Edited by
Dirk Lehmhus, Matthias Busse,
Axel S. Herrmann, and
Kambiz Kayvantash

Structural Materials and
Processes in Transportation

Related Titles

Schmitz, G. J., Prahl, U. (eds.)

Integrative Computational Materials Engineering

Concepts and Applications of a Modular Simulation Platform

2012

ISBN: 978-3-527-33081-2

Mittal, V. (ed.)

Modeling and Prediction of Polymer Nanocomposite Properties

Series: Polymer Nano-, Micro- and Macrocomposites (Volume 4)

2013

ISBN: 978-3-527-33150-5

Thomas, S., Joseph, K., Malhotra, S.K., Goda, K., Sreekala, M.S. (eds.)

Polymer Composites

3 Volume Set

2014

ISBN: 978-3-527-32985-4

Kainer, K. U. (ed.)

Magnesium

8th International Conference on Magnesium Alloys and their Applications

2010

ISBN: 978-3-527-32732-4

Wagg, D., Bond, I., Weaver, P., Friswell, M. (eds.)

Adaptive Structures

Engineering Applications

2007

ISBN: 978-0-470-05697-4

Gaudenzi, P.

Smart Structures

Physical Behaviour, Mathematical Modelling and Applications

2009

ISBN: 978-0-470-05982-1

*Edited by Dirk Lehmhus, Matthias Busse, Axel S. Herrmann,
and Kambiz Kayvantash*

Structural Materials and Processes in Transportation

WILEY-VCH

The Editors

Dr.-Ing. Dirk Lehmhus
University of Bremen
ISIS Sensorial Materials
Scientific Centre
Wiener Straße 12
28359 Bremen
Germany

Prof. Dr.-Ing. Matthias Busse
University of Bremen
ISIS Sensorial Materials Scientific
Centre and Fraunhofer IFAM
Wiener Straße 12
28359 Bremen
Germany

Prof. Dr.-Ing. Axel S. Herrmann
Faserinstitut Bremen e.V.
Am Biologischen Garten 2
28359 Bremen
Germany

Prof. Dr. Kambiz Kayvantash
Cranfield University
School of Applied Sciences
Building 61, Cranfield campus
Cranfield, MK43 0AL
UK

and

CADLM Sarl
43 rue du Saule Trapu
91300 Massy
France

Library of Congress Card No.:
applied for

British Library Cataloguing-in-Publication Data
A catalogue record for this book is available from the British Library.

Bibliographic information published by the Deutsche Nationalbibliothek
The Deutsche Nationalbibliothek lists this publication in the Deutsche Nationalbibliografie; detailed bibliographic data are available on the Internet at <http://dnb.d-nb.de>.

© 2013 Wiley-VCH Verlag GmbH & Co. KGaA, Boschstr. 12, 69469 Weinheim, Germany

Print ISBN: 978-3-527-32787-4
ePDF ISBN: 978-3-527-64987-7
ePub ISBN: 978-3-527-64986-0
mobi ISBN: 978-3-527-64985-3
oBook ISBN: 978-3-527-64984-6

Cover Design Adam-Design, Weinheim
Typesetting Laserwords Private Limited, Chennai, India
Printing and Binding Strauss GmbH, Moerlenbach

Foreword

From Mechanics and Materials Science to Engineering

Transportation has at all times given engineers, craftsmen, and builders a challenge: making it easier to move. This means: make it lighter, stronger, more reliable, more comfortable, easier to build, less expensive . . . all with available materials.

Although means of transportation have been designed and built over millennia, truly engineered products have arisen recently. The first engineering book, *La Science des Ingénieurs*, by Belidor, was published in 1729. A. Wöhler developed the theory of mechanical fatigue and used it to improve the rolling stock of Northern German Railways in the mid-nineteenth century. A.N. Krylov applied modern stress analysis methods to ship design, based on his initial publication in 1906. A.A. Griffith published his theory of the strength of glass fiber, underlying structural composite performance, in 1920. The commercial production of glass fibers, by Owens Corning, dates back to no further than 1935 Ref. [1]. Also the field of polymer chemistry for technical plastics – the molecular design of new materials – emerged in the second half of the twentieth century. This paved the way for the widespread replacement of natural, renewable materials for strong lightweight structures.

Challenges of the Past Century and the Breakthroughs They Have Provoked

We are only a century away from the major roots of our current design and materials science for transportation! But the new scientific tools developed since then have allowed engineers to tackle some big challenges during the twentieth century.

Big strides were made, for example, in protecting metal structures from corrosion using organic coatings, and where relevant, using noncorroding materials such as stainless steel (patented in 1912) and plastics.

Sometimes problems with fabrication processes led to improved design rules. During World War II, the massive failure of a number of Liberty Ships practically gave birth to the discipline of fracture mechanics. More than 2700 Liberty Ships were built between 1941 and 1945 at unprecedented war-time rates. "One shipyard built a Liberty ship in five days. The massive increase in production was possible in large part because of a change from riveted to electric-arc welded construction . . . saving a thousand tons of weight in the hull" Ref. [2]. The better understanding of crack propagation led to durable progress in ship – and other – production thereafter.

Air transport, with its literally sky-high requirements, stimulated a race for energy efficiency and mass reduction, and thus prompted great strides in the development of new grades of aluminum, titanium, and carbon-fiber composites. A significant step in modern aircraft design was the De Havilland Mosquito, whose airframe largely consisted of sandwich structures built up of thin skins of laminated plywood over a balsa core. Its lightweight and aerodynamics gave it a speed superior to any other airplane of the day. Sandwich panels are now widespread in all kinds of vehicles, including trains.

The emergence of powerful and affordable computers quickened the pace of development by providing a means to create and "test" designs numerically. It thus provided a new way to meet the challenge of improving the robustness of material models and predictive simulation. This led to significant progress in reliability and safety. A second effect was the capacity to imagine and realize structures with much more complex geometries. Ensuring, in particular, the safety of ever more sophisticated structural systems became intimately linked to finding the compromise between strength and formability in new-generation materials, be they steel alloys, nonferrous alloys, or structural plastics.

Airplanes, boats, trains, trucks, and cars today are thus built of very different materials than they were even 50 years ago.

Challenges of the New Century and the Breakthroughs They Will Require

In this twenty-first century, armed as we are with numerical tools and a rich knowledge of materials and process science – to which this book is contributing – the new challenge is applying our technological base to make our prosperity last and share it more widely. We have become obsessed with finding the optimal materials to push the performance envelope, improve safety, reduce costs, and preserve our resources – in a competitive global market environment.

From the point of view of materials and mechanical engineering, a number of standards have gained preeminence for various transportation applications such as plastic bumpers, magnesium steering wheels, composite train cabins, aluminum bus frames, titanium ship propeller shafts, and carbon-fiber composite structural components for aircraft. We have mastered the most pressing trade-offs so far, but the uncomfortable realization remains, that our best solutions are local, rather than global optima, increasingly subject to extrinsic factors.

Mass reduction for the improvement of energy consumption and emissions has become one of the principal drivers in material selection for transportation. The debate has evolved from the primitive level of asking which material – polymer composites or metals – future vehicles will be made of, to an understanding that significant progress can only be made by thinking in terms of pragmatic, fully optimized multimaterial designs used in appropriate vehicle architectures.

In the automotive sector, performance improvement has finally become synonymous with downsizing, for higher efficiency in use. The newest generations of internal combustion engines, with outputs upward of 90 kW l^{-1} and 200 Nm l^{-1}, rely on increasingly sophisticated materials with high temperature and dynamic wear resistance. A typical engine consists of more than 60 grades of metallic alloys and 20 types of organic materials, plus a variety of coatings and surface treatments. Electric vehicle (EV) motors, running at 20 000 rpm, will require yet different structural materials with improved creep and magnetic resistances.

Although very promising, however, advanced materials such as carbon-fiber composites are not suited for some of the applications with the strongest projected growth. For both individual and shared means of transportation, the access to mobility in new markets is opened under different economic constraints. Here the conditions for success are low initial and operation costs, local availability of materials, damage tolerance under loads that are more severe than in countries with mature infrastructures, good repairability, and adaptability to a widely varying, sometimes minimal industrial system.

Durable development requires not only robustness and durability of the means of transportation itself but also a design that favors repair, reuse, and recycling. The recycling pathways for many material classes are not yet reliable or even inexistent in many regions. Designs that favor easy separation for maximal recovery rates will become an essential element of a durable mobility for all.

Finally, for a global economy that is growing well above 3% per year according to the World Bank, using recycled materials will not suffice to cover our industries' needs – no matter how intensely we recycle. At such growth rates, we will have to continue injecting primary materials into the production streams. That means that we must pay increased attention to our resources, with the aim of reducing the footprints in minerals and metals, energy, and water. Given the expected long-term upward trend in energy costs, getting a grip on our material consumption will soon be synonymous with protecting ourselves against cost impacts, and thus finding a better path toward durable business.

Toward a Society Working by the Principles of Durability and Material Efficiency

Several trends are outlining the way ahead for structural materials:

- mass reduction, with, for example, a goal of -20% in production cars within 10 years;
- a pronounced bifurcation toward more/less "technicality," with high tech for a growing premium segment, and ultralow cost, frugally engineered low tech (but high appeal!) for the booming new markets;
- the integration of functions such as decoration, optical effects, and information (sensors) in structural parts;

- design-for-*x*, where "*x*" stands for drivers such as cost, global production, logistics, disassembly, recycling, and the environment;
- more focus on process to require less raw materials and energy.

We are indeed shifting our paradigm from one of pure technical efficiency to one of a broader sustainable material management (SMM). As a recent European report remarked, "the EU is the world region that outsources the biggest part of resource extraction required to produce goods for final demand" Ref. [3]. SMM is thus not just a catchphrase for durability – be it environmental or business – but in fact one of raw material security. In several ways, beyond the pure technical challenge, mass reduction is becoming a strategic society issue.

The general term of ecodesign is gaining a strong, concrete footing as a credible practice to make "green" good for business. Michael Ashby has dedicated a recent book to the issue, extending his popular material selection charts to cover environmental criteria for broad classes of materials and applications. Ref. [4]. A metric of particular relevance, in relation with the mainstream considerations of CO_2 emissions, is the energy trade-off between manufacturing and in-use consumption. Life cycle analysis (LCA) is thus finally becoming established as an economic tool, with strong bearing on not only corporate social responsibility, but also on mid-to-long-term competitiveness. It helps us understand where, in the supply chain, we can reduce our energy impacts, and therefore costs. Such approaches aim at guiding economical material selection in a broader perspective. Beyond the traditional steel versus aluminum versus composite choices, they allow us to construct optimized material–process–localization combinations or aim for the right pace of introduction of recycled materials.

The conditions of success are multiple. Starting with reliable and comparable data, we need to define new value equations that integrate extended design drivers such as sustainability, and allow arbitration between them. In some cases, we will need to make hard choices between adapting the requirements to available technologies that may be more economical or durable and developing new technologies when more sophisticated materials are considered readily accessible over the longer term. Third, management needs to provoke and actively support more interdisciplinary research and improved cross-functional interactions within and between companies. Underlying all this is an integrated approach that gives appropriate simultaneous consideration to design, materials, and process: a truly production- and life-cycle-oriented engineering. And finally, we may simply need to have more courage and willingness to make decisions that could pay off only in the midterm – to truly build a vision.

Outlook on Structural Materials for Transportation

The future of structural materials lies in optimized hybrid structural systems that conform to a larger set of requirements and constraints. A commonly heard expression is "the right material at the right place, at the right time." In order to develop durable solutions to the technical challenges of safe and efficient transportation, we will need to solve equations that integrate macro-economic and policy factors, trade considerations and externalities, and customer appeal and reassurance.

To do this successfully, the necessary starting point is a comprehensive knowledge of the options, of the potentials, and limitations of the different classes of structural materials. This is precisely what this book aims to give you. Beyond the basic yet extensive design and process reference data, it aims to bring you new insights based on recent, first-hand information from some of the top research centers in the EU.

As we look toward the future, we will want to keep our scope wide: state-of-the-art research is continuously extending the horizon for novel, sometimes surprising applications of classical materials. This is the spirit behind concrete canoes, carbon-fiber-reinforced bridges, plastic engines, and cars made of castor-oil-based polymers. Having ready access to comparable data will, it is hoped, promote synergies between fields of application.

We are confident that this book will be useful to you to orient your current material selection and also as a starting point to imagine the means of transportation of tomorrow.

Dr Patrick Kim
VP R&D Benteler Automotive
formerly VP Materials Engineering
Renault

References

1. Timoshenko, S.P. (1953) *History of Strength of Materials*, McGraw-Hill.
2. matdl.org/failurecases/Other_Failure_Cases/Liberty_Ship.
3. Wuppertal Institute et al. (2010) Sustainable Materials Management for Europe – from Efficiency to Effectiveness. Report for the Belgian Government, March 2010, *http://www.euractiv.com/sites/all/euractiv/files/SMMfor%20EuropeStudy_0.pdf*.
4. Ashby, M. (2009) *Materials and the Environment*, Butterworth-Heinemann.

Contents

Preface

This book is meant to provide an introduction to current developments in the field of structural materials for the transportation industry. This includes rail, maritime, automotive, and aerospace industries, with a focus on the last two. Deliberately excluded from the scope are purely functional materials.

Quite literally, structural characteristics of materials are the backbone of any engineering design. They provide self-supporting capabilities to components where mechanical stability is a secondary concern. Whenever the bearing of mechanical loads becomes the primary role, and materials are optimized in view of this demand, we speak of structural materials. Solutions that address this challenge are what this book revolves around. The perspective chosen to deal with the topic is that of materials science and engineering. We have structured our work accordingly by dedicating the central parts to the main material classes.

However, any structural material we see in a specific application is in fact a combination of material and process. Its properties are defined by both, and thus, treating one aspect while neglecting the other is not a viable option. Besides, it is development tools which are built on similarly advanced modeling and simulation techniques that finally enable usage of emerging materials by allowing their evaluation in diverse application environments. With this in mind, we have included these aspects in our book. In all of them, the perspective is forward-facing: We do not intend to comprehensively cover the fundamentals of the various fields. Instead, we have attempted to identify major trends and highlight those that we see at the threshold to practical application.

Materials are evolving. So are the processes associated with them, as well as the tools and methodologies that allow their development and application. The rate of change in material development is dictated by external pressure. We have defined the major periods in the development of early mankind by the structural materials that dominated them. When bronze technology evolved, stone had to yield. The same occurred to bronze once iron became available on a larger scale. In transportation, we have seen change from wood and other natural materials to metal, and nowadays to composites. The rate of change steps up once pressure rises. The period of time that one material prevails appears to become shorter and shorter. On the other hand, since the shift from natural to technical materials, we mostly observe additions to the spectrum of materials rather than complete replacement. This may not be true for individual exponents of a material class, but definitely so for the classes themselves. Transportation, in all its width, is currently under considerable pressure to increase resource efficiency. One major handle to achieve this is lightweight design. This affords either new structural concepts or new materials offering improved performance. Very often, both go hand in hand. While such general pressure strengthens the motivation to search for entirely new approaches, it will also fuel inter-area competition. The past has shown that this may significantly speed up developmental processes within one class of materials. A good example in this respect is the recent evolution of high-strength steels, which took place at least partly in response to aluminum-centered automotive body designs entering volume production. A comparable situation can be observed in the commercial aircraft industry, where large-scale introduction of carbon-fiber-reinforced composites challenges the established status of aluminum alloys. New production processes support such tendencies are enablers of cross-fertilization between modes of transport: considering their properties, automotive design could profit greatly from application of carbon-fiber-based composites, too, but the sheer cost of state-of-the-art aerospace materials and processes forbids immediate takeover. Adaptation of processes to match another industry's needs, like transition from single part to large-scale series production, can help diminish such barriers. As a result, we currently see an extremely high rate of change in the range of available materials for load-bearing structures in the transport industry. With this in mind, it is the conviction of editors and authors of this book that a work is needed that familiarizes materials scientists, design engineers, and innovation managers in industry with developments in structural materials science and engineering that are likely to find their way into high-technology products within the next 5–10 years. We do give some background on the various materials and technologies, but the major focus is on what is currently on the verge of application.

Besides our primary target audience, we are confident that students and graduates in mechanical engineering, as well as academic researchers in the field, will find this compilation helpful to first get and then adjust their bearings through a highly dynamic field of research. In this sense, we intend our book to serve as a guideline for both groups. As such, it is meant to give them the first idea of the respective material class as well as a clear vision of where the present focus of developmental work will lead it within the near to mid-term future. This knowledge base shall allow them to decide which material classes and subclasses to study in more detail in view of their specific interest. Suggestions on where to search for in-depth fundamental information and keep track of future advances shall complete the picture.

The book is structured along the major material classes relevant for transport industry structural applications. All of these are treated in separate parts, starting with metals (Part I) and proceeding via polymers (Part II) and composites (Part III) to cellular materials (Part IV). Each part covers associated processes on the level of its individual chapters, that is, for the exemplary case of metals separately for iron-, aluminum-, magnesium-, and titanium-based materials (Chapters 1–4). In a further section (Part V), selected aspects of modeling and simulation techniques are being treated. Highlights have been set here in terms of modeling approaches covering multiple scales of material description (Chapter 13) and adaptation of artificial intelligence (AI) techniques to material modeling (Chapter 14). The use of fundamental ab initio techniques in designing new metallic material compositions and states is treated in yet another subsection (Chapter 15). Finally, specific trends that go beyond an individual class of materials are discussed in Part VI. An example are hybrid design approaches, which attempt to locate the optimum material for a purpose at the place where best use can be made of its properties, thus leading to complex, multimaterial structures (Chapter 16). In extrapolating trends already discussed in terms of structural health monitoring for composite materials (Chapter 8), material-integrated sensing and intelligence, summarized under the descriptive term of sensorial materials, are covered in Chapter 17. Additive manufacturing as an approach with promise for highly versatile production and structural complexity that in some respects cannot be reached by other processes is presented in the final chapter (Chapter 18).

We have attempted to organize each of the main chapters in Parts I–IV and Part VI in a similar way. In these predominantly material-related chapters, we start with some fundamentals and go on to detail new developments. In this, we do not separate material and process because of the close link between both. However, we do subdivide the chapters according to distinctions that are already established for the respective class of materials. An example is the distinction between wrought and cast alloys realized in the chapters on aluminum and magnesium. A similar approach, though adapted to the specifics of such composites, is reflected in the separation between processes involving thermoplastic versus those employing thermoset matrices in the chapter on polymer matrix composites. The major chapters are concluded with a section on further reading for intensified study and a hint at major organizations, conferences, or other events dedicated to the respective topic.

We are extremely grateful to the many authors who have shouldered the task of providing the content to this work. We are indebted to Dr. Martin Preuss (Wiley-VCH) who encouraged us to venture this endeavor, which was originally based on two symposia organized in the course of the Euromat 2009 Conference held in Glasgow (UK) from 7 September 2009 to 10 September 2009 (www.euromat2009.fems.eu). Finally, our thanks go to Lesley Belfit, again of Wiley-VCH, who helped us steer our course through all the hindrances of the editorial process with grace and patience.

Dirk Lehmhus
Matthias Busse
Axel S. Herrmann
Kambiz Kayvantash

List of Contributors

Claus Aumund-Kopp
Fraunhofer Institute for
Manufacturing Technology and
Advanced Materials (IFAM)
Shaping and Functional Materials
Wiener Straße 12
28359 Bremen
Germany

Jorge Barcena
TECNALIA
Industry and Transport Division
Mikeletegi Pasealekua 2
E-20009
Donostia-San Sebastian
Spain

Daniele Bassan
Centro Ricerche Fiat S.C.p.A.
Sede legale e amministrativa
Strada Torino, 50
10043 Orbassano (TO)
Italy

Joachim Baumeister
Fraunhofer IFAM
Powder Technology
Wiener Straße 12
28359 Bremen
Germany

Banu Berme
RWTH Aachen University
Institut für Eisenhüttenkunde
(IEHK)
Department of Ferrous
Metallurgy
Intzestrasse 1
52072 Aachen
Germany

Stefan Bosse
University of Bremen
Department of Mathematics and
Computer Science
Working Group Robotics
Robert Hooke Str. 5
28359 Bremen
Germany

Christian Brauner
University Bremen
Faserinstitut Bremen e.V.
Am Biologischen Garten 2
28359 Bremen
Germany

Matthias Busse
University of Bremen
ISIS Sensorial Materials Scientific
Centre and Fraunhofer IFAM
Wiener Str. 12
28359 Bremen
Germany

Sandro Campos Amico
Rio Grande do Sul Federal
University (UFRGS)
Programa de Pós-Graduação em
Engenharia de Minas
Metalúrgica e de Materiais
(PPGEM)
Av. Bento Gonçalves,
9500, Agronomia
Porto Alegre
RS 91501-970
Brazil

Aravind Dasari
Nanyang Technological
University
School of Materials Science and
Engineering
Blk N4.1, 50 Nanyang Avenue
639798 Singapore
Singapore

Pedro Egizabal
TECNALIA
Industry and Transport Division
Mikeletegi Pasealekua 2
E-20009
Donostia-San Sebastian
Spain

Luiz Antonio Ferreira Coelho
Santa Catarina State University
(UDESC)
Center of Technological Sciences
Department of Mathematics
Campus Universitário Prof.
Avelino Marcante s/n
Bom Retiro
Joinville
SC 89219-710
Brazil

Martin Friák
Max-Planck-Institut für
Eisenforschung GmbH
Max-Planck-Strasse 1
40237 Düsseldorf
Germany

Maider García de Cortázar
TECNALIA
Foundry and Steelmaking Unit
Mikeletegi Pasealekua 2
E-20009
Donostia-San Sebastian
Spain

Axel von Hehl
Stiftung Institut für
Werkstofftechnik (IWT)
Badgasteiner Str. 3
28359 Bremen
Germany

Axel S. Herrmann
University Bremen
Faserinstitut Bremen e.V.
Am Biologischen Garten 2
28359 Bremen
Germany

Jörg Hohe
Fraunhofer-Institut für
Werkstoffmechanik IWM
Wöhlerstr. 11
79108 Freiburg
Germany

Norbert Hort
Helmholtz-Zentrum Geesthacht
Magnesium Innovation Centre
(MagIC)
Max-Planck-Straße 1
21502 Geesthacht
Germany

Juan F. Isaza P.
Fraunhofer Institute for
Manufacturing Technology and
Advanced Materials (IFAM)
Shaping and Functional Materials
Wiener Straße 12
28359 Bremen
Germany

Kambiz Kayvantash
Cranfield University
School of Applied Sciences (SAS)
Centre for Automotive
Technology
Building 61, Cranfield Campus
Cranfield MK43 0AL
UK

and

CADLM Sarl
43 rue du Saule Trapu
91300 Massy
France

Peter Krug
Cologne University of Applied
Sciences
Institute of Automotive
Engineering (IFK)
Betzdorfer Strasse 2
50679 Köln
Germany

Dirk Lehmhus
University of Bremen
ISIS Sensorial Materials
Scientific Centre
Wiener Str. 12
28359 Bremen
Germany

Yann Le Petitcorps
LCTS UMR 5801
3 Allée de la Boétie
33600 Pessac
France

Jörg Neugebauer
Max-Planck-Institut für
Eisenforschung GmbH
Max-Planck-Strasse 1
40237 Düsseldorf
Germany

James Njuguna
Cranfield University
School of Applied Sciences
Cranfield
Bedfordshire MK43 0AL
UK

Luiz Cláudio Pardini
Centro Técnico Aeroespacial
Comando da Aeronáutica
Instituto de Aeronáutica e Espaço
Pça Marechal Eduardo Gomes
Vila das Acácias
Sao Jose dos Campos
SP 12228-904
Brazil

Christian Peters
University Bremen
Faserinstitut Bremen e.V.
Am Biologischen Garten 2
28359 Bremen
Germany

Sérgio Henrique Pezzin
Santa Catarina State University
(UDESC)
Center of Technological Sciences
Department of Chemistry
Campus Universitário Prof.
Avelino Marcante s/n
Bom Retiro
Joinville
SC 89219-710
Brazil

Krzysztof Pielichowski
Cracow University of Technology
Department of Chemistry and
Technology of Polymers
ul. Warszawska 24
31-155 Kraków
Poland

Ulrich Prahl
RWTH Aachen University
Institut für Eisenhüttenkunde
(IEHK)
Department of Ferrous
Metallurgy
Intzestrasse 1
52072 Aachen
Germany

Dierk Raabe
Max-Planck-Institut für
Eisenforschung GmbH
Max-Planck-Strasse 1
40237 Düsseldorf
Germany

Ali Ramazani
RWTH Aachen University
Institut für Eisenhüttenkunde
(IEHK)
Department of Ferrous
Metallurgy
Intzestrasse 1
52072 Aachen
Germany

Miguel A. Rodriguez-Perez
University of Valladolid
CellMat Laboratory
Department of Condensed Matter
Physics
Facultad de Ciencias
Paseo de Belén 7
47011 Valladolid
Spain

Patrick Schiebel
University Bremen
Faserinstitut Bremen e.V.
Am Biologischen Garten 2
28359 Bremen
Germany

Wim H. Sillekens
European Space Agency – ESTEC
Keplerlaan 1
2201 AZ Noordwijk
The Netherlands

Eusebio Solórzano
Universidad de Valladolid
CellMat Laboratory
Department of Condensed Matter
Physics
Facultad de Ciencias
Paseo de Belén 7
47011 Valladolid
Spain

André Stieglitz
University Bremen
Faserinstitut Bremen e.V.
Am Biologischen Garten 2
28359 Bremen
Germany

Lothar Wagner
Clausthal University of
Technology
Institute of Materials Science and
Engineering
38678 Clausthal-Zellerfeld
Germany

Jörg Weise
Fraunhofer IFAM
Powder Technology
Wiener Straße 12
28359 Bremen
Germany

Manfred Wollmann
Clausthal University of
Technology
Institute of Materials Science and
Engineering
38678 Clausthal-Zellerfeld
Germany

Part I
Metals

Axel von Hehl

Metals belong to the eldest engineering materials of the humankind. The history can be traced back to the Copper Age when metallic products were firstly made of native metals. However, in contrast to natural materials, such as wood or stone, native metals were very rare. The rising needs of metallic products implied the invention and control of metallurgical processes, such as ore smelting and casting. In the period of the Bronze Age (2200–800 BC), when copper was firstly alloyed with tin, followed by the Iron Age (1100–450 BC), when advanced smelting and working techniques were introduced, the cornerstones for the industrial production were put in place. Since the beginning of the industrial production in the late eighteenth century, the demand for metals has been increased steadily (see Figure P1.1) [1]. The rapid technological development has led simultaneously to today's alloys and production processes. Nowadays, metals are spread in all technical applications of human life. The present means of transportation would have been inconceivable without metals.

On microscale, metals are characterized by a crystalline structure. Electrons of the outer shell being freely movable between the atomic cores keep the crystal together. The electron's free mobility is responsible for both the high electrical and thermal conductivity of metals. Furthermore, the formation of close atomic packing within the crystal [2] enables a ductile atomic sliding under mechanical shear load. The siding proceeds on the densely packed planes along the close packed directions without changing the bonding conditions. The spatial arrangement of these sliding planes is shown in Figure P1.2. The number of sliding systems is the product of the number of sliding planes and the number of close packed sliding directions. While body centered cubic (BCC) and face centered cubic (FCC) structures, which can be found in ferritic steel and aluminum alloys exhibit respectively 12 and more systems for atomic sliding, the number of sliding systems in hexagonal crystal structures, which are typical for magnesium and β-titanium alloys, is dependent on the c/a ratio of the unit cell. Besides three basal sliding systems, alloys with c/a below 1.63 exhibit up to nine additional pyramid and prism sliding systems. In polycrystals, a number of more than five enables any deformation [3]. This fact explains the outstanding plasticity of most metals and, consequently, their excellent workability.

Structural Materials and Processes in Transportation, First Edition.
Edited by Dirk Lehmhus, Matthias Busse, Axel S. Herrmann, and Kambiz Kayvantash.
© 2013 Wiley-VCH Verlag GmbH & Co. KGaA. Published 2013 by Wiley-VCH Verlag GmbH & Co. KGaA.

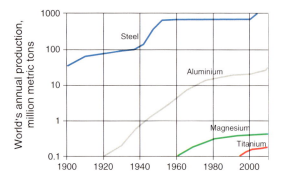

Figure P1.1 World's annual production during the last century (following [1]).

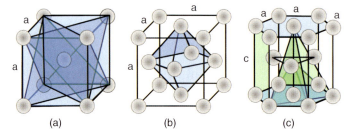

(a)   (b)   (c)

Figure P1.2 Sliding planes of (a) body centered cubic (BCC), (b) face centered cubic (FCC), and (c) hexagonal crystal structure.

The mass and the diameter of atoms as well as how close atoms are packed determine the density of a material [2]. Metals with a density of below 5 g cm^{-3} belong to the class of light metals [4]. In this class, magnesium is the lightest metal with a density of about 1.74 g cm^{-3} followed by aluminum with 2.70 g cm^{-3}, and titanium with 4.50 g cm^{-3}. The density difference between the pure metal and its alloys are usually very low. With a density of 7.86 g cm^{-3}, iron, as the base metal of steel, belongs to the class of heavy metals. Because of its very high Young's modulus, and the variety of technical and metallurgical means to reach highest strength properties, nevertheless, this metal is still among the materials being attractive for light-weight designs.

Weight reduction is an important subject for the transportation industry to improve their products' energy efficiency, particularly during the operation time. Low energy consumption benefits the reduction of CO_2 emission and helps the electric mobility by increasing the transportation range. The main lightweight drivers are illustrated in Figure P1.3 [5].

Generally, the potential of materials for weight savings can be evaluated by regarding their density-specific properties [6]. Compared to fiber-reinforced polymers (FRPs), metallic materials, such as steel, aluminum, and magnesium alloys as well as titanium alloys, provide an excellent impact toughness, a good wear, and thermal

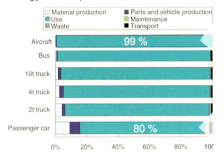

Energy consumption

CO$_2$ emission

Fleet limits:
- 2012: 130 g/km (acc. to EU legislation)
- 2020: 90 g/km (target)

Electric mobility

Transportation range:
- Weight of battery units
- Payload

Figure P1.3 Drivers of weight reduction and multimaterial design as a strategy for weight reduction (following [5, 7]).

resistance as well as a high life cycle fatigue along with moderate materials and processing costs and, furthermore, an outstanding recyclability.

Within this material class, steel is distinguished by its high specific tensile stiffness combined with a good fracture toughness and low material costs. Light alloys such as aluminum and magnesium alloys are characterized by high specific bending stiffness and strength values along with excellent compression stability. Titanium alloys are more expensive, but advantageous when, for example, a very high specific tensile strength and a good wet corrosion resistance is required. Figure P1.4 gives an overview of the mechanical properties for different loading conditions relative to those of steel as the reference material.

However, owing to the ongoing need of weight reduction, alternative structural materials, such as FRP, penetrate more and more into transportation applications (see Figure P1.3). Compared to metallic materials, for example, carbon-fiber-reinforced polymers (CFRPs) are still very expensive, but on the other hand they exhibit specific strength and stiffness values, which are significantly higher than those of metallic materials. Consequently, effective weight reduction calls for advanced multimaterial designs (Figure P1.3) [7]. Thus, in order to ensure a cost efficient weight reduction, a strong position of metals is required. This necessitates a multidisciplinary approach that combines materials science and production technology with designing and dimensioning.

The following sections shall give an overview of recent advances in the development of steel, aluminum, and magnesium alloys as well as titanium alloys, which are intended for application in the transportation sector. Basically, the alloy development is closely related to process development, comprising a variety of production techniques associated with the development of advanced metallic

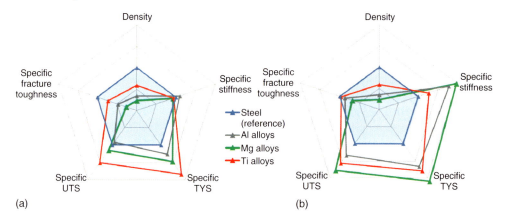

Figure P1.4 Density and relative mechanical properties of metals (a) for tension/compression load and (b) for bending/buckling load (reference: steel) (values according to [6]).)

products. Therefore, descriptions of selected process developments are also part of these sections.

References

1. Degischer, H.P. and Lüftl, S. (eds) (2009) Leichtbau–Prinzipien, Werkstoffauswahl und Fertigungsvarianten, Wiley-VCH Verlag GmbH, Weinheim, ISBN: 978-3-527-32372-2, p. 80.
2. Ashby, M.F. and Jones, D.R.H. (2005) *Engineering Materials 1*, 3rd edn, Butterworth–Heinemann as an imprint of Elsevier, Oxford, ISBN: 978-0750663809.
3. Gottstein, G. (2010) *Physical Foundations of Materials Science*, 1st edn, Springer-Verlag, Berlin, Heidelberg, ISBN: 978-3-642-07271-0.
4. Roos, E. and Maile, K. (2004) Werkstoffkunde für Ingenieure: Grundlagen, Anwendung, Prüfung, Springer-Verlag, ISBN: 978-3540220343.
5. Kasai, J. (2000) *Int. J. Life Cycle Assess.*, **5** 5316.
6. Ashby, M.F. (1999) *Materials Selection in Mechanical Design*. Butterworth-Heinemann Verlag, Oxford, ISBN: 978-0750643573.
7. Goede, M., Dröder, K., Laue, T. (2010) Recent and future lightweight design concepts–The key to sustainable vehicle developments. Proceedings CTI International Conference–Automotive Lightweight Design, Duisburg, Germany, November 9–10, 2010.

1
Steel and Iron Based Alloys

Ali Ramazani, Banu Berme, and Ulrich Prahl

1.1
Introduction

As a consequence of global warming the demand for transport vehicles with lower emissions and higher fuel efficiency is increasing [1]. Reduction of travel time and CO_2 emissions per passenger is also noticed in transport applications. Different methods have been utilized to improve the fuel economy, which is also demanded by the governments. For instance, the Obama administration in the United States has recently proposed a law requiring a 5% increase in fuel economy during 2012–2016 and the development of the Corporate Average Fuel Economy (CAFE) to obtain an average automotive fuel efficiency of 35.5 miles per gallon by 2016 [1–3]. Since the development of new power generation system in hybrid-electric vehicles is expensive, new technologies aim to improve fuel economy through improving aerodynamics, advanced transmission technologies, engine aspiration, tires with lower rolling resistance, and reduction of vehicle weight [4]. Usage of new kinds of lightweight steels in vehicle design for automotive and transport applications is of great importance nowadays for enhancing safety, improving fuel economy, and reducing lifetime greenhouse gas emissions.

Sheet and forged materials for transport applications require both high strength and formability. High strength allows for the use of thinner gauge material for structural components, which reduces weight and increases fuel efficiency [5]. High strength also improves the dent resistance of the material, which is esthetically important. Another advantage of higher strength material is improved passenger safety due to the higher crash resistance of the material. Therefore, the high strength material must be formable to allow economical and efficient mass-produced automotive parts.

New tools and technologies such as modeling and simulations are being innovated for the design of steels for future ecologically friendly vehicles. There are also several different organizations that work on the evaluation of new kinds of lightweight steels. One of them is the World Auto Steel organization that has developed the latest version of the so-called advanced high strength steels (AHSSs) with the newest technology to evolve body structures with higher strength, lower

Structural Materials and Processes in Transportation, First Edition.
Edited by Dirk Lehmhus, Matthias Busse, Axel S. Herrmann, and Kambiz Kayvantash.
© 2013 Wiley-VCH Verlag GmbH & Co. KGaA. Published 2013 by Wiley-VCH Verlag GmbH & Co. KGaA.

weight, and consequently lesser CO_2 emissions compared to the other body materials. Another program, new steel body (NSB), of ThyssenKrupp made a point of
weight-saving potential of steel; it provides the benefits of tubular components with
conventional stamped parts using advanced multiphase steels. Also, the future
steel vehicle (FSV) program worked on the steel body structure designs. Over 35%
mass reduction in the benchmark vehicle and approximately 70% reduction in the
whole life cycle emissions are achievable in this program [6–10]. Finally, the most
extensive research is however conducted by the ultralight steel auto body-advanced
vehicle concepts (ULSAB-AVC) Consortium. This program was established about
a decade ago by the collaborative efforts of 35 steel companies in the world to
find steel solutions to the problems faced by automotive companies all around the
world. The ULSAB-AVC program follows the ultralight steel auto body (ULSAB)
program (results announced worldwide in 1998). Here, two important drivers for
the ULSAB-AVC were the US Partnership for a New Generation of Vehicles (PNGV)
and the EUCAR (The European CO_2 reduction program) projects, which provided
references for setting ULSAB-AVC targets [8–11]. This program focused on the
development of steel applications for future vehicles, considering the increasing
demands shown in Figure 1.1. Weight reduction, crash safety, compatibility with
the environment, and economic features can be addressed as the main goals of this
program [6–13].

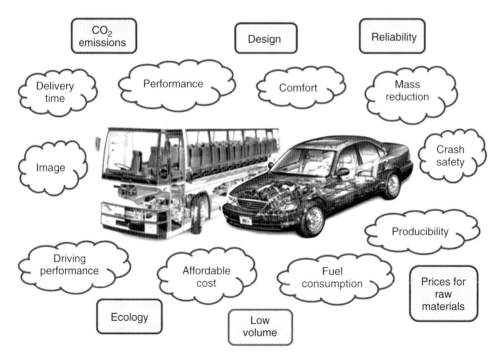

Figure 1.1 Increasing demands on the transportation [13].

1.2
Sheet Steels

1.2.1
Development Strategy and Overview

The ULSAB program deals with both high strength steels (HSSs) and ultrahigh strength steels (UHSSs), which were mostly conventional microalloyed sheet grades. Within the ULSAB-AVC program, the application of newer types of high-strength sheet steels, the so-called AHSSs, has been considered. Therefore, as the first task, a consistent classification of the various grades of steels has been identified in order to manage the goals of the program [4, 6–11]. In this terminology, steels are identified as "XX aaa/bbb," where the first digits (XX; these are not constricted to be two digits) represent the type of the steel, the next digits (aaa) are responsible for the minimum yield strength (YS) of the material in MPa, and the last digits (bbb) depict the minimum achieved ultimate tensile strength (UTS) of the investigated steel. The world wide steel classification developed by the ULSAB-AVC program is reported in Table 1.1 [11]. The identified mechanical properties of the steels chosen for the ULSAB-AVC body structure, closures, ancillary parts, suspension, and wheels are reported in Table 1.2.

There are various mechanisms available to strengthen steels. HSSs such as BH (bake hardening), HSLA (high strength low alloy), CMn (carbon manganese), or ISs (isotropic steels) have been optimized by the application of classical concepts such as solid solution strengthening, precipitation strengthening, and grain refinement. In addition, AHSS concepts such as DP (dual phase), TRIP (Transformation-Induced Plasticity), and CP (complex phase), or Mart (martensitic) use the multiphase character of the microstructure, which may also include phase transformation and additional deformation and strengthening concepts [1]. The special class of MnB (manganese boron) steels in some sense stands outside of this logic as it is based on martensitic structure (to achieve maximum strength, while accepting poor deformation values), which is reasonable only by application of the very specific processing concept of press hardening. Figure 1.2 shows a comparison between the tensile strength and total elongation of HSS and AHSS steels in comparison to HSS. Classical press forming steels such as mild steels and IF (interstitial free)

Table 1.1 The universal classification introduced by the ULSAB-AVC Consortium [11].

Designator	Steel type	Designator	Steel type
Mild	Mild steel	DP	Dual phase
IF	Interstitial free	SF	Stretch flangeable
BH	Bake hardenable	TRIP	Transformation induced plasticity
CMn	Carbon manganese	CP	Complex phase
HSLA	High strength low alloy	Mart	Martensite
IS	Isotropic	MnB	Hardenable manganese boron

Table 1.2 Mechanical properties of steels chosen for the ULSAB-AVC components [11].

Steel grade	YS (MPa)	UTS (MPa)	Total EL (%)	n value (5–15%)	r (bar)	K value (MPa)
BH 210/340	210	340	34–39	0.18	1.8	582
BH 260/370	260	370	29–34	0.13	1.6	550
DP 280/600	280	600	30–34	0.21	1	1082
IF 300/420	300	420	29–36	0.2	1.6	759
DP 300/500	300	500	30–34	0.16	1	762
HSLA 350/450	350	450	23–27	0.14	1.1	807
DP 350/600	350	600	24–30	0.14	1	976
DP 400/700	400	700	19–25	0.14	1	1028
TRIP 450/800	450	800	26–32	0.24	0.9	1690
DP 500/800	500	800	14–20	0.14	1	1303
CP 700/800	700	800	10–15	0.13	1	1380
DP 700/1000	700	1000	12–17	0.09	0.9	1521
Mart 950/1200	950	1200	5–7	0.07	0.9	1678
Mart 1250/1520	1250	1520	4–6	0.065	0.9	2021

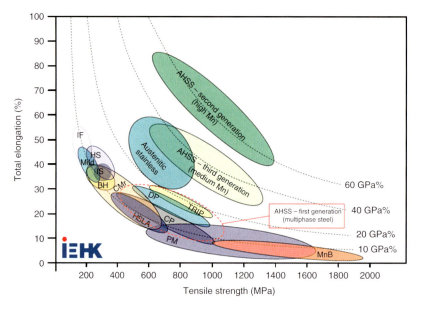

Figure 1.2 Strength-formability relationships for mild, conventional high strength steels (HSS), and three generations of advanced high strength steels (AHSS). Source: After Ref. [9].

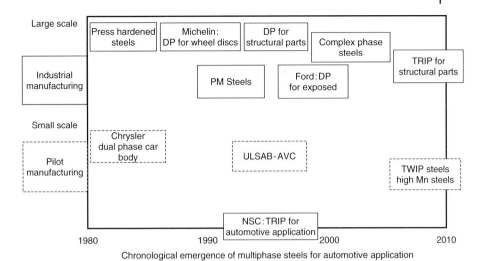

Chronological emergence of multiphase steels for automotive application

Figure 1.3 The rapid development of multiphase steels for automotive applications since 1980s.

are included, as well as austenitic steels for comparison, but will not be discussed in detail.

While HSS typically shows 10 GPa% ecoindex, first generation AHSS shows 15–20 GPa% and second generation AHSSs, namely TWIP steels (twining induced plasticity) with Mn content, show between 15 and 25 wt%, providing high strength–ductility combinations with up to more than 60 GPa%. Third generation AHSSs are currently being developed as an alternative with medium Mn (8–12 wt%) content aiming at properties between those of first and second generation AHSS 30–45 GPa% [2].

Some exemplary steps in the application of multiphase steels within automotive industry can be found in Figure 1.3. The development as well as the industrial implementation started with DP steels for wheel disc application in 1990 [14, 15]. Nowadays, TRIP and CP steel qualities have been documented in several applications at least within the advanced quality segment [15, 16], and TWIP steel is still on pilot-scale development stage.

1.2.2
Multiphase Microstructure Design

The microstructure of AHSS steels is the main reason for the difference in the mechanical properties of conventional HSS and AHSS steels. The presence of martensite, bainite, and/or retained austenite in the microstructure of first generation AHSS steels leads to the production of unique mechanical properties compared to the conventional microalloyed steels. These steel grades also demonstrate a superior combination of high strength and good formability because of their low YS and high UTS, which results in higher work hardening rates [2,

Table 1.3 Typical microstructure and tensile strength for first generation AHSS [18].

First generation AHSS	Microstructure	UTS (MPa)
DP	Ferrite + martensite	400–1000
TRIP	Ferrite + martensite/ bainite + retained austenite	500–1000
CP	Ferrite + bainite + pearlite	400–1000
MART	Martensite	700–1600

9–13]. The target microstructures of the third generation AHSS are ultrafine ferrite matrix with bainite/martensite, stabilized high austenite fractions, composites, and nanoprecipitates [17].

AHSS are classified in three categories based on the microstructure of these steels [4]. Ferrite-based steels are the first generation of AHSS and have proved to be economically feasible replacements for conventional HSS (Table 1.3). The main deformation mechanisms are homogeneous or inhomogeneous dislocation gliding (Figure 1.4).

A desire to produce materials with significantly higher strength and high formability in parallel has led to the evolution of the second generation AHSS, which were developed on the basis of an austenitic microstructure with excellent formability, including TWIP steel, lightweight steel with induced plasticity (L-IP), and shear band formation-induced plasticity (SIP) steel.

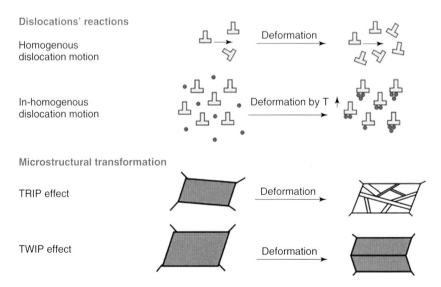

Figure 1.4 Deformation mechanisms in steels.

The TRIP and TWIP effects are based less on each specific change in shape – the shape and volume keep changing – as in the generation of many new interfaces that arise continuously during plastic deformation and thus continuously form new barriers to the movement of dislocations and contribute to an exceptionally strong bonding. In the structure description, therefore, it is particularly important to know the stability of the austenite phase, which is controlled by the stacking fault energy (SFE). Thus, apart from the homogeneous dislocation movement, the deformation mechanisms that are usable in steels are the TRIP effect, the TWIP effect, and a recently discussed inhomogeneous dislocation movement, which is sometimes also referred as SBIP or MBIP (shear-band-induced plasticity or micro-band-induced plasticity, respectively, Figure 1.4).

Nevertheless, the very high Mn content leads to specific disadvantages during processing (mainly, but not only, segregation). On the other hand, the impressive combination of strength and ductility give specific challenges to tools and dies during component manufacturing. Thus, the development of a third generation AHSS has been aimed to fill the gap between the acceptable balance of the strength and ductility of the first generation AHSS and the complex and costly manufacturing of the second generation AHSS using intermediate Mn content. It is represented by a region in Figure 1.2, which shows that the new generation of steels should offer ecoindex at 40 GPa% by presenting ultimate tensile strength in the range 600–1600 MPa and total elongation in the range 30–50%.

Since microstructure determines the behavior and performance of AHSS, design and control of the microstructure becomes essential in order to achieve these desired values [17]. The data in Table 1.4 show the UTS and uniform elongation for single phases of steel. It is evident that austenite presents high ductility and martensite shows high strength [1]. As reported by Matlock *et al.* by application of the model of Mileiko [19], the mechanical properties required for the third generation of AHSS could be realized by producing AHSS with austenite and martensite microstructure [1, 17] (Figure 1.5).

1.2.3
HSLA Steel

The development of high-strength low alloy (HSLA) steels over the past 50 years has been and continues to be of great importance. Advances in properties such as formability and weldability, and increased performance with tougher and

Table 1.4 Mechanical properties of single phases in steel [1].

Constituent	UTS (MPa)	Uniform elongation (–)
Ferrite	300	0.3
Martensite	2000	0.08
Austenite	640	0.6

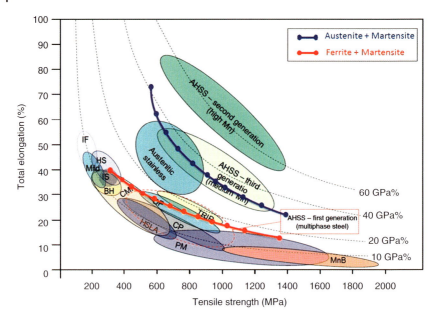

Figure 1.5 Superposition of the strength/ductility predictions using Mileiko's model on the strength/ductility map shown in Figure 1.2. For this figure, the austenite was assumed stable in contradiction of martensitic transformation with deformation.

stronger grades enable these HSSs to remain competitive with other materials in transportation applications. All these desirable properties are obtained by a judicious adjustment of its chemical composition and the implementation of a controlled rolling schedule [20].

In fact, strengthening in these steel grades takes place through dislocation glide, grain refinement, solid-solution, precipitation, and phase transformation strengthening mechanisms. Here, grain refinement, which can be controlled by advanced precipitation design, is the most important mechanism as it increases both strength and toughness [21].

HSLA steels contain a very small percentage of carbon, medium manganese level, and relative low levels of phosphorus and sulfur. It may also contain small amounts of copper, nickel, niobium, nitrogen, vanadium, chromium, molybdenum, silicon, or zirconium. Therefore, HSLAs are referred to as *microalloyed*.

There are basically four main steps in HSLA processing [21, 22]. First, the slabs are reheated, leading to complete austenitization and dissolution of precipitates. Second, they are rolled at high temperature in the roughing phase, resulting in recrystallization of the large austenite grains. Third, a final hot-rolling process follows at lower temperatures, and finally cooling processes will incorporate the microstructure [21, 23]. The alloying elements can act as solutes or they may transform into precipitates, such as carbide or nitrides, creating barriers to dislocation and defect motion during deformation. The hot-rolling processes can affect the austenite microstructure, which controls the phase transformation

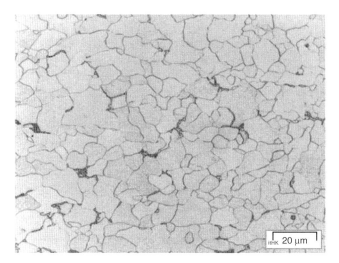

Figure 1.6 A microstructure of HSLA steel.

kinetics. Finally, the transformation temperature and cooling rates from austenite to ferrite determine the final microstructure [21, 22].

In Figure 1.6 a microstructure of HSLA steel consisting of a relatively fine grained ferritic matrix with a small amount of pearlite is demonstrated. Carbonitride precipitates of microalloying elements can also be observed. Pearlite is an undesirable strengthening agent because it reduces toughness.

As already mentioned, the alloying elements are added to these steels for strengthening purposes. These elements are intended to alter the microstructure of carbon steels, which is usually a ferrite–pearlite aggregate, to produce a very fine dispersion of alloy carbides in an almost pure ferrite matrix. This eliminates the toughness-reducing effect of a pearlitic volume fraction, yet maintains and increases the material's strength by refining the grain size, which in the case of ferrite increases YS by 50% for every halving of the mean grain diameter. Meanwhile, precipitation strengthening plays a minor role.

HSLA steels present a YS of 340–420 MPa and UTS of 410–510 MPa, while their uniform elongation ranges between 10 and 25%. Consequently, because of their higher strength and toughness, HSLA steels usually require 25–30% more power to form, as compared to carbon steels [21]. The nominal stress–strain curve of HSLA 350/450 steel is shown in Figure 1.7.

HSLA sheets are used in trucks, construction equipment, off-highway vehicles, mining equipment, and other heavy-duty vehicles for constructing chassis components, buckets, grader blades, and structural members outside the body [22, 24].

1.2.4
BH Steel

BH is an advanced technique to produce low carbon steels with high strength, used for car bodies, which require a good combination of formability and high strength.

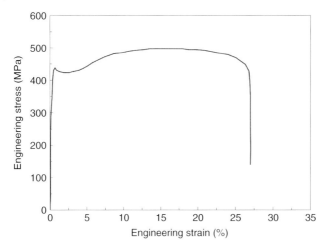

Figure 1.7 Engineering stress–strain curves of HSLA 350/450 steel.

This method enables the usage of low carbon, ultralow carbon, and IF steels instead of using expensive high alloy steels. An optimized batch annealing treatment of cold-rolled sheet is necessary to have enough carbon in solution for the later BH, which gives additional strength for automotive bodies and panels [25–27]. Thus, BH steels get an increase of strength by carbon aging during heat treatment. In the initial state, dissolved carbon atoms have a concentration of approximately 10–20 ppm, and are homogeneously distributed in the ferritic matrix; the material is resistant to aging at room temperature.

By deformation of the sheet new dislocations are introduced, at which carbon atoms attach at the subsequent heat treatment at 170 °C. Solute carbon and nitrogen atoms diffuse and interact with mobile dislocations, resulting in the Cottrell atmosphere occurrence. Therefore, the BH strengthening increases with increasing solute carbon. On account of the Cottrell clouds, further dislocation movements are more difficult, and the material gets an increase in yield by dislocation blocking in addition to the work hardening. This process is shown schematically in Figure 1.8 [26–29].

BH steels are mainly used in the automotive industry for structural and exterior panels. The advantages of these steels are that they have a good cold formability because of the relatively low YS and high *r*-and *s*-values during the manufacturing state and that the heat treatment is carried out anyway during the paint backing. The yield increase is about 40–60 MPa during annealing at 170 °C for 20 min. Also, the BH effect is the return of a distinct yield point [27].

1.2.5
DP Steel

DP steels are characterized by a microstructure consisting of hard martensite particles in a soft and ductile ferrite matrix. This morphology presents remarkable

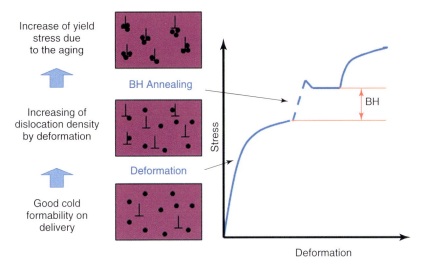

Figure 1.8 Strain- and bake-hardening index and increase in the yield strength during bake hardening.

mechanical properties. It can be attributed to high work hardening rate, good ductility, and superior formability in comparison to other HSLA steels [30–32]. Besides the parallel existence of a ductile matrix-like ferritic phase and island martensitic grains, the properties are imposed by transformation-induced geometrically necessary dislocations (GNDs). Owing to this GND effect continuous yield behavior, low YS without decrease in tensile strength, high initial hardening rate, and finally retarded necking behavior are caused. There are two main methods of producing DP structures: cold rolling plus intercritical annealing and quenching, or hot rolling with controlled quenching as shown in Figure 1.9a.

The intercritical annealing process involves heating the cold-rolled steel (with ferrite–pearlite microstructure) between A_{c1} and A_{c3} temperatures (ferrite–austenite phase region), where a certain amount of austenite is formed and then cooled rapidly to room temperature. Subsequently, the steel may be aged at low temperatures of up to 400 °C for 1.5 min so that the quenched-in solute carbon precipitates increase the ductility. In some process lines, quenching is interrupted to galvanize the sheet (hot dip galvanizing line). In this case, the steel is held at the temperature of the zinc bath, which is approximately 460 °C in hot dip galvanizing lines.

For hot-rolled DP steels, the hot-rolling process takes place in the austenitic region, and steel is then cooled into the intercritical region where ferrite nucleates and grows from unstable austenite, which results in a fine two-phase morphology. The material is then quenched sufficiently fast to transform the austenite into martensite [33–35]. Figure 1.9b demonstrates a microstructure of DP steel with 20% martensite.

DP steels offer an improved combination of strength and ductility. These steels have high strain hardening capacity. This gives them good strain redistribution

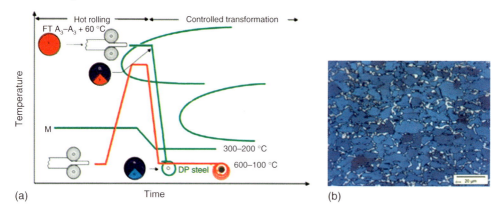

(a)

(b)

Figure 1.9 (a) Schematic production routes of cold-rolled and hot-rolled DP steels and (b) microstructure of DP steel with 20% martensite.

capacity, and thus ductility. As a result of strain hardening, the mechanical properties of the finished part are superior to those of the initial blank. High mechanical strength of the finished part causes excellent fatigue strength and good energy absorption capacity in these steels, making them suitable for use in structural parts and reinforcements. Strain hardening and strong BH effect give these steels excellent potential for reduced skin and structural part weight. The other attractive properties of DP steels for automotive applications are continuous yielding, high YS to tensile strength ratio, reduced cost, and excellent surface finishing because of the elimination of the yield point elongation [36–40].

Figure 1.2 shows the tensile strength and total elongation of ferrite–martensite DP steels compared to low alloy steels strengthened by solid solution and participation hardening [41]. According to this figure, DP steels with tensile strength and total elongation in the range 250–1000 MPa and 10–35%, respectively, show a superior combination of strength and ductility compared to other steels. The application of DP steels in automotive components such as bumpers, wheels, wheel discs, pulleys, and springs has demonstrated up to 30% weight reduction with an increase in the component life. DP steels display crashworthiness features, because they also have a very good post-uniform elongation. Therefore, DP steels are used in the crash-sensitive parts in the front and rear rails of automobiles [42].

Ferrite is the softest phase of steel, which has a body-centered cubic (BCC) crystal structure that can contain only a maximum of 0.02% carbon. The primary phase in low carbon steels is ferrite [43]. The flow stress of ferrite depends on its chemical composition and its microstructural features [44]. It can be expressed in terms of various strengthening components. It is important to evaluate these strengthening mechanisms, the factors controlling them, and their effect on other properties such as toughness and ductility [45–47]. Different contributions to the flow stress of ferrite can be addressed as lattice resistance, dislocation hardening, grain size strengthening, solid solution strengthening, and precipitation or dispersion hardening.

Martensite is a nonequilibrium phase that develops when austenite phase is rapidly quenched, so that there is no diffusion of the carbon. It is a supersaturated solid solution of carbon in ferrite with a generally body-centered tetragonal crystal structure (BCT). Since martensite is generally very hard, it can be considered as the reinforcing phase of DP steel, and thus, its main role in DP steel is to carry a significant part of the applied load. Accordingly, to describe the mechanism of the strengthening effect of martensite in DP steels, recognition of the parameters controlling the martensite strength is important [48].

In some cases, there is a small amount of bainite and/or retained austenite within the martensite structure depending on the chemical or process parameter and sheet thickness, and thus, the hard constituent in DP steels should be considered as a second phase containing martensite with or without bainite and/or retained austenite [49]. The strength of the hard phase in the as-quenched state depends primarily upon the carbon content, and increasing the carbon content in martensite leads to an increase in the YS [47, 49]. The strength of martensite depends also on the substitution alloying elements such as manganese and silicon that cause the solid solution strengthening, but it is considered that this effect plays a secondary role compared to the strong strengthening effect of carbon atoms [50, 51].

Owing to the combination of high work hardening rate and excellent elongation in DP steels, this grade of steel depicts much higher UTS compared to the conventional steel with similar YS. Figure 1.10 illustrates the quasi-static stress–strain response of DP600 steel compared to HSLA steel with similar yield strength (YS = 350 MPa). The DP steel exhibits higher initial work hardening rate, uniform and total elongation, UTS, and lower YS/TS ratio in comparison to the HSLA steel with the same YS [2].

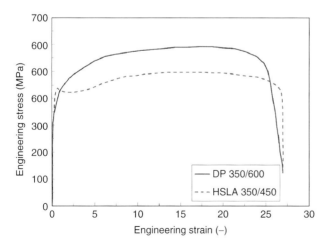

Figure 1.10 Comparison of engineering stress–strain curves of HSLA 350/450 and DP 350/600 steels.

According to the thickness, dual-phase steel products can be broadly categorized into three groups. These are products with thickness more than 2.5 mm, between 2.5 and 2 mm, and less than 1.5 mm. The thickest one is used for wheel disc application. They are hot-rolled and the most important property for this application is hole-expansion ability. The medium thickness grades are hot-rolled products too, and are used as bumpers in cars or chassis in larger vehicles, such as trucks. The chassis of the larger vehicles are generally gas metal arc welded. Automakers are interested in these grades for their high strength and excellent crash-worthiness. The thinnest sections are cold-rolled-annealed products and are joined by resistance spot welding. They are chosen for their formability combined with high strength and moderate hole-expansion ability [52].

1.2.6
TRIP Steel

TRIP steels are currently the widespread steel options because of their high mechanical properties, weight reduction, and improved safety performance. These steels present high strength levels with high ductility, even higher than comparable DP steels. TRIP phenomenon can account for the ductility of the steels, which can be outstandingly improved by the strain-induced phase transformation in which the retained austenite plays a significant role. Here, carbon-enriched austenite is a metastable phase at room temperature. Therefore, the processing routes of TRIP steels are designed to maintain austenite in the multiphase microstructure at room temperature [53–55].

There are two procedures to generate TRIP steels. The first process includes an intercritical annealing of cold-rolled sheet, as for DP steels. The multiphase steel is initially heated to a temperature of 780–800 °C between A_{c1} and A_{c3} in austenite plus ferrite range, and during this time, a controlled volume fraction of austenite is formed inside ferrite grains and at ferrite grain boundaries. This intercritical annealing is following a fast cooling to the isothermal holding stage in the bainite treatment. The cooling rate is approximately 15–25 °C s^{-1} to avoid ferritic microstructure formation. The holding stage is kept in the range 350–500 °C and during this time, part of austenitic microstructure transforms to bainite; however, carbon redistributes from bainitic ferrite to the surrounding residual austenite islands. As a consequence, M_s temperature is reduced to below room temperature; in the meantime, the stability of retained austenite is notably increased by carbon enrichment. The higher silicon and carbon content of TRIP steels also result in significant volume fractions of retained austenite in the final microstructure. This procedure is represented schematically in Figure 1.11 [54, 55].

Hot rolling is the second procedure used to develop TRIP steel, in which two stages are applied to develop the hot-rolled TRIP steel. The first stage is austenitic hot rolling, combined with a holding period in the austenite plus ferrite range. Afterwards, this step is followed by coiling of the material in the bainite formation (about 500 °C), during which partial austenite transforms to bainite.

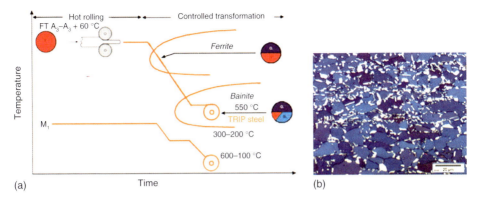

Figure 1.11 (a) Schematic production routes of cold-rolled and hot-rolled TRIP steels and (b) typical micrograph of TRIP steel.

During the hot-rolling process, the pearlite formation and cementite precipitation are suppressed. This procedure is represented schematically in Figure 1.11a [56].

The alloying elements play a major role for the production route and for the properties of the steels. TRIP steels have higher carbon content than DP steels. And also manganese, aluminum, and silicon are the most important elements for low alloy TRIP steels. Although, silicon has a great effect on the development of TRIP steels as it retards cementite precipitation during isothermal bainite holding, in some cases it is replaced by other elements because of its negative impact in coating [56–59]. The typical microstructure of TRIP steels consists of ferrite (50–60%), retained austenite (5–15%), and bainite (25–40%) (Figure 1.11b). The retained austenite transforms into martensite during deformation, resulting in a higher work hardening rate [60]. This transformation propagates dynamically and is strain-induced.

Figure 1.12 shows a comparison between the stress–strain response of HSLA, DP, and TRIP steels with a similar yield stress. TRIP steels show lower initial work hardening rate compared to DP steel. Because higher martensite content in DP steel is more, free dislocations are generated in the ferrite grains by the formation of surrounding martensite. In TRIP steels, the microstructure of martensite is displaced by the bainite. As a consequence, free dislocations are less effective in ferrite matrix and the yield processes with less initial hardening [61, 62]; nevertheless, the hardening rate persists at higher strains accompanied by TRIP effect where the work hardening of the DP steel starts to diminish. TRIP steel also shows higher tensile strength and ductility in comparison to DP and HSLA steels with similar YS.

In TRIP steels, strength–ductility balance can be adjusted by controlling the stability and content of the retained austenite. Therefore, TRIP steels can be used in current steel components reducing the volume against a mild steel formed component by 20%, while maintaining the same stiffness [15, 16]. Furthermore, with the weight savings, the amount of inertial losses could be reduced by 11%.

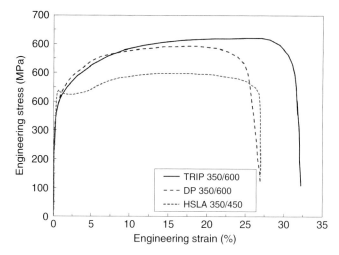

Figure 1.12 Comparison between the stress–strain response of HSLA 350/450, DP 350/600, and TRIP 350/600 steels.

Typical thicknesses of TRIP steels are 2–5 mm for hot-rolled grades and 0.6–1.5 mm for cold-rolled grades. These steels are used to construct the complex parts of automobiles, where the best formability is required and the part cannot be manufactured by DP and CP steels. For instance, applications of DP and TRIP steels can be for car body panels, pillars, and crash-relevant parts such as side impact beams, under body parts, and wheels. The energy absorption during the TRIP effect also enables an application to be used for crash-relevant components (e.g., B-pillar) [63, 64].

1.2.7
CP Steel

High-strength CP steel as an AHSS is characterized by a multiphase microstructure consisting of a fine ferrite matrix and a significant volume fraction of hard constituents, such as bainite, martensite, and small amounts of retained austenite, which is further strengthened by fine precipitates [65–69]. CP steels include many of the same alloy elements found in DP and TRIP steels such as manganese (Mn) and silicon (Si); however, small quantities of niobium (Nb), titanium (Ti), and/or vanadium (V) are additionally added to these steel grades in order to get fine strengthening precipitates. The microalloying elements postpone the austenite to ferrite transformation and correspondingly lead to the formation of bainite and martensite. An example of CP steel microstructure is shown in Figure 1.13. It is evident that CP steel has an extremely fine microstructure.

CP steels have complex structures in their matrix based on bainitic phases, and the production procedure of these steels depends strongly on the processing conditions. They are mostly produced by hot-rolling and controlled cooling processes. The exposed coiling temperature used in this process must be chosen carefully in order

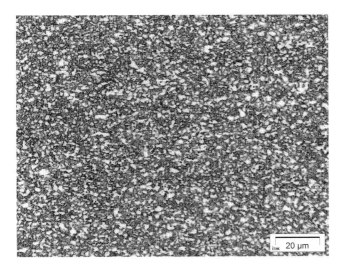

Figure 1.13 A microstructure of CP steel.

to achieve the desired bainite fraction and distribution in the matrix and thereby to achieve the desired mechanical properties [70]. Complex steels are suitable for hot forming at temperatures below 700 °C [71]. However, cold-rolled CP steels have recently been preferred. According to the flow chart of the process, the cold-rolled steel is intercritically annealed and then it is held isothermally at a temperature higher than the martensite start temperature to control the bainitic transformation and also the related mechanical properties [69, 72].

Since CP steels consist of a significant volume fraction of bainite, these steel grades represent very high UTSs and higher level of YS at tensile strength levels equal to those of DP steels. CP steels show a tensile strength of approximately 800 MPa and above with an elongation of approximately 10–15% (total elongation) [73, 74]. Also, by balancing the percentage of ferrite, bainite, martensite, and precipitation hardening phases in the microstructure, CP steels will provide an attractive combination of high strength, wear resistance, good cold formability, and weldability. CP steels are also characterized by continuous yielding and high uniform elongation. CP steels with the bainitic matrix have superior formability because the difference between the hardness of bainite and martensite is quite insignificant. Comparison of the stress–strain characteristics of conventional DP600 with CP steel demonstrates a clearly higher YS level of CP steel, as schematically shown in Figure 1.14 [75].

CP is characterized by high energy absorption and high residual deformation capacity. Therefore, CP steel is an excellent choice for applications where a combination of high strength and ductility together with high energy absorption and high residual deformation capacity is required. Consequently, these steels are utilized in the automotive industry for constructing structures in the high-energy absorption area of car body such as bumpers and B-pillar reinforcements [65, 72].

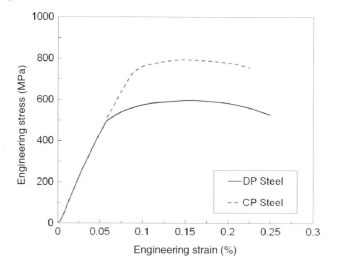

Figure 1.14 Comparison of quasi-static stress–strain behavior of CP 780 and DP 600 steels.

1.2.8
PM Steel

Martensite transformation hardening is considered as a convenient mechanism, which is employed to achieve further increase in the yield and tensile strengths. However, martensitic steels present generally very little uniform elongation. And consequently, their applications become limited. Therefore, these steels are often subjected to post-quench tempering to improve ductility, and can provide sufficient formability even at extremely high strengths [4].

In martensitic steels, austenite to martensite transformation takes place during quenching on the run-out table or in the cooling section of the annealing line, which leads to a microstructure including principally lath martensite, as shown in Figure 1.15. Steels with a lath martensite microstructure provide an excellent toughness at yield strengths of about 1000 MPa. In this regard, martensitic steels provide the highest strengths among the AHSS, up to 1700 MPa UTS. This high strength–high toughness combination is obtained in the as-quenched matter without any additional tempering process; therefore, the heat treatment process of steels with lath-martensitic microstructure is always more economical, compared to the heat treatment process of conventional tempering steels [76]. Martensitic steels follow a similar cooling pattern as CP steels; nevertheless, the chemistry of Martensitic steel (MS) steels is adjusted to produce less retained austenite and form fine precipitates to strengthen the martensite and bainite phases. These steels consist of martensitic matrix and a small amount of ferrite and/or bainite.

The increased strength, hardness, and brittleness of steels after quenching from austenite are the result of martensite formation. The strength of martensite depends on the carbon content and the strengthening mechanism is mostly

Figure 1.15 Microstructure of hot-rolled martensitic steels.

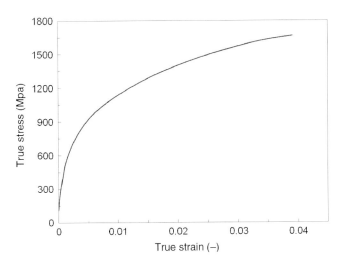

Figure 1.16 Flow curve of PM steel MS-W 1200.

activated by solid solution hardening because of the interaction between carbon and dislocations. Increasing carbon content in the matrix also increases the yield and tensile strengths of the martensitic steels [72, 76]. An exemplary flow curve can be found in Figure 1.16.

Applications of Partial martensite steel (PM) steels in transport industry can be addressed to the construction of structures such as removable chassis components (frame cross members, suspension control arms, etc.), side impact intrusion beams, vehicle seat frames and support, entire chassis frame rails, and entire vehicle space frames [76–79]. As an example, the application as a bumper beam in a sheet thickness of 1.75 mm may yield a weight saving of more than 40% in comparison with conventional HSLA steel [80].

1.2.9
TWIP Steel

TWIP steels offer an extraordinary combination of mechanical properties in terms of strength and ductility far beyond the combinations being found in the above discussed low alloyed sheet steel concepts [81, 82].

TWIP steels have high manganese content (15–30%) that causes the steel to be fully or mostly austenitic at room temperature. In addition to the above discussed deformation mechanisms, in these steels deformation is controlled by the formation of deformation twins causing high hardening rate by dynamic grain refinement (Figure 1.5). The resultant twin boundaries act as additional grain boundaries and strengthen the steel [83]. TWIP steels usually have high uniform elongation and high work hardening rate. Some microstructures that show the twin formation after deformation are given below, in Figure 1.17a. In the electron backscatter diffraction (EBSD) data in Figure 1.17b,c the strong epsilon martensite and twin formation can be seen due to deformation.

SFE, which depends on chemical composition, is the most important parameter influencing deformation mechanism type. Manganese is the main alloying element in TWIP-steels. It preserves the austenitic structure based on the ternary system

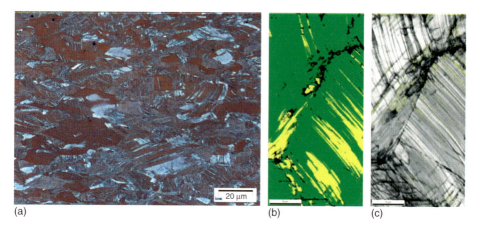

(a)   (b)   (c)

Figure 1.17 (a) Micrograph of TWIP steel after deformation, (b) EBSD data indicating phases green austenite, yellow epsilon martensite, and (c) EBSD data indicating orientation changes (grain boundaries and twins).

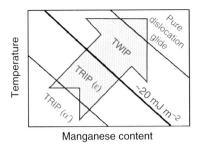

Figure 1.18 Schematic presentation of deformation mechanisms related to the SFE value as a function of chemical composition and temperature.

of the Fe–C–Mn system. As a consequence of evolution of SFE with changing chemical composition, the deformation mode changes from TRIP to TWIP to SLIP, while at low Mn content the TRIP effect is dominant [84]. High manganese steels are plastically deformed through strain-induced martensite formation, mechanical twinning, and pure dislocation glide, because of increases in the SFE value. If SFE is very low (≤ 20 mJ m^{-2}), martensitic induced plasticity is favored. However, higher SFE (20 mJ m^{-2}) suppresses martensitic phase transformation and favors mechanical twinning until 60 mJ m^{-2} (Figure 1.18). And at high SFE (≥ 60 mJ m^{-2}) the glide of perfect dislocations is dominant because the partition of dislocations into the Shockley partial dislocations is difficult [84].

Current alloy development focuses on fine tuning mechanical properties and suppressing the risk of delayed cracking. Here, Al, which increases ductility by increasing TWIN effect by decreasing SFE, is under investigation [85].

According to the design criteria, TWIP steels are mostly used at seat cross member parts to prevent passengers from front crashing and side crashing, at A- and B-pillar parts for strength, and at crash-relevant parts (e.g., longitudinal beam) and complex geometry parts or parts with special properties (e.g., suspension dome that requires good fatigue properties) [85].

1.2.10
MnB Steels for Press Hardening

Classical ferrite-based sheet concepts are either good formability but low strength as in mild steels or high strength with poor formability as in manganese steels. Crash-relevant structural components strength is the most relevant property being offered in an effective and cheap way by martensitic microstructure. Nevertheless, classically martensitic steels might show acceptable toughness after optimized tempering, but deformability is too low for complex shaping. Here, a combined process has been introduced – press hardening or hot stamping that combines good formability in the as-received ferritic pearlitic state, and high strength with acceptable toughness after specific annealing and quenching in the die [86–90].

The press hardening concepts are based on boron steels that have both good formability in ferrite pearlite state and high strength in quenched state. In press

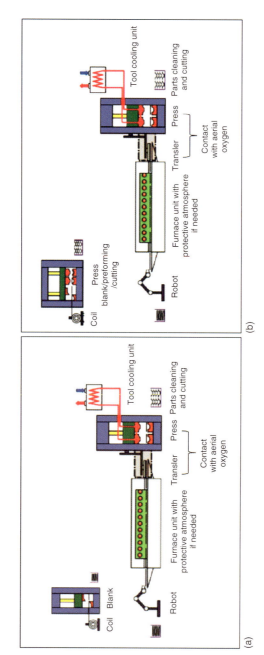

Figure 1.19 Schematic illustration of (a) direct and (b) indirect hot stamping [93].

hardening or hot stamping process, the hot blank is placed in a tool that is cooled by water. The tool remains cool for approximately 15 s. Then, the part is taken from the tool at a temperature of 80 °C and air cooled. The production rate is 2 or 3 stampings per min. As can be seen in Figure 1.19, there are two different hot stamping methods; direct and indirect hot stamping. In direct hot stamping, the blank is austenized at a temperature of 900–950 °C, then placed in a die and formed at high speeds. In indirect hot stamping, the parts are first cold formed, austenized in a furnace, and then quenched in the cooled die [89–93].

After rolling, quenchable boron steels provide good blanking and cold forming ability. After annealing and quenching and tempering, they are well suited for wear- and abrasion-resistant applications, and lead to large weight savings up to 50% compared to an HSLA grade.

High mechanical properties of boron steels are the result of a 10–30 ppm boron addition. Boron segregates to austenite grain boundaries and retards the ferrite nucleation and austenite decomposition. The hardenability depends on the austenization time and temperature, grain size, heat treatment, and alloying elements, mainly carbon content [92]. As can be seen in Figure 1.20, 27MnCrB5 contains pearlite and high amounts of ferrite in as-received condition. After stamping, a fully martensitic microstructure could be seen. 37MnB4 steel has more homogeneous pearlite/ferrite distribution in the matrix. It also undergoes full martensitic transformation after hot stamping. However, the morphologies of the formed martensite are different. By increasing the cooling rate, it might be possible to obtain a fully martensitic microstructure in Martensitic-phase steel (MSW) and TRIP steels also [93].

The comparison of the true strain–true stress curves of high Mn steel and low Mn steel are given in Figure 1.21. In as-received condition, both of them have almost the same stress–strain values. After stamping, the steels with higher Mn content have higher stress, but lower elongation values [92, 93]. The final properties in the product depend on additional tempering parameters.

The application of boron steels has been developed for the automotive industry. They are mostly used in structural, crash relevant parts such as door beams, A- and B-pillars, front and rear bumpers, and side impact beams [93].

1.2.11
LH Steel

Air-hardened (low-hardenability (LH)) steels are developed by Salzgitter to create a sheet steel quality with good formability and the option of a mild strengthening after deformation by an annealing process without quenching. Here, the quenching process shall be avoided in order to overcome the distortion problem. Thus, this steel quality has to be discussed separately in the delivered and the annealed state after deformation [94–96].

Their excellent hardenability is not only due to carbon and manganese alloying elements but also due to chrome, molybdenum, vanadium, boron, and titanium. They have ferritic microstructure with carbonitride precipitates, small amount of retained austenite in soft as-delivered condition (Figure 1.22). The presence of

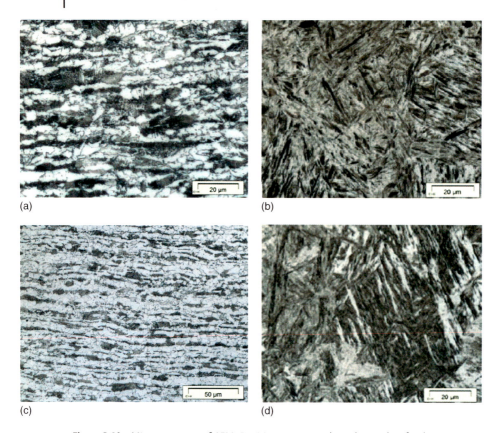

Figure 1.20 Microstructures of 37MnB4 (a) in as-received condition, (b) after hot stamping and 27MnCrB5, (c) in as-received condition, and (d) after hot stamping.

carbides in the fine-grained ferritic matrix can be observed in the SEM image clearly. Meanwhile, this steel shows good coatability and weldability when it is treated and tempered [95].

The excellent formability–high strength combination is considered as a result of the so-called post forming heat treatment(PFHT) process that offers an alternative to the press hardening of UHS components. The heat treatment is conducted in a furnace in inert gas atmosphere, cooled in air or protective gas, and finally annealed. During hardening, the microstructure is heated to austenite region (∼950 °C) in inert gas atmosphere and then cooled in air or inert gas, leading to martensitic structure and high strength values. As a result of the following annealing process, the residual stresses inside the microstructure decrease, resulting in low hardness, but high ductility. A comparison of strength level before and after forming and annealing is shown in terms of flow curves in Figure 1.23. The low mechanical properties in the untreated soft condition increase after air hardening to tensile strength values in the range 800–1000 MPa, while the YS increases in the range 600–800 MPa [95–99].

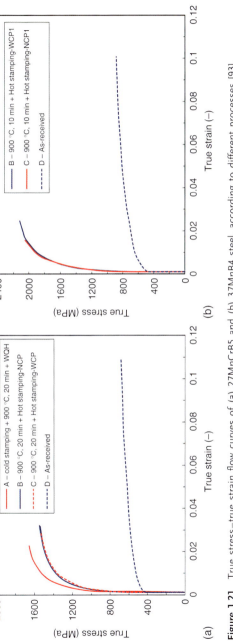

Figure 1.21 True stress–true strain flow curves of (a) 27MnCrB5 and (b) 37MnB4 steel, according to different processes [93].

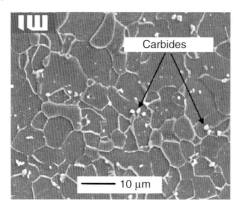

Figure 1.22 Scanning electron micrograph of the initial structure of LH800 with primary carbides in different grains [95].

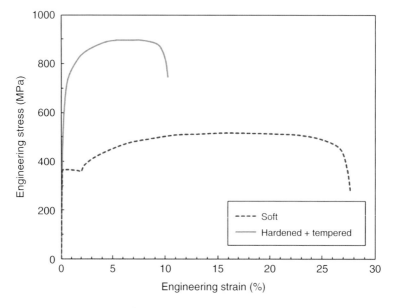

Figure 1.23 Stress–strain flow curve of LH900 untreated and after hardening and tempering [99].

The characteristics of these steels are that they show excellent formability after cold rolling, and are therefore very suitable for the production of complex components and usage under cyclic loads. They possess higher strength than HSS; therefore, the plate thickness can be reduced easily. They can be formed both by deep-drawing and by other forming methods. They were developed with the aim of reducing the vehicle weight and providing crash safety, and first used by Mercedes-Benz, in the new E Class, in 2009. In particular, they can also be used in welded elements that are subjected to high loads, and in those that perform safety-related support functions, such as cardan shafts, hybrid, and transverse control arms, roll bars, trailer axles, and A-/B-pillars [94–99].

1.2.12
Q&P Steel

Quenching and partitioning (Q&P) steels have been currently introduced by Speer [100], to fill the gap between the first and second generation of AHSS in the sheet properties diagram (Figure 1.2). Q&P procedure is primarily developed for the production of austenite-containing steels, based on the assumption that carbide formation can be suppressed by suitable alloying elements, and austenite can be highly enriched in carbon because of the rejection of carbon from supersaturated martensite under para-equilibrium conditions. Therefore, the key point for developing Q&P steel is to achieve a significant amount of retained austenite in the steels, which can transform to martensite at higher strains, so that the work hardening rate of the steel will increase. Because austenite is not an equilibrium phase in the steel at room temperature, the main challenge of producing Q&P steel is the stabilization of a high content of retained austenite in the final structure [100–105].

In order to produce Q&P steel a special thermal treatment is required, which consists of three steps, as shown in Figure 1.24. First, a sample with a fully austenitic microstructure is rapidly cooled to a specific quench temperature (QT), between the martensite start temperature M_s and the martensite finish temperature M_f, to create a controlled volume fraction of martensite. The samples are then held at a partitioning temperature (PT) to allow the carbon to diffuse from martensite regions to surrounding austenite regions. This step increases the austenite stability,

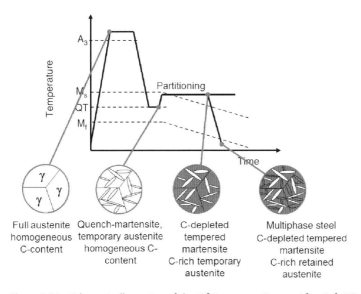

Figure 1.24 Schematic illustration of the Q&P process. Source: After Ref. [98].

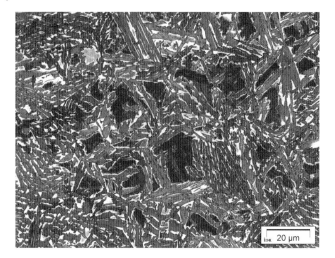

Figure 1.25 An optical microstructure of Q&P.

and results in a higher austenite fraction. Finally in the third step, the steel is cooled from PT to room temperature, resulting in the formation of additional martensite [103, 104, 106]. These three steps of heat treatment can be reduced to two steps by applying the same temperature for Q&P [107, 108].

An Light optical microscopy (LOM) microstructure of typical Q&P steel microstructure is depicted in Figure 1.25, including ferrite, martensite, and retained austenite.

The Q&P steels generally consist of carbon-depleted martensite and carbon-enriched stabilized austenite, which has the direct attraction of achieving higher strength levels. Since quenching is done between the martensite start temperature M_s and the martensite finish temperature M_f, a much higher amount of retained austenite is achievable, compared to low alloyed TRIP steels. Thus, the TRIP effect may be stronger than in low alloyed TRIP steel, and it can be seen in the flow curve with a strong increase in hardening during loading (Figure 1.26).

The tensile strength of Q&P steel increases by decreasing the amount of ferrite. The ductility and toughness of these steels increase by increasing the amount of retained austenite [109, 110]. Since the carbide formation consumes carbon in martensite, formation of carbide has to be avoided to get high amounts of retained austenite [104, 110]. Therefore, the value of carbon in the chemical composition of steel is a critical factor for the stabilization of retained austenite during cooling from partitioning temperature to room temperature and usually should be higher than 0.2%. Consequently, carbide-forming elements such as Ti, Nb, and Mo have to be avoided and some elements such as Al and Si should be added to suppress the precipitation of carbides and stabilize austenite [111–115].

Figure 1.27 shows the strength–ductility combinations characteristic of the Q&P steels in comparison to DP, TRIP, and MART steels. Mechanical properties of Q&P steels depend significantly on their microstructures [103]. As shown in this figure,

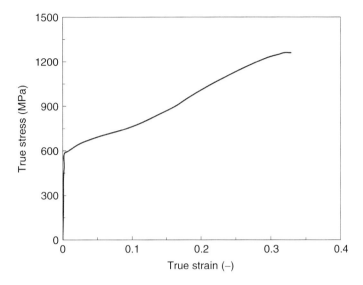

Figure 1.26 Flow curve of Q&P steel.

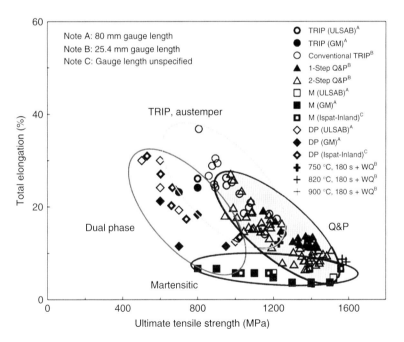

Figure 1.27 Total elongation vs. ultimate tensile strength for sheet steels processed with microstructures characteristic of TRIP, DP, MART, and Q&P steels. Source: After Ref. [103].

Q&P steels show very high strength levels combined with sufficient ductility. These mechanical properties make these grades of steels suitable for use in automotive applications such as pillars and anti-intrusion barriers [116]. For instance, the component of B-pillar is constructed successfully by Q&P 1000 steel [107].

Nevertheless, the preferred times and temperatures for the partitioning treatment and the optimum final quenching temperature that yields the maximum amount of retained austenite are not yet clearly established, and significantly more work is required to optimize Q&P processing and design alloys with regard to the current and future production constraints [101, 103].

1.3
Forging Steels

Concerning forging steels, running development projects aim on the one hand for improvement of properties, but more effort is spent later to increase process efficiency by shortening the process chain. Besides alloy development, based on conventional forging and hardening methods, some new technologies and steel qualities are developed to respond to the demands of the industry in a most economical way. This literature review is about the precipitation hardened microalloyed ferrite–pearlite steels, forging of high strength ductile bainitic steels, case-hardening steels with Nb-alloying, and TRIP steels, which are the results of these new strategies.

1.3.1
AFP Steel

These new kinds of forging steels are one of the results of developments in the automobile industry. They have two subgroups: carbon steels and microalloyed steels [117]. This concept is based on precipitation hardening of ferrite–pearlite steel (in German called *AFP steel*). The processing of this kind of steel eliminates the steps of quenching, straightening, and tempering after forging, which results in major energy savings and reduction of quench cracks. The requirement for higher strength, toughness, formability, and fatigue resistance in the automotive industry leads to the development of microalloyed forging steels [13].

The principle of the AFP steel is a controlled cooling process (e.g., slow cooling after air-cooling) after forging, which results in ferritic–pearlitic microstructure. However, as can be expected by carbon steels, such a pure microstructure will not yield the desired better mechanical properties, and therefore microalloying additions are offered to control the strength mechanisms. Especially, low fractions of titanium (Ti), vanadium (V), and niobium (Nb), and their interactions with carbon (C) and nitrogen (N) play an important role on the strength mechanisms of microalloyed steels by supporting the precipitation. These carbides, nitrides, and carbonitrides are required to obtain the strength of tempered martensite microstructure. They make a major contribution to the precipitation strengthening

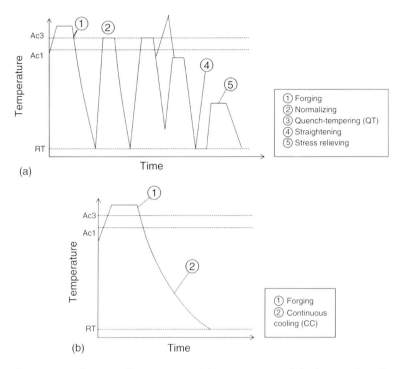

Figure 1.28 Schematic of (a) conventional heat treatment and (b) direct-cooling after forging.

by acting as obstacles for the dislocation, interface movements. By increasing the temperature the precipitates of Ti, V, and Nb are very stable. TiN, which is stable at around 1200 °C, controls the austenite grain size at high forging temperatures. Vanadium-Nitrides (VN) dissolves at low temperatures; therefore, it supports precipitation hardening during cooling, and afterwards the forging [118, 119].

As mentioned before, the process steps of these steels are quite simple. It is based on a rather slow cooling rate after forging. In Figure 1.28 the conventional heat treatment and direct cooling after forging are compared. By direct cooling after forging, cost and time savings can be achieved. The as-rolled or as-forged steels are reheated in order to control the dissolution of the microalloying elements in the steel composition, and then the reheated steels are deformed at high temperatures before transformation during cooling. At lower reheating temperature the solubility is reduced, which limits Nb addition, while V addition is getting important. By higher reheating temperatures, the solubility of Nb increases and therefore high amounts of Nb and also V can be used. Here, Nb refines the final microstructure [120, 121].

Nb has three important roles such as grain refinement, reduction of pearlite interlamellar spacing, and the related precipitation strength. Considering the 0.03% alloying content, Nb provides higher strength (150 MPa) than V (50 MPa). To obtain the similar strength value by V, a vanadium level of 0.07% is required [121].

Many vehicle components, especially in power train and suspension systems that require especially high fatigue resistance, are manufactured by these microalloyed forged steels. They are used just after direct cooling after forging without any heat treatment [118–120]. Forging gives a number of advantages such as enhancement of lower weight, less machining, and no heat treatment costs [117]. The application areas of forged steels in a vehicle are gears, crankshafts, connecting rods, axles, spindles, wheel hubs, and steering links. Crankshafts that are hot forged are mostly manufactured from medium to high carbon grades with high manganese contents. These higher strength steels provide lighter crankshafts that respond to the higher torque demands of the customers. Wheel hubs and spindles that are also hot forged, induction hardened, and machined have 0.5% C [122]. The final strength level reached within the AFP concept is shown in Figure 1.29 in comparison to the properties available from recently developed Nb microalloyed AFP (AFP-M) steel and high strength and ductile bainite (HDB) steel.

1.3.2
HDB Steel

Steels with bainitic microstructures show an excellent strength–toughness combination without any additional heat treatment. HDB steels have a tensile strength of >1200 MPa, YS of >850 MPa and elongation of 10% at room temperature.

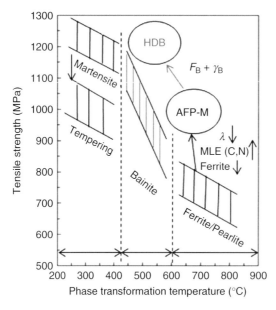

Figure 1.29 The strength of different kind of steels in different temperatures with the related microstructural features.

Figure 1.29 shows the strength of HDB steels compared to the classical AFP and Nb microalloyed AFP-M forged steels. To enhance a bainitic phase field and retard a diffusion controlled ferrite/pearlite phase transformation, alloying elements such as chrome (Cr), molybdenum (Mo), boron (B), and titanium (Ti) are added. Also niobium (Nb) is used to increase the YS.

HDB steels have a higher martensite fraction than the AFP steels. The cooling for employing a bainitic microstructure can be conducted not only continuously but also with isotherm holding steps in the bainitic field. The secondary phase in bainite is retained austenite that enhances better toughness.

The lamellar structure in bainite shows differences according to the transformation temperature. By increasing isothermal transformation temperature, the width of the lamellar increases, thereby decreasing both YS and tensile strength. However, elongation shows no dependency on this lamellar width [123].

1.3.3
Nb Microalloyed Case-Hardening Steels

Case-hardening steels are very common in the application of power transmission gears in vehicles because of their high strength in the case and high toughness in the core related to their carbon contents. The larger the case depth the higher is the fatigue resistance, which also strongly depends on the austenite grain size in the carburized case.

Nb has an important role on the mechanism of the case-hardening steels, such as retardation of recrystallization, suppression of austenitic grain size, and precipitation hardening [124]. Suppression of austenite grain size is needed during austenitization where Nb(C,N) precipitates act as obstacles at grain boundaries. Figure 1.30 shows the effect of Nb clearly; the average grain size of the Nb-free steel is 26 μm, while the average grain size of 0.04 Nb steel is 15 μm.

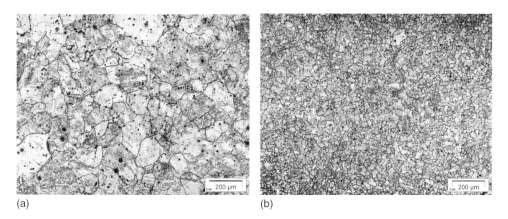

(a) (b)

Figure 1.30 Micrographs of the carburized case of (a) Nb-free and (b) 0.04 Nb.

Even in small quantities Nb delays the austenite to ferrite transformation because it has a low solubility in iron that leads to Nb segregation at grain boundaries. The sample with Nb possesses finer lath martensite in the core, finer needle type martensite in the case, and finer grain size, which provides a high fatigue limit, compared to Nb-free steels [124–127].

1.3.4
Nb Microalloyed TRIP Steels

The retained austenite amount in TRIP steels plays an important role in the improvement of the formability. Especially, bainitic ferrite lath structure provides a high stretch–flangeability combination [128]. Forging is generally conducted to destroy the cast structure, and the released high energy levels cause deformation [129].

As mentioned before, Nb has major effects on the microstructure and mechanical properties of the steels. Its grain refinement and precipitation hardening effects increase the strength values. It increases the stability of austenite resulting in the retention of austenite in the final microstructure. And also, it supports bainitic reaction. Especially, at faster deformation rates it enriches the retained austenite with carbon. Therefore, the retained austenite can be found in the final microstructure. Nb also enhances better bake hardenability [130].

Besides Nb, other alloying elements can also have some effects on the TRIP structure, as can be seen in Figure 1.31. Here, bainite (B), retained austenite (Ar), and some precipitates in the ferrite grain boundaries have been found. Addition of Nb and V provides the formation of (Nb,V)-carbides, instead of the coarse Nb carbides [131].

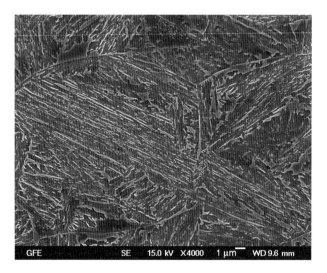

Figure 1.31 SEM images of Nb/V steel.

1.4
Casting Steel

1.4.1
Austempered Ductile Cast Iron (ADI)

Austempered ductile iron (ADI) is a concept aiming to strengthen the fatigue properties of cast iron. It can be interpreted as an answer of the casting industry to the optimized fatigue properties from forged components. ADI was introduced in the mid-1970s in commercial transportation applications in Europe. It consists of a microstructure containing spherical graphite inserted in a matrix, which in general is a mixture of bainitic ferrite and austenite plus small amounts of martensite and/or carbides. The bainitic ferrite is generated during isothermal transformation of austenite within an austempering process, with the bainite start temperature in the range 250–400 °C. This morphology presents excellent mechanical fatigue properties in comparison to classical cast [132]. ADI also shows good castability, lower processing cost, higher damping capacity, and 10% lower density in comparison with forged steels [133–137]. Therefore, ADI becomes a good candidate for many structural and wear applications in the automotive industry because of its high mechanical properties such as ductility, strength, fatigue strength, fracture toughness, wear resistance combinations, and design flexibility at low cost.

After casting, the austempering process of a component consists of two stages, as schematically shown in Figure 1.32. The first step is austenization, where the casted component is heated to temperatures between 850 and 950 °C with holding time of 15 min to 2 h. The second step is austempering. Therefore, after austenization, the

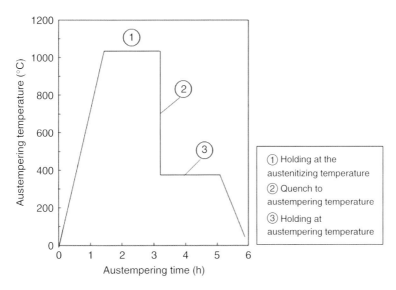

Figure 1.32 Schematic illustration of the austempering process.

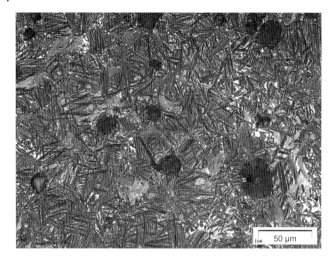

Figure 1.33 A characteristic microstructure of ADI [129].

component is quenched within a salt bath at a temperature in the range 450–250 °C and held isothermally for 30 min to 3 h, followed by controlled cooling to room temperature [134, 136, 138]. The challenge in this concept is the accurate temperature control that is needed to control the final properties.

ADI normally has the chemical composition of Fe-3.6 C-2.5 Si-0.5 Mn-0.05 Mg in wt%. Some additional alloying elements such as Mo, Ni, and Cu can be added to these cast iron grades. These alloying elements suppress the pearlite reaction so that austenite can transform into bainite. Cr and V may be added to increase the hardenability [139].

The microstructure of ADI is a mixture of bainitic ferrite and high carbon retained austenite. Other constituents such as martensite, carbides, and pearlite can be seen in the microstructure. However, the presence of these constituents reduces ductility. A typical micrograph of ADI is shown in Figure 1.33. As shown in this figure, ADI microstructure consists of upper bainitic ferrite (dark sheaves), retained austenite (light phase), and some martensite inside of the retained austenite [140].

However, the mechanical properties of ADI can be varied over a wide range by a suitable choice of heat treatment and mechanical treatment after annealing (Figure 1.34). A high austempering temperature, 410 °C, produces ADI with high ductility and YS in the range of 500 MPa with good fatigue and impact strength. When subjected to surface treatments such as rolling or peening after heat treatment, the fatigue strength of ADI is increased significantly and is competitive to gas-nitrided and case-carburized steels [141]. On the other hand, a lower transformation temperature, for example, 370 °C, results in ADI with very high YS (750 MPa), while 260 °C may yield to 1400 MPa, high hardness, good wear resistance, and contact fatigue strength. This high strength ADI has lower fatigue strength as-austempered, but it can be improved with rolling or grinding. Thus,

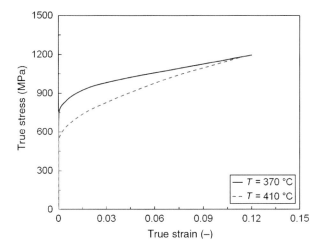

Figure 1.34 Flow curve of ADI for two annealing temperatures. Source: After Ref. [144].

through relatively simple control of the austempering conditions, ADI can be given a range of properties [142, 143].

Furthermore, ADI has good low temperature properties and can work at temperatures as low as −80 °C. In general, ADI maintains at least 70% of its room temperature impact strength at −40 °C [145, 146]. ADI also shows higher damping capacity compared to steel, which makes the parts absorb energy two to five times more than steels. Therefore, the level of noise decreases approximately 8–10 decibels by using this material in gear boxes.

The combination of these characteristics, and presenting good mechanical properties, high wear resistance, and low cost design flexibility introduce ADI as a new class of automotive materials [147, 148]. ADI is mostly used in crankshafts, camshafts, hypod pinion, and ring gear pairs, spring hanger brackets, clip plates, impellers, valve bodies, compressor housing, gears, and drilling heads [139, 149].

References

1. Qu, H. (2011) Advanced High Strength Steel Through Paraequilibrium Carbon Partitioning and Austenite Stabilization.

2. Bhattacharya, D. (2006) An overview of advanced high strength steels (AHSS). Advanced High Strength Steel Workshop, Arlington, October 22–23, 2006.

3. Burns, T.J. (2009) Weldability of a dual-phase sheet steel by the gas metal arc welding process. Master of Applied Science, University of Waterloo Library, Canada.

4. http://www.ncms.org/index.php/ programs/lightweight-automotive-materials/

5. Hulka, K. (2003) Modern multi-phase steels for the automotive industry. *Mater. Sci. Forum*, **414–415**, 101–110.

6. World Auto Steel http://www.worldautosteel.org/projects/ AHSSGuidelines/AHSS-application-guidelines-version-4.aspx (accessed 21 December 2012).

7. Steel Market Development Institute http://www.autosteel.org/en/Programs/

Future%20Steel%20Vehicle.aspx (accessed 21 December 2012).

8. AISI (2002) ULSAB-Advance Vehicle Concepts, AISI, Michigan, January 2002.

9. (2001) ULSAB-Advance Vehicle Concepts, Technical Transfer Dispatch, May 6, 2001. *http://www.autosteel.org/~/media/ Files/Autosteel/Programs/ULSAB-AVC/avc_ttd6.pdf*

10. (1998) ULSAB Program Phase 2 Final Report, Porsche Engineering Services, March 1998.

11. (2001) ULSAB-AVC-PES Engineering Report, October 2001. *http://c315221.r21.cf1.rackcdn.com/ ULSAB-AVC_EngRpt_ Ch1ProgramBackground.pdf*

12. Zuidema, B.K., Denner, S.G., Engl, B., and Sperle, J. (2001) New high strength steels applied to the body structure of ULSAB-AVC. *SAE Trans.*, **42**, 984–992.

13. Matlock, D.K., Krauss, G., and Speer, J.G. (2005) New microalloyed steel applications for the automotive sector. *Mater. Sci. Forum*, **500–501**, 87–96.

14. Papaefthymiou, S. (2004) *Failure Mechanisms of Multiphase Steels*, Shaker Verlag GmbH, Aachen.

15. Anderson, D. (2009) New Study Finds Increased Use of Advanced High-Strength Steels Helps Decrease Overall Vehicle Weight. AISI News Release.

16. Stodolsky, F., Vyas, A, and Cuenca, R. (1995) Lightweight Materials in the Light-Duty Passenger Vehicle Market: Their Market Penetration Potential and Impacts, Vienna.

17. Hempowitz, H. (2009) Ressourceneffizienter Einsatz von Stahlwerkstoffen und Stahltechnologien, Wiesbaden.

18. Heimbuch, R. (2006) An overview of the auto/steel partnership and research needs. Advanced High-Strength Steels: Fundamental Research Issues Workshop, Arlington, Virginia.

19. Mileiko, S.T. (1969) The tensile strength and ductility of continuous fibre composites. *J. Mater. Sci.*, **4**, 974–977.

20. Tanakat, T. (1981) *Int. Met. Rev.*, **4**, 185.

21. Skobir, D. (2011) High strength low alloy (HSLA) steels. *MTAEC9*, **45**(4), 265–301.

22. Patel, J., Klinkenberg, C., Hulka, K. (2003) Hot Rolled HSLA Strip Steels for Automotive and Construction Applications. Niobium Products Company, Düsseldorf.

23. Facco, G. (2005) Effect of cooling rate and coiling temperature on the final microstructure of HSLA steels after HSM and/or laboratory TMP processing. Master thesis. University of Pittsburgh, Pennsylvania

24. Takahashi, M. (2003) Development of High Strength Steels for Automobiles. Technical Report 88, Nippon Steel (UDC 699. 14. 018. 295-415: 629. 11. 011). (*http://www.nssmc.com/en/tech/report/ nsc/pdf/n8802.pdf*)

25. Momeni, A., Dehghani, K., Abbasi, S., and Torkan, M. (2011) Bake Hardening of a Low Carbon Steel for Automotive Applications, Association of Metallurgical Engineers of Serbia, pp. 131–138.

26. Trimberger, R., Fleischman, E., and Salmon-Cox, P. (2003) New Ultra Low Carbon High Strength Steels with Improved Bake Hardenability for Enhanced Stretch Formability and Dent Resistance. Final Report DE-FC07-01ID14045 (University of Pittsburgh, Office of Research 350 Thackeray Hall, Pittsburgh, PA 15260). (*http://www.osti.gov/bridge/servlets/purl /820518-SoQi6S/native/820518.pdf*)

27. Kawasaki Steel 21st Century Foundation (2003) Applications-quest for new breakthrough in steel materials. *An Introduction to Iron and Steel Processing*, Kawasaki Steel 21st Century Foundation, *http://www.jfe-21st-cf.or.jp/chapter_1/1b_2.html* (accessed 21 December 2012).

28. Buahombura, P. (2009) High Strength Low Alloy Steel (HSLA), Ultra Low Carbon Steel, Advance High Strength Steel, Suranaree University of Technology.

29. (2002) Hot Rolled and Cold Rolled Steels, Automotive Steel Design Manual, pp. 2.4.1–2.4.16.

30. Bag, A., Ray, K.K., and Dwarahadasa, E.S. (1999) Influence of martensite content and morphology on tensile and impact properties of high-martensite dual-phase steels. *Metall. Mater. Trans. A*, **30**, 1193–1202.

31. Mediratta, S.R., Ramaswamy, V., and Rama Rao, R. (1985) Low cycle fatigue behaviour of dual-phase steel with different volume fractions of martensite. *Int. J. Fatigue*, **7**(2), 101–106.

32. Skare, T. and Krantz, F. (2003) Wear and frictional behaviour of high strength steel in stamping monitored by acoustic emission technique. *Wear*, **225**, 1471–1479.

33. Bhadeshia, H.K.D.H. and Honeycombe, R.W.K. (1995) *Steels: Microstructure and Properties*, Butterworth-Heinemann, London.

34. Rocha, R.O., Melo, T.M.F., Pereloma, E.V., and Santos, D.B. (2005) Microstructural evolution at the initial stages of continuous annealing of cold rolled dual-phase steel. *Mater. Sci. Eng. A*, **391**(1–2), 296–304.

35. Baudin, T., Quesnea, C., Jura, J., and Penelle, R. (2001) Microstructural characterization in a hot-rolled, two-phase steel. *Mater. Charact.*, **47**, 365–373.

36. Rigsbee, J.M. and VanderArend, P.J. (1979) *Formable HSLA and Dual-Phase Steels*, TMS–AIME, Warrendale, pp. 56–86.

37. Speich, G.R. and Miller, R.L. (1979) *Structure and Properties of Dual-Phase Steels*, TMS–AIME, Warrendale, pp. 145–182.

38. Rashid, M.S. (1981) Dual phase steels. *Annu. Rev. Mater. Sci.*, **11**, 245–266.

39. Speich, G.R. (1981) *Fundamental of Dual-Phase Steels*, TMS–AIME, Warrendale, pp. 3–45.

40. Balliger, N.K. and Gladman, T. (1981) Work hardening of dual-phase steels. *Met. Sci.*, **15**, 95–108.

41. International Iron and Steel Institute (2002) Ultra Light Steel Auto Body – Advanced Vehicle Concepts (ULSAB–AVC). Overview Report.

42. Abdalla, A.J., Hein, L.R.O., Pereira, M.S., and Hashimoto, T.M. (1999) Mechanical behavior of strain aged dual phase steels. *Mater. Sci. Technol.*, **15**, 1167–1170.

43. Bleck, W. (2007) *Material Science of Steel*, Mainz, Aachen.

44. Thomser, C. (2008) *Modeling of the Mechanical Properties of Dual-Phase Steels Based on Microstructure*, Aachen. PhD Thesis, RWTH-Aachen University, Germany, 2009

45. Pickering, F.B. (1978) *Physical Metallurgy and Design of Steels*, Applied Science Publishers, London.

46. Gladman, T. (1997) *The Physical Metallurgy of Microalloyed Steels The Institute of Materials*, Institute of Materials, London.

47. Leslie, W.C. (1981) *The Physical Metallurgy of Steels*, McGraw-Hill, New York.

48. Smallman, R.E. and Bishop, R.J. (1999) *Modern Physical Metallurgy and Materials Engineering*, Science, Process, Application, 6th edn, Butterworth-Heinemann, Oxford.

49. Leslie, W.C. and Sober, R.J. (1967) The strength of ferrite and martensite as functions of composition, temperature and strain rate. *Trans. ASM*, **60**, 459–484.

50. Pickering, F.B. (1981) Advances in the physical metallurgy and applications of steels. *Met. Soc.*, 5–25.

51. Sinha, A.K. (1989) *Ferrous Physical Metallurgy*, Butterworth-Heinemann, Boston, MA.

52. Mukherjee, K. (2009) *Grain Refinement in Dual Phase Steels*, Springer, Vancouver.

53. Sakuma, Y., Matsumura, O., and Akisue, O. (1991) Influence of C content and annealing temperature on microstructure and mechanical properties of 400 °C transformed steel containing retained austenite. *ISIJ Int.*, **31**(11), 1348–1353.

54. Jacques, P.J., Delannay, F., Cornet, X., Harlet, P., and Ladriere, J. (1998) Enhancement of the mechanical properties of a low-carbon, low-silicon steel by formation of a multiphased microstructure containing retained austenite. *Metall. Mater. Trans. A*, **29**(9), 2383–2393.

55. Srivastava, A.K., Jha, G., Gope, N., and Singh, S.B. (2006) Effect of heat treatment on microstructure and mechanical properties of cold rolled C–Mn–Si TRIP-aided steel. *Mater. Charact.*, **57**(2), 127–135.

56. Bleck, W. (2000) Using the TRIP effect- the dawn of a promising group of cold formable steels. International Conference on TRIP-Aided High Strength Ferrous Alloys, Ghent, pp. 13–23.

57. Van der Zwaag, S., Zhao, L., Krujiver, O., and Sietsma, J. (2002) Thermal and mechanical stability of retained austenite in aluminium-containing multiphase TRIP steels. *ISIJ Int.*, **42**(11), 1565–1570.

58. Baik, S.C., Kim, S., Jin, Y.S., and Kwon, O. (2001) Effects of alloying elements on mechanical properties and phase transformation of cold rolled TRIP steel sheets. *ISIJ Int.*, **41**(3), 290–297.

59. Jacques, P.J., Girault, E., Martens, A., Verlinden, B., van Humbeeck, J., and Delannay, F. (2001) The developments of cold-rolled TRIP-assisted multi-phase steels. Al-alloyed TRIP assisted multiphase steels. *ISIJ Int.*, **41**(9), 1068–1074.

60. Bleck, W. (2007) *Materials Science of Steel*, Mainz, Aachen.

61. Sakuma, Y., Matlock, D.K., and Krauss, G. (1992) Intercritically annealed and isothermally transformed 0.15Pct C steels containing 1.2 Pct Si-1.5Pct Mn and 4 Pct Ni: part 1. transformation, microstructure, and room-temperature mechanical properties. *Metall. Mater. Trans. A*, **23**, 1221–1232.

62. Sakuma, Y., Matlock, D.K., and Krauss, G. (1992) Intercritically annealed and isothermally transformed 0.15Pct C steels containing 1.2 Pct Si-1.5Pct Mn and 4 Pct Ni: part 2. Effect of testing temperature on stress–strain behavior and deformation-induced austenite transformation. *Metall. Mater. Trans. A*, **23**, 1233–1241.

63. Li, W. and Al, W. (2004) Application of TRIP steel to replace mild steel in automotive parts. International Conference on Advanced High Strength Sheet

Steels for Automotive Applications Proceedings, June 2004, pp. 31–36.

64. Papaefthymiou, S. (2004) *Failure Mechanisms of Multi Phase Steels*, Shaker Verlag GmbH, Aachen.

65. Edward, G.O. (2006) Advanced High Strength Steel (AHSS) Application Guidelines, Version 3, International Iron and Steel Institute, Committee on Automotive Application, September, 2006.

66. Mesplont, C., De Cooman, B.C., and Vandeputte, S. (2002) Dilatometric study of the effect of soluble boron on the continuous and isothermal austenite decomposition in 0.15C-1.6Mn steel. *Ironmaking Steelmaking*, **29**, 39.

67. Mesplont, C., Vandeputte, S., and De Cooman B.C. (2001) Proceedings of the 43rd Conference on Mechanical Working and Steel Processing, Charlotte, NC, 2001, p. 359.

68. Sarkar, S., Militzer, M., Poole W.J., and Fazeli, F. (2007) Proceedings of International Symposium Advanced High Strength and other Specialty Sheet Steel Products for the Automotive Industry, MST'07, Michigan 2007, Materials Science & Technology Conference, p. 61.

69. ThyssenKrupp Steel (2009) Complex-Phase Steels CP-W and CP-K.

70. Heller, T. and Nuss, A. (2005) Effect of alloying elements on microstructure and mechanical properties of hot rolled multiphase steels. *Ironmaking Steelmaking*, **32**(4), 303–308.

71. Mesplont, C. and De Cooman, B.C. (2003) Effect of austenite deformation on crystallographic texture during transformations in microalloyed bainitic steel. *Mater. Sci. Technol.*, **19**, 875.

72. Kleiner, L.M. and Simonov, Y.N. (1999) Structure and properties of low-carbon martensitic steels. *Met. Sci. Heat Treat.*, **41**, 366–368(accessed 25 September 2009). (*http://link.springer.com/article/ 10.1007%2FBF02474887?LI=true*)

73. Shaw, J., Engl, B., Espina, C., Oren, E.C., and Kawawoto, Y. (2002) *New Steel Sheet and Steel Bar Products and Processing* (eds D.W. Anderson *et al.*), SAE-SP 1685, Society of Automotive

Engineers (SAE), Warrendale PA, pp. 63–71.

74. Kuziak, R., Kawalla, R., and Waengler, S. (2008) Advanced high strength steels for automotive industry. *Arch. Civil Mech. Eng.*, **8**, 103–117.

75. Schaeffler, D.J. (2005) Introduction to advanced high strength steels, part 1: grade overview. *Fabricator* (Online Journal). *http://www.thefabricator.com/MetalsMaterials/MetalsMaterials_Article.cfm?ID=1139*

76. Leslie, W.C. (1966) *Proceedings of the 12th Sagamore Army Materials Research Conference*, Syracuse University Press, Syracuse, NY, p. 43.

77. Codd, D. (2005) *Seam Welded Air-Hardenable Corrosion Resistant Steel Tubing: Automotive Applications Overview*, KVA Inc., Escondido.

78. Schulz-Beenken, A.S. (1997) Martensite in steels: its significance, recent developments and trends. *J. Phys.*, **4**, 359–366.

79. Arcelor Mittal (2009) Martensitic Steels.

80. ThyssenKrupp Steel (2009) Product Information Martensitic-Phase Steel.

81. SAE International *http://www.sae.org/mags/aei/7584* (accessed 21 December 2012).

82. Worldsteel Association (2008) Report for an Advanced High Strength Steel Family Car.

83. Bleck, W. and Fischer, M. (2009) Gefüge und Eigenschaften neuer kaltumformbarer Stähle, in *Metallographie Tagung*, Vol. **43**, Aachen, Germany.

84. Aydin, H. (2009) Effect of Microstructure on Static and Dynamic Mechanical Properties of Third Generation Advanced High Strength Steels, Vancouver.

85. Kriangyut, P. (2008) *Deformation Mechanisms and Mechanical Properties of Hot Rolled Fe-Mn-C-(Al)-(Si) Austenitic Steels*, Aachen. PhD Thesis, RWTH-Aachen University, Germany, 2008.

86. Vaissiere, L., Laurent, J.P., and Reihardt, A. (2003) Development of pre-coated boron steel for applications on PSA Peugeot Citroen and Renault bodies in white. *J. Mater. Manuf.*, **111**, 909–917.

87. Kolleck, R., Steinhoefer, D., Feindt, J.A., and Bruneau, P. (2004) Manufacturing methods for safety and structural body parts for lightweight body design. Proceedings of the IDDRG, Conference, 2004, pp. 167–173.

88. Garcia Aranda, L., Ravier, P., and Chstel, Y. (2003) Hot stamping of quenchable steels: material data and process simulations. Proceedings of the IDDRG, Conference, 2003, pp. 155–164.

89. Philip, T.V. (1991) *Metals Handbook*, 10th edn, ASM International, Vol. **1**, pp. 428–430.

90. Klein, M., Spindler, H., Luger, A., Rauch, R., Stiaszny, P., and Eigelsberger, M. (2005) Thermomechanically hot rolled high and ultra high strength steel grades–processing, properties and application. *Mater. Sci. Forum*, **500–501**, 543–550.

91. Wilsius, J., Hein, P., and Kefferstein, R. (2006) Status and future trends of hot stamping of USIBOR 1500P. Proceedings of the 1st Erlangener Workshop Warmblechumformung, 2006, pp. 82–101.

92. Haga, J., Miziu, N., Nagamichi, T., and Okamoto, A. (1998) *ISIJ Int.*, **38**, 580.

93. Naderi, M. (2007) Hot stamping of ultra-high strength steels. PhD thesis. RWTH-Aachen University, Aachen.

94. Salzgitter Flachstahl *http://www.salzgitterflachstahl.de/de/Produkte/warmgewalzte_produkte/stahlsorten/Lufthaertende_Staehle/* (accessed 21 December 2012).

95. Schaper, M. (2010) Mikrostrukturelle Vorgänge bei der Verformung verschiedener höher- und höchstfester Stahlblechwerkstoffe. Habilitation thesis. Hanover.

96. Noman, M., Clausmeyer, T., Barthel, C., Svendsen, B., Huetink, J., and Van Riel, M. (2010) Experimental characterization and modeling of the hardening behavior of the sheet steel LH800. *Mater. Sci. Eng. A*, **527**, 2515–2526.

97. Flaxa, V. and Schoettler, J. (2007) Entwicklung: Produktion und Eigenschaften hochfester Staehle für den Automobilbau. Proceeding of 14th

Saxony's Metal Forming Meeting
"Werkstoffe und Komponenten für
den Fahrzeugbau", Freiberg, Germany,
December 4–5, 2007, pp. 127–136.

98. Lund, T., Ölund, P., Larsson, S.,
Neuman, P., and Rösch, O. (2005)
Air-hardening, low to medium carbon
steel for improved heat treatment. US
Patent 6902631, International Clas-
sification: C23C008/80; C23C008/22;
C21D001/18, June 7, 2005.

99. (2006) Air-Hardening Steels,
Material Data Sheet, Salzgitter.
*http://www.salzgitterflachstahl.de/MDB/
News/2006/LH800/SZFG_Material_
Data_Sheet_LH.pdf*

100. Speer, J.G., Streicher, A.M., Matlock,
D.K., Rizzo, F.C., and Krauss, G. (2003)
in *Austenite Formation and Decomposition* (eds E.B. Damm and M. Merwin),
TMS/ISS, Warrendale, pp. 505–522.

101. Matlock, D.K. and Speer, J.G. (2006)
Design considerations for the next
generation of advanced high strength
sheet steels. Proceedings of the 3rd
International Conference on Advanced
Structural Steels (ICASS), Geongju,
2006, pp. 774–781.

102. Matlock, D.K. and Speer, J.G. (2009)
in *Microstructure and Texture in Steels
and Other Materials* (ed H. Arunansu),
Springer, London, pp. 185–205.

103. Speer, J.G., Matlock, D.K., De Cooman,
B.C., and Schroth, J.G. (2003) Carbon
partitioning into austenite after martensite transformation. *Acta Mater.*, **51**,
2611–2622.

104. Cao, W., Shi, J., Wang, C., Wang, C.,
Xu, L., Wang, M., Weng, Y., and Dong,
H. (2010) in *Proceedings of International
Conference on Advanced Steels* (eds Y.
Weng, H. Dong, and Y. Gan), Beijing,
pp. 196–215.

105. Clarke, A.J., Speer, J.G., Miller, M.K.,
Hackenberg, R.E., Edmonds, D.V.,
Matlock, D.K., Rizzo, F.C., Clarke,
K.D., and De Moor, E. (2008) Carbon
partitioning to austenite from martensite or bainite during the quench and
partition (Q&P) process: a critical
assessment. *Acta Mater.*, **56**, 16–22.

106. Wang, L. and Feng, W. (2010) in *Proceedings of International Conference on*
Advanced Steels (eds Y. Weng, H. Dong,
and Y. Gan), Beijing, **101**, pp. 242–245.

107. Wang, C.Y., Shi, J., Cao, W.Q., and
Dong, H. (2010) Characterization of
microstructure obtained by quenching
and partitioning process in low alloy
martensitic steels. *Mater. Sci. Eng. A*,
527, 3442–3449.

108. Speer, J.G., Rizzo Assungcao, F.C., and
Matlock, D.K. (2005) The quenching
and partitioning process: background
and recent progress. *Mater. Res.*, **8**(4),
417–423.

109. Speer, J.G., Edmonds, D.V., Rizzo,
F.C., and Matlock, D.K. (2004) Partitioning of carbon from supersaturated
plates of ferrite, with application to
steel processing and fundamentals of
the bainite transformation. *Curr. Opin.
Solid State Mater. Sci.*, **8**, 219.

110. Edmondsa, D.V., He, K., Rizzo, F.C.,
De Cooman, B.C., Matlock, D.K., and
Speer, J.G. (2006) Quenching and partitioning martensite—a novel steel heat
treatment. *Mater. Sci. Eng. A*, **438–440**,
25–34.

111. Clarke, A., Speer, J.G., Matlock, D.K.,
Rizzo, F.C., Edmonds, D.V., and He,
K. (2005) Proceedings of the Inter-
national Conference on Solid–Solid
Phase Transformations in Inorganic
Materials, Arizona 2005, pp. 99–108.

112. Gerdemann, F.L.H., Speer, J.G., and
Matlock, D.K. (2004) *Materials Sci-
ence & Technology (MS&T)*, TMS, New
Orleans, LA and Warrendale, PA,
September 26–29, 2004, pp. 439–449.

113. De Cooman, B.C. and Speer, J.G.
(2006) Quench and partitioning steel:
a new AHSS concept for automotive
anti-intrusion applications. *Steel Res.
Int.*, **77**, 634–640.

114. Zhong, N., Wang, Y., Zhang, K., and
Rong, Y.H. (2011) Microstructural
evolution of a Nb-microalloyed advanced high strength steel treated by
quenching-partitioning-tempering process. *Steel Res. Int.*, **82**(11), 1332–1337.

115. Streicher, A.M., Speer, J.G., Matlock,
D.K., and De Cooman, B.C. (2004)
Proceedings of the International Con-
ference Advanced High-Strength Sheet
Steels for Automotive Applications,
Warrendale, pp. 54–62.

116. Cristinance, M., Milbourn, D.J., and James, D.E. (1998) *Corus Engineering Steels*, Prod/EP5, pp. 1–13. (*http://www.tatasteeleurope.com/file_source/StaticFiles/Business%20Units/Engineering%20steels/PRODEP5.PDF*)

117. Matlock, D.K. (2010) *Advanced Steel Processing and Products Research Center Department of Metallurgical and Materials Engineering*, Material & Manufacturing 2010, ARAI Forging Industry Division, Pune, India, pp. 51–56.

118. Van Tyne, C.J., Matlock, D.K., and Speer, J.G. (2008) *Advanced Steel Processing and Products Research Center*, 19th International Forging Congress, FIA, Cleveland, OH, USA, pp. 189–197.

119. Huchtemann, B. and Schüler, V. (1990) Entwicklungsstand der ausscheidungshärtenden ferritisch-perlitischen (AFP_) Stähle mit Vanadiumzusatz für eine geregelte Abkühlung von der Warmformgebungstemperatur, Krefeld.

120. Tither, G. (2001) *Niobium in Steel Castings and Forgings*, Reference Metals Company, Inc., Bridgeville.

121. Bayer, J. (2003) Steel bars for automotive applications. *Adv. Mater. Processes*, **161**, 46–49.

122. Keul, C., Urban, M., Back, A., Hirt, G., and Bleck, W. (2010) Entwicklung eines hochfesten duktilen bainitischen (HDB) Stahls für hochbeanspruchte Schmiedebauteile. *Schmiede J.*, **9**, 28–31.

123. Mougin, J., Dierickx, P., Robat, D., and Vernis, J.R. (2005) Gears and springs in niobium microalloyed steels for automotive applications. *Mater. Sci. Forum*, **500–501**, 753–760.

124. Ma, L., Wang, M.Q., Shi, J., Hui, W.J., and Dong, H. (2008) Influence of niobium microalloying on rotating bending fatigue properties of case carburized steels. *Mater. Sci. Eng. A*, **498**, 258–265.

125. Woods, J.L., Daniewicz, S.R., and Nellums, R. (1999) Increasing the bending fatigue strength of carburized spur gear teeth by presetting. *Int. J. Fatigue*, **21**, 549–556.

126. Farfan, S., Rubio-Gonzalez, C., Cervantes-Hernandez, T., and Mesmacque, G. (2004) High cycle fatigue, low cycle fatigue and failure modes of a carburized steel. *Int. J. Fatigue*, **26**, 673–678.

127. Sugimoto, K., Muramatsu, T., Hashimoto, S., and Mukai, Y. (2006) Formability of Nb bearing ultra-high-strength TRIP-aided sheet steels. *J. Mater. Process. Technol.*, **177**, 390–395.

128. Wietbrock, B., Bambach, M., Seuren, S., and Hirt, G. (2010) Homogenization strategy and material characterization of high-manganese TRIP and TWIP steels. *Mater. Sci. Forum*, **638–642**, 3134–3139.

129. Bhattacharya, D. and Fonstein, N., (2005) Niobium Bearing Advanced Sheet Steels for Automotive Applications at ArcelorMittal, Global R & D, East Chicago, Indiana, pp. 91–102.

130. Wang, X.D., Huang, B.X., Wang, L., and Rong, Y.H. (2008) Microstructure and mechanical properties of microalloyed high-strength transformation-induced plasticity steels. *Metall. Mater. Trans. A*, **39 A**, 1–7.

131. Alan, P.D. and David, C.F. (2000) Lightweight Iron and Steel Castings for Automotive Applications SAE 2000 World Congress Detroit, Michigan.

132. Owen, W.S. (1954) *Trans. ASM*, **46**, 812–829.

133. Yang, J. and Putatunda, S. (2005) *Effect of Microstructure on Abrasion Wear behavior of Austempered Ductile Cast Iron (ADI) Processed by a Novel Two-Step Austempering Process*, Elsevier B.V., Detroit.

134. Kim, Y., Shin, H., Park, H., and Lim, J. (2008) Investigation into mechanical properties of austempered ductile cast iron (ADI) in accordance with austempering temperature. *Mater. Lett.*, **62**(3), 357–360.

135. Putatunda, S. and Gadicherla, P. (1999) Influence of austenitizing temperature on fracture toughness of a low manganese austempered ductile iron (ADI) with ferritic as cast structure. *Mater. Sci. Eng. A*, **268**, 15–31.

136. Meena, A. and El Mansori, M. (2011) Study of dry and minimum quantity

lubrication drilling of novel austempered ductile iron (ADI) for automotive applications. *Wear*, **271**, 2412–2416.

137. Putatunda, S., Kesani, S., Tackett, R., and Lawes, G. (2006) Development of austenite free ADI (austempered ductile cast iron). *Mater. Sci. Eng. A*, **435-436**, 112–122.

138. Putatunda, S. (2001) Development of austempered ductile cast iron (ADI) with simultaneous high yield strength and fracture toughness by a novel two-step austempering process. *Mater. Sci. Eng. A*, **315**, 70–80.

139. Putatunda, S.K. and Gadicherla, P.K. (2000) Effect of austempering time on mechanical properties of a low manganese austempered ductile iron. *J. Mater. Eng. Perform.*, **9**, 193–203.

140. Yescas-Gonzalez, M.A. (2011) Modelling the microstructure and mechanical properties of austempered ductile irons. PhD thesis. University of Cambridge, Cambridge, UK.

141. Jokipii K. (1984) Austempered ductile iron as material for gears and other applications. Proceedings of the 1st International Conference on Austempered Ductile Iron, Chicago, April 2–4, 1984.

142. Gundlach, R.B. and Janowak, J. (1985) Austempered ductile irons combine strength with toughness and ductility. *Met. Prog*, **128**, 19–26.

143. Lin, C.K. and Hung, T.P. (1995) *Int. J. Fatigue*, **18**, 309–320.

144. Aranzabal, J., Gutierrez, I., Rodriguez-Ibabe, J.M., and Urcola, J.J. (1997) Influence of the amount and morphology of retained austenite on the mechanical properties of an austempered ductile iron. *Metall. Mater. Trans. A*, **28**, 1143–1156.

145. Hayrynen, K.L., Moore, D.J., and Rundman, K.B. (1993) Tensile and fatigue properties of relatively pure ADI. *AFS Trans.*, **101**, 119–129.

146. Keough, J.R. (ed) (1998) Austempered ductile iron, in *Ductile Iron Data for Design Engineers* Chapter IV, Published by Rio TINTO.

147. Rossi, F.S. and Gupta, B.K. (1981) Austempering of nodular cast iron Automobile component. *Met. Prog*, **119**, 25–31.

148. Salonen, P. (1997) International Conference on Engineering Design (ICED) 97, Tampere, pp. 637–640.

149. British Cast Iron Research Association (1991) Selected Case Studies of Austempered Ductile Iron Components, BCIRA.

2
Aluminum and Aluminum Alloys

Axel von Hehl and Peter Krug

2.1
Introduction

Aluminum accounts for about 8% of the Earth's elements and is the third most abundant element in the Earth's crust. Although as a chemical element aluminum was discovered in 1807, the essential steps for the commercial use of today's most common light metal were taken around 80 years later. The Bayer process was developed, enabling efficient extraction of alumina (Al_2O_3) from the ore (bauxite), while fused-salt electrolysis (that is still applied today) enabled the industrial production of molten pure aluminum from alumina. Since then, the worldwide primary production has been continuously increasing, particularly after World War II. Today's annual production of primary aluminum is far beyond 30 million metric tons (Figure 2.1) [1, 2]. The world's main producers are located in Asia, with the Asian production volume being around the sum of European and American production volumes [3]. The expensive production of primary aluminum is reflected by the high price that ranged between $1100 and $3300 per metric ton during the last decade [4].

An important attribute of aluminum is its recyclability. Since the commercial production started, approximately three-quarters of the primary aluminum ever produced is estimated to be still in use. Around one-third of today's aluminum products available on the market is sourced from recycled metal [5, 6]. In comparison to the production of primary aluminum by fused-salt electrolysis – still one of the most energy-intensive smelting processes, remelting of aluminum, because of its low melting point, requires much less energy consumption. Thus, recycling is a supporting pillar of aluminum production, benefiting the material's life cycle performance [7].

The physical properties of aluminum, which make the lightweight material so successful, are primarily its low density of 2.7 kg m^{-3} combined with a sufficiently high Young's modulus of about 72 GPa. In addition, aluminum provides a high electrical conductivity of $37.8 \text{ m } \Omega^{-1} \text{ mm}^{-2}$, a high thermal conductivity of $238 \text{ W m}^{-1} \text{ K}^{-1}$ and an excellent inherent corrosion resistance. Besides, aluminum has a melting point of 660 °C, a specific heat of $0.917 \text{ J g}^{-1} \text{ K}^{-1}$, and a thermal

Structural Materials and Processes in Transportation, First Edition.
Edited by Dirk Lehmhus, Matthias Busse, Axel S. Herrmann, and Kambiz Kayvantash.
© 2013 Wiley-VCH Verlag GmbH & Co. KGaA. Published 2013 by Wiley-VCH Verlag GmbH & Co. KGaA.

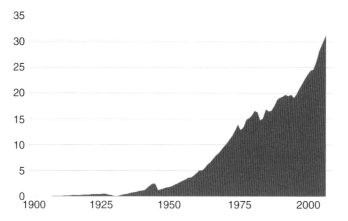

Figure 2.1 Primary aluminum production per year (million tons) [1].

expansion of 23.8×10^{-6} l K^{-1}. Its processing characteristics cover, among others, a good formability and an excellent weldability. A further outstanding attribute of aluminum is its ability to harden when combined with certain alloying elements, which precipitate out of their supersaturated solid solution. This strengthening phenomena, also called the age-hardening effect, was first observed by Alfred Wilms in copper alloyed aluminum, known as *Duralumin*, at the beginning of the last century. The first aluminum alloy system was born [8]. Since then, many further alloy systems have been explored, promoting the development of a huge number of different alloy grades.

With a current global share of more than one-quarter the main field of application is the transportation sector, followed by packaging and construction [3]. The outstanding position of aluminum in the transportation sector is based on a long tradition. In 1899, the first car with an aluminum body was displayed at the Berlin international car exhibition. Then, in 1903, the Wright brothers' historical aircraft was propelled by an engine made with aluminum components. After the great demand for application of aluminum in military aircraft during World War II, the production volume had increased rapidly, driven by the still rising civil aviation. The beginning of serial application in the automotive industry was indicated by the large-volume production of aluminum engine blocks in the mid-1960s. The need for reducing fuel consumption induced by the oil crisis in the 1970s had created the necessity for weight reduction, enabling the further penetration of aluminum in car production. In 1994, Audi rolled out the first serial upper class limousine providing a self-supporting aluminum body, registered as "Audi Space Frame" (ASF$^{\circledR}$) [9]. Today, the average penetration of aluminum in cars has reached more than 130 kg per vehicle, whereas the highest growth has to be recorded for car body components [10, 11].

Besides the growing demands for energy efficiency in transportation, future legal regulations for CO_2 emission control have increased the need for weight reduction and promote advanced lightweight designs considering the variety of materials

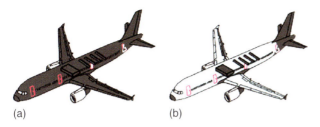

(a)   (b)

Figure 2.2   Different scenarios of material designs in short-range aircraft. (a) CFRP predominated design and (b) aluminum predominated design.

available at present. These materials provide an extended range of lightweight potentials with regard to in-service characteristics and life cycle costs. Hence, the permanent penetration of new lightweight materials and lightweight designs leads partially to considerable material replacements. However, it also intensifies the material development of established lightweight materials, improving their competitiveness.

The challenges of the aluminum industry comprise all the fields of the transportation sector. Driven by the huge replacement of aluminum alloys by carbon-fiber-reinforced polymers (CFRPs) in long-range aircraft, considerable material improvements are needed in order to retain the dominating position of aluminum alloys in short-range aircraft, and also for future aircraft generations (Figure 2.2). The major defiances facing the material development lie in the improvement of fatigue and damage tolerance properties, impact and corrosion resistance, strength properties also at elevated temperatures, and structure integrity without affecting the inherent cost advantages of aluminum alloys regarding material costs, as well as costs of component manufacturing and assembly [12–16].

In the field of automotive manufacturing, material development of aluminum alloys is mostly directed to components for the car body followed by components for structure and chassis. Current activities comprise the improvement of structural integrity by manufacturing (i.e., forming, casting, and joining), and also improvements of in-service characteristics (such as strength properties and corrosion resistance). Apart from development strategies focusing on the all-aluminum body in white, where the aluminum sheet material proportion is particularly high, the current development of aluminum alloys goes toward application in advanced multimaterial designs (Figure 2.3) for high-volume cars [11, 17–19].

Using selected examples, this chapter gives an overview of recent advances in the development of aluminum alloys intended for application in the transportation sector. The focal point is mainly on aviation and the automotive industry as major fields of high-volume applications. Thus, niche applications are not considered.

Sections 2.2 and 2.3 distinguish between developments of wrought alloys and casting alloys, respectively, divided into products classes. Generally, the alloy

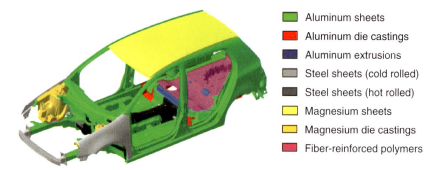

Aluminum sheets
Aluminum die castings
Aluminum extrusions
Steel sheets (cold rolled)
Steel sheets (hot rolled)
Magnesium sheets
Magnesium die castings
Fiber-reinforced polymers

Figure 2.3 "Super Light Car" approach for multimaterial designs in high-volume cars [19] (53 mass% aluminum, 36 mass% steel, 7 mass% magnesium, and 4 mass% FRP [17]).

development is closely related to process development, comprising a variety of production technologies associated with the development of advanced aluminum products. Therefore, descriptions of selected process developments are also part of these sections. Section 2.4 is focused on selected developments regarding secondary processes, such as joining techniques, heat treatment and measures for surface protection. Section 2.5 deals with case studies for future applications divided into wrought products and castings. In this context, the huge lightweight potential of age-hardenable alloys for high-strength applications, processed by semi-hot deep drawing (HDD), is highlighted and compared with press-hardened steels. In addition, the example of gas injection is given, along with another case study from powder metallurgy. Section 2.6 gives a summary and outlook, and the section "Further Reading" rounds off the chapter by giving an overview for the further reading.

2.2
Wrought Alloys and Associated Processes

2.2.1
Alloy Classes and Their Basic Constitution

Wrought aluminum alloys are usually registered in accordance to the alloy classification of the Aluminum Association. The used term is a four digit number, where the first digit indicates the alloy class, the second its variant, and the last two digits the alloy's record number. This classification has also been adapted into the European standard EN 573. While using the designation according to this standard, the acronym AW (aluminum wrought) has to be added in front of the four digit number. Table 2.1 gives a short overview of the alloy classes and their basic constitution.

The delivery conditions of non-heat-treatable alloys are designated as follows:

Table 2.1 Alloy classes and their basic constitution.

Alloy class (international designation)	Alloy system (basic elements)	Basic constitution (range in wt%)	Age-hardenable
1xxx	Al	≥99.0% Al	No
2xxx	Al–Cu	0.7–6.8% Cu	Yes
3xxx	Al–Mn	0.05–1.8% Mn	No
4xxx	Al–Si	0.6–21.5% Si	No
5xxx	Al–Mg	0.2–6.2% Mg	No
6xxx	Al–Mg–Si	0.2–1.6% Mg 0.2–1.7% Si	Yes
7xxx	Al–Zn–Mg	0.8–12.0% Zn 0.5–3.7% Mg	Yes
8xxx	Al (other elements, e.g., Fe and Li)	—	—

Condition	Description
F	As-fabricated (no mechanical property limits specified)
O	Annealed (soft tempered)
H1x	Strain-hardened (cold formed)
H2x	Strain-hardened and partially annealed
H3x	Strain-hardened and stabilized (e.g., for corrosion)
H4x	Strain-hardened and painted or lacquered

While the first digit after H indicates the type of processing, the second digit after H indicates the degree of strain hardening (1–9).

The temper designations of heat-treatable alloys are as follows:

Condition	Description
O	Annealed (soft tempered)
W	Solution annealed and quenched
T1	Hot formed, quenched, and naturally aged
T2	Hot formed, quenched, cold formed, and naturally aged
T3	Solution annealed, quenched, cold formed, and naturally aged
T4	Solution annealed, quenched, and naturally aged
T5	Hot formed, quenched, and artificially aged
T6	Solution annealed, quenched, and artificially aged
T7	Solution annealed, quenched, and artificially overaged
T8	Solution annealed, quenched, cold formed, and artificially aged

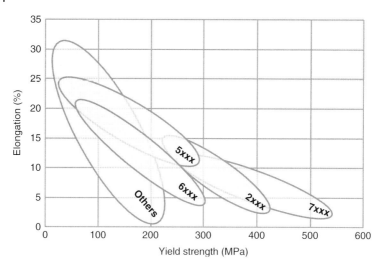

Figure 2.4 Usual strength/ductility range of wrought alloy classes.

Usually, the first digit after T is followed by further digits defining additional requirements. For more details see [20–23].

The usual strength/ductility range of the alloy classes is given in Figure 2.4.

2.2.2
Plates and Sheets

Aluminum plate and sheet material are available in various dimensions. Coiled strip material, for example, for automotive panels, can be produced with a thickness <1 mm and a width of more than 2 m, while several millimeters thick wide sheets, for example, for aircraft panels, can be produced with a width of more than 3 m and a length of more than 12 m [13].

As "state of the art" today's industrial aluminum plate and sheet production consists of a sequence of different single processes starting with direct chill (DC) casting of large rectangular aluminum ingots followed by homogenization treatment, hot and cold rolling, interannealing between the cold-rolling operations and final annealing [24]. Final annealing gives the material the desired microstructure and properties for subsequent sheet-metal forming processes, such as deep drawing, stretch, and roll forming, as well as creep forming or other special forming techniques, and specifies the in-service characteristics of the final product.

2.2.2.1 Alloy Development

Age-hardenable Mg containing 2xxx alloys show high potential for application in high-strength aircraft components operating at higher temperatures. Additions of Ag [25] as well as Zr and Sc [26] are proven to be promising approaches to enhance the material properties at elevated temperatures. While thermally stable $Al_3(Sc,Zr)$ dispersoids (modified $L1_2$ structure) improve the in-service creep resistance and delay static recrystallization (SRX) up to 500 °C [27], Ag stimulates the precipitation of a uniform dispersion of the thermally stable Ω phase during age-hardening [28, 29]. Latter precipitations increase the material strength significantly at temperatures up to 200 °C [27–29]. Regarding the development of future alloy variants, it can be expected that a superposition of $Al_3(Sc,Zr)$ dispersoids and Ω phase particles will lead to an optimum balance between strength, creep resistance, and crack resistance, meeting the demands for high-strength aircraft structures operating at elevated temperatures [27].

Li-containing aircraft alloys are well known for decades for their benefits in reducing density (3% per 1 mass% Li) as well as increasing Young's modulus (5% per 1 mass% Li). Despite their extraordinary low density compared to usual alloys of similar strength, these alloys of the first two generations containing 2–2.5% Li, such as alloy 2090 or alloy 8090, received limited success because of various technical disadvantages such as low fracture toughness, poor resistance against stress corrosion cracking (SSC), insufficient ductility, and high anisotropy [12, 30]. With the development of the third generation alloy in the last decade, focusing on the Al–Cu–Li alloy system, these problems were broadly solved; however, a decrease in density lowering the Li content to <2% had to be tolerated. With the growing demands on the balance between high strength and high damage tolerance of Li-containing 2xxx and 8xxx, aircraft alloys require a precise alignment of chemical composition and thermomechanical processing (TMP). In this context, novel generations of Al–Cu–Li alloys for plates and sheets were recently developed by Alcan Aerospace [12]. The new plate alloy 2050-T8 exhibits fracture toughness and yield strength values (being at 64 MPa $m^{1/2}$ and 530 MPa, respectively) ranging between those of well-established high damage-tolerant plate alloy 2x24-T3 and those of the high-strength plate alloy 7x49-T7. The new sheet alloy, for example, for fuselage skins, that is registered as alloy 2198-T8 offers toughness values exceeding by far those of the well-established high damage-tolerant sheet alloys 2x24-T3 and 6x56-T6, along with strength values slightly higher than those of the high-strength alloy 7x75-T6 [12]. As a further Al–Cu–Li alloy variant for fuselage and outer wing skin panels, a new 2199 sheet alloy with higher Li content was developed by Alcoa [16]. This novel alloy is characterized by improved corrosion resistance because of the addition of Zn. As a result of the higher Li content (between 1.5 and 2%) and a slightly decreased Cu content (<3%) weight savings of around 20% at moderately elevated strength can be offered [16].

With the density of present day Al–Cu–Li alloys in the range of about 2.65 g cm^{-3}, Al–Mg–Sc alloys are highly predestined for sheets being dedicated for the production of fuselage skins or other aircraft applications such as wing flaps, where high strength, good notch toughness, improved fatigue, and damage tolerance properties along with a good impact and corrosion resistance as well as an excellent weldability are needed. Today's commercial Al–Mg–Sc alloys contain up to 4.5% Mg and, because of the low solubility of Sc, not more than 0.4% Sc. The strength properties of sheet material mainly result from solute hardening by means of Mg as well as work hardening during cold rolling. Thermally stable primary $Al_3Sc(Zr)$ dispersoids precipitating nearly completely during conventional DC slab casting promote a fully non-recrystallized and highly polygonized microstructure during processing. However, owing to the small Sc proportion retained in supersaturated solid solution the age-hardening effect is very low [31, 32]. Thus, today's commercial alloys are rated among the non-age-hardenable 5xxx series. Nevertheless, under the labels KO8242 (already registered as alloy 5024) and KO8542, Aleris recently developed advanced 5xxx alloy variants for roll-formed and creep-formed fuselage skins exceeding the toughness/strength balance of that of age-hardenable reference alloys such as 2x24-T3 or 6x56-T6 [13, 14]. Providing comparable or even higher yield strength properties [13] these advanced Al–Mg–Sc alloys exhibit a 4.3–4.7% lower density and a 2.2–5.8% higher Young's modulus, and compared to 2x24-T3 alloys [13, 14] a better damage tolerance, a very low corrosion susceptibility, and an excellent weldability. However, the high strengthening effect of Al_3Sc precipitations (of about 45–50 MPa per 0.1 mass% Sc [33]) introduced by artificial aging is far from being exhausted. To face this challenge, EADS Deutschland GmbH (together with RSP Technology Inc.) developed a prototype alloy in 2003 that was later patented under the registered trademark "Scalmalloy" [31]. Strip-casting technologies such as "ScalmalloySC," which were adapted to this novel second generation of Al–Mg–Sc alloys, enable an extraordinary high solidification rate that keeps a high amount of Sc and Zr retained in supersaturated solid solution, as a prerequisite for a subsequent precipitation of fine disperse secondary $Al_3Sc(Zr)$ phase during hot rolling or age-hardening. Depending on the alloy's Sc content ranging between 0.7 and 1.4% Sc these precipitations effect very high tensile strength levels ranging from 500 to 600 MPa (i.e., typical values of today's high-strength 7xxx alloys) combined with sufficient elongations of at least 10%. The current research on these future Al–Mg–Sc alloys focuses on an accurate adjustment of temperature and time parameters during semiproduct processing in order to receive optimal age-hardening results [32]. However, the huge age-hardening potential of Sc hasn't been completely explored yet. This necessitates further fundamental investigations [34].

Regarding the application of sheet alloys in the automotive industry, non-heat-treatable alloys from the 5xxx series entered the automotive sector in the mid-1980s [18] as a serious alternative to low carbon steel grades. While Mg, as the main alloying element, enhances strain hardening significantly, an addition of Mn mainly reduces recovery, and thus, both alloying elements are often used in combination for automotive sheet applications [24]. Besides the high strengthening effect Mg additionally increases the cold formability and, the occurrence of flow lines during deep drawing; wherefore, today's 5xxx alloys such as 5049 and 5454 are often used for nonvisible hang-on parts (such as door inner panels) or chassis parts (e.g., seam welded tubes) with a high demand on formability and lower demand on surface appearance [17, 24]. For high-strength structural applications where intergranular corrosion can be excluded, alloys with higher Mg content of about 4.5% such as alloy 5182 (AlMg4.5Mn0.4) are used. Novel alloy developments characterized by an Mg content up to 6.5% provide a significant increase in tensile strength of more than 320 MPa and elongations of around 26%, while formability (Figure 2.5) benefiting the cold forming of complex structural components is substantially improved [11, 35].

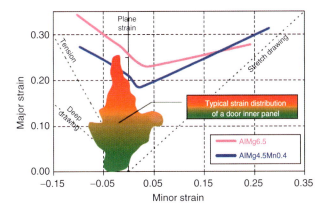

Figure 2.5 Forming limit curves of AlMg6.5 compared to AlMg4.5Mn0.4 [11, 35].

Age-hardenable Al–Mg–Si sheet alloys belonging to the 6xxx alloy series provide good strength/toughness balance and excellent corrosion resistance in condition T6, and additionally, a good cold formability along with an excellent weldability in condition T4 before age-hardening. Driven by the rising demand for weight savings in aircraft production by rivetless joining, 6xxx sheet alloys, such as 6010-T6, 6013-T6, or 6x56-T6, have already replaced a considerable amount of conventional 2xxx alloys, such as 2x24-T3, for sheet applications such as skin panels being used in the lower part of the fuselage. However, in future, 6xxx sheet alloys may compete increasingly with advanced Al–Cu–Li and Al–Mg–Sc alloys, provided the manufacturing restrictions of these advanced alloys concerning weldability and formability, respectively, can be solved.

Since the requirements for automotive panels have diversified in the last decade, 6xxx alloys are preferentially used in visible exterior panels as a consequence of their optimum surface appearance after forming and their age-hardenability during paint baking [17, 18]. Furthermore, 6xxx alloys, such as 6016 and 6111 [36], provide an in-service corrosion resistance and compatibility with adhesive bonding techniques [17]. The excellent dent resistance of 6xxx panels compared to that of 5xxx panels results from the sequential precipitation of Mg–Si containing metastable β' dispersoids during bake hardening at $170–190\,^{\circ}$C. Thus, 6xxx panels are usually cold formed in condition T4. However, compared to 5xxx alloys, the formability of today's 6xxx automotive alloys is constrained. Therefore, by means of metallurgical improvements Hydro recently developed an advanced 6016-T4 alloy variant, labeled as "6/30+," for more sophisticated forming parts [37]. Providing elongation values of more than 30% in all directions, this new alloy shows a forming limit curve with nearly 20% higher major true strain values than those of the conventional alloy 6016-T4 [37].

Plate alloys from the 7xxx series, providing tensile strength values up to 600 MPa, are needed for huge ultra-high-strength applications. With respect to aircraft structures this includes a high compressive strength and high cycle fatigue strength combined with high fracture toughness, good corrosion resistance, and sufficient

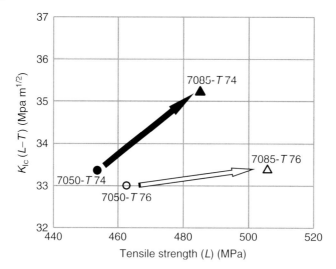

Figure 2.6 Strength and toughness properties of alloy 7085-T7x versus incumbent alloy 7050-T7x in the longitudinal direction of a 152 mm thick plate [38].

resistance against SSC. Recent developments are focused on Cu-containing alloy variants with an increased Zn content, exhibiting a higher strengthening potential for ultra-thick section parts. In this context, Alcoa recently developed an advanced plate and forging alloy, which is registered as alloy 7085 [38]. As a result of the raised Zn content up to 8.0% along with the reduced Cu and Mg content, alloy 7085-T7 possesses enhanced strength and toughness properties compared to incumbent thick-section alloys, such as alloy 7050-T7 (Figure 2.6). In addition, the quench sensitivity of the properties during aging is considerably low [38], benefiting the age-hardening results of plates with enlarged thickness as well as the compliance of distortion limits. Facing the same property targets, a similar alloy, registered as alloy 7081, was also developed by Aleris [14]. Besides, more academic investigations have been focused on the effect of further alloying elements on the peak hardening results of Al–Zn–Mg-based sheet alloys. For example, the addition of Ag obviously enables an increase of tensile strength in a similar manner as the addition of a comparable amount of Cu, promoting an increased number of fine spherical η' precipitations [39].

Regarding automotive applications, recent 7xxx alloy developments are directed to safety-relevant substructures, such as B-pillars or transmission tunnels, usually made of ultra-high-strength steels (e.g., hot-stamped boron steel grades). This includes Cu-free (engineering) alloys, such as 7021, offering an excellent weldability as well as Cu-containing (aerospace) alloys, such as 7081, providing maximum strength properties. In this context, Aleris developed both a modified 7021 automotive alloy, offered as "Structurelite 400," and a modified 7081 automotive alloy, labeled as "Structurelite 5xx," providing yield strength values in aged conditions of more than 400 and 500 MPa, respectively, which are achievable by common paint bake operations [40] (see also Section 2.5).

2.2.2.2 Process Development

With the objective of enhancing the sheet production efficiency as an alternative to conventional sheet production starting with DC ingot casting, a couple of continuous process routes for smaller quantities or particular sheet products have been developed in the last decades. Coupling a continuous casting process of thin aluminum strips directly from the melt with a subsequent cold-rolling operation can render hot rolling redundant. One of the best known continuous casting processes is doubtless the twin-roll casting (TRC) [24, 41]. Characterized by a very quick solidification between the two rolls, TRC enables strip production with a strip thickness of 2–12 mm [24]. Despite an enormous productivity potential, various microdefects (e.g., surface pores and center segregations) and inappropriate texture may arise during casting and limit the range of commercial application [41, 42]. Thus, in the field of TRC, there is still a great need for research [43].

As a novel strip-casting technology, spray-rolling combines TRC and spray forming (Figure 2.7) [44]. Tensile properties of strips obtained after spray-rolling including subsequent solution heat treatment and cold-rolling meet or exceed those of strips processed via the conventional ingot process route, thus eliminating DC ingot casting, homogenization, and hot-rolling operations. Owing to the very high solidification rate of the droplets during flight, impact, and consolidation between the mill rolls, spray-rolling enables a broader production range of alloys with wide solidification ranges than conventional TRC [44].

This process attribute, while being typical for the spray-forming technology, also benefits the production of age-hardenable Al–Mg–Sc alloys by keeping a huge amount of Sc retained in supersaturated solution. However, the precipitation of Al_3Sc dispersoids cannot be avoided completely. Therefore, rapid cooling of the deposit is necessary to suppress precipitation coarsening after spray forming. Subsequent flat rolling at moderately elevated temperatures leads to extraordinary aging results, with tensile strength values and elongations close to those of today's 7xxx-T6 sheet alloys [33]. Furthermore, spray forming enables the deposition of multiple alloy layers within a single sheet by transversally moving a flat substrate

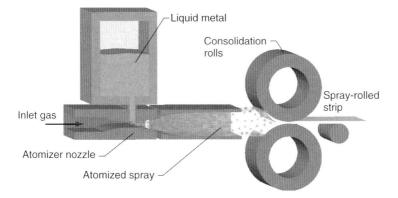

Figure 2.7 Schematic of the spray-rolling approach [44].

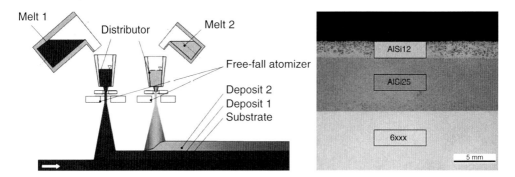

Figure 2.8 Micrograph of spray-formed multilayer sheet [45].

through two or more serially connected spray-forming atomizers bearing various alloy types (Figure 2.8). As an alternative to commercially produced multialloy aluminum ingots, labeled as Novelis Fusion [41], multilayer spray-forming allows a combination of various alloys with different characteristics (e.g., formability, toughness, corrosion resistance, and wear resistance), but without limitation in the thickness ratio of the layers [45].

Although cold workability, as one of the most crucial processing characteristics, is adjusted by final annealing after cold rolling, the inherent forming properties are significantly influenced by the thermomechanical conditions of the upstream process stages. On this issue, a couple of investigations were done in the last decade. Particularly, the interaction between hot-rolling and cold-rolling conditions has a great impact on the forming behavior (Figure 2.9) [46]. While incomplete recrystallization after hot rolling intensifies a distinctive anisotropy after cold rolling, a fully recrystallized (cube) texture after hot rolling promotes a cold-rolling texture that leads to a more equalized earing formation, for example, after cup deep drawing as the result of a less anisotropic forming behavior of such processed sheets [24, 46].

Further improvements of the formability can be realized by applying advanced forming techniques.

In order to enable the production of particularly complex and difficult structures, which are usually made of easy formable deep drawing steel grades, hot deep drawing (HDD) can be applied to non-heat-treatable alloys, such as 5xxx grades [11, 47, 48], and heat-treatable alloys, such as 7xxx alloys [49, 50]. Concerning 5xxx alloys, the best drawing limits (up to 2.6) are obtainable by applying different temperatures locally (i.e., heating the flange up to 250 °C and cooling the punch to room temperature) [51]. Regarding sheets made of ultra-high-strength age-hardenable 7xxx alloys, for example, for B-pillars or transmission tunnels, HDD helps overcome their inherent poor cold formability restrictions. Usually, HDD of 7xxx sheets is performed at high temperatures during solution annealing, raising the material's formability to an optimum. However, subsequent quenching may lead to distortion that necessitates an additional calibration operation and, thus, reduces the productivity. Therefore, semi-HDD that is performed during

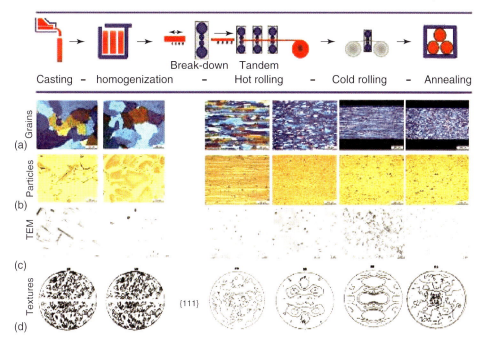

Break-down Tandem

Casting – homogenization – Hot rolling – Cold rolling – Annealing

(a) Grains

(b) Particles

(c) TEM

{111}

(d) Textures

Figure 2.9 Evolution of the microstructure during the sheet production according to Hirsch [24]: (a) grain structure, (b) precipitation structure, (c) TEM, and (d) texture as {111} pole figures.

age-hardening [49] seems to be a promising approach for the efficient production of complex-shaped components made of 7xxx alloys (Section 2.5).

As further approaches for sheet-metal forming of complex-shaped structures with detailed features that cannot be produced by conventional deep drawing, sheet-metal hydroforming (SHF) [52, 53], viscous pressure forming (VPF) [54, 55], superplastic (blow) forming (SPF) [56, 57], or electromagnetic forming (EMF) [58] can be applied. The punching pressure during SHF and VPF is generated by viscous media, while SPF works with gas pressure and EMF with electromagnetic forces.

SHF is a cold-working technology for the manufacture of lightweight components with improved structural integrity, increased strength, and stiffness compared to conventional deep drawing parts. Hollow components are usually made of longitudinal welded tubes enabling tighter wall thickness and diameter tolerances than components made of extruded tubes [20]. The final shape is obtained through multiple stages comprising bending, pre-forming, and finishing, where pre-forming has a great impact on typical failure modes such as folding back, wrinkling, buckling, and fracture [59].

EMF is a cold-forming approach to overcome inherent formability limitations of aluminum alloy sheets [58, 60]. Owing to the high strain rate and an inertial stabilization of material failure modes, EMF can promote significant increases in strain-to-failure values [61, 62]. Thus, compared to conventional deep drawing the frictionless electromagnetic punching of the workpiece toward a complex-shaped counter die enables the cold forming of parts with more complicated geometry [58]. Furthermore, EMF is also applicable for compression of cross sections of hollow components, for example, tubes [63].

VPF, as a novel flexible cold-forming technology, uses a semisolid macromolecular polymer [54, 55, 64]. Due to the feasibility of a nonuniform pressure field being adaptable to the strain rate of the sheet material, VPF is proved to reduce the spring-back behavior as a result of an improved stress distribution after forming particularly a sheet material with lower ductility [54].

SPF, as a hot-forming technique, is characterized by the feasibility of extraordinary large strain-to-failure response of more than 300% that makes SPF attractive for the production of complex three-dimensional automotive panels. However, because of the long-term production cycles SPF is currently only restricted to the manufacturing of low-volume cars. In addition, typical extremely fine-grained alloys that are suitable for SPF are mostly too expensive and therefore rarely appropriate for high-volume automotive applications. To face these challenges, a concept to promote SPF as a cost-competitive alternative to traditional forming techniques has been recently developed [56]. In order to decrease the materials cost penalty this concept envisages the use of conventional or slightly modified alloys. Despite the coarser grain size resulting from the lack of grain growth inhibitors, experimental results of conventional 5182 sheets show a strain-to-failure response of considerably more than 100%, which is enough for the one-piece production of many automotive applications, which are conventionally manufactured with an assortment of stamped, extruded, and cast parts. In order to compensate for the production time penalty, the contrived concept proposes a one-piece SPF of integral structures, such as inner doors, from a single blank. An associated overall cost reduction may promote the competitive position of SPF, particularly in terms of medium-volume automotive applications [56].

As a sheet-metal forming technique for prototype manufacturing or low-quantity production of complex-shaped components, incremental sheet forming (ISF) is a cost-efficient alternative to conventional deep drawing. Along with simple CNC (computer numerical controlled) forming tools, ISF makes the investment of deep drawing presses using costly dies redundant [65]. The desired sheet-metal part contour is formed by a locally operating tool (forming pin), moving along a predefined tool path and effecting a gradual plastic deformation with large plastic strains (logarithmic strains of up to 1.8) [66]. The most common process variants are single-point incremental forming (SPIF) and two-point incremental forming (TPIF) [67], where TPIF uses a partial or full die as a contour-supporting tool benefiting the contour accuracy. As an advanced process variant duplex incremental forming (DPIF) offers both a high shape flexibility and accuracy without the need for any contour-specific tools, by using a movable counter tool moving simultaneously on the opposite side of the blank [68]. A current advancement of TPIF is characterized by an in-process support by means of stretch-forming and local laser heating, effecting both a more uniform thickness distribution and a reduced production time [69].

Figure 2.10 Profile intensive space-frame architecture applied to Audi R8 [19].

2.2.3
Extrusions

Aluminum extrusions are available with a variety of differently shaped cross sections ranging from base profiles, such as rods and tubes, to multiple hollow extrusions, which are predominantly used in rail applications. Open profiles such as stringers are typical high-volume applications in aircraft, which are needed for enhancing the bending stiffness of fuselage skins and for discharging the forces. In automotive applications, such as space-frame body-in-whites (Figure 2.10), hollow extrusions prevail, offering a high moment of inertia of area [70].

Today's industrial aluminum extrusion production consists of a sequence of several single processes starting with DC ingot casting followed by homogenization treatment, (hot) extrusion, stretching, and final annealing [71]. The last process gives the material the desired in-service characteristics, or, if a subsequent hydroforming is intended, the required forming properties. The cross section and accuracy of the exiting profile is specified by the extrusion die geometry; wherefore particular attention has to be paid to the die design. Concerning the extrusion process variants it has to be distinguished between direct extrusion and indirect extrusion, whereas the more elaborately equipped indirect extrusion is characterized by a lower pressing force demand due to reduced friction [20]. The hydrostatic extrusion and the so-called Comfom extrusion are further process variants, which are not very common for the mass production of structural components and, therefore not considered in the following.

2.2.3.1 Alloy Development

Novel generations of Al−Cu−Li alloys, which are predominantly developed for aircraft sheet applications such as fuselage skin panels (Section 2.2.2), are also available for aviation extrusions, such as stringers. New Al−Cu−Li alloys developed by Alcan Aerospace, such as 2195-T8 and 2196-T8, provide an improved toughness/strength balance by far exceeding that of conventional stringer alloys such as 2024-T3, 2027-T3, or even 6056-T6 [12]. As a result of the increased Li content alloy 2196 exhibits a distinctly reduced density. Despite the decreased Cu content this alloy provides both high-strength and higher toughness values, while the high Cu (and low Li) containing alloy variant 2195 achieves the same strength properties as the high-strength stringer alloy 7349-T76 (Figure 2.11) promoting the potential for replacement of 7xxx alloys.

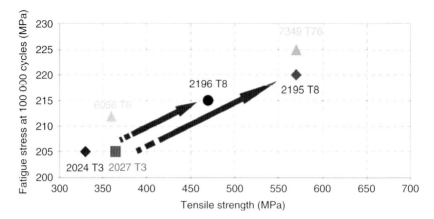

Figure 2.11 Properties of extruded stringers made of Al–Cu–Li alloys versus conventional stringer alloys [12].

Owing to their good damage tolerance and excellent weldability, extrusions made of 6xxx alloys, such as stringers, are increasingly used in aircraft designs. Furthermore, as an approach for improving the fuselage skin integrity, the extrusion of stringer-integrated panels is conceivable. Driven by the increasing demand of weight savings in the automotive industry, aluminum extrusions are more and more applied for safety-relevant substructures, such as bumper beams, side skirts, or crash boxes [17, 72]. Concerning the requirements for crash characteristics, these kinds of extrusions provide inherent functions, such as high specific bending stiffness and excellent energy absorbability, respectively. Medium-strength age-hardenable alloys from the 6xxx series with tensile strength values of about 300 MPa, such as 6061, exhibit a considerable extrudability of more than 60, allowing both the productivity target of a high-volume production and the design target of load-optimized cross sections with wall thicknesses ranging from 1.8 to 3.0 mm to be met. Furthermore, the approach of an intensive in-process quenching applied directly after extrusion enables the elimination of the solid-solution treatment, and thus, improves energy and cost efficiency [72]. The final ductility and strength is controlled by subsequent precipitation hardening, considering the required structural characteristics. Bumper beams are mainly peak hardened (T6) [72], while, for example, crash boxes have to provide a distinctive ductility affecting the resulting buckling behavior proved during axial crushing testing [17].

As an advanced solution for complex-shaped space-frame extrusions, Trimet and Honsel have currently developed a novel 6xxx alloy called *Trimal 52* [73, 74]. Considering the whole manufacturing chain (from casting to final age-hardening), multihollow space-frame extrusions with a tensile strength above 300 MPa and elongations of more than 11% could be produced, providing excellent crash characteristics indicated by a very regular wrinkling (Figure 2.12) without any cracks [73, 74].

Extrusions from the 7xxx alloy series are mainly used when ultra-high tensile strength with values between 300 and 600 MPa is required. In the field of aircraft

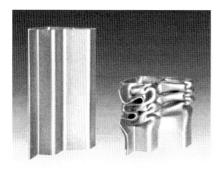

Figure 2.12 Wrinkling of a multihollow space-frame extrusion after axial crushing [73].

applications this includes components such as seat tracks or stringers. With Zn and Mg as the main alloying elements, high-strength age-hardenable 7xxx alloys have proved to be very difficult to extrude. Particularly, when the cross-sectional geometry is complex, the limited extrudability of below 10 constrains the productivity significantly [20, 72]. In addition, micro cracks may occur during hot extrusion if extrusion speed and temperature are not accurately adjusted. Stringers made of copper containing 7xxx alloys, such as 7075-T7, 7136-T7, or 7349-T7, are predominantly used in highly compressed aircraft sections [13]. However, contrary to Al–Mg–Sc alloys or alloys from the 6xxx series, which are appropriate for laser beam welding (LBW), stringers made of those 7xxx alloys are still integrated into the structure by riveting. Thus, in order to overcome the weldability restrictions, addition of Sc, which has already been applied to 2xxx alloys [27], seems to be a promising approach both for reducing hot-cracking during welding and for retaining the required strength properties in the welding zone.

Regarding automotive applications, 7xxx extrusions show a high potential for complex high-strength space-frame structures. While quenching occurs during the extrusion process the required strength/ductility balance is achieved by adjusting the T7 condition. However, the major challenge concerning the application is to overcome the already mentioned extrudability restrictions benefiting both the process productivity and the feasibility of sophisticated cross section designs.

2.2.3.2 Process Development

Aluminum and its alloys are characterized by a high stacking-fault energy that promotes microstructural mechanisms such as dynamic recovery (DRV) during hot-working at high temperatures and also dislocation-based subgrain formation over discontinuous dynamic recrystallization (DRX) [71, 75]. Furthermore, geometric dynamic recrystallization (GRX) occurs in regions with high plastic shear deformation [76–79], while SRX may proceed after plastic deformation [76, 78, 80]. All these interacting mechanisms are proved to be responsible for the local subgrain structure during hot-working being crucial for the final strength and ductility of the material. Regarding age-hardenable alloys, it is also recognized that the subgrain morphology influences the precipitation behavior during quenching and age-hardening [20]. Therefore, a lot of research activities have been launched in the field of microstructure modeling in order to enable the prediction of microstructure evolution during hot extrusion [81, 82]. However, some phenomena, such as peripheral coarse grain and grain growth, have to be investigated more deeply [81, 83].

An intensive quenching immediately after extrusion in order to suppress precipitation before age-hardening is a presupposition for optimal material properties. Therefore, knowledge of the temperature and time-dependent precipitation behavior during quenching is of great importance. These interrelations can be displayed by continuous cooling precipitation (CCP) diagrams recorded by courtesy of a novel method using the differential scanning calorimetry (DSC) [84–86]. Current investigation results by means of quenching 6xxx alloys prove that the precipitation kinetic is strongly dependent on the alloy content. Both the lower critical cooling rate (start of supersaturation of solid solution) and the upper critical cooling rate (completion of supersaturation of solid solution) increase with rise in the alloy content. Even if the same alloy type is used, a deviation of the chemical composition may cause significantly differing material properties after subsequent precipitation hardening [84]. Thus, further investigations concerning the optimization of the quenching process are still being carried out.

Profiled automotive components can also be produced by sheet-metal forming. However, in contrast to the conventional deep drawing of double half-shell components (such as cross members or roof rails) the extrusion of profiles requires considerably lower tool costs, even in the case of low-volume production [87]. Otherwise, deep drawing allows the production of complex three-dimensional workpieces with different cross sections, while the shape of extruded profiles is always straight and their cross sections are constant along their longitudinal axis. The production of curved profile shapes usually requires an accurately adjusted and thus cost-intensive cold stretch-bending operation considering spring-back behavior, internal stress distribution, and deformation of the cross section [88]. In order to avoid these disadvantages and to enable smaller bending radii, several warm bending techniques (performed directly during extrusion) have been developed in the last decade [88, 89]. The use of a guiding tool that deflects the exiting strand, for example, curved profile extrusion (CPE), enables the production of three-dimensional bent and twisted profiles (Figure 2.13), whereas the plasticity

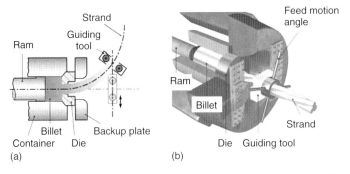

Figure 2.13 Schematic of (a) curved profile extrusion and (b) twist profile extrusion [88, 90].

results from the extrusion process itself and not from lateral forces, benefiting both the material properties and the accuracy of the curved profile shape [88, 90]. Furthermore, curved hollow extrusions are potentially well suited for hydroforming, enabling the production of curved hollow structures, for example, top side rails [20], without cold pre-bending.

As a further profile-forming approach twist profile extrusion (TPE) and helical profile extrusion (HPE) allow the manufacturing of straight, but accurately twisted profiles. While TPE uses an external tool for twisting the profile (Figure 2.13), the helical profile shape generated by HPE is internally formed by using a helical cut die geometry [90, 91]. For example, by HPE it is possible to extrude near-net-shaped components such as helical gear rods [91] or screw rotors [90], for example, for automotive applications.

With the objective of reducing the energy consumption for the production of aluminum extrusions, a new hot extrusion approach has been developed in recent years using compacted milling and turning chips as billet material instead of conventional DC-cast solid billets [92]. By means of extrusion trails with alloy 6060, it could be demonstrated that a sufficient extrudability and surface quality is achievable when using chips of the same alloy type. Furthermore, the microstructure did not exhibit pores or inclusions at the welded chip boundaries, and the yield strength was comparable with those of conventional profiles in extruded condition [92].

2.2.4
Forgings

High-quantity aluminum forgings are commonly produced by die forging, enabling a near-net shape manufacturing of components with high integrity and geometrical complexity nearly similar to castings. Furthermore, on account of their dense fiber structure, forgings are predestinated for applications where maximum possible safety against misuse or in case of impact or cyclic load is required [20]. In the automotive industry, die forgings, such as control arms, are predominately used in the chassis, and also in combustion engines, where die-forged pistons or con-rods are employed [20, 93]. Large die forgings are also used in the aircraft industry. This comprises components such as vertical stabilizer fittings, pylon diaphragm brackets [94], or huge wing spars as today's largest die forgings in the world (Figure 2.14) [95]. In addition, large open die forgings dedicated, for example, for frame segments, and large seamless forged or rolled rings are also needed for aircraft applications, the latter primarily for smaller sized short-range jets (e.g., from Embraer).

Figure 2.14 Inner center spar of Airbus A380 [95].

The industrial aluminum die forging production of today comprises of a sequence of several single processes starting with DC ingot casting and homogenization treatment followed by heating, pre-forging (drawing out, upsetting), die forging, solution treatment, cold calibrating, and, finally, precipitation hardening. The last process gives the material the desired in-service characteristics. The material flow and, thus, the final shape of the forging are specified by the geometries of the upper and the lower dies; wherefore, particular attention has to be paid to the die engraving [20]. If the desired shape necessitates a sequence of die forging operations, an accurate adjustment of the dies employed is of great importance. As forging machines, hydraulic presses or drop hammers are usually applied for the manufacturing of both die forgings and open die forgings. Large seamless rings, also classified as forgings, are more efficiently produced by means of ring-rolling machines (Figure 2.18).

2.2.4.1 Alloy Development

For the production of die-forged automotive parts, such as control arms, suspension arms, or even wheels, alloys from the Al–Mg–Si system are very common because of their excellent formability. By adjusting a stoichiometric Si/Mg ratio, alloys from this alloy system, such as 6xxx alloys, are hardenable because of the precipitation of metastable partially coherent β' phase during artificial aging. With the objective of using the considerable strengthening effect of finely dispersed incoherent β (Mg_2Si) precipitations, novel alloys with more than 20 mass% Mg_2Si were developed. Such alloys can hardly be produced by casting. Otherwise, the hardening effect was only negligible as a consequence of the usually low solidification and cooling rate during casting, generating coarse Mg_2Si precipitations. Therefore, the application of spray forming, proved as a rapid solidification and cooling technique, enables a considerably finer dispersion of the particles. Owing to the extraordinarily large amount of fine Mg_2Si particles embedded in the Al matrix, these alloys exhibit an increased Young's modulus of around 80 GPa, and a distinctly decreased density below 2.5 g cm^{-3} combined with a significantly enhanced strength of 450 MPa at room temperature and 150 MPa at 300 °C. The obtained properties are well suited for advanced high-temperature applications, such as die-forged pistons (Figure 2.15) in high-performance combustion engines [93, 96, 97].

Despite the comparatively low formability of 7xxx alloys [20], forgings made of these alloys, such as 7050-T7, 7075-T7, or 7085-T7, are needed for huge ultra-high-strength aircraft applications, which can be found, for example, in the wing unit [95], where high compressive strength and high cycle fatigue strength combined with high fracture toughness, good corrosion resistance, and sufficient resistance against SSC are required.

Equivalent to plate alloys, recent developments of 7xxx forging alloys are focused on very high-strength alloy variants with an increased Zn content along with a reduced Cu and Mg content and the minor addition of elements such as Sc, Zr, and Ag for ultra-thick section three-dimensional parts. As already mentioned in Section 2.2.2, Alcoa developed the plate and forging alloy 7085-T7 as an enhanced alternative to thick-section alloys, such as alloy 7050-T7 or 7010-T7. Extraordinary

Quelle: Mahle GmbH

Figure 2.15 Die-forged high-performance piston made of spray-formed alloy with more than 20 mass% Mg_2Si. (Courtesy of Mahle GmbH.)

large die-forged parts made of this advanced alloy, such as, spar components, are already used in the Airbus A380 [38].

A further thick-section alloy variant, developed by Otto Fuchs and registered as alloy 7037-T7, enables achieving the properties, strength, ductility, and fracture toughness, similar to conventional alloys, such as 7050-T7 or 7010-T7, but, the alloy's lower quenching sensitivity makes feasible these properties with considerably enlarged cross sections up to 270 mm. Typical examples for forgings made of this alloy are, die-forged pylon brackets or vertical stabilizer fittings, used in the Airbus A380 (Figure 2.16) [94].

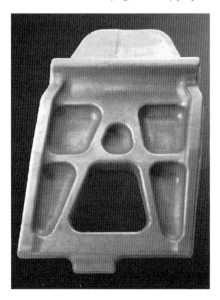

Figure 2.16 Vertical stabilizer fitting of Airbus A380 (mass 246 kg). (Courtesy of Otto Fuchs KG.)

2.2.4.2 **Process Development**

Following the near-net shape philosophy, precision forging (PF) is a more recent, advanced process variant of die forging that is applied in order to reduce the dimensional oversize and to obtain an excellent dimensional accuracy. PF was intensively investigated for steel applications [98], but it is also applied to aluminum applications [99] providing narrow tolerances, thin cross sections, small radii, 0–1° draft angles, and excellent surface finishes [100]. Furthermore, the existence of the weaker short transverse (S-T) direction that is typical for die forgings can be avoided by adjusting the material flow.

Combining the advantages of die forging and casting both thixoforming (TF) and thixocasting (TC) are further process variants for the production of complex near-net-shaped components, which are used in automotive applications, such as steering knuckles or suspension arms [101]. Compared to TC, TF requires a liquid fraction that is much lower (usually below 30%) and can be regarded as a lubricant during the forming process. Nevertheless, aluminum alloys, commercially used for TF, are usually Al–Si casting alloys, such as A356 and A357 [102]. However, particularly complex-shaped workpieces (Figure 2.17) tend to segregations during TF, which significantly depends on the local material filling velocity and local flow length inside the die. Thus, the challenge for future research is focused on optimizing the forming parameters in order to minimize segregations inside complex TF components [103]. Further research activities concentrate on TF of high-strength wrought alloys, such as alloy 7075, which are typically used for forgings, but until now regarded as unsuited for TF because of their wide solidification interval promoting the tendency for hot tearing. By using, for example, feedstock in the extruded state that utilizes recrystallization and partial melting (RAP), the generation of a thixoformable microstructure is largely possible [102].

As an advanced forming process, ring rolling enables the production of huge seamless rings up to 10 m diameter (Figure 2.18). In contrast to the discontinuously working open die forging process, the continuously operating ring-rolling process

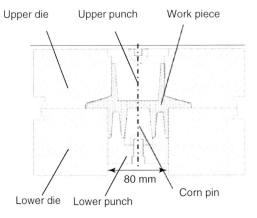

Upper die   Upper punch   Work piece

80 mm

Corn pin

Lower die   Lower punch

Figure 2.17 Schematic of complex forming dies and the components cross section after thixoforming [103].

Figure 2.18 Production of large seamless rolled rings (left, axial rolling gap and right, radial rolling gap). (Courtesy of Rothe Erde GmbH.)

based on radial-axial multi-pass-rolling enables high-quantity production. Besides conventional rings exhibiting rectangular cross sections, it is also possible to produce rings with complex-shaped cross sections by using profiled rolls or even die rolls. However, a thermomechanical rolling that is well established in the field of flat rolling is extremely difficult to apply to ring rolling on account of its highly unsteady process behavior, which hampers the controllability of the microstructural evolution, during and between the rolling passes. To overcome these restrictions, an advanced simulation model has recently been developed for implementation into the machine control system, enabling an in-process simulation of the local microstructure evolution in the cross section [104, 105]. This model originally developed for ring rolling of steels is also applicable for aluminum alloys.

2.3
Casting Alloys and Associated Processes

In comparison to the development intensity within wrought alloys, only few new casting alloys for transportation applications have been developed during the last years. This is due to several reasons, that is, enormous cost pressure, enhancement of mechanical performance of wrought alloys accompanied with innovative forming processes, limited cooperation among casting alloy producers to share development cost, and concentration on processes instead of materials. Although these reasons can be well understood, they have, in all, led to a limited number of innovative casting alloys, which have found their way into mass production. Foundry men have to focus on liquid metal handling and treatment such as grain refinement, degassing, modification, and melt cleanliness, which will not be treated here because of limited space, and the reader is kindly referred to the references given in the section "Further Reading." From a performance point of view, all the points mentioned earlier must be handled with care to produce a reliable and long-enduring cast component. Even when high-performance alloys are used, inclusions, porosity, and trace elements will destroy all good properties if not controlled in a specific and proper manner. Therefore, process improvements have helped to produce sound castings, but the complete interactions of trace elements, impurities with minor and major alloying elements still remain unclear and are not fully understood to date. Application of unusual alloying elements often fails because of a lack of appropriate master alloys. In addition a single-edge obsession to cost structure ignoring system cost and focusing on part-to-part comparison still drives "development" of cast components. Nevertheless, some new developments have come along, which will be treated in more detail. Only limited focus is set on process improvements, while emphasis is on the enhancement of the alloys.

2.3.1
Alloy Classes and Their Basic Constitution

Similar to wrought alloys, there is a designation system for cast aluminum alloys. Unfortunately, it is only similar and not straightforward and parallel to the wrought designation. A good overview is given in [106]. The designation can be seen in Table 2.2. Please note that alloys are only designated if a certain amount per year is produced. Therefore, special alloys do not have a number, and in the text only their nominal composition is given. A comparison to all national and international standards is given in [107].

In this system, major alloying elements and certain combinations of elements are indicated by the number series in Table 2.2. The 6XX.X and 9XX.X series are not currently in use, but they are being held open for possible use in the future. The digit following the decimal in each alloy number indicates the form of the product:

- "0" (zero) following the decimal indicates the chemistry limits applied to an alloy casting;
- "1" (one) following the decimal indicates the chemistry limits for ingot used to make the alloy casting;

Table 2.2 Alloy classes and their basic constitution [106].

Alloy class (international designation)	Alloy system (basic elements)	Basic constitution (range in wt%)	Age-hardenable
1xx.x	Al	≥99.0% Al	No
2xx.x	Al–Cu	3.7–9.0% Cu	Yes
3xx.x	Al–Si–Mg	4.5–23.0% Si	Yes
	Al–Si–Cu	0.03–5.0% Cu	
	Al–Si–Cu–Mg	0.05–1.5% Mg	
4xx.x	Al–Si	3.3–13.0% Si	No
5xx.x	Al–Mg	2.5–10.6% Mg	No
	Al–Mg–Si	0.1–2.2% Si	
6xx.x	Not in use	n.a.	n.a.
7xx.x	Al–Zn	2.7–8.0% Zn	Yes
	Al–Zn–Mg	0.2–2.4% Mg	
	Al–Zn–Mg–Cu	0.1–1.0% Cu	
8xx.x	Al–Sn	5.5–7.0% Sn	No
9xx.x	Not in use	n.a.	n.a.

- "2" (two) following the decimal also indicates ingot but with somewhat different chemical limits (typically tighter, but still within the limits for ingot).

Generally, the XXX.1 ingot version can be supplied as a secondary product (remelted from scrap), whereas the XXX.2 ingot version is made from primary aluminum (reduction cell). Some alloy names include a letter. Such letters, which precede an alloy number, distinguish between alloys that differ only slightly in percentages of impurities or minor alloying elements (for example, 356.0, A356.0, B356.0, and F356.0). The temper designations of heat-treatable alloys are the same, as shown in Table 2.1, with the exception of strain hardening, which does not apply to casting alloys.

2.3.1.1 Alloy Development

Alloys from the 1xx series contain more or less only aluminum, no real alloy development can be done here. The main focus is on the concentration of trace elements influencing casting performance and physical properties. Therefore, no further comments are made here.

Alloys of the 2xx series are well known for their hot-cracking tendency because of their long solidification range; therefore, special care must be taken to avoid unsound castings by grain refinement or by thermal management of dies [106]. In sand casting, sand molds with reduced rigidity should be used and – where applicable – steel chills to install high temperature gradients [108]. In contrast, Sigworth [109–111] showed that lowering the titanium content will have beneficial effects on hot tearing although it contradicts common grain-refining theory at first sight. Oxide-free melt is mandatory to avoid hot tearing and this is the reason that most molds have integrated filters to catch oxide skins before they are taken out of the melt and before they enter the die cavity [108]. 2xx alloys exhibit the highest strength among aluminum castings in the heat-treated state and are therefore used in high mechanical performance application in aerospace or automotive industry, for example, alloy 201, also known as KO1, a silver containing and therefore very expensive material [112]. Silver enhances the precipitation sequence (in a similar way as described in Section 2.2). Owing to their copper and magnesium content, these alloys are heat treatable, but with strength, a high sensitivity for SSC arises, which usually affords corrosion protection or overaged T7 conditions [113, 114].

Additions of zirconium and strontium to an Al–2Cu–1Si alloy will increase strength and ductility by changing the precipitation sequence and grain size. Titanium and strontium concentration have to be controlled carefully, as there is a deleterious interaction between titanium and zirconium in their grain-refining effect. Improvement in creep resistance by a factor of 3–5 could be achieved by adding rare earth elements, especially 1 wt% lanthanum, to an Al–6Cu alloy. This is achieved by the precipitation of $Al_{11}La_3$, which inhibits grain boundary migration [115]. Another rare earth element, cerium, is added together with nickel to an Al–5Cu–Mn base alloy to increase hot strength and creep resistance for power train applications, mainly cylinder heads. Long-term application temperatures above 250 °C and with thermal treatment up to 400 °C are claimed for this type of alloy, but castability still remains a challenge [116, 117].

Most of the aluminum castings are made from one or the other alloy of the 3xx family. Owing to their good castability, fluidity, and low or no hot tearing tendency (also similar to 4xx alloys), especially when approaching eutectic Al–Si compositional range, these alloys are widely used in sand, gravity die, and high-pressure die casting (HPDC).

If there is contact between metallic dies and the melt, die sticking is problematic and reduces die life dramatically in HPDC. Therefore, certain iron content is installed to avoid die attack, but some countermeasures have to be taken to avoid the formation of needle-shaped, embrittling AlSiFe phases. Usually, manganese is present in these alloys to alter the undesirable morphology to a more convenient, globular one. Cobalt and cerium have the same effect but are limited in use because of their high price. Manganese addition also requires good temperature control to avoid precipitation and settling of the so-called sludge particles [118, 119]. Nickel counterstrikes the action of manganese, and titanium has to be present to a certain extent to keep good die release behavior [120]. Most challenging is the modification of the eutectic silicon, due to the hindered nucleation of faceted silicon onto nonfaceted aluminum. Usually, sodium or strontium is used but antimony can also be added. There are several publications on the modifying effect of other elements and/or compounds such as calcium, calcium carbide, or sulfur but are still not commonly agreed yet, which may be due to a certain phosphorus content, which may or may not be analyzed correctly [121–123]. In hypereutectic alloys such as 390, refinement of primary silicon is necessary besides the modification of the eutectic silicon. A new approach was launched by Zak and Tonn [124] with the addition of rapidly solidified master alloys of Zr and TiC-containing grain refiner, resulting in a much better distribution and size of primary silicon phase in comparison with only phosphorus. The main application of hypereutectic alloys, such as A390, is still in monolithic engine blocks, but cast-in-place cylinder inserts for the BMW magnesium crankcase [125] are also still in manufacture. In North America, this type of alloy is also used for pistons, sometimes with a reduced silicon content [126]. Higher silicon contents up to 30 wt% can be cast only when steel and not sand molds are used. Otherwise, coarse primary silicon will precipitate out of the melt, which causes increased cost in machining. An Al–30Si–Cu–Ni–Mg alloy was squeeze-cast, exhibiting an average primary silicon size of 70 μm. Therefore, applications as cylinder liners or pistons could be possible. HONDA developed a near eutectic, Al–13Si–4Cu alloy with a high nickel concentration of about 5.5 wt% for HPDC. The high cooling velocity of casting is necessary to achieve a fine microstructure. In addition, the Fe/Mn ratio and absolute Mn content were adopted to avoid large crystallization of Al-(Ni,Fe) plates. Strength could be increased as well as wear resistance, resulting in prolonged microwelding time between ring grove and piston ring material. Fatigue strength was improved at 250 °C by a factor of 1.5 and by a factor of 1.8 at 300 °C, respectively.

3xx alloys often contain magnesium for heat treatability and strength. Heat treatment response of 319-type alloys cast in permanent molds can be improved by strontium and/or magnesium, the latter especially above the specification limit, leading to a change in aging behavior caused by different precipitation sequences and forced segregation of copper, which must be counteracted with the grain refiner [127, 128]. Copper additions usually improve strength by solid-solution hardening and by precipitation hardening during heat treatment. Copper, iron, and zinc may be present to a certain extent when secondary (remelted from scrap) alloys are used or alloyed on purpose with iron to reduce its activity in the melt, thereby avoiding melt attack and die soldering in steel dies.

An interesting, nearly copper- and magnesium-free alloy AlSi9MnMoZr, with an unusually high zirconium content (0.3 wt%) and molybdenum as an unusual alloying element, was introduced, which is traded under the brand marks Castasil-37 or Trimal 37, mainly for HPDC thin-walled body parts. Owing to its composition it shows no undesired aging and should be used in the F state to maintain good ductility, which is important for casting nodes in space-frame design or for internal door parts as shown in Figure 2.19 [129, 130].

As alloys from the 4xx series are binary no significant alloy development can be done here. The main focus is the modification of silicon eutectic and the concentration of trace elements that influence the solidification morphology. Therefore, no further comments are made here.

In alloys from the 5xx series, magnesium improves corrosion resistance markedly, and even in seawater applications alloys with up to 9 wt% magnesium are used but they are not heat treatable. Aluminum–magnesium melts always contain beryllium to overcome excessive dross formation. However, BeO is a toxic compound, but no substitute has been found to date. Vanadium may reduce the beryllium content but still approximately 60 ppm Be has to be present in such alloys [131]. Alloying with manganese leads to microstructures with outstanding performance. As the problem of adjusting the right microstructure in AlMgMn alloys has been overcome, there was, and still is, some renaissance of this type of alloy. Owing to the fact that the microstructure is influenced by the cooling rate

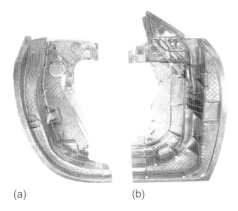

(a)   (b)

Figure 2.19 Internal door parts for Jaguar sports car, Castasil-37 High-pressure die casting 620 × 340 × 170 mm, weight: 1.2 kg (a) and 700 × 340 × 170 mm, weight: 2.1 kg (b) [130].

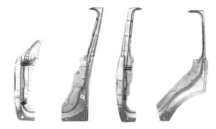

Figure 2.20 Internal door parts for off-road vehicle, Magsimal-59, as-cast state, high-pressure die casting, suited to welding, wall thickness 1.8–2.0 mm, 1400 × 500 mm to 1000 × 240 mm, weight: 2.0–2.2 kg.

during solidification [132–135] besides the exact control of trace elements, they have gained importance in HPDC. AlMg5Si2Mn has been introduced under the brand name Magsimal or Maxxalloy [136] with the same principal composition for several applications in the automotive industry such as internal door panels (Figure 2.20), strut mountings, bracket for stabilizer rods, rear cross members and oil pans.

The main reason is the high ductility in the as-cast state with elongation to fracture up to 15% [137, 138] with reasonable strength. Some improvement of hot strength could be achieved by adding chromium or rare earth elements [136, 139]. Further enhancement of ductility can be reached by lowering the magnesium content to 3% and thus reaching elongation values in the as-cast state of about 20–25%. Even higher elongation values up to 27% could be gained with magnesium contents of about 1 wt%, but yield stress and ultimate tensile strength will decrease from 170/300 MPa (yield stress/tensile strength) at 5 wt% magnesium to 120/200 MPa at 1 wt% magnesium. This outstanding ductility allows untypical joining methods to be applied for cast components, such as self-piercing riveting (SPR), clinching, and flanging. Beyond those, alloys with 3 and 1 wt% magnesium exhibit an unusual deformation behavior because elongation (A_5) and reduction of area (Z) increases while yield stress remains nearly constant during dynamic loading. Raising the strain rate from 0.0004 to 200 s^{-1} will force A_5 by 50% and Z by 40%, making these alloys particularly suitable for crash-relevant parts where high energy absorption is mandatory (Figure 2.21) [140, 141].

Alloying an AlMg3 base alloy with silicon and copper leads to superior creep resistance, retained aging, and good hot strength properties, but hot-cracking tendency must also be taken into account. One possibility to counteract against hot cracking is to replace copper by scandium and zirconium. Zirconium reduces the amount of scandium needed, which is helpful because of the high scandium price of approximately US$2500 kg^{-1} [116, 142–144]. Especially, the price makes the proposed application in cylinder heads for mass production unlikely.

Alloys of the 7xx series are based on Al–Zn–Mg and provide intermediate to high strength accompanied with high elongation values. They are heat treatable in principle but will age at room temperature because of a tremendous retrograde solubility of zinc with temperature, thus eliminating artificial aging.

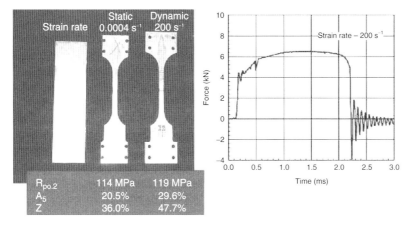

Strain rate	Static 0.0004 s⁻¹	Dynamic 200 s⁻¹
$R_{po.2}$	114 MPa	119 MPa
A_5	20.5%	29.6%
Z	36.0%	47.7%

Figure 2.21 Results of dynamic tensile tests of Magsimal-25, as-cast state, performed at Fraunhofer Institut für Werkstoffmechanik, Freiburg, Germany. Values shown are an average of five specimens [140].

In contrast, these alloys are very difficult to cast, even more difficult than 2xx alloys. To improve castability manganese content should be kept as low as possible. For corrosion resistance an increase in chromium is helpful, and higher magnesium content will lead to better strength [145]. Incorporating these points into alloy design led to an AlZn3Mg3Cr alloy, usually cast in permanent molds. As heat treatment and especially quenching is not necessary for this alloy type, no additional cost for heat treatment and straightening operations will be incurred. A reduction of titanium led to a reduced hot-cracking tendency and SSC is no issue as the zinc to magnesium ratio is far below the critical value of 2–3 [145]. Optimizing AlZnMgCu type alloys for permanent mold castings led to a composition of Al–7.4Zn–2.5 Mg–2.9Cu, which achieved 500 MPa UTS and 460 MPa yield stress after a T7 temper. In parallel, hot-cracking tendency was reduced, too [146].

2.3.1.2 Process Development
Casting alloys are processed by all typical casting techniques such as sand, investment, gravity die, semipermanent mold, low-pressure die casting (LPDC), high-pressure die casting (HPDC), and squeeze casting. Besides this a great variety of TC methods are available, operating in a temperature range where the alloys used are in a semisolid state. In the following, some of the processes mentioned earlier are discussed [143].

2.3.1.3 Sand Casting
Most emphasis was placed on the development of improved binder systems for automatic core making or to replace organic with inorganic binder systems [147–149]. Especially, the ecological aspect gained more and more attention as legislative restrictions and, last but not least, cost for depositing of used sand led to systems that can be refurbished in house more easily and to reduce the

amount of waste. The most well-known example is given by crankcases made by the core package method with cast-in-place gray cast iron liners. A very interesting sand casting method is introduced as "ablation technology" where the sand is washed away by a water shower, thus improving cooling of the partially solidified cast part by increasing solidification velocity of the residual liquid. Refinement of microstructure and improved mechanical properties are shown in some cast samples. The biggest advantage is to cast thin- and thick-walled sections and control the uniformity of microstructure by well-directed cooling with water sprays [150].

2.3.1.4 Low-Pressure Die/Sand Casting

In the manufacture of the linerless so-called monolithic crankcases (Figure 2.22), this process is a prominent method to handle hypereutectic Al–Si alloys up to 18% silicon and even higher. Owing to the high melting temperature, no other casting method will lead to satisfactory results in terms of the lifetime of the dies, microstructure, and gas content. Owing to the latter, parts which have to be welded or which are remelted for tribological reasons (cylinder walls) have to be cast in this way, although HPDC was greatly improved during the past decades concerning gas content; but mold filling in this process is purely lamellar and therefore avoids the entrainment of hydrogen or oxide skins.

An interesting enhancement of LPDC is the assistance of vacuum during the mold-filling procedure. The melt is not transferred to the die by applying nitrogen gas pressure to the melt but by evacuating the die cavity and sucking in the melt from the holding furnace. A very innovative component is the Porsche Panamera rear subframe (Figure 2.23). The cast part replaces a mounted steel sheet design with more than 30 single parts. It is cast with a single sand core and saves 4 kg compared to steel in the final part [152].

2.3.1.5 Investment Casting

This process is used mainly for high-performance parts for aerospace, military, and racing applications, such as rudders, hatches, hand wheels, cargo loading boxes, exhaust ducts, struts, venting door frames, housings, and tracks. Usually,

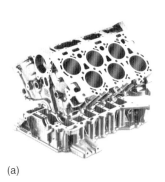

(a)   (b)

Figure 2.22 VW/AUDI W12-engine, low-pressure die cast (a). View of opened die (b) [151].

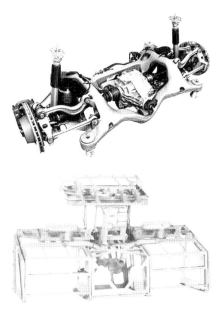

Figure 2.23 Rear subframe assembled with rear axle [152].

Figure 2.24 Cockpit instrumentation panel, 11 investment cast parts assembled together instead of 300 single components and 600 rivet connections [156].

a wax model of the cast component is produced first. Onto this wax pattern a ceramic shell is built up by dipping the wax pattern into a ceramic slurry and subsequently sprinkling with solid particles. After several runs and drying operation the wax is remelted and removed from the shell, which subsequently is fired and cast. For further information refer to [153]. This method allows parts to be cast from a few grams to several kilograms with high accuracy, surface quality, undercuts, and even with complicated cores placed in the wax model and held by metallic pins. Especially, the number of single parts needed for big components can be reduced to a small number of pieces with a high degree of integration, thus reducing assembly time and cost dramatically. In Figure 2.24 a cockpit instrumentation panel is shown, which was originally assembled from 300 single parts and 600 rivets. Investment casting reduced the number of parts down to 11 and assembly takes only 10% of the initial time. Typical alloys for investment castings are Al–5Cu–1.5Ni–Mn–Sb–Co–Ti–Zr (RR350), Al–7Si–0.6Mg, Al–3Mg, Al–5Si–1Cu–Mg, and Al–4Cu–Ti–Mg–(Ag). Process development in terms of solidification modeling and directional solidification control leads to a certain improvement of casting quality, especially of components with differing wall thicknesses. These methods are known under the name HERO Premium Casting process [154] or Sophia Process [155].

2.3.1.6 **High-Pressure Die Casting (HPDC)**

With HPDC a variety of components can be produced. It ranges from small and simple parts to huge-, thick-, and thin-walled components with high complexity, produced in sophisticated dies with several movable cores and squeeze pins. Some decades ago, HPDC parts were said to be of low quality, which was caused by poor process control. High gas contents, endrained oxides, presolidified silicon, and sludge particles led to poor mechanical properties. In the mean time a close control of the whole process such as shot control, die venting, and thermal management of die and melt had led to a process capable of producing weldable and heat-treatable components at a high productivity level. In particular, the reduction of gas porosity – starting from a properly degassed melt to a well-dried but still lubricated mold – and an evacuated die cavity have changed the character of HPDC completely. Melt supply beneath the melt bath level by sucking the melt into the shot sleeve through complete evacuation of the die cavity improved mechanical properties significantly. Melt heating from the top of the melt chamber, which is an usual arrangement in these types of casting units, is not recommended for alloys with higher magnesium content because of the enhancement of oxidation. Another way of transferring the melt is through the use of a liquid metal pump to avoid oxide films in the cast piece [157].

2.3.1.7 Semisolid Processes

After its rollout Thixocasting was mainly done with Al–7Si–Mg alloy and research and development concentrated on the production of appropriate feedstock material by either chemical (grain refiner) or physical (electromagnetic stirring) means. By lowering the melt temperature to the liquidus–solidus interval of the alloy reduces the hydrogen content significantly [158]. In this state, the material behaves thixotropic, similar to, for example, tomato ketchup. It is usually processed in HPDC or squeeze casting machines with an upstream heating unit. In the semisolid state a billet can be cut in two pieces without any effort but keeps its shape for a while, which allows a robot to transfer it to the shot sleeve. In addition to the very low gas content, a reduced shrinkage makes this technique favorable in comparison to normal HPDC. During the last decade, this principle was modified and improved in several ways, mostly the semisolid metal supply, because preproduction of specific thixo-castable billets is expensive. A cost effective route is to produce the slurry shortly before casting, by cooling it down from liquid to semisolid temperatures. This can be done by cooling down and tempering (NRC™, New Rheocast process). After reaching the right portion of solid the slurry will be injected by a squeeze casting unit. Other processes include Slurry on Demand (SoD™) and Semisolid Rheocasting (SSR™). Worth mentioning are the rheocasting variations that will provide potential for exceptional cost-effectiveness such as Sub Liquidus Casting (SLC™) from THT Presses, Inc. and Continuous Rheoconversion Process (CRP™) from WPI [159].

Other processes deal with the improvement of the feedstock metal such as the deformation semisolid casting (D-SSC) process where the semisolid slurry is directly supplied into the shot sleeve after the material has undergone a marked deformation to refine microstructural features such as crystallite and intermetallic precipitation size. In this way, the undesired β-Al_5FeSi phase can be distributed homogeneously and the morphology altered to a more globular shape [160].

2.3.2
Powder Metallurgy of Aluminum Alloys

2.3.2.1 Liquid Phase Sintering

Another way to produce net shape parts is related to powder metallurgical processes. In comparison to castings, there is no need for risers and complicated gating, but higher cost for the production of base powders and a certain limit to extruded 2D geometries have to be taken into account. Powders may be elemental or already alloyed, where compositions are mainly taken from 2xxx, 6xxx, and 7xxx series. Besides that some Al–Si alloys and also some metal matrix composites are processed. As shaping process is done in steel or hard metal dies, a minimum number of parts are necessary (approximately 10 000) so that this manufacturing method is competitive. Also, some lubricants and binders are necessary to provide good ejection behavior after the shaping process and an appropriate strength to the green part after shaping. These auxiliary agents and different powders will be blended and the right powder size distribution will be installed. The shaping process is done in mechanical or hydraulically driven uniaxial molding presses, which can be supported by vibration or ultrasound to achieve a higher green density. The final consolidation process is launched with a first debindering step to remove or burn the organic-based shaping agent, followed by a sintering step in which the formation of a liquid phase is promoted. The occurrence of a liquid phase leads to a quick rearrangement of the particles accompanied with a significant shrinkage. Some dissolving of small particles or elemental powder particles will follow, and in a last step normal diffusion sintering or solid-state sintering will take place. As a result, one will obtain a nearly dense part that may be transferred in most cases to a subsequent cold calibration after cooling to ambient temperature to ensure dimensional accuracy. Nevertheless, a certain fraction of porosity will remain in the part. In contrast to most Fe-based powder metallurgical parts, aluminum powders require liquid phase sintering because of the inevitable presence of aluminum oxide on the surface of the powder particles and only limited shear forces acting during the production of the green parts. Special care has to be taken that hydrogen pickup during sintering is avoided or at least restricted, which otherwise will result in additional porosity. Typical automotive parts are cam phasers, bearing caps, drive pullies, sprocket wheels, pistons, guide sleeves, oil pump gears, and rotors. Powders used for those parts will have compositions similar to wrought aluminum alloys and are produced by nitrogen or water atomization. The main target of the atomization is not to install a certain microstructure due to rapid solidification but to create small droplets, thus leading to powders for further processing and shaping [161]. The reader may also refer to the ''Further Reading'' section.

2.3.2.2 Advanced Powder Metallurgical Alloys and Processes

Owing to the presence of a liquid phase and the high temperatures during sintering it is obvious that parts produced by the earlier mentioned process will not have extraordinary mechanical properties. Diffusion processes, recrystallization, and Ostwald ripening will lead to a more or less equilibrium microstructure, while the residual porosity is detrimental to the endurance limit. To achieve better properties more sophisticated manufacturing methods have to be applied. First of all, the solidification rate will be increased by high-pressure gas atomization or by changing atomizing gas to helium, which will increase the cost of powder generation significantly. But higher solidification rates allow unusual alloy compositions with high concentrations of dispersoid-forming elements such as Fe, Ni, Co, Mn, and Cr or the use of elements usually not present in aluminum alloys, such as refractory or rare earth metals [162–165]. All these elements have a low diffusivity in the aluminum matrix and are therefore not prone to ripening, exhibiting high solvus temperatures and contributing to hot strength and thermal stability. High amounts of silicon are responsible for reduced thermal expansion and a higher Young's modulus up to 115 MPa [166]. Manganese-containing powders in the Al–Zn–Cu–Mg system are said to achieve tensile strength values of above 900 MPa because of the precipitation of the Q-phase and Al_6Mn, which influences DRX behavior [167].

2.3.2.3 Consolidation and Extrusion

To improve the strength and to overcome the problems due to the occurrence of liquid phases during sintering, another way of consolidation is necessary. Usually, rapidly solidified, gas atomized (nitrogen, helium, or argon) powder is used, which will be canned and extruded afterward. By keeping a high deformation ratio and thereby applying high shear forces between the particles, one can achieve a breakup of oxide skin into small pieces and weld all particles together, resulting in extrusion semis with a density close to the theoretical value. The powders are filled in a container that will be evacuated. After being heated to an appropriate temperature the canned powder is extruded to the desired shape. In comparison to net shape parts from liquid phase sintering, further machining is required. The additional handling of powder for canning, extrusion, and machining as well as the higher cost of gas atomized powders is the reason that those parts are only used in highly sophisticated applications or where higher requirements (strength, endurance limit, or thermal expansion) have to be met. Instead of extrusion hot isostatic pressing (HIP) can be used to consolidate to nearly 100% of the theoretical density, but because of the limited shear forces during the HIP cycle, mechanical properties are lower than those obtained from deformed material.

2.3.2.4 Melt Spinning

To improve properties further, higher cooling rates could be applied, but this can only be achieved with very fine powders or by making use of another technique, that is, melt spinning. This method produces thin ribbons with solidification rates that exceed those of normal gas atomizing by a factor of 10–100. The received ribbon must be chopped to flakes, but can be handled similar to powders. The additional process step will not contribute significantly to overall cost, but as melt spinning does not need atomizing gas, marked savings can be achieved. By applying melt spinning to an Al–Fe–Ni alloy room temperature strength will reach values above 600 and 400 MPa at 200 °C in the non-heat-treated state. This alloy is hardened solely by dispersoids and therefore aging does not alter precipitation size and distribution [168, 169]. A scandium-containing Al–Mg–Zr alloy group (Scalmalloy) produced by melt spinning, hipping, and hot extrusion exhibits up to 650 MPa yield strength without further heat treatment. Strength is achieved by fully coherent Al_3Sc particles. The alloy composition is not prone to SSC and may replace 2xxx or 7xxx alloys. High scandium price and challenging TMP with manufacturing temperatures below 350 °C to avoid loss of coherency among scandium precipitates make production of these alloys expensive and difficult [170, 171].

2.3.2.5 Spray Forming

Spray forming of aluminum is a process that will overcome partially the competitive disadvantages of the powder extrusion process. An alloyed aluminum melt is atomized (usually with nitrogen) and accelerated to a turning substrate, which will be withdrawn downward in the same rate at which the mixture of solid particles and semisolid/liquid droplets will consolidate on the substrate, resulting in a cylindrically shaped billet with a density of 90–97% of theoretical density – depending on the spraying conditions and the alloy's composition. The extrusion or hot isostatic consolidation process is needed to close residual porosity, which results from atomizing gas and not from hydrogen content (if removed properly before atomizing although a certain degassing takes place during atomizing). As droplets and particles exist only within a closed chamber with inertial atmosphere no contamination due to oxygen will occur. In the transportation sector the most important products are cylinder liners (Al-25–4Cu–Mg) for gasoline engines followed by spool valves (Al–17Si–4Fe–3Cu–Mg–Zr) for camshaft control [172, 173]. Typical parts produced by spray forming and subsequent forming processes are shown in Figure 2.25.

As the spray-forming process allows much higher concentrations of alloying elements, a number of promising new alloys have been developed in the last 10 years. Hata *et al.* [175] had shown that an Al–8.1Cr–9.9Fe–3Ti alloy spray formed to disk shape and subsequently forged shows better properties than only hipped material. High-strength and low-density alloys are produced via spray forming by [176] alloying up to 20 wt% magnesium with additions of vanadium, manganese, and chromium. The spray-formed billets were extruded and heat treated, resulting in yield stresses up to 450 MPa and UTS of 585 MPa, while elongation still reaches 19%. With high magnesium contents, density is reduced markedly, especially when silicon is also present, as the lightweight Mg_2Si (density approximately 1.95 g cm^{-3}) contributes according to its content [177, 178]. Spray forming does not

Figure 2.25 Automotive parts manufactured from spray-formed aluminum alloys [174].

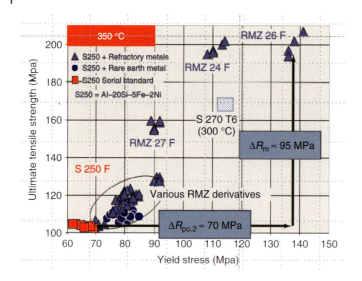

Figure 2.26 Hot strength at 350 °C. Comparison between standard S250 (=Al–20Si–5Fe–2Ni) and derivatives with rare earth and/or refractory metals [174].

provide such high cooling rates as melt spinning. Nevertheless quasi-crystalline or nanocrystalline phases can be obtained with spray forming [179]. Al–Si–Li alloys show favorable microstructure with low density and good mechanical properties by using this method [180, 181]. Alloying Al–20Si–5Fe–2Ni with rare earth and/or refractory metals leads to outstanding hot strength (Figure 2.26) but makes hot-forming processes such as extrusion difficult [174].

2.4
Secondary Processes

2.4.1
Joining

The huge variety of today's joning techniques and the rapidly increasing development of new techniques demonstrate the significance of joining for modern lightweight design. Nowadays, welding is still one of the most common joining methods for joining automotive parts, however, adhesive bonding as well as mechanical joining techniques are getting more and more important. The developemant of so called high-speed joining (HSJ) techniques are mainly driven by the pressure to reduce production costs. In this context the development of the self-piercing riveting (SPR) (Figure 2.27) as a HSJ technique alternatively to traditional resistance spot welding (RSW) has promoted punch riveting in the automobile industry in recent years, also driven by the trend toward multimaterial designs [182–185]. Current research activities in the field of aluminum applications

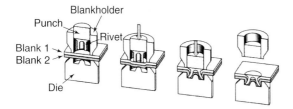

Figure 2.27 Schematic of the self-piercing riveting (SPR) process [183].

focus, among others, on the comparative investigation of SPR and RSW [185], the distortion characteristics of SPR joints [184], and the suitability of aluminum rivets instead of common steel rivets, benefiting the recyclability of riveted aluminum panels [183].

Being the traditional method in aircraft production and because of its proven reliability, riveting is still the dominant method for joining aircraft structures. However, the growing demands for cost savings during production and weight savings during operation as today's major drivers for technological development have promoted alternative joining methods such as the joining of metallic fuselage components by LBW (Figure 2.28) [186]. Since its introduction in the beginning of the last decade, LBW has continuously replaced the cost- and weight-intensive conventional riveting, supported by the increasing use of sheets and stringers made of excellent weldable and corrosion resistant alloys from the 6xxx series, such as alloy 6056-T6, being applied in the lower part of the fuselage. Today's LBW process is performed with high-performance CO_2 lasers by using filler wire made of AlSi12 [187]. The welding speed is about 8 m min^{-1} for 2–3 mm thick sheets [20]. Conventional L stringers are welded to fuselage skins from both sides by moving two lasers simultaneously along the longitudinal axis of the stringer [186]. Up to now, for example, Airbus has produced more than 1000 shells for aircraft types A318, A340, and A380 [95]. Current investigations comprise, among others, the weldability of structures, such as stringers, skin panels, or flaps, made of advanced Al–Cu–Li and Al–Mg–Sc alloys, and the design of advanced stringer profiles, such as U and Y profiles, adjusted to both enhanced weldability and loading performance [188].

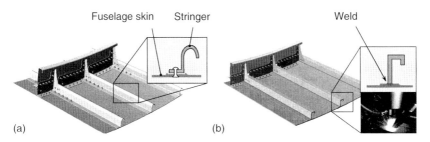

Figure 2.28 Joining of fuselage skins by (a) conventional riveting and (b) laser beam welding (LBW) [189].

As a solid-state, hot-shear joining method [190–192] potentially suitable for joining sheets (without filler metal and inert gas) friction stir welding (FSW) has received considerable attention in recent years offering various advantages such as mechanical properties of the weld similar to the base material, weldability of different metals, little distortion, low residual stresses, and few weld defects without porosity [193, 194]. FSW applies a rotating tool moving along the butting surfaces of two rigidly clamped sheets placed on a backing plate. While heat (being necessary for reducing the material's flow stress) is generated by friction, plasticization and transport of the plasticized material into the joint occurs during translation of the tool along the welding direction (Figure 2.29) [194, 195].

However, because of the high press and feed forces FSW requires a strong machine design. Furthermore, compared to other welding techniques the welding speed is very low (e.g., 0.8 m min^{-1} during welding of 5 mm thick sheets [20]). Thus, to date FSW has only been established in a small segment of aircraft applications, where the use of tailored welded blanks offers advantages. Recent research activities comprise, among others, the influence of different tool designs, such as pinless scroll tools or bobbin tools, on the resulting weld quality [193, 196] and the suitability of serial industrial robots for enhancing automation [197].

Friction-stir processing could also be used to rework the microstructure to achieve better mechanical properties and fatigue behavior. Possible applications would be pistons and cylinder heads. Parts treated in this way withstand the high thermomechanical loadings, especially in engines with high output power/displacement ratios, or will reduce the cost of complicated casting technology and alloys to achieve the same level of thermomechanical performance [198, 199]. Trials with spray-formed Al–25Si–4Cu–1Mg and Al–35Si material showed better properties at room and at elevated temperatures in the weld nugget than in the base material and, on account of high shear forces, a reduced silicon particle and grain size [174].

In case of materials that are not in equilibrium as in parts manufactured by powder metallurgical means, the occurrence of liquid phases has to be avoided. Therefore, a solid-state process is preferred over all other conventional welding techniques except electron beam welding. In this type of welding, only a very small portion of the material is remelted, and solidifies in a rapid manner because of a high self-quenching effect [200].

Figure 2.29 Schematic of the friction-stir welding (FSW) process [194].

2.4.2
Heat Treatment

Heat treatment of aluminum casting and wrought alloys is done mainly for increasing strength. This is done by a solution treatment near solidus temperature to dissolve and to distribute homogeneously precipitation-forming elements in the aluminum lattice. This is followed by a rapid quench to keep those elements in a supersaturated solid solution. Precipitation of hardened particles is done either by natural aging when the alloy system is appropriate (Al–Zn–Mg) or by an aging treatment to the desired strength level. Especially, quenching is always accompanied by distortion, and therefore, research and development had focused on two main issues. One is to provide an appropriate quench rate and another is to quench uniformly. The first needs reliable information on quench rates and the last an appropriate set up of chilling media supply. Quench sensitivity is usually detected by the so-called C-factor analysis [201] but a more precise method was developed by Keßler *et al.* [85] and Milkereit *et al.* [202–205] through the use of a high velocity calorimeter leading to time–temperature–precipitation diagrams similar to time–temperature–transition diagrams used in steel heat treatment [84]. Simple immersion into cold water is replaced by immersion in water–polymer mixtures to either reduce quench intensity or enhance it by forced air or process gas in different arrangements (Figure 2.30) [206, 207]. The next step would be the consequent introduction of spray cooling. Owing to geometrical reasons control of uniform quench intensity at different wall thicknesses is very difficult, and this method still needs further investigation before being implemented into production [211].

2.4.3
Surface Protection

Aluminum has the advantage of having a good surface protection due to its oxide skin, which usually forms immediately when pure aluminum is exposed to

 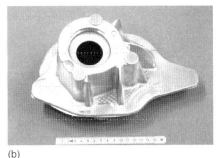

(a)   (b)

Figure 2.30 Nozzle array for high-pressure air quenching of suspension strut, Al–7Si–Mg (a). Cast piece (b) [207].

normal air. Unfortunately, additions of copper or iron will alter its good corrosion resistance and therefore measures have to be taken, such as paints or cladding to prevent contact with aggressive liquids. Oxide skin, always present when aluminum surface is exposed to oxygen-containing atmosphere, can be artificially thickened by anodizing. This process gives the opportunity to color the oxide skin in a variety of different colors. This provides a reasonable protection and is done mainly for design reasons and/or corrosion resistance. This artificial oxide skin is rather robust, but a few challenges can hardly be tolerated. For abrasive attack hard anodizing should be used to avoid wear. Spark plasma assisted processes act in a similar way in producing a thick alumina layer. For more information about standard surface techniques, the reader is referred to the Section "Further Reading."

The most challenging loading of oxide layer came along with brushless car wash shops. In this type of car cleaning, alkaline solutions of pH 12.5 or even higher will be used to remove dust and dirt without touching. As aluminum oxide will dissolve in such solutions additional surface treatment has to be done. As can be seen in Figure 2.31, this could be done by a modification of the whole anodizing process, which provides a certain short-time protection (Sealomax) [208].

For better long-term protection, a transparent, hydrophobic silicon oxide layer of a few microns thickness will be placed on top of the anodization layer (Figure 2.32) by a nano sol–gel process (Trimcotal).

2.5
Case Studies

2.5.1
Advancement on Deep Drawing of 7xxx Sheets for Automotive Applications

Ultra-high-strength alloys from the 7xxx series are well established in the aircraft industry. However, despite their enormous lightweight potential, indicated by tensile strength values up to 650 MPa, these alloys are rarely used in the automotive sector. Today, their application is mainly restricted to a few extruded components such as bumper beams, while sheet-metal parts can be neglected. As an attempt to figure out the potential of 7xxx alloys for substitution of ultra-high-strength boron steel grades for B-pillars, a high-strength aerospace sheet alloy, registered as alloy 7081, was chosen for a performance study. By means of a side-impact simulation (according to EuroNCAP specifications) of VW Golf V it could be demonstrated that a B-pillar made of a 3.5 mm thick 7xxx-T6 sheet matches the performance of 2 mm thick boron steel B-pillar. In addition, the aluminum structure is 2.4 kg lighter and, thus, enables weight savings of around 40% (Figure 2.33) [11, 210].

However, 7xxx alloys are more expensive than boron steel grades and even than typical automotive aluminum alloys, such as those from the 5xxx or 6xxx series. In addition, with Zn and Mg as the main alloying elements, 7xxx alloys are demonstrably difficult to form. As today's standard, cold deep drawing (CDD) is carried out in T4 condition. However, because of the rapidly rising flow stress levels

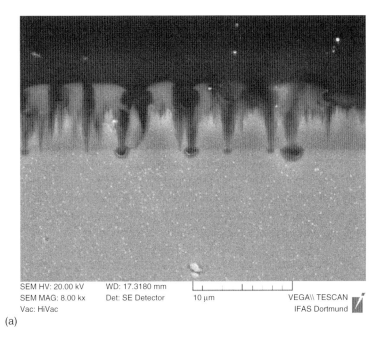

SEM HV: 20.00 kV   WD: 17.3180 mm
SEM MAG: 8.00 kx   Det: SE Detector   10 μm   VEGA\\ TESCAN
Vac: HiVac   IFAS Dortmund

(a)

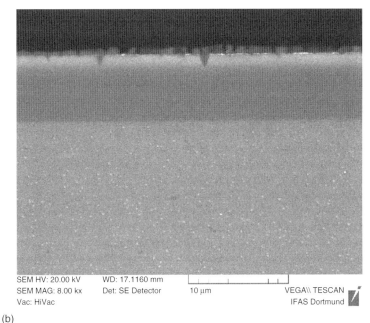

SEM HV: 20.00 kV   WD: 17.1160 mm
SEM MAG: 8.00 kx   Det: SE Detector   10 μm   VEGA\\ TESCAN
Vac: HiVac   IFAS Dortmund

(b)

Figure 2.31 Anodized surface after alkaline attack. Conventionally anodized: massive degradation and crater formation reaching down to the substrate (a) and Sealomax treated surface: only small penetration on surface of layer (b) [209].

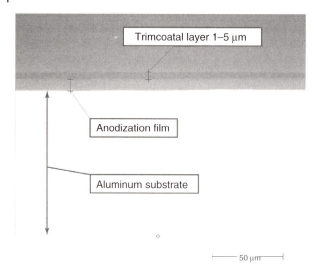

Figure 2.32 Hydrophobic silicon oxide layer (Trimcoatal layer) on top of a anodization film of a trim part [209].

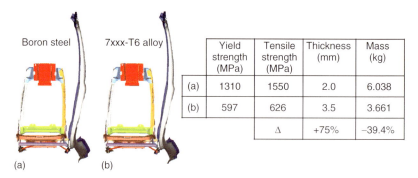

	Yield strength (MPa)	Tensile strength (MPa)	Thickness (mm)	Mass (kg)
(a)	1310	1550	2.0	6.038
(b)	597	626	3.5	3.661
		Δ	+75%	−39.4%

Figure 2.33 Deformation results of a side-impact simulation: (a) B-pillar made of incumbent hot-stamped boron steel grade versus (b) B-pillar made of 7xxx-T6 sheet alloy available as "Aleris Structurlite™ 550" [11] (Aleris).

during natural aging the deep draw ability at ambient temperature is hardly limited [49]. Moreover, arising spring-back behavior after CDD has to be considered, when designing the deep drawing process. In addition, attention has to be paid to internal stresses, induced by cold forming. Inhomogeneous stress distribution may lead to excessive distortion after stress-release during upstream processes, such as age-hardening and machining, reducing the productivity of CDD significantly.

As an advanced forming technique, HDD helps overcome the productivity restrictions of conventional CDD. Being performed during solution annealing, HDD utilizes the prevalent high temperatures, ranging between 350 and 500 °C, which leads to considerably reduced flow stress levels along with enlarged plastic strains, benefiting the formability significantly, and thus, the feasibility of more complex-shaped structures [47]. However, with rapid quenching during or after processing being necessary, artificial aging usually promotes the tendency for distortion [211]. Thus, in order to ensure the requested shape accuracy additional calibration operations may be required, which may also impair the productivity, even if this impairment is certainly lower than that due to the application of conventional CDD.

With the objective to promote the attractiveness of 7xxx sheet alloys for high-strength automotive applications, Aleris recently developed advanced modifications of aerospace alloys, such as alloy 7021 and the Cu-containing alloy 7081, labeled as "Structurelite 400" and "Structurelite 5xx," respectively, being well-adjusted to the demands of the automotive production. In contrast to conventional CDD, deep drawing has to be performed immediately (i.e., within 15 min) after solution treatment and quenching in the so-called fresh W condition (instead of conventional T4 conditions), when the age-hardened elements are still in supersaturated solution. This condition, indicated by very low flow stress values (around 100 MPa in case of "Structurelite 400"), significantly enhances the cold workability as well as the feasibility of more complex-shaped structures. This also lowers spring-back effects and a shape deviation induced by internal stresses and, thus, promotes the shape accuracy without falling back on additional calibration operations. Furthermore, compared to the above-mentioned aerospace alloys the developed automotive alloy modifications allow achieving the in-service yield strength targets of more than 400 and 500 MPa, respectively, during paint baking that makes conventional precipitation annealing (for adjusting condition T6) redundant [40].

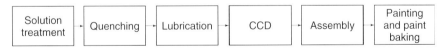

As a further promising approach to overcome the forming limits of 7xxx alloy sheets benefiting the deep drawing productivity, Austria Metall AG, in cooperation with LKR Ranshofen, has currently started the development of a novel deep drawing technology that is called *semi-HDD* [49]. Being performed during age-hardening, semi-HDD utilizes elevated temperatures limited up to the material's recrystallization temperature; these operation temperatures promote recovery, which leads to low flow stress levels and enlarged flow strain values, raising the

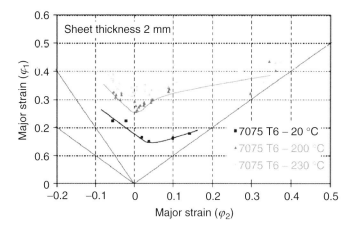

Figure 2.34 Forming limit curves of 7075-T6 alloy sheets at different forming temperatures [49].

forming limits significantly and benefiting the feasibility of more complex-shaped structures (Figure 2.34).

Furthermore, the intensive plasticization of the forming zone, as a result of the low flow stress level, leads to a small spring-back effect, considerably benefiting the shape accuracy. In addition, in contrast to HDD that is carried out during solution annealing, semi-HDD takes place at lower temperatures during or after artificial aging, followed by air cooling. Therefore, the level of thermal stresses arising after semi-HDD is expected to be far lower than those after HDD that necessitates a subsequent rapid quenching. Thus, the shape accuracy after semi-HDD is potentially higher. These interrelations are part of current investigations.

As a further advantage, semi-HDD is even suitable for the production of structures made of conventional 7xxx aerospace alloys, as the investigated alloy 7075-T6 exemplifies. Parts produced by semi-HDD usually leave the forming process in overaged condition (T7x) in order to achieve an optimal balance between high strength and enhanced resistance against SSC. The deep drawing itself is carried out under isothermal conditions within 1 min (including the time for conductive heating of the blanks) [49]. The use of sheets in peak-hardened condition (T6) along with the short processing time enables the utilization of elevated temperatures, which are slightly higher than conventional aging temperatures, benefiting both low forming limits and short overaging times. Figure 2.35 shows the impact of the forming temperature and time on the final tensile strength of alloy 7075-T6 at room temperature [49].

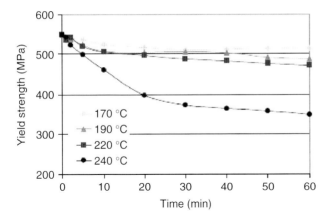

Figure 2.35 Impact of the forming temperature and time on the final tensile strength of alloy 7075-T6 at room temperature [49].

Parts produced by semi-HDD are also suitable for paint baking. However, as a result of the continuation of the overaging process (i.e., 20 min at 185 °C), the strength properties after paint baking are slightly lower than those directly after semi-HDD.

As Figure 2.36 shows, high-strength 7xxx alloys provide specific tensile strength values, which are equal to or even higher than those of incumbent ultra-high-strength boron steels, while the offered elongations are nearly doubled.

Besides the high potential of 7xxx alloys for automotive applications this case study additionally exemplifies the permanent competition between both different process variants and different material classes.

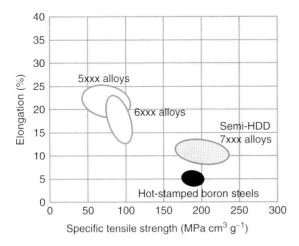

Figure 2.36 Specific tensile strength and elongations of 7xxx alloys versus incumbent hot-stamped boron steel grades [49].

2.5.2
Advancement on High-Pressure Die Casting with Gas Injection

Complicated parts with tiny small sections and cavities of extreme length to diameter ratios can hardly be produced by HPDC. As cavities can only be formed by movable cores, complex design or undercuts cannot be cast by this method. To reduce weight and to eliminate long movable cores from HPDC tooling, a new method was developed by Kallien and Weidler [212] not only for aluminum but also for zinc and magnesium based alloys.

It makes use of gas pressure usually applied with injection molding of polymer parts. During solidification, a gas is injected into the die to produce defined hollow structures (Figure 2.37). In polymer processing this is mainly done to avoid surface sink marks at the thick section because of the long solidification time in comparison to metal melts. In HPDC, parts can be produced with cavities but without cores, especially when very long bores have to be cast where conventional tooling reaches its limits. The short time available for gas injection and the resistance of the injector nozzle against aluminum melts pose a challenge [212]. In comparison to polymer melts, aluminum will solidify 10 times quicker, which needs a fast and precise pressure supply and control. The positioning of the injection nozzle also plays an important role in getting the gas transferred in all the desired regions [213]. Therefore, solidification modeling is mandatory for optimum die design. In principle, the smoothness of the inner surface depends on the total length of the hollow section, and opens the opportunity to use such parts as ducts for fluid flow without additional machining (Figure 2.38).

Inner surface roughness is dependent on the geometry and length of the hollow section, and additional research for optimum mold and gas injection design is

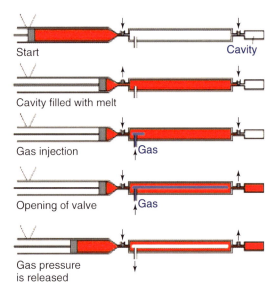

Figure 2.37 Principle of gas injection during HPDC [212].

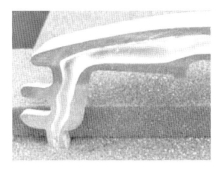

Figure 2.38 Smooth inner cavity produced by gas injection during HPDC, enabling undisturbed fluid flow [212].

Figure 2.39 Complicated internal cavity. Inner surface smoothness is dependent on shape and length [212].

necessary (Figure 2.39). Another advantage is that the pressure applied counteracts against solidification shrinkage, which gives a better accuracy in geometry.

The use of salt or sand cores in HPDC is limited because the high melt velocity entering the die and the high pressure applied to get a dense cast piece, will destroy such cores. This rough condition necessitates additional measures to be taken when hollow parts are cast in place, for example, filling the cavities with sand or another filler material to avoid collapsing. The method of gas injection seems to be quite an interesting way to produce lightweight parts with an internal complicated hollow structure, providing additional functionality to cast parts without excessive machining. Costs are only slightly higher with this method, but one can easily think about several applications where the benefits will compensate for that issue.

2.5.3
Advancement on Combined Method: Powder Metallurgy, Casting, Forging, and Extrusion

Producing a functional layer could be done with an unusual combination of casting with the powder metallurgical technique of spray forming and subsequent forming

Figure 2.40 V8-engine with cast-in-place spray-formed liners [166].

operations. As explained above, spray forming is used to produce millions of aluminum-based hypereutectic cylinder liners. These liners are cast in place in HPDC crankcases similar to the Daimler V8 engine shown in Figure 2.40.

Owing to the increased thermomechanical loadings because of engine downsizing and the introduction of chill water channels in the bore bridge regions, a perfect metallurgical bond between HPDC crankcase and spray-formed liner is mandatory. A good bonding transfers into a uniform cylinder wall temperature avoiding hot spots and keeping bore bridge temperature lower than with iron-based liners. To assure the desired amount of substance-to-substance bond, an outer layer around the liners is necessary, which exhibits a good melting behavior during casting and has itself a perfect bonding to the spray-formed liner material. Among gray cast iron liners for high output power engines, a coating by thermal spraying process with an arc remelted Al–12Si wire is "state of the art."

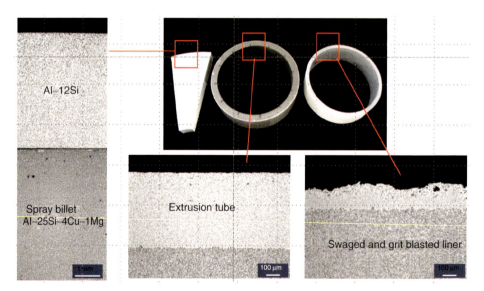

Figure 2.41 Production sequence of coated spray-formed liners (from left to right). Al–12Si cast around billet using LPDC, extrusion leads to perfect welding of cast material onto the spray-formed material, final hot swaging and grit blasting do not affect the joint quality [174].

Coating piece by piece is not very advantageous; therefore, a method for coating 500 liners simultaneously has been developed by an exceptional combination of manufacturing processes (Figure 2.41). For this purpose the spray-formed Al–25Si–4Cu–1Mg billet is placed into an LPDC machine and a thin layer of Al–12Si is cast around it. As the billet functions as a tremendous chill iron, preheating and special die design are necessary to provide a sound "cast around" shell. In this state, only a few spots have a good bonding. Perfect bonding is achieved in the subsequent step of coextrusion where the billet and the Al–12Si shell together will be transformed to seamless tubes. As shown in Figure 2.42 cast and spray-formed material are completely fused together. The fine microstructure of the cast layer resulting from the high cooling action of the billet during casting is the reason that it can easily be extruded. As accuracy during the extrusion of spray-formed high-strength material is not sufficient, the tubes covered with Al–12Si are hot round swaged to its final diameter. The good bonding of the layer permits this deformation process with a ratio of approximately 1–2 and layer and liner thickness are halved while the tube/layer composite is stretched twice without creating any defects at the interface.

Liners produced with this method were cast in and the bonding was checked by ultrasonic inspection and by metallographic methods. By comparing a non-coated liner and a coextruded liner – both cast in place in a V6 engine – it becomes apparent that coextrusion enhances metallurgical bonding. As Al–12Si has a higher liquidus temperature than the liner material, the reason for this enhancement lies in the consistency of the oxide skin, which can be washed away much more easily than the oxides from a "naked" liner. This process also gives the opportunity to cover other extrusion profiles especially when the surface is machined subsequently, thus saving the high-performance material only milling away the layer as a cheap sacrifice or using different casting materials, for example, magnesium alloys for other cast-in-place concepts.

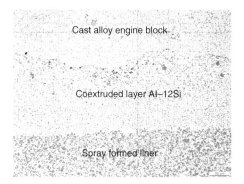

Figure 2.42 Microstructure of cast-in-place liner with Al–12Si coextruded layer showing good bonding between cast material and layer and as well between layer and liner [174].

2.6
Summary and Outlook

Being a pioneer in the field of transport applications aluminum has been developed to be the most important light alloy. Both its outstanding processing and in-service characteristics are exemplified by a huge number of available alloys, which are used in high-volume quantities, particularly in the aviation and automotive industries today.

In the field of aircraft applications, the major challenges are focused on the improvement of fatigue and damage tolerance properties, impact and corrosion resistance, strength properties also at elevated temperatures, and structure density without affecting the inherent cost advantages of aluminum (i.e., low material costs, cost-efficient component manufacturing and assembly). This refers particularly to fuselage structures such as skin panels, extrusions, and forgings. In this context advanced aircraft alloys such as Li-containing 2xxx alloys and Sc-containing 5xxx alloys from the current generation are of great interest. While Li-containing 2xxx alloys possess a reduced density and increased Young's modulus combined with toughness values exceeding by far those of high damage-tolerant 2xxx and 6xxx reference alloys along with strength values close to high-strength 7xxx alloys, Sc-containing 5xxx alloys provide an extraordinary weldability along with a toughness/strength balance equal to that of the above-mentioned reference alloys, although the outstanding strengthening potential of Sc at room temperature and elevated temperatures has not been exhausted yet. To face this challenge a couple of strip-casting technologies have been developed in the last decade in order to retain a huge amount of Sc as far as possible in supersaturated solution for subsequent age-hardening. Besides, recent developments of ultra-high-strength aircraft alloys from the 7xxx series are focused on alloy variants with an increased Zn content along with a reduced Cu and Mg content for huge plates and forgings with considerably enlarged cross sections.

In the field of automotive production, the material development is mostly directed to cost-efficient structural integrity (e.g., by forming, casting, and joining) combined with functional attributes (such as improved crash characteristics, increased dent, as well as corrosion resistance). Characterized by an elevated Mg content, novel work-hardenable 5xxx sheet alloys exhibit tensile strength and elongation values similar to those of today's incumbent deep drawing steels promoting the application for complex-shaped inner parts. Besides, by metallurgical measures the cold workability of bake hardenable 6xxx sheet alloys being used for automotive panels has also been substantially improved. In order to promote the formability of complex-shaped structures a couple sheet-metal forming approaches, such as (semi-)HDD, SHF, VPF, SPF, EMF, or ISF, have been focused on in recent years. In addition, a couple of forming approaches being applied during extrusion have been developed in order to enable the production of twisted or three-dimensional curved profiles. In the field of forgings, the development of manufacturing technologies following the near-net shape philosophy, for example, such as PF, TF, and TC, have been intensified.

As advanced joining techniques, a variety of processes have been established in aircraft production. While LBW as a cost- and weight-efficient alternative to riveting has widely been established for joining stringers/fuselage skin panels made of 6xxx alloys, FSW is currently restricted to a small segment, where tailored welded blanks are needed. In the field of automotive production SPR as an HSJ alternative technique to traditional resistance sport welding has received considerable attention in recent years, mainly driven by the trend toward multimaterial designs.

The huge lightweight potential of high-strength sheet alloys from the age-hardenable 7xxx series for safety-relevant automotive substructures was highlighted by means of a B-pillar being usually made of ultra-high-strength boron steels. Despite weight savings of around 40% the poor cold workability and the resulting bad productivity of the employed conventional CCD process hamper the use of 7xxx sheet alloys. To overcome these restrictions several advanced deep drawing approaches have been developed within the last few years. As a promising approach that even enables the use of sheets made of Cu-containing high-strength aerospace alloys, semi-HDD proves the feasibility of efficient production of sophisticated automotive components indicated by complex shapes and excellent in-service characteristics.

This case study exemplifies the permanent competition between both different process variants and different material classes. Even if aluminum is assumed to take up a dominant position particularly in future multimaterial designs, this perspective necessitates a continuous and intensive alloy and process development in order to compete with emerging lightweight materials, such as FRP.

2.7
Further Reading

This section is addressed to readers who need in-depth information to selected topics and also to those who need assistance to deepen their knowledge in the field of aluminum alloys and their associated processes. The following reference selection is only a small extract from the literature to be recommended for further reading.

Aluminum and Aluminum Alloys

- Gottstein, G. (2010) *Physical Foundations of Materials Science*, 1st edn, Springer-Verlag, Berlin, Heidelberg, (Original edition 2004), ISBN: 978-3-642-07271-0
- Hesse, W. 2010 Key to Aluminium Alloys, 9th edn, Alu Media, 628 pp., ISBN: 978-3-942-486-05-7 (English/German).
- Aluminum Reference Library DVD (2011) 2011 edn, ASM International, 15000 pp., ISBN: 978-1-61503-723-0
- Kaufman, J.G. (1999) Properties of Aluminum Alloys: Tensile, Creep and Fatigue Data at High and Low Temperatures, ASM International, 311 pp., ISBN: 978-0-87170-632-4

- Kaufman, J.G. (2008) Properties of Aluminum Alloys: Fatigue Data and Effects of Temperature, Product Form, and Processing, ASM International, 574 pp., ISBN: 978-0-87170-839-7.
- Mazzolani, F.M. (1995) Aluminium Alloy Structures, 2nd revised edn, E & FN Spon, an imprint of Chapman & Hall, 693 pp., ISBN: 978-0-419-17770-8

Associated Processes

- Bauser, M., Sauer, G., Siegert, K. (eds) (2006) Extrusion, 2nd edn, ASM International, ISBN: 978-0-87170-837-3.
- Gilbert Kaufman, J., Rooy, E.L. (2004) Aluminum Alloy Castings: Properties, Processes, and Applications, ASM International, Ohio, ISBN: 978-0-87170-803-8.
- Hirsch, J. (ed.) (2006) Virtual Fabrication of Aluminium Products – Microstructural Modeling in Industrial Aluminium Fabrication Processes, John Wiley & Sons, Inc., ISBN: 978-3-527-31363-1.
- Hirt, G. and Kopp, R. (2009) Thixoforming: Semi-Solid Metal Processing, Wiley-VCH Verlag GmbH, Weinheim, ISBN: 3527322043.
- Müller, K. (ed.) (2004) Fundamentals of Extrusion Technologies, 1st edn, Giesel Verlag, Isernhagen 278 pp., ISBN: 3-87852-016-6.
- Schatt, W.,Wieters, K.P. and Kieback, B. (eds) (2006) Pulvermetallurgie: Technologien und Werkstoffe: Technologie und Werkstoffe Springer Berlin, Heidelberg; 2nd Neu bearbeitete und erweiterte Aufl., November 10, 2006, ISBN: 354023652X.
- Semiatin, S.L. (ed.) (2006) ASM Handbook, Metalworking: Sheet Forming, Vol. 14B, ASM International, Materials Park, OH 924 pp., ISBN: 978-0-87170-710-9.
- Yamagata, H. (2011) The Science and Technology of Materials in Automotive Engines, Woodhead Publishing Limited, Cambridge, ISBN: 1 85573 742 6 ISBN. Further advisable literature, such as [214–217], are listed in the section ''References''; however, these comprehensive works are unfortunately not available in English yet.

Acknowledgment

The authors gratefully acknowledge A. Barr, B. Striewe, O. Karsten and K.Schimanski for their assistance and H.-W. Zoch for his support.

References

1. European Aluminium Association (ed.) Aluminium: The History, *http://www.eaa.net/upl/4/default/doc/Fact%20Sheet_Alu%20History.pdf* (accessed 2 March 2011).
2. Gesamtverband der Aluminiumindustrie e.V (ed.) Gesamtbedarf an Aluminium, *http://www.aluinfo.de/index.php/bedarf-weltwit.htm* (accessed 2 March 2011).
3. The London Metal Exchange (ed.) Industry Usage, *http://www.lme.com/aluminium_industryusage.asp* (accessed 2 March 2011).
4. The London Metal Exchange (ed.) Price Graphs, *http://www.lme.com/*

aluminium_graphs.asp (accessed 7 April 2011).

5. International Aluminium Institute (ed.) Aluminium for Future Generations/2009 Update, *http://www.world-aluminium.org/cache/fl0000336.pdf* (accessed 2 March 2011).

6. Huber, P.H.foreword (2009) Life cycle performance of aluminium applications. *Int. J. Life Cycle Assess.,* **14** (Suppl. 1), S1.

7. de Schrynmakers, P. (2009) Life cycle thinking in the aluminium industry. Life cycle performance of aluminium applications. *Int. J. Life Cycle Assess.,* **14**(Suppl. 1), S2–S5.

8. Starke Jr, E.A., Precipitation hardening: from Wilm to the present. in: *Aluminium Alloys – Their Physical and Mechanical Properties*, Vol. 1, Hirsch, J., Skrotzki, B, Gottstein, G. (eds), Wiley-VCH Verlag GmbH, 2008, ISBN: 978-3-527-32367-8, pp. 3–13.

9. European Aluminium Association (ed.) Aluminium in transport *http://www.eaa.net/upl/4/default/doc/Fact%20Sheet_Transport.pdf* (accessed 2 March 2011).

10. European Aluminium Association (ed.) Aluminium in cars, *http://www.eaa.net/upl/4/en/doc/Aluminium_in_cars_Sept2008.pdf* (accessed 2 March 2011).

11. Hirsch, J., Bassan, D., Lahaye, C., and Goede, M. (2009) Aluminium in innovative light-weight car design. Proceedings SLC Conference Latest Developments for Lightweight Vehicle Structures, Wolfsburg, Germany, May 26–27, 2009.

12. Lequeu, V., Eberl, F., Rhenalu, A., Jambu, S., Aviatube, A., Warner, T., Danielou, A., and Bes, B. (2008) Latest generation of Al-Li alloys developed by Alcan Aerospace. Proceedings EUCOMAS, Berlin, Germany, May 26–27, 2008.

13. Miermeister, M., Bürger, A., Spangel, S., Kröpfl, I., and Heinz, A. (2008) Aleris' advanced aluminium sheet and plate products for high-performance aircraft applications. Proceedings EUCOMAS, Berlin, Germany, May 26–27, 2008.

14. Spangel, S., Bürger, A., Miermeister, M., and Kröpfl, I. (2009) New developments on aluminium products for advanced aircraft applications. Proceedings EUCOMAS, Augsburg, Germany, September 8–10, 2009.

15. Gradinger, R. (2009) Herausforderungen bei der industriellen umsetzung des Legierungskonzeptes Scalmalloy für Flugzeuganwendungen. *Berg- Huettenmaenn. Monatsh.,* **154**(9), 403–406.

16. Rioja, R. (2006) Aerospace alloys advance at Alcoa. Bei Alcoa weiterentwickelte luftfahrtlegierungen. *Adv. Mater. Process.,* **164**(12), 30–34.

17. Hirsch, J. and Laukli, H.I. (2010) Aluminium in innovative light-weight car design. Proceedings of the 12th International Conference on Aluminium Alloys, Yokohama, Japan, September 5–9, 2010, pp. 46–53.

18. Aiura, T. and Sakurai, T. (2010) Development of aluminum alloys and new forming technology for automotive parts. Proceedings of the 12th International Conference on Aluminium Alloys, Yokohama, Japan, September 5–9, 2010, pp. 62–67.

19. Goede, M., Dröder, K., and Laue, T. (2010) Recent and future lightweight design concepts – the key to sustainable vehicle developments. Proceedings CTI International Conference – Automotive Lightweight Design, Duisburg, Germany, November 9–10, 2010.

20. Ostermann, F. (ed.) *Anwendungstechnologie Aluminium*, 2 Neu bearbeitete und aktualisierte Aufl., Springer-Verlag, Berlin, Heidelberg, 2007, ISBN: 978-3-540-71196-4.

21. J.G. Kaufman, *Introduction to Aluminum Alloys and Tempers*, ASM International, 2000, ISBN: 0-87170-689-X.

22. W. Hesse, *Aluminium-Schlüssel / Key to Aluminium Alloys*, 7th edn, Aluminium-Verlag, Düsseldorf, 2006, ISBN: 978-3-87017-282-4.

23. The Aluminum Association (ed.) (2009) Registration Record Series–Teal Sheets: International Alloy Designations and Chemical Composition Limits for Wrought Aluminum and Wrought

Aluminum Alloys, Revised February 2009, *http://www.alueurope.eu/wp-content/uploads/2012/03/Teal_Sheets.pdf.*

24. Hirsch, J. (2010) in *Fundamentals of Aluminium Metallurgy: Production, Processing and Applications, Part 3 Processing and Applications of Aluminium and Its Alloys* (ed R. Lumley), Woodhead Publishing Limited, pp. 721–748.

25. Polmear, I.J., Pons, G., Barbaux, Y., Octor, H., Sanchez, C., Morton, A.J., Borbidge, W.E., and Rogers, S. (1999) After Concorde: evaluation of creep resistant Al-Cu-Mg-Ag alloys. *Mater. Sci. Technol.*, **15**, 861–868.

26. Dudko, V.A., Kaibyshev, R.O., and Salakhova, E.R. (2009) *Phys. Met. Metallogr.*, **107**, 90–95.

27. Lee, Y.Y. (2010) Scandium effect on mechanical and physical properties for 2x19 Al alloy. Proceedings of the 12th International Conference on Aluminium Alloys, Yokohama, Japan, September 5–9, 2010, pp. 2281–2286.

28. Vladislav, A., Roman, V., Tatyana, P., and Eva, L. (2010) Structure and properties of semiproducts from Al-Cu-Mg-Ag V-1213 alloy. Proceedings of the 12th International Conference on Aluminium Alloys, Yokohama, Japan, September 5–9, 2010, pp. 2405–2410.

29. Nikulin, I., Kipelova, A., Gazizov, M., Teleshov, V., Zakharov, V., and Kaibyshev, R. (2010) Novel Al-Cu-Mg-Ag alloy for high temperature applications. Proceedings of the 12th International Conference on Aluminium Alloys, Yokohama, Japan, September 5–9, 2010, pp. 2303–2308.

30. Ovsyannikov, B.V. and Popov, V.I. (2010) Development of a new aluminium-lithium alloy of Al-Cu-Mg-Li (Ag, Sc) system intended for manufacturing sheets, thin-walled sections and forgings. Proceedings of the 12th International Conference on Aluminium Alloys, Yokohama, Japan, September 5–9, 2010, pp. 441–446.

31. Gradinger, R., Schneider, R., and Palm, F. (2008) 2nd generation of AlMgSc materials dedicated for short manufacturing chains in the aerospace industry. Proceedings EUCOMAS, Berlin, Germany, May 26–27, 2008.

32. Herding, T., Keßler, O., and Zoch, H.W. (2007) Spray formed and rolled aluminium-magnesium-scandium alloys with high scandium content. *Materialwiss. Werkstofftech.*, **38**(10), 855–861.

33. Seidman, D.N., Marquis, E.A., and Dunand, D.C. (2002) Precipitation strengthening at ambient and elevated temperatures of heat-treatable Al(Sc) alloys. *Acta Mater.*, **50**, 4021–4035.

34. von Bargen, R., von Hehl, A., and Zoch, H.W. (2011) Precipitation hardening behaviour of Al-2Sc micro sheets. *Mater. Sci. Forum*, **690**, 327–330.

35. Wagner, P., Brinkman, H.J., Bruenger, E., and Keller, S. (2008) Aluminium in the automobile – light, safe, and sustainable, in *Proceedings of the New Developments in Sheet Metal Forming* (ed. M. Liewald), Fellbach, Stuttgart, June 3–5, 2008, DGM, ISBN: 978-3-88355-305-8.

36. Hirsch, J. (2009) Aluminium in innovative light-weight car design. *Z. Aluminium*, **85**, 28–33, ISSN: 0002–6689.

37. Steffen, M.P. (2010) Aluminium in der Anwendung: Dauerhaft, Umformbar, Erneuerbar. VDI-Konstruktion, 9, pp. 6–7.

38. Boselli J., Chakrabarti D.J. and Shuey, R.T. Aerospace applications: metallurgical insights into the improved performance of aluminium alloy 7085 thick products, in *Aluminium Alloys – Their Physical and Mechanical Properties*, Vol. 1, Hirsch, J., Skrotzki, B, Gottstein, G. (eds.), Wiley-VCH Verlag GmbH, 2008, ISBN: 978-3-527-32367-8 pp. 202–208.

39. Tokugawa, H., Iida, K., Nakamura, J., Matsuda, K., Kawabata, T., Ikeno, S., Yoshida, T., and Murakami, S. (2010) Effect of additional elements on the age harding behavior of Al-Zn-Mg-Si alloys. Proceedings of the 12th International Conference on Aluminium Alloys, Yokohama, Japan, September 5–9, 2010, pp. 2075–2078.

40. Smeyers, A., Schepers, B., Braunschweig, W., Bürger, A., and Vieregge, K. (2011) 7xxx Grades for Automotive Applications. *Aluminium International Today* (Jan./Feb. 2011), pp. 37–39.

41. Gras, C., Meredith, M., and Hunt, J.D. (2005) Microstructure and texture evolution after twin roll casting and subsequent cold rolling of Al–Mg–Mn aluminium alloys. *J. Mater. Process. Technol.*, **169**, 156–163.

42. Gras, C., Meredith, M., and Hunt, J.D. (2005) Microdefects formation during the twin-roll casting of Al–Mg–Mn aluminium alloys. *J. Mater. Process. Technol.*, **167**, 62–72.

43. Sarkar, S., Wells, M.A., and Poole, W.J. (2006) Softening behaviour of cold rolled continuous cast and ingot cast aluminum alloy AA5754. *Mater. Sci. Eng., A*, **421**, 276–285.

44. McHugh, K.M., Lin, Y., Zhou, Y., Johnson, S.B., Delplanque, J.-P., and Lavernia, E.J. (2008) Microstructure evolution during spray rolling and heat treatment of 2124 Al. *Mater. Sci. Eng., A*, **477**, 26–34.

45. Uhlenwinkel, V., Meyer, C., Ristau, R., Jahn, P., Müller, H.R., Krug, P., Trojahn, W., and Hesse, D. (2010) Spray Forming of Multilayer Materials. PMTEC, Fort Lauderdale, June 27–30, 2010, Paper No. 2010-01-0041.

46. Hirsch, J. (2008) Texture evolution during rolling of aluminium alloys, Light metals; Proceedings of the Technical Sessions Presented by the TMS Aluminum Committee at the TMS 2008 Annual Meeting & Exhibition. *Miner. Met. Mater. Soc.*, 1071–1077.

47. van den Boogaard, A.H. and Huétink, J. (2006) Simulation of aluminium sheet forming at elevated temperatures. *Comput. Meth. Appl. Mech. Eng.*, **195**, 6691–6709.

48. Palumbo, G. and Tricarico, L. (2007) Numerical and experimental investigations on the warm deep drawing process of circular aluminum alloy specimens. *J. Mater. Process. Technol.*, **184**, 115–123.

49. Sotirov, N., Simon, P., Waltenberger, T., Uffelmann, D., and Melzer, C. (2010) Towards high strength 7xxx aluminium sheet components through warm forming. Proceedings of the 12th International Conference on Aluminium Alloys, Yokohama, Japan, September 5–9, 2010.

50. Sotirov, N., Simon, P., Waltenberger, T., Uffelmann, D., and Melzer, C. (2010) Erweiterte Einsatzmöglichkeiten von höchstfesten 7xxx-Aluminiumblechen für Leichtbaukomponenten durch Halbwarmumformen. 66. Kolloquium für Wärmebehandlung, Werkstofftechnik, Fertigungs- und Verfahrenstechnik, Wiesbaden, Germany, October 13–15, 2010.

51. van den Boogaard, H., Bolt, P.J., and Werkhoven, R.J. (2001) Modeling of AlMg sheet forming at elevated temperatures. *Int. J. Form. Processes*, **4**, 361–375.

52. Lang, L., Li, T., An, D., Chi, C., Nielsen, K.B., and Danckert, J. (2009) Investigation into hydromechanical deep drawing of aluminum alloy – complicated components in aircraft manufacturing. *Mater. Sci. Eng., A*, **499**, 320–324.

53. Wu, C.-T., Li, M.-F., Wang, C.-C., Wang, K.-S., and Zheng, W.-W. (2010) Use of resin tooling with sheet metal hydroforming for the manufacture of aluminum alloy cases. Proceedings of the 12th International Conference on Aluminium Alloys, Yokohama, Japan, September 5–9, 2010, pp. 2069–2074.

54. Wang, Z.J., Li, Y., Liu, J.G., and Zhang, Y.-H. (2007) Evaluation of forming limit in viscous pressure forming of automotive aluminum alloy 6 k21-T4 sheet. *Trans. Nonferrous Met. Soc. China*, **17**, 1169–1174.

55. Hussain, M.M., Rauscher, B., and Tekkaya, A.E. (2008) Working media based manufacturing of hybrid metal-plastic-compounds by using thermoplastic melt as pressurized media. *Materialwiss. Werkstofftech.*, **39**(9), 627–632.

56. Friedman, A. and Luckey, S.G. (2004) On the expanded usage of superplastic forming of aluminum sheet for automotive applications. *Mater. Sci. Forum*, **447–448**, 199–204.

57. Sun, P.-H., Wu, H.-Y., Lee, W.-S., Shis, S.-H., Perng, J.-Y., and Lee, S. (2009) Cavitation behavior in superplastic 5083 Al alloy during multiaxial gas blow forming with lubrication. *Int. J. Mach. Tool Manuf.*, **49**, 13–19.

58. Oliveira, D.A., Worswick, M.J., Finn, M., and Newman, D. (2005) Electromagnetic forming of aluminum alloy sheet: free-form and cavity fill experiments and model. *J. Mater. Process. Technol.*, **170**, 350–362.

59. Lang, L., Li, H., Yuan, S., Danckertc, J., and Nielsen, K.B. (2009) Investigation into the pre-forming's effect during multi-stages of tube hydroforming of aluminum alloy tube by using useful wrinkles. *J. Mater. Process. Technol.*, **209**, 2553–2563.

60. Jimbert, P., Eguia, I., Perez, I., Gutierrez, M.A., and Hurtado, I. (2011) Analysis and comparative study of factors affecting quality in the hemming of 6016T4AA performed by means of electromagnetic forming and process characterization. *J. Mater. Process. Technol.*, **211**(5), 916–924.

61. Oliveira, D.A. (2002) Electromagnetic forming of aluminum alloy sheet: experiment and model. Master thesis. Department Of Mechanical Engineering, University of Waterloo, Waterloo, Ontario.

62. Oliveira, D.A. and Worswick, M.J. (2003) Electromagnetic forming of aluminum alloy sheet. *J. Phys. IV*, **110**, 293–298.

63. Demir, O.K., Psyk, V., and Tekkaya, A.E. (2009) Simulation of wrinkle formation in free electromagnetic tube compression. Proceedings of the 3rd International Conference on Accuracy in Forming Technology ICAFT2009 in Association with 16. Sächsische Fachtagung Umformtechnik SFU, Chemnitz, Germany, November 10–11, 2009.

64. Wang, Z. and Li, Y. (2008) Formability of 6 k21-T4 car panel sheet for viscoelastic–plastic flexible-die forming. *J. Mater. Process. Technol.*, **201**, 408–412.

65. Maidagan, E., Zettler, J., Bambach, M., Rodríguez, P.P., and Hirt, G. (2007) A new incremental sheet forming process based on a flexible supporting die system. *Key Eng. Mater.*, **344**, 607–614.

66. Bambach, M. and Hirt, G. (2005) Performance assessment of element formulations and constitutive laws for the simulation of incremental sheet forming (ISF), in *8th International Conference on Computational Plasticity COMPLAS VIII* (eds E. Onate and D.R.J. Owen), CIMNE, Barcelona September 5–7, 2005.

67. Bambach, M., Taleb Araghi, B., and Hirt, G. (2009) Strategies to improve the geometric in asymmetric single point incremental forming. *Prod. Eng. Res. Devel.* doi: 10.1007/s11740-009-0150-8.

68. Meier, H., Smukala, V., Dewald, O., and Zhang, J. (2007) Two point incremental forming with two moving forming tools. *Key Eng. Mater.*, **344**, 599–605.

69. Taleb Araghi, B., Göttmann, A., Bambach, M., and Hirt, G. (2010) Hybride inkrementelle blechumformung (IBU) für leichtbaustrukturen, in *Proceedings of the 25th Aachener Stahlkolloquium Umformtechnik Stahl und NE-Werkstoffe* (eds G. Hirt and R. Kopp), Aachen March 11–12, 2010.

70. Mayer, H., Venier, F., and Koglin, K. (2002) *The New Audi A8 – Special Edition of ATZ/MTZ-Fachbuch*, Vieweg + Teubner Verlag.

71. Bauser, M., Sauer, G., and Siegert, K. (eds.), *Extrusion*, 2nd edn, ASM International, Materials Park, OH 2006, ISBN: 978-0-87170-837-3 pp 141–194.

72. Cho, H., Kim, S., Kim, C., Kim, J., and Lee, C. (2010) Development of high-strength aluminum extrusion alloy for automobile bumper. Proceedings of the 12th International Conference on Aluminium Alloys, Yokohama, Japan, September 5–9, 2010, pp. 2079–2083.

73. Fuchs, H. (2010) Struktureller Leichtbau mit integralen Leichtmetallbauteilen. *Lightweightdesign*, **3**, 25–29.

74. Ehrke, J., Rosefort, M., Koch, H., Schnapp, D., and Gers, H. (2009) Eine neue aluminiumlegierung für hochbeanspruchte space frame profile. Proceedings EUCOMAS, Augsburg, Germany, September 8–10, 2009.

75. McQueen, H.J. and Blum, W. (2000) *Mater. Sci. Eng., A,* **290**, 95–107.

76. Humphreys, F.J., and Hatherly, M. *Mater. Recrystallization and Related Annealing Phenomena,* 2nd edn, Elsevier, Oxford, 2004, ISBN: 0-08-044164-5.

77. van Geertruyden, W.H., Browne, H.M., Misiolek, W.Z., and Wang, P.T. (2005) *Metall. Mater. Trans. A,* **36A**, 1049–1056.

78. Doherty, R.D., Hughes, D.A., Humphreys, F.J., Jonas, J.J., Juul Jensen, D., Kassner, M.E., King, W.E., McNelly, T.R., McQueen, H.J., and Rollet, A.D. (1997) *Mater. Sci. Eng., A,* **238**, 219–274.

79. Kassner, M.E. and Barrabes, S.R. (2005) *Mater. Sci. Eng., A,* **410–411**, 152–155.

80. Gottstein, G. *Physical Foundations of Materials Science,* Springer-Verlag Berlin, Heidelberg, 2004, ISBN: 978-3-540-40139-1.

81. Schikorraa, M., Donatib, L., Tomesanib, L., and Tekkaya, A.E. (2008) Microstructure analysis of aluminum extrusion: prediction of microstructure on AA6060 alloy. *J. Mater. Process. Technol.,* **201**, 156–162.

82. Velay, X. (2009) Prediction and control of subgrain size in the hot extrusion of aluminium alloys with feeder plates. *J. Mater. Process. Technol.,* **209**, 3610–3620.

83. Kaysera, T., Klusemann, B., Lambers, H.-G., Maier, H.J., and Svendsen, B. (2010) Characterization of grain microstructure development in the aluminum alloy EN AW-6060 during extrusion. *Mater. Sci. Eng., A,* **527**, 6568–6573.

84. Milkereit, B., Schick, C., and Keßler, O. (2010) Continuous cooling precipitation diagrams depending on the composition of aluminum-magnesium-silicon alloys. Proceedings of the 12th International Conference on Aluminium Alloys, Yokohama, Japan, September 5–9, 2010, pp. 407–412.

85. Keßler, O., von Bargen, R., Hoffmann, F., and Zoch, H.W. (2006) Continuous cooling transformation (CCT) diagram of aluminum alloy Al-4.5Zn-1 Mg. Proceedings of the 10th International Conference on Aluminium Alloys, Vancouver, Canada, July 9–13, 2006, pp. 1467–1472.

86. von Bargen, R., Keßler, O., and Zoch, H.W. (2007) Kontinuierliche zeit-temperatur- ausscheidungsdiagramme der aluminiumlegierungen 7020 und 7050. *HTM J. Heat Treat. Mater.,* **62**, 285–293.

87. Zengen, K.H. and von Possehn, T. (2002) Aluminum – the light body material. Proceedings of New Advances in Body Engineering, Aachen, Germany, November 27–28, 2002.

88. Jeswiet, J., Geiger, M., Engel, U., Kleiner, M., Schikorra, M., Duflou, J., Neugebauer, R., Bariani, P., and Bruschi, S. (2008) Metal forming progress since 2000. *CIRP J. Manuf. Sci. Technol.,* **1**, 2–17.

89. Müller, K.B. (2006) Bending of extruded profiles during extrusion process. *Int. J. Mach. Tool Manuf.,* **46**, 1238–1242.

90. Ben Khalifa, N., Becker, D., Schikorra, M., and Tekkaya, A.E. (2008) Recent developments in the manufacture of complex components by influencing the material flow during extrusion. *Key Eng. Mater.,* **367**, 55–62.

91. Takatsuji, N., Takarada, Y., Aida, T., and Tanpa, K. (2010) Forming of aluminum helical gear by twist extrusion processing. Proceedings of the 12th International Conference on Aluminium Alloys, Yokohama, Japan, September 5–9, 2010, pp. 2057–2062.

92. Tekkayaa, A.E., Schikorraa, M., Beckera, D., Biermannb, D., Hammerb, N., and Pantke, K. (2009) Hot profile extrusion of AA-6060 aluminum chips. *J. Mater. Process. Technol.,* **209**, 3343–3350.

93. Stelling, O., Ellendt, N., Uhlenwinkel, V., von Hehl, A., Keßler, O., and Zoch, H.W. Influence of content, distribution and size of Mg2Si-precipitates on properties of spray formed aluminum alloys. in: *Aluminium Alloys – Their Physical and Mechanical Properties,* Vol. 1, Hirsch, J., Skrotzki, B, Gottstein, G. (eds.), Wiley-VCH Verlag GmbH, Weinheim 2008, ISBN: 978-3-527-32367-8, pp. 2231–2237.

94. Hilpert, M., Terlinde, G., and Witulski, T. (2009) A new aluminum alloy for high strength heavy aerospace forgings Proceedings EUCOMAS, Augsburg, Germany, September 8–10, 2009.

95. Rendigs, K.H. and Knüwer, M. (2010) Metal materials in Airbus A380. 2nd Izmir Global Aerospace and Offset Conference, Izmir, Turkey, October 6–8, 2010.

96. Stelling, O., Ellendt, N., Uhlenwinkel, V., von Hehl, A., Krug, P., and Zoch, H.W. (2009) Coarsening of spray formed aluminium alloys with high Mg2Si content during subsequent processing. Proceedings of the 4th International Conference on spray deposition and melt atomization and 7th International Conference on Spray Forming, Bremen, Germany, September 7–9, 2009.

97. Mahle GmbH (2005) Werkstoff auf der Basis einer Aluminium-Legierung, Verfahren zu seiner Herstellung sowie Verwendung hierfür. Offenlegungsschrift DE 10 2004 007 704 A1, Bundesrepublik Deutschland – Deutsches Patent- und Markenamt.

98. Behrens, B.-A., Doege, E., Reinsch, S., Telkamp, K., Daehndel, H., and Specker, A. (2007) Precision forging processes for high-duty automotive components. *J. Mater. Process. Technol.*, **185**, 139–146.

99. Shan, D.B., Wang, Z., Lu, Y., and Xue, K.M. (1997) Study on isothermal precision forging technology for a cylindrical aluminium-alloy housing. *J. Mater. Process. Technol.*, **72**, 403–406.

100. Aluminum Precision Products (ed.) Capabilities–Precision Aluminium Forgings, *http://www. aluminumprecision.com/capabilities/ precision-aluminum-forgings* (accessed 12 March 2011).

101. Fuganti, A. and Cupitò, G. (2000) Thixoforming of aluminium alloy for weight saving of a suspension steering knuckle. *Metall. Sci. Technol.*, **18**(1), 19–23.

102. Chayong, S., Atkinson, H.V., and Kapranos, P. (2005) Thixoforming 7075 aluminium alloys. *Mater. Sci. Eng., A*, **390**, 3–12.

103. Wang, K., Kopp, R., and Hirt, G. (2006) Investigation on forming defects during thixo-forging of aluminum alloy AlSi7Mg. *Adv. Eng. Mater.*, **8**(8), 724–730.

104. von Hehl, A. (2010) *Thermomechanische Behandlung beim Ringwalzen – Modelle zur schnellen Simulation der Gefügeevolution*, Shaker Verlag, Aachen 2010, ISBN: 978-3-8322-8926-3.

105. von Hehl, A. and Hirt, G. (2010) Fast simulation models for the microstructure simulation during ring rolling. *HTM J. Heat Treat. Mater.*, **65**(6), 287–298. ISSN: 1867–2493.

106. Apelian, D. (2009) Aluminum Cast Alloys: Enabling Tools for Improved Perform, North American Die Casting Association, Wheeling, IL.

107. Datta, J. (1997) *Key to Aluminium Alloys*, 5th edn, Aluminium-Verlag, Dusseldorf.

108. Fasoyino, F.A., Thomson, J.P., Sahoo, M., Burke, b., and Weiss, D. (2007) Permanent mold casting of aluminum alloys A206.0 and A535.0. *AFS Trans.*, **115**, 207–220, Paper 07-095(02).

109. Sigworth, G. (1996) Hot tearing of metals. *AFS Trans.*, 1053–1062.

110. Sigworth, G. (2001) Grain refining of aluminum casting alloys. Proceedings of the 6th International AFS Conference – Molten Aluminum Processing, Orlando, Florida, November 11–13, 2001.

111. Sigworth, G.K., Koch, H., and Krug, P. (2001) High strength, natural aging aluminum casting alloys for automotive applications. Light metals Symposium in Proceedings of the 40th – Annual Conference of Metallurgists and Electrometallurgy, Toronto, Canada, August 26–29, 2001, pp. 349–358.

112. Chien, K.-H., Kattamis, T.Z., and Mollard, F.R. (1973) Cast microstructure and fatigue behavior of a high strength aluminum alloy (KO-1). *Metall. Mater. Trans. B*, **4**(4), 1069–1076.

113. Jean, D., Major, J.F., Sokolowski, J.H., Warnok, B., and Kasprzak, W. (2009) Heat treatment and corrosion resistance of B206 aluminum alloy. *AFS Trans.*, **117** Paper 09–040, 113–129.

114. Nabawy, A.M., Samuel, A.M., Samuel, F.H., and Doty, H.W. (2009) Effect of Zr addition and aging treatment on the performance of Al-2%Cu base alloys. *AFS Trans.*, **117**, Paper 09–082, 209–223.

115. Yao, D., Zhao, W., Zhao, H., Qui, F., and Jiang, Q. (2009) High creep resistance behavior of the casting Al-Cu alloy modified with La. *Scr. Mater.*, **61**, 1153–1155.

116. Lenczowski, B., Koch, H., Eigenfeld, K., Plege, B., Franke, A., and Klan, S. (2004) Neue entwicklungen auf dem gebiet der warmfesten aluminium-gusswerkstoffe. *Giesserei*, **91**(08), 32–38.

117. Lenczowski, B. (2005) Neue warmfeste Aluminiumlegierungen für Hochleistungsmotoren. Abschlussbericht BMBF Verbundvorhaben Förerkennzeichen 03N3097A, 01.04.2001-30.09.2004.

118. Jorstad, J.L. (1987) Understanding sludge. Proceedings of the 14th SDCE International Die Casting Congress and Exposition, Toronto, Ontorio, May 11–14, 1987, Paper No. G-T87-011.

119. Gobrecht, J. (1975) Gravity-segregation of iron, manganese and chromium in an aluminum-silicon casting, part I. *Giesserei*, **61**(10), 263–265.

120. Tanihata, A., Sato, N., Katsumata,K., and Shiraishi, T. (2006) Development of high-strength piston material with high pressure die casting. SAE 2006 World Congress and Exhibition, Detroit, Michigan, April, 2006, Paper No 2006-01-0986.

121. Zak, O., Zak, H., and Tonn, B. (2008) Gefügebeeinflussung der legierung AlSi17Cu4Mg durch calciumcarbidzugabe. *Druckguss-Praxis*, (3), S.109–112.

122. Loper, C.R. Jr, Park, J., Fournelle, J., Seong, H.-G., and Cho, J.-I. (2001) Interaction of phosphorus and bismuth in A356Alloy. Transactions of the American Foundry Society and the One Hundred Fifth Annual Castings Congress, Dallas, Texas, pp. 1–31.

123. Khalifa, W., Samuel, A.M., Samuel, F.H., Chicoutimi, A., Doty, H.W., and Valtierra, S. (2007) Effect of Bismuth and Calcium Additions and Heat Treatment on the Microstructure and Mechanical Properties of

B319 Al Cast Alloys, Fonderie Fondeur d'aujourd'hui, No. 270, December 2007 pp. 13–23.

124. Zak, O. and Tonn, B. (2008) Effect of different micro alloying elements on the structure of hypereutectic aluminium-silicon alloys used to produce monolithic engine blocks. Proceedings of the International Conference Aluminium Alloys ICAA11, Aachen, Germany, September 22–26, 2008, pp. 270–275.

125. Schöffmann, W., Beste, F., Atzwanger, M., Sorger, H., Feuikus, F.J., and Kahn, J. (2005) *VDI-Tagung Gießereitechnik im Motorenbau, VDI Wissensforum GmbH, Magdeburg, February 1–2, 2005* VDI-Bericht Nr. 1830, VDI Verlag GmbH, Düsseldorf, pp. 93–114.

126. Jorstad, J.L. (2009) The progress of 390 alloy: from inception until now (20090184). *AFS Trans.*, **117**, 241–249 Paper 09-152SA.

127. Tavitas-Medrano, F.J., Gruzleski, J.E., Samuel, F.H., Valtierra, S., and Doty, H.W. (2007) Artificial aging behavior of 319-type cast aluminum alloys with Mg and Sr additions. *AFS Trans.*, **115** Paper 07-015(02), 135–150 (One Hundred Eleventh Annual Metalcasting Congress, Houston, Tex, May 15–18, 2007).

128. Samuel, A.M., Oullet, P., Samuel, F.H., and Doty, H.W. (1997) Microstructural Interpretation of Thermal Analysis of Commercial 319 Al Alloy with Mg and Sr Additions. *AFS Trans.*, **105**, 951–962.

129. Rheinfelden Alloys (2010) Primary Aluminium Casting Alloys, Rheinfelden Alloys GmbH & Co. KG, product catalogue, Version 7-01/2010.

130. Dragulin, D., Franke, R., Hoffmann, O., Zovi, A., and Casarotto, F. (2007) Al-Si-Druckgusslegierungen – Theoretische und praktische Aspekte-neue Entwicklungen. *Druckguss-Praxis*, **4**.

131. Koch, H. (1996) WO001996015281A1, Aluminium Gusslegierung.

132. Koch, H. and Franke, A.J. (1999) Development of a super ductile high-pressure die casting alloy for crash relevant parts. Transactions of the 20th International Die Casting

Congress & Exposition Cleveland, OHIO, November, 1999, pp. 269–272.

133. Trenda, G., Kraly, A., Pabel, T., and Rockenschaub, H. (2006) Die optimierung der mechanischen eigenschaften von AlMgSi-druckgusslegierungen. *Gießerei-Rundschau*, **53**(3/4), 48–49.

134. Trenda, G. (2006) Die casting alloy for ductile thin-walled structural parts. *Cast. Plant Technol.*, **22**(1), 28.

135. Dragulin, D., Franke, R., Hoffmann, O., Schwerin, C., and Gundlach, C. (2006) Al-Mg-Si-Druckgusslegierungen–Praktische und Theoretische Aspekte. *Druckguss-Praxis*, **8**, 318–320.

136. Trenda, G. (2006) Die vorteile von chrom in aluminium-magnesium-legierungen. *Giesserei*, **93**(9), 32–35.

137. Wuth, M.C., Koch, H., and Franke, J.A. (1999) Producing steering wheel frames with an AlMg5Si2Mn-type alloy. Proceedings of the TMS Annual Meeting, San Diego, California.

138. Koch, H., Krug, P., and Franke, A.J. (2000) Ductile high pressure die casting parts for automotive application – alloy development and experience from practice. Proceedings of the 11th Biennial Die Casting Conference of the Australian Die Casting Association, Melbourne, Australia, September 3–6, 2000, Paper 26, 26/1-25/5.

139. Leis, W. (2007) Druckguss, teil 1: werkstoffe. *Giesserei*, **94**(02), 48–59.

140. Krug, P., Koch, H., and Klos, R. (2000) Two Diecasting Alloys Make Their Debut. Diecasting World (März 2000), pp. 32–34.

141. Krug, P., Koch, H., and Klos, R. (2000) Magsimal-25 – a new high-ductility die casting alloy for structural parts in automotive industry. International Conference Advanced Materials and Their Processes and Applications, Materials Week, München, Germany, September 25–28, 2000.

142. Eigenfeld, K., Franke, A., Klan, S., Koch, H., Lenczowski, B., and Plege, B. (2004) New development in heat resistant aluminium castings. *Cast. Plant Technol. Int.*, **20**(4), 4–9.

143. Feikus, F.J. (2009) Leichtmetallguss. *Giesserei*, **96**(05), 40–52.

144. Röyset, J. (2007) Scandium in aluminium alloys: physical metallurgy, properties and applications. *Metall. Sci. Technol.*, **25**(2), 11–21.

145. Sigworth, G. (2004) Development Program on Natural Aging Alloys, Final Technical Report DE-FC07-02ID14230/DE-FC36-02ID14230, United States Department of Energy (DOE), pp. 1–32.

146. Zak, H. and Tonn, B. (2009) Optimierung von höchstfesten AlZnMgCu-legierungen für den kokillenguss. *Giesserei Praxis*, **60**(11), 349.

147. Bischoff, U. and Georgi, B. (2005) *VDI-Tagung Gießereitechnik im Motorenbau, VDI Wissensforum GmbH, Magdeburg, February 1–2, 2005*, VDI Verlag GmbH, Düsseldorf, pp. 267–374 VDI-Bericht Nr. 1830.

148. Kato, Y., Zenpo, T., and Asano, N. (2005) New core binder system for aluminum casting based on polysaccharide. *Trans. AFS*, **113**, 327–332, Paper No. 05-038(04).

149. Rigel, J.A. and Sturtz, G.P. (2004) Developing new PUCB binders for aluminum casting applications. *Trans. AFS*, **112**, 575–586, Paper No. 04–029.

150. Grassi, J., Campbell, J., Hartlieb, M., and Major, F. (2009) The ablation casting process. *Mater. Sci. Forum*, **618–619**, 591–594.

151. Kolbenschmidt Product Brochure *http://www.kspg-ag.de/pdfdoc/kspg_produktbroschueren/2007/at06_vw_audi_w12_e.pdf* (accessed 17 April 2011).

152. European Aluminium Association European Aluminium Award 2010. *http://www.aluminium-award.eu/2010/news/winners-european-aluminium-award-2010/* (accessed 8 April 2011).

153. Aue, H., Blank, W., Feikus, F., Finke, D.. Gottschalk, J., Hanke, K., Hippler, P., Jahn, J., Knaus, J., Kohlgrüber, K., Präfke, K., Schädlich-Stubenrauch, J., Scholz, G., Schütt, K.-H., and Weihnacht, W. (2008) Feingießen–Herstellung, Eigenschaften, Anwendung,

konstruieren+giessen, 33 Jahrgang, Nr. 1.

154. Tital GmbH *http://www.tital.de/e/ technologies/design-to-cost/* (accessed 8 April 2011).

155. Hornung, B. (2010) High strength aluminium investment casting at Zollern, using the Sophia process. New Frontiers in Light Metals – Proceedings of the 11th International Aluminium Conference INALCO 2010, Eindhoven, The Netherlands, pp. 289–293.

156. Tital GmbH *http://www.tital.de/cms/ upload/images_big/design_to_cost-big.jpg* (accessed 8 April 2011).

157. Denso, K.R. (2004) *J. Korean Foundry Soc.*, **24**(1), 3–9.

158. Granath, O., Wessen, M., and Cao, H. (2010) Porosity reduction possibilities in commercial aluminium A380 and magnesium AM60 alloy components using the RheoMetal process. *Metall. Sci. Technol.*, **28**(1), 1–11.

159. Jorstad, J.L. (2006) Aluminum Future Technology in Die Casting. Die Casting Engineer (Sept. 2006), pp. 18–25.

160. Sato, T. (2008) Development of deformation-semisolid-casting (D-SSC) process and applications to some aluminum alloys. *J. Mater. Sci. Technol.*, **24**(1), 12–16.

161. Verlinden, B. and Froyen, L. (1994) Aluminium Powder Metallurgy, TALAT Lecture 1401, EAA – European Aluminium Association, *http://www.eaa.net/eaa/education/ talat/lectures/1401.pdf* (accessed 17 April 2011).

162. Abramov, V.O. and Sommer, F. (1994) Structure and mechanical properties of rapidly solidified Al-(Fe,Cr) and Al-Mg-(Fe,Cr) alloys. *Mater. Lett.*, **20**(5/6), 251.

163. Rizzi, P. and Battezzati, L. (2004) *Mechanical Properties of Al Based Amorphous and Devitrified Alloys Containing Different Rare Earth Elements*, Fracture and Flow of Advanced Glasses, Elsevier, pp. 94–100.

164. Battezzati, L., Baricco, M., Kusy, M., Palumbo, M., Rizzi, P., and Ronto, V. (2004) *Amorphization and Devitrification of Al-Transition Metal-Rare Earth Alloys*, *Amorphous and Nanocrystalline Metals*, MRS, Warrendale, PA, pp. 21–32.

165. Kaji, T., Hattori, H., Hashikura, M., Tokuoka, T., Takikawa, T., Yamakawa, A., and Takeda, Y. (2001) Development of Tough and Heat-resistant Nanocrystalline Aluminum Alloy, SEI Technical Review, No. 51 (Jan. 2001), pp. 114–120.

166. Krug, P. (2006) in *Neue Pulvermetallurgische Al-Werkstoffe und deren Anwendung* (ed H. Kolaska) Hrsg. 25. Hagener Symposium Pulvermetallurgie, November 23–24, 2006, Heimdall Verlag, Witten, pp. 135–151.

167. Morimoto, Y., Adachi, H., Osamura, K., Kusui, J., and Okaniwa, S. (2006) Effect of Mn intermetallic particle on microstructure development of P/M Al-Zn-Mg-Cu-Mn alloy. *Mater. Sci. Forum*, **519–521**, 1623–1628.

168. Hummert, K., Schattevoy, R., Broada, M., Knappe, M., Muller, R., Beiss, P., Klubberg, F., Schubert, T., and Weissgarber, T. (2008) Nanostructured high-strength aluminium for automotive applications advances in powder metallurgy and particulate materials. Proceedings of the International Conference and Exhibition on Powder Metallurgy and Particulate Materials, Princeton, New Jersey, Vol. 3, pp. 09-161–09.171.

169. Hummert, K., Schattevoy, R., Broda, M., Knappe, M., Beiss, P., Klubberg, F., Schubert, T.H., and Leuschner, R. (2009) Nano-crystalline P/M aluminium for automotive applications, in *Rapidly Quenched and Metastable Materials (RQ13)*, IOP, Bristol.

170. Bosch, A.J., Senden, R., Entelmann, W., and Palm, F. (2008) Scalmalloy-A unique high strength and corrosion insensitive AlMgScZr material concept. Proceedings of the International Conference Aluminium Alloys ICAA11, Aachen, 22–26 September, 2008, pp. 270–275.

171. Tsivoulas, D. and Robson, J.D. (2006) Coherency loss of Al3Sc precipitates during ageing of dilute Al-Sc alloys. *Mater. Sci. Forum*, **519–521**, 473–478.

172. Krug, P., Kennedy, M., and Foss, J. (2006) New aluminium alloys for

cylinder liner applications. SAE-World Congress, Detroit, April 3–6 2006, Technical Paper, 2006-01-0983.

173. Beiss, P., Klubberg, F., Schäfer, H.J., Krug, P., and Weiss, H. (2008) Fatigue behaviour of high performance Al alloys. Proceedings of the International Conference Aluminium Alloys ICAA11, Aachen, Germany, September 22–26, 2008, pp. 1581–1588.

174. Krug, P., Commandeur, B., Dang, T., and Przeorski, T. (2009) What shall we do with spray formed aluminum alloys? 4th International Conference on Spray Deposition and Melt Atomization and 7th International Conference on Spray Forming–SDMA 2009-ICSF VII, Bremen, September 7–9, 2009.

175. Hata, H., Kajihara, K., Takagi, T., Takahara, T., and Ehira M. (2006) Spray forming of Al alloy of high temperature strength. Proceedings of the 3rd International Conference on Spray Deposition and Melt Atomization and 6th International Conference on Spray Forming–SDMA 2006-ICSF VI, Bremen, Germany, September 4–6, 2006.

176. Sun, T., Sun, Y., Wang, F., Xin, H., Guo, M., and Zhu, X. (2006) The research of microstructures and mechanical properties of the spray-deposited high magnesium aluminum alloy. Proceedings of the 3rd International Conference on Spray Deposition and Melt Atomization and 6th International Conference on Spray Forming, SDMA 2006-ICSF VI, Bremen, September 4–6, 2006.

177. Stelling, O., Irretier, A., Keßler, O., Krug, P., Commandeur, B. (2006) New light-weight aluminium alloys with high Mg2Si-content by spray forming. Proceedings of the 10th International Conference on Aluminum Alloys-ICAA10, Vancouver, Canada, July 9–13, 2006; *Mater. Sci. Forum*, **519–521**, 1245–1250.

178. Stelling, O., Ellendt, N., Uhlenwinkel, V., Krug, P., Kessler, O., Zoch, H.-W., and Commandeur, B. (2008) Einfluss des primär ausgeschiedenen Mg2Si-anteils sowie dessen verteilung auf die eigenschaften sprühkompaktierter

AlMgSiCu-legierungen. *J. Heat Treat. Mater.*, **63**(5), 276–283.

179. Grant, P. and Krug, P. (2009) Spray forming of nanostructured materials. 4th International Conference on Spray Deposition and Melt Atomization and 7th International Conference on Spray Forming–SDMA 2009-ICSF VII, Bremen, Germany, 7–9 September 2009.

180. Hogg, S.C., Palmer, I.G., Thomas, L.G., and Grant, P.S. (2007) Processing, microstructure and property aspects of a spray cast Al-Mg-Li-Zr alloy. *Acta Mater.*, **55**, 1885–1894.

181. Hogg, S.C., Palmer, I.G., and Grant, P.S. (2006) An investigation of novel spraycast Al-Mg-Li-Zr-(Sc) alloys. *Mat. Sci. Forum*, **519–521**, 1629–1633.

182. He, X., Pearson, I., and Young, K. (2008) Self-pierce riveting for sheet materials: state of the art. *J. Mater. Process. Technol.*, **199**, 27–36.

183. Hoang, N.H., Porcaro, R., Langseth, M., and Hanssen, A.-G. (2010) Self-piercing riveting connections using aluminium rivets. *Int. J. Solids Struct.*, **47**, 427–439.

184. Cai, W., Wang, P.C., and Yang, W. (2005) Assembly dimensional prediction for self-piercing riveted aluminum panels. *Int. J. Mach. Tool Manuf.*, **45**, 695–704.

185. Han, L., Thornton, M., and Shergold, M. (2010) A comparison of the mechanical behaviour of self-piercing riveted and resistance spot welded aluminium sheets for the automotive industry. *Mater. Des.*, **31**, 1457–1467.

186. Kocik, R., Vugrin, T., and Seefeld, T. (2006) Laserstrahlschweißen im flugzeugbau: stand und künftige anwendungen. Proceedings 5, Laseranwenderforum, Bremen, September 13/14, 2006, pp. 15–26.

187. Cui, C., Schulz, A., Schimanski, K., Zoch, H.W., Baumgart, P., Syassen, F., and Kocik, R. (2007) Development of new filler materials for welding of aluminum structures by spray forming. Proceedings of the International Conference on Applied Production Technology, Bremen, Germany, September 17–19, 2007.

188. Dittrich, D., Brenner, B., Kirchhoff, G., and J. Hackius (2010) New stiffener designs and Al-alloys for laser welded integral structures to sustain higher load at reduced weight. Proceedings EUCOMAS, Berlin, Germany, Juni 7–8, 2010.

189. Schumacher, J. (2002) *Laserfügen: Prozesse, Systeme, Anwendungen, Trends, Strahltechnik*, Vol. 19, Bias Verlag, pp. 5–14.

190. Thomas, W.M., Nicholas, E.D., Needham, J.C., Murch, M.G., Temple-Smith, P., and Dawes, C.J. (1991) Friction stir butt welding. International Patent Application No. PCT/GB92/02203.

191. Dawes, C.J. and Thomas, W.M. (1996) Friction stir process welds aluminum alloys. *Weld. J.*, **75**(3), 41–45.

192. Thomas, W.M. and Dolby, R.E. (2003) *Proceedings of the Sixth International Trends in Welding Research, April 15–19, 2002*, ASM International, pp. 203–211.

193. Tozaki, Y., Uematsu, Y., and Tokaji, K. (2010) A newly developed tool without probe for friction stir spot welding and its performance. *J. Mater. Process. Technol.*, **210**, 844–851.

194. Sheikhi, S. Zettler, R. dos Santos, J.F. Fortschritte beim rührreibschweißen von aluminium, magnesium und stahl, *Materialwiss. Werkstofftech.* **37** (2006) **9** 762–767.

195. Nandan, R., DebRoy, T., and Bhadeshia, H.K.D.H. (2008) Recent advances in friction-stir welding – process, weldment structure and properties. *Prog. Mater. Sci.*, **53**, 980–1023.

196. Deloison, D., Marie, F., Guérin, B., and Aliaga, D. (2008) Multi-physics modelling of bobbin-tool friction stir welding–Applications to a latest generation aluminium alloy. Proceedings EUCOMAS, Berlin, Germany, May 26–27, 2008.

197. Monsarrat, B., Dubourg, L., Fortin, Y., Bres, A., Guérin, M., Banu, M., Perron, C., and Wanjara, P. (2009) Friction stir welding of aerospace structures using low-cost serial industrial robots. EUCOMAS, Augsburg, Germany, September 08–10, 2009.

198. Grant, G., Hovanski, Y., Beardsley, M.B., and Veliz, M. (2009) Tailored materials for high efficiency CIDI engine. 2009 DOE Hydrogen Program and Vehicle Technologies Program Annual Merit Review and Peer Evaluation Meeting, May 18–22, 2009.

199. Santella, M.L., Engstrom, T., Storjohann, D., and Pan, T.-Y. (2005) Effects of friction stir processing on mechanical properties of the cast aluminum alloys A319 and A356. *Scr. Mater.*, **53**(2), 201–206.

200. Zenker, R., Krug, P., Buchwalder, A., Dickmann, T., Frenkler, N., and Thiemer, S. (2006) *3. VDI-Fachtagung Zylinderlaufbahn, Kolben, Pleuel, Böblingen, März 7–8, 2006* (Hrsg. VDI Wissensforum GmbH) VDI-Bericht 1906, VDI Verlag GmbH, Düsseldorf, pp. 259–274.

201. Bernadin, J.D. and Mudawar, I. (1995) Validation of the quench factor technique in predicting hardness in heat treatable aluminium alloys. *Int. J. Heat Mass Transfer*, **38**(5), 863–873.

202. Milkereit, B. (2011) Kontinuierliche Zeit-Temperatur-Ausscheidungs-Diagramme von Al-Mg-Si-Legierungen, Dissertation Universität Rostock, Shaker Verlag, Aachen, ISBN: 978-3-8322-9993-4.

203. Milkereit, B., Beck, M., Reich, M., Kessler, O., and Schick, C. (2011) Precipitation Kinetics of an aluminium alloy during Newtonian cooling simulated in a differential scanning calorimeter. *Thermochim. Acta*, **522**, 86–95.

204. Milkereit, B., Jonas, L., Schick, C., and Kessler, O. (2010) Das kontinuierliche zeit- temperatur- ausscheidungs- diagramm einer aluminiumlegierung EN AW-6005A. *HTM J. Heat Treat. Mater.*, **65**(3), 159–171.

205. Milkereit, B., Kessler, O., and Schick, C. (2009) Recording of continuous cooling precipitation diagrams of aluminium alloys. *Thermochim. Acta*, **492**, 73–78.

206. Rose, A., Kessler, O., Hoffmann, F., and Zoch, H.-W. (2005) Quenching distortion of aluminium castings – improvement by gas cooling. Proceedings of the 1st International Conference on Distortion Engineering, Bremen, Germany, September 14–16, 2005, pp. 487–494.

207. Rose, A., Keßler, O., Hoffmann, F., Zoch, H.-W., and Krug, P. (2007) Age hardening of forged aluminum components - distortion behavior after gas quenching. *HTM Z. Werkst. Wärmebeh. Fertigung*, **62**, 58–61.

208. Steins, R. (2011) Sol–gel-beschichtung für dekorative automobilteile stabile wanderung auf schmalem grat. *J. Oberflächentechnik*, **5**, 14–17.

209. Steins, R. (2011) Innovative aluminium surfaces – meet today's challenges. 11th International Car Symposium, Bochum, Germany, January 27, 2011.

210. Lahaye, C., Hirsch, J., Bassan, D., Criqui, B., Sahr, C., Goede, M. in: *Aluminium Alloys – Their Physical and Mechanical Properties*, Vol. 1, Hirsch, J., Skrotzki, B, Gottstein, G. (eds.), Wiley-VCH Verlag GmbH, 2008, ISBN: 978-3-527-32367-8 pp 2231–2237.

211. Karsten, O., Schimanski, K., von Hehl, A., and Zoch, H.W. (2011) Challenges and solutions in distortion engineering of an aluminium die casting component. 5th International Light Metals Technology Conference, Lüneburg, Germany, July 19–22, 2011.

212. Kallien, L.H. and Weidler, T. (2009) Using gas injection in high pressure die casting technology. 113th Metalcasting Congress, Las Vegas, Nevada, April 7–10, 2009 in Die Casting Engineer, 5/2009, pp. 50–52.

213. Kallien, L.H., Weidler, T., Hermann, C., and Stieler, U. (2006) Pressure die castings with functional cavities produced by gas injection. *Gießereiforschung*, **58**(4), 2–9.

214. Ostermann, F. (ed.) (2007) *Anwendungstechnologie Aluminium*, Springer Verlag, Berlin Heidelberg, 2007, (Original edition 1998), ISBN: 978-3-540-71196-4.

215. Kammer, C. (2009) *Aluminium-Taschenbuch*, Vol. 1: Grundlagen und Werkstoffe, 16th edn, Alu Media, ISBN: 978-3-87017-292-3.

216. Drossel, G., Friedrich, S., Huppatz, W., and Kammer, C. (2009) *Aluminium-Taschenbuch*, Vol. 2: Umformung von Aluminiumwerkstoffen, Gießen von Aluminium-Teilen, Oberflächenbehandlung von Aluminium, Recycling und Ökologie, 16th edn, Alu Media, ISBN: 978-387017-293-0.

217. Kammer, C. (2011) *Aluminium-Taschenbuch*, Vol. 3: Weiterverarbeitung und Anwendung, 16th edn, Alu Media, ISBN: 978-394248-608-8.

3
Magnesium and Magnesium Alloys

Wim H. Sillekens and Norbert Hort

3.1
Introduction

Magnesium is the lightest engineering metal and is abundantly available as a resource from the Earth's crust and from seawater. It – with atomic number 12 – was discovered as an element in 1774 and named after the ancient city Magnesia in East Thessalia, where the raw material magnesite was being found. The metal is being produced on an industrial scale since 1886. The extraction of the metal from minerals such as dolomite ($MgCO_3 \cdot CaCO_3$) and magnesite ($MgCO_3$) is done by thermal reduction or electrolysis.

The assets of magnesium for use as a construction material are its low density, good specific strength (i.e., strength relative to density), high vibration-damping capacity, good electromagnetic shielding, excellent castability, good machinability, and weldability, as well as its durability and recyclability. The drawbacks are its corrosion susceptibility, moderate elastic moduli, and limited room-temperature ductility and formability, as well as the safety hazard involved in certain manufacturing activities. Pure magnesium has a density of 1740 kg m^{-3}, Young's modulus of 44 GPa, melting point of 650 °C, specific heat of 1026 J kg$^{-1} \cdot$K^{-1}, and linear thermal expansion of 27.1 μm m$^{-1} \cdot$K^{-1}. Notably, its low density renders magnesium about 75% lighter than ferrous and 35% lighter than aluminum alloys. The mechanical properties of magnesium alloys are typically in the same ballpark as those of common aluminum alloys.

Magnesium usage has seen some periods of strong growth over the past century. The metal was widely used in Germany during both World Wars. During the latter, it was applied extensively in (bomber and combat) aircraft of German as well as of Allied make. In the 1960–1970s, the VW Beetle included up to 20 kg of magnesium components especially in the drivetrain, making Volkswagen the single largest global user at that time. After this period of bloom, however, interest declined again with pricing and increasing technical requirements – following the introduction of water-cooled engines with more horsepower – acting in favor of aluminum. As of the early 1990s, there has been an accelerated worldwide growth in the use of magnesium following the growing demand for lightweight products

Structural Materials and Processes in Transportation, First Edition.
Edited by Dirk Lehmhus, Matthias Busse, Axel S. Herrmann, and Kambiz Kayvantash.
© 2013 Wiley-VCH Verlag GmbH & Co. KGaA. Published 2013 by Wiley-VCH Verlag GmbH & Co. KGaA.

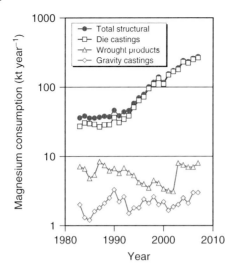

Figure 3.1 Development of global primary magnesium use for structural applications.

for transport applications, aiming to increase fuel efficiency and reduce emissions. Volkswagen in Germany was again one of the drivers of this renaissance in a joint venture with Dead Sea Works in Israel for material supply. Car manufacturers are presently using several magnesium components in high-volume production and in amounts that typically range between 0.5 and 20 kg per vehicle. Apart from that, magnesium has found widespread application in other sectors, notably in consumer electronics (casings for computers, cell phones, and cameras, as well as for hand tools).

Figure 3.1 shows how global primary magnesium shipments for structural use have developed during the past decades, coming from a recorded 36 100 metric tons in 1983 to 276 000 metric tons in 2007 (statistical data from [1, 2]). Noting that magnesium consumption is represented in the graph on a logarithmic scale, die castings are by far the prevailing application, with wrought products and gravity castings being only marginal in terms of volume. Coming from a steady situation in the period before 1990, growth in magnesium die castings has been driving overall growth since – averaging to an annual 33% increase during the 1990s and a sustained annual 25% during the 2000s. Die castings are now the largest overall application area for magnesium (45% of global consumption in 2007), followed by aluminum alloying (30%) and desulfurization of steel (17%) [2]. With magnesium being a small market as compared to other materials for transport, however, supply and demand are volatile and susceptible to economic tendencies as with the recent downturn, when global primary production dropped from 2007 to 2009 by 22%, yet to recover to the earlier high by 2010 [1]. This is also reflected by the metal's base price, which varied between $6000 and $2000 t^{-1} for commercially pure magnesium during this particular period [3].

Currently, 70–80% of the primary production of magnesium is based in China, with the remainder coming from the United States, Israel, Russia, and some other minor contributors. Magnesium parts are mainly used in the automotive sector in the United States and Europe and in the electronics sector in Asia; the exception to this is China, where they are used in both application areas alike.

For structural applications, magnesium is employed in an alloyed form. Workability and properties – and hence application areas – depend on alloy constitution and alloying concentrations. Generic alloys are commonly designated according to the ASTM classification, in which each of the main alloying elements is indicated by a letter and its contents in weight percent by a number following the concerned letter combination (A denotes aluminum, E rare earth elements, J strontium, K zirconium, M manganese, W yttrium, Z zinc, etc.). Additional letters following this designation are used to label the alloying variant (A, B, C, etc. for subsequent generations). For example, AZ91E is the fifth generation of an alloy with about 9 wt% aluminum and 1 wt% zinc. The temper designations (e.g., F for as fabricated, O for annealed recrystallized, H for strain hardened, and T for thermally treated to stable tempers) known from aluminum alloys are also used as addendums. Alternatively the DIN classification is used, comprising a similar element–content combination in accordance with the alloy's main composition (in this sense, the previous example is designated MgAl9Zn1). Proprietary alloys are often indicated by a trade name.

Some examples of early magnesium applications are given in Figure 3.2. The vehicle body structure shown in Figure 3.2a is a sheet-metal assembly that has been in series production in the 1930s as a lightweight substitution for steel [4], while the drivetrain components shown in Figure 3.2b are castings for the VW Beetle prototype dating back to 1936.

As indicated earlier, the pursuit of vehicle weight reduction in the transport sector has propelled the most recent upheaval of magnesium use. Meanwhile, roadmapping studies of the automotive industry identified a variety of components for redesign and their associated technical challenges [5–7]. Components for the interior (steering column and wheel core, pedal brackets, seat frame, etc.) and the drivetrain or powertrain (intake manifold, transmission case, crankcase, etc.) have meanwhile become state of the art, while components for the body (door frame and panel, A and B pillars, etc.) and chassis (engine cradle, suspension arm, wheel, etc.) are rather in the phase of exploration for mass-produced cars.

(a)

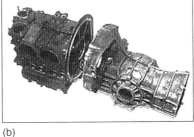
(b)

Figure 3.2 Early magnesium applications: (a) bus–trailer superstructure and (b) crankcase with gearbox.

In addition to weight saving, the substitution of mainly steel and aluminum is facilitated if other favorable attributes of magnesium can be addressed as well, for instance, if a consolidation of parts in an assembly can be realized because of the excellent castability or if an improvement in noise, vibration, and harshness (NVH) characteristics can be realized because of the high vibration-damping capacity. Transport applications beyond the automotive industry exist as well, not that much in terms of volume but quite so as an arena for development. One specific category is sports and leisure, ranging from high-end applications such as racing cars and motorcycles (wheels) to bicycles for daily use. Bicycles contain a variety of parts that are of interest; of these parts, the magnesium suspension fork is already common while wrought products (frames, wheel rims) are introduced as well [8]. For civil aviation, regulatory restrictions still apply concerning the use of magnesium for aircraft interiors, but with its substantial potential (seats, luggage racks, trolleys, galleys, etc.), stakeholders are now working together to relieve these [9]. Military vehicles for ground and air are a traditional area of application for magnesium; with the peaks in its use being directly linked to periods of tension and escalation, this remains to be a market with a dynamics of its own and with present applications in jeeps, helicopters, and so on [10].

Critical for any application are the costs of magnesium components versus other material solutions, more in particular any permissible additional costs that are obviously to be offset by the benefits during their life cycle such as fuel savings, emission reductions, and deployment enhancement. For the cost-driven automotive sector, margins are at the lower end (roughly 0–1 times the magnesium base-metal price), while truck and rail tend to be in the middle (1–10 times), and military and (aero)space at the higher end (up to 100 times).

Figure 3.3 presents a contemporary wrought product: a bicycle crank set that was developed by Timminco Corporation and Kikusui Forging Company for SR Suntour. This high-end product received the 2007 "excellence for design" award

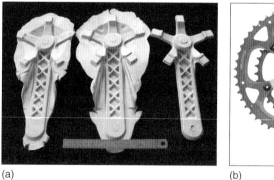

(a)   (b)

Figure 3.3 Current magnesium applications: SR Suntour forged AZ80A alloy bicycle crank set; (a) manufacturing sequence and (b) finished and mounted component. (Courtesy of Kikusui Forging Co., Yokohama, Japan, and Applied Magnesium International, Denver CO, USA.)

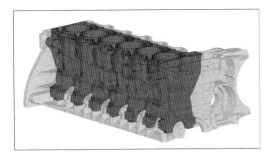

Figure 3.4 Current magnesium applications: BMW NG6 high-pressure die-cast composite crankcase with an AlSi17Cu4 aluminum alloy insert and an AJ62 magnesium alloy sleeve. (Courtesy of BMW AG, Landshut, Germany.)

of wrought products from the International Magnesium Association (IMA). The forged cranks, featuring a tailored mesh design, are made of the AZ80A alloy and are 30% lighter than the original aluminum version.

As a contemporary cast product, Figure 3.4 presents the world's first magnesium–aluminum composite crankcase that has been in production by BMW since 2004. As of this writing, more than 1.5 million units have been manufactured for the concerned 3.0 l six-cylinder in-line gasoline engine. The crankcase is made by overmolding the premanufactured aluminum casting, that is, by using it as an insert in the magnesium high-pressure die-casting mold. The aluminum insert integrates the cylinder linings, the cooling system, and the crankshaft's main bearings, while the magnesium sleeve forms the mounting structure for the attachments and engine cradle. For this high-temperature application, the creep-resistant AJ62 magnesium alloy is used. To obtain an intricate interface between the aluminum and magnesium castings, the insert is pretreated with an arc spraying process to provide a porous, rough AlSi12 coating with high specific surface area. As compared to its aluminum predecessor, the composite crankcase features 10 kg of weight saving and a simultaneous increase in engine performance.

One specific technological road-mapping study for the period 2005–2020 targets a potential increase from 5 to 160 kg of magnesium for the average US carmakers' vehicle, which could bring 130 kg of weight saving over current ferrous and aluminum parts [11]. This aspiration is based on an inventory of components that have already been demonstrated to be technically feasible. Critical in this best-case scenario are cost/quality issues, engineering/manufacturing challenges, and an enabling infrastructure. By means of illustration, Figure 3.5 shows selected possible applications of magnesium parts. Notable in this respect is also that the primary focus for weight saving is in the front and upper car areas as well as in the unsprung mass because this yields better comfort and driving characteristics in addition to improved mileage.

In this chapter, recent advances in the development of magnesium alloys and their processing for (high-volume) transport applications are introduced from an end user's perspective, assuming a basic understanding of metallurgy and

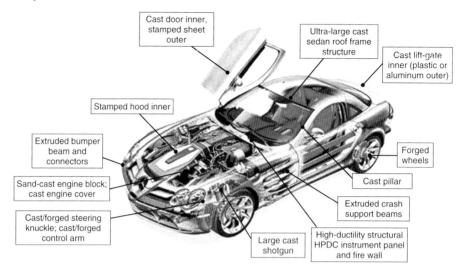

Figure 3.5 Possible applications of new magnesium components in the front and upper car areas. (Courtesy of LightWeight Strategies LLC, Franklin, MI, USA.)

manufacturing. The emphasis is on technological rather than scientific aspects. For the sake of brevity, a number of specific developments such as bulk-metallic glasses, metal–matrix composites, powder metallurgy, superplastic, and severe plastic deformation processes are left out of consideration. As the scope of this work dictates, application areas other than transport – both established (electronics) and emerging (biomedical) – are omitted as well. Further, magnesium base-metal production and some secondary manufacturing processes such as joining and machining are not treated. The same holds for (computer) modeling and simulation as supportive tools for a variety of research and engineering topics. Sections 3.2 and 3.3 deal with alloys and processes for wrought and cast products, respectively. The former category covers development routes for alloys, extrusion, forging, and sheet processing. The latter category covers development routes for alloys, die casting, and gravity casting. After that, Section 3.4 briefly addresses corrosion and its prevention, safety hazards, and sustainability issues. Section 3.5 introduces some prototypes that have resulted from recent research and development efforts and can thus be regarded as an outlook for what is ahead in vehicle construction. Finally, Section 3.6 recapitulates the current situation and gives some thoughts on prospective future developments, while Section 3.7 provides suggestions as to the gathering of information for further exploration of this subject.

3.2
Wrought Alloys and Associated Processes

With magnesium having only a minor market, the development of wrought alloys is lagging behind that of competing materials such as aluminum. Further, manufacturing processes for wrought magnesium products have a technological as well as a technical backlog. Academic and industrial interest, however, has increased in recent times on account of the favorable mechanical attributes (strength and ductility) as well as the shapes that can be manufactured complementary to castings. General themes in this are the development of alloys with better workability and in-service properties and of processes with better performance in terms of productivity and resulting product quality. Meanwhile, the number of industrial suppliers that are active in this field is still limited.

The following subsections outline the state of affairs in the field of wrought alloys and their processes. This follows the usual classification into extrusion, forging, and sheet processing, but starts with an overview of alloy development. Accompanying (research) examples are drawn from the immediate awareness spheres of the authors; owing to the variety of ongoing efforts, they can merely serve as illustrations to provide a taste of the issues that are being dealt with.

3.2.1
Alloys

Mechanical properties of wrought magnesium alloys are listed in Table 3.1, compiled from [12–15] but noting occasional deviations in data between individual sources. Alloys from the AZ series (with aluminum and zinc as the main alloying elements) have already been around for several decades and are the mainstream, with AZ31 representing a general-purpose alloy with moderate strength and good workability, and AZ61 and AZ80 featuring higher strength at the expense of extrudability and forgeability. Other regular alloy series are ZM (with zinc and manganese) and ZK (with zinc and zirconium), the most common representatives being the general-purpose ZM21 alloy and the high-strength ZK60 alloy. These alloys are generally not heat-treatable, with only AZ80 and ZK60 having a significant age-hardening response. A particular aspect that is to be mentioned here is the directionality of mechanical properties. This shows, for instance, a disparity in yield stress between compression and tension (as is obvious from the table), but also in strength differences between different directions in the material (e.g., longitudinal vs transverse) and in the plastic anisotropic behavior in forming operations. These effects are associated with the anisotropy of the crystal lattice of magnesium (hexagonal close-packed) in conjunction with the often pronounced crystallographic textures as they develop during (thermo)mechanical working.

Development strategies for new alloy compositions often rely on the mechanisms of chemical grain refinement and precipitation in order to enhance mechanical properties. An example of the former is the conventional grain-refining addition of zirconium; yet, it is ineffective in alloys containing aluminum and/or manganese. An example of the latter is by micro-alloying using (such elements as) calcium, silver, and tin, which induce a substantial hardening response on aging. The addition of rare earth elements including yttrium is especially to be mentioned as this favorably affects both quoted strengthening mechanisms and has a randomizing effect on textures, which effectively provides a route for developing both stronger

Table 3.1 Basic properties of magnesium alloys for wrought applications (underlined: proprietary alloys).

Alloy (temper)[a]		Typical mechanical properties[b]				Workability[c]	Corrosion resistance[c]
		TYS (MPa)	UTS (MPa)	El. (%)	CYS (MPa)		
Extrusions	AZ31B/C (F)	200	255	12	97	G	G
	AZ61A (F)	205	305	16	130	F	G
	AZ80A (T5)	275	380	7	240	P	E
	ZM21 (F)	155	235	8	—	E	F
	MRI-301 (F)	220	290	17	225	—	—
	Elektron 675	310	410	9	—	E	—
	AM30	165	—	12	—	E	—
Forgings	AZ31B (F)	195	260	9	85	E	G
	AZ61A (F)	180	295	12	115	G	G
	AZ80A (T6)	250	345	5	185	F	E
	ZK60A (T6)	270	325	11	170	G	P
	MRI-301 (F)	220	290	17	225	—	—
	Elektron 675	310	410	9	—	E	—
Sheet	AZ31B (H24)	220	290	15	180	F	G
	ZM21 (O)	120	240	11	—	G	F
	ZM21 (H24)	165	250	6	—	F	F

[a]Temper condition affects mechanical properties.
[b]TYS, tensile yield stress; UTS, ultimate tensile strength; El.; tensile elongation; CYS, compressive yield stress – values for longitudinal direction.
[c]Relative ratings: E, excellent; G, good; F, fair; P, poor.

and less anisotropic alloys. Hence, this class of alloys recently attracted a lot of research interest. Although initially a mixture of rare earth elements in its naturally occurring proportions was used (so-called misch metal, which mainly consists of cerium, lanthanum, neodymium, and praseodymium), meanwhile the insight has rooted that specific rare earth elements induce specific effects so that now rather individual elements such as neodymium or gadolinium are added.

Further, a particularly exploratory development for high-temperature alloys is the addition of zinc and yttrium and is characterized by very high strength resulting from the so-called long-period stacking order (LPSO) phase. Another line of development is the so-called lean alloys; these are relatively low in alloying contents to give favorable workability (e.g., extrudability) but without compromising the mechanical properties. Finally, a less common option is to use additions that change alloy constitution: high amounts of lithium (>10 wt%) change the crystal lattice of magnesium to body-centered cubic, which increases the number of available slip systems at room temperature and thus improves formability. These alloys have some distinct drawbacks though.

Examples of alloy developments that recently made it to commercialization are as follows. MRI-301 as developed by the Magnesium Research Institute, Israel, contains neodymium and yttrium as its main alloying elements and is reported to have high strength and creep resistance [13]. Elektron 675 is a proprietary alloy of Magnesium Elektron, England, with substantial amounts of gadolinium, yttrium, and some zirconium that offers exceptional strength [14]. Finally, a modification explored by General Motors, USA, is based on the AM series; for instance, the lean alloy AM30 – which is essentially the AZ31 alloy in which zinc content is reduced – shows improved extrudability as well as subsequent formability at similar strength [15].

In addition to chemical composition, an important aspect is the processing of the base metal into feedstock for extrusion and forging, and into thin-gauge sheet for sheet-metal forming. Casting is the first step, where the so-called (semi-continuous) direct-chill casting is the prevailing method for series production of strands and slabs. Here, process engineering is concerned with such issues as reaching high productivity while simultaneously avoiding hot cracking and realizing a fine and homogeneous microstructure. Next, the remaining unfavorable (macro)segregations and eutectic structures in the as-cast material can (partly) be resolved by homogenization/solutionizing heat treatment. Feedstock for extrusion and forging is then mostly produced by hot pre-extrusion of the strands (followed by cutting into billets and slugs of the required size), with the associated advantage of a further physical grain refinement as a result of dynamic recrystallization. Thin-gauge sheet for sheet-metal forming is conventionally manufactured from the slabs in a sequence of hot-rolling operations, generally with frequent reheating between passes to account for the magnesium's low heat capacity. These upstream thermomechanical operations are also relevant in a sense that they affect textural development and thus have a bearing on anisotropic effects.

3.2.2
Extrusion

Hot extrusion is used for manufacturing long, straight (semi)finished products, commonly called *sections*, *extrusions*, or *profiles*. The process basically consists of placing a heated billet in a container and then pressing it through a die into the desired cross-sectional geometry. Products range from basic shapes such as bar, wire, and tube to intricate, multihollow sections in a wide variety of dimensions. The required force is usually applied mechanically by a ram that moves the billet in a forward direction; this is the so-called direct extrusion process. Magnesium can in principle be processed with the same equipment as that used for aluminum. Process settings such as billet temperatures (typically 300–400 °C for magnesium alloys vs 400–500 °C for aluminum alloys) and working procedures need to be adapted though. A major impediment is in the limited productivity: in general, the extrusion speed for magnesium alloys is 5–10 times lower than that of comparable aluminum alloys. As a consequence, the extrusion operation constitutes the single largest share of the product's cost price.

A principal limitation of extrusion in general and of magnesium extrusion in particular is the occurrence of hot shortness, also known as hot cracking. This defect appears as recurring circumferential surface cracks, and is obviously detrimental to product quality, especially for structural and decorative applications. The defect is initiated by incipient melting of the material in the plastic zone because of excessive temperature rise caused by the conversion of mechanical work into heat. Therefore, the phenomenon is not only alloy dependent, but is also related to the extrusion speed: the higher the speed, the more adiabatic–like the process becomes. Effectively, hot shortness fixes the speed for obtaining sound products to an upper level.

As hot shortness is such a critical aspect in the extrusion of magnesium alloys, considerable research effort has been dedicated to stretching this limit. For example, it is demonstrated that the indirect extrusion process (in which the material flows backward through a die that is incorporated in a hollow ram) and the hydrostatic extrusion process (in which the force is applied by a fluid that is pressurized by a ram that moves into the container) are especially advantageous for magnesium alloys in terms of productivity and product quality as a result of the reduced redundant work and/or friction. However, these alternative processes have some practical drawbacks (more elaborate equipment and processing, and restrictions on product shapes and dimensions) so that their current industrial use is limited. Hydrostatic extrusion holds clear potential for basic shapes, but conventional direct extrusion will remain the primary option for mainstream production.

Figure 3.6 identifies the diverse aspects that affect the process and the resulting product in hot direct extrusion. Feedstock quality has already been addressed in the previous subsection on (wrought) alloys; this can be recapitulated in that low-melting-point constituents should be avoided and a homogeneous, fine equiaxed

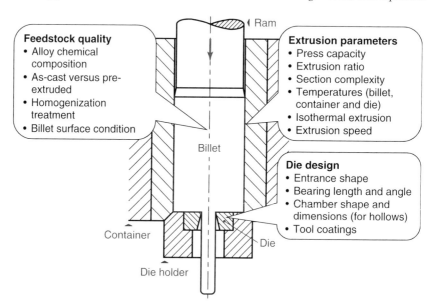

Figure 3.6 Hot direct extrusion: process principle and focus areas for development.

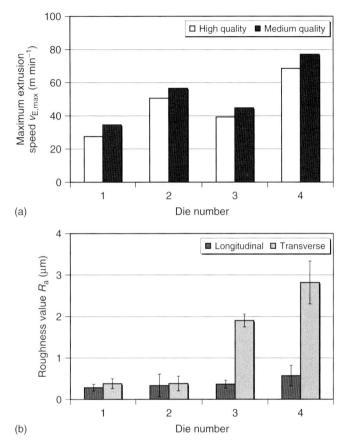

Figure 3.7 Hot direct extrusion: (a) effect of die geometry on extrudability and (b) surface condition of ZM21 6.5 mm $\times$ 2.0 mm strip; extrusion ratio $R = 40$, and billet temperature $T_b = 375\,^{\circ}$C.

microstructure without segregations is preferred. The selection of appropriate extrusion parameters is generally a complex task and is often still empirically based. For example, the fixing of the billet temperature for extrusion-speed optimization is usually a trade-off between press capacity and hot shortness. In a wider sense, this does also include more advanced concepts such as isothermal extrusion. Further, the fact that besides surface quality processing conditions also affect the mechanical properties of the sections needs to be taken into account. Die design obviously starts from known solutions for aluminum extrusion, yet needs to be adapted because of the higher susceptibility of magnesium alloys to hot shortness.

The following research example is concerned with the productivity of the direct extrusion process and the surface quality of the extruded sections depending on the die design [16]. Extrusion experiments involved trials on a laboratory press (make Loire/ACB, 50 t capacity, Ø25 mm container) with pre-extruded ZM21 billets. Dies of different shapes were used to extrude identical strip sections: dies

1, 2, and 3 were dies with a square entrance and bearing lengths (or die lands) of 0, 2, and 4 mm, respectively, while die 4 had a shaped entrance and a bearing length of 2 mm. By adopting the presence of microcracks (as observed with an optical microscope) and macrocracks (as observed with the bare eye) as some arbitrary criteria for "high" and "medium" surface quality, respectively, results can be compared as shown in Figure 3.7a. Speed values are on the high side as compared to regular values on account of the simplicity of the investigated section shape. The differences in the maximum allowable extrusion speed between the dies are notable, where the shaped die performs better than any of the square dies. The surface roughness as characterized by the arithmetic mean roughness value R_a is shown in Figure 3.7b (average values for all explored extrusion speeds yielding sound extrusions). Roughness in the transverse direction is generally higher because of the nature of surface generation in contact with the die bearing (often referred to as *die lines*). For the square dies, the directional dependence of the roughness increases with bearing length. The most pronounced difference between the longitudinal and the transverse directions, however, is found for the shaped die. To put the presented results in perspective, roughness values for extruded sections are commonly in the range $0.8 \leq R_a \leq 3.2\,\mu m$.

3.2.3
Forging

In forging, products are shaped from slugs by applying compressive forces through various tools and dies. Where closed-die forging is the most common variant, this metal-working category also includes such processes as swaging and impact extrusion. Part shapes can be intricate, and, as such, it is a competitive process to casting. Magnesium can in principle be processed with the same equipment as used for other metals, although high-speed presses should be avoided, dies and slugs need to be heated, and magnesium-related working procedures are to be followed. Issues that hamper a more widespread use of magnesium forgings are the lack of analytical capabilities on process design and product performance and the unfamiliarity with the material throughout the forging sector.

Because of the limited formability of magnesium alloys at room temperature, forging has to be conducted at elevated temperatures. Depending on the alloy, implied deformation, press speed, and so on, temperatures have to be kept within a certain process window. Cold cracking and press capacity define the lower temperature limit. Conversely, the upper limit is determined by hot cracking. Another important feature of magnesium forging is of the anisotropic flow. The phenomenon is visualized in Figure 3.8b, showing the footprints (bottom surfaces) of some cylindrical slugs that were axially forged into a conical shape similar to the one sketched in Figure 3.8a. As these bottom areas are not confined by the tooling and thus tend to flow in the direction of the lowest resistance, they give an indication of the directionality of the plastic behavior. The slugs were machined from different directions in the feedstock, their length direction coinciding with either the axial or the radial orientation thereof. Notably, the footprint for the

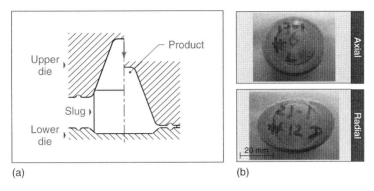

(a)   (b)

Figure 3.8 Closed-die forging: (a) process principle and (b) anisotropic flow of pre-extruded AZ80A.

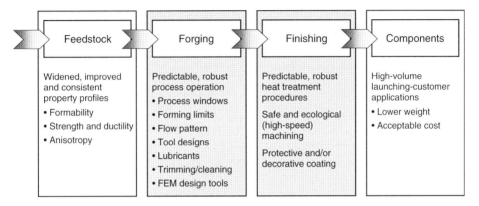

Figure 3.9 Closed-die forging: innovation objectives, arranged according to the processing chain (from left to right).

former (Figure 3.8b, top photograph) remains basically circular, while that for the latter (Figure 3.8b, bottom photograph) has developed into an elliptical shape. Such anisotropic effects are typical for pre-extruded feedstock as a result of their crystallographic textures.

The manufacturing of magnesium forgings involves a sequence as visualized in Figure 3.9. Perceived items for advancement are specified for each area as well. The feedstock area covers the specifics of the alloying (chemical compositions) and production of the slugs. The forging area entails the processing of the feedstock into semifinished parts; this embraces the preparation of the slugs including any required preforming such as upsetting and bending, one or more forging steps, and trimming of the flash. The finishing area covers all additional working operations to arrive at the finished parts; this may include heat treatment to modify mechanical properties, machining for the provision of (bore) holes as well as for reasons of accuracy and surface condition, and coating to enhance resistance to corrosion and wear and/or visual appearance. Finally, the components area refers to such aspects as appropriate designs that meet functionality as well as any other requirements.

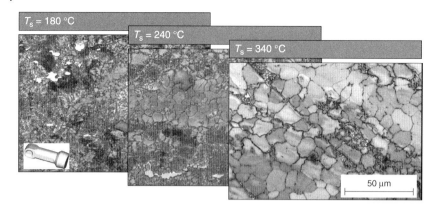

Figure 3.10 Closed-die forging: microstructures of a forged AZ80A shock-absorber head (inset); slug size Ø47 mm × 140 mm, slug temperature T_s varied, die temperatures $T_d \sim$ 100 °C, and press speed $v_R \sim$ 0.3–0.4 m s^{-1} – same magnification for all light-optical micrographs.

It is noteworthy that each of the involved unit operations is characterized by a diversity of (process) parameters and variables, which also interact with the upstream and downstream operations. Manufacturing productivity and product quality are thus the result of a complex interplay within the processing chain.

The following is a research example of the influence of the forging parameters on the structural quality of the component [17]. Figure 3.10 shows light-optical images that reveal how the microstructure of an AZ80A forging alters when slug temperature is varied within a wide range. The grain size increases with forging temperature. The mechanism behind this is that plastic deformation in the warm-working regime induces a recrystallization in which substantially smaller grains are formed, followed by grain growth until the parts have cooled to a certain threshold value. A higher forging temperature thus implies more extensive grain growth and, in the end, a coarser microstructure. This also affects the mechanical properties. In the example at hand, tensile testing along the longitudinal axis of the forgings revealed a drop in tensile yield stress (TYS) from 305 to 240 MPa (−20%) on raising the slug temperature T_s from 180 to 340 °C. Simultaneously, ultimate tensile strength (UTS) dropped from 390 to 330 MPa (−15%) while tensile elongation increased from 8 to 13.5% (+70%).

3.2.4
Sheet Processing

Sheet-metal parts are generally manufactured in a chain of operations in which the upstream processing consists of the already outlined direct-chill casting and (hot) rolling of sheet and the downstream processing consists of processes such as deep drawing and stretching to produce thin-walled three-dimensional shapes such as body panels. Magnesium can in principle be processed along this route with proper adaptations. Owing to the more elaborate working and the limited size of

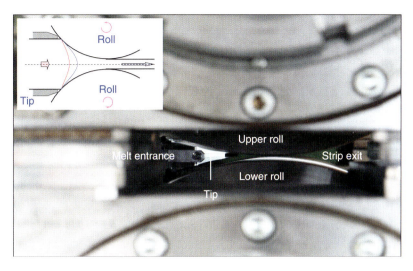

Figure 3.11 Twin-roll casting: side view of the tip arrangement between the rolls and process principle (inset). (Courtesy of HZG MagIC Department Wrought Magnesium Alloys, Geesthacht, Germany.)

scale, magnesium sheet and sheet-metal parts are still much more expensive than their steel and aluminum counterparts. Their manufacturability and functionality, however, have meanwhile been demonstrated in several studies.

With respect to the upstream processing, the trend to integrate the conventional processes in the so-called twin-roll casting process is now attracting ample attention for magnesium as well. Here, the melt is directly transferred from a tip or nozzle between two rotating water-cooled rolls and the metal solidifies while being simultaneously reduced in thickness to the final gauge. Figure 3.11 clarifies the principle, with the material flow being from left to right. With typical gauge thicknesses of 4–8 mm, this makes several hot-rolling operations redundant. Except for an increased cost-effectiveness, technical assets are in the high solidification rates, which are claimed to lead to finer microstructures, reduced segregation, and extended solid solubility. Twin-roll casting of magnesium is under development at several institutes and companies worldwide, with industrial production being in the start-up phase and twin-roll cast sheet becoming commercially available [18]. One particular challenge is the manufacturing of wide strip – with widths of up to 2000 mm suitable for automotive sheet – in which tip design for a balanced flow of the melt between the rolls is a critical aspect. Initial assessments of the deep-drawability of twin-roll cast AZ31 sheet in comparison with conventionally manufactured equivalents suggest a somewhat different behavior in terms of their processing window, obviously related to the differences in microstructures (including textures) resulting from both manufacturing routes [19].

With respect to the downstream processing, any operations involving substantial plastic straining (deep drawing, stretching, hydroforming, but also piercing) are to be done at elevated temperature to bypass the material's limited room-temperature

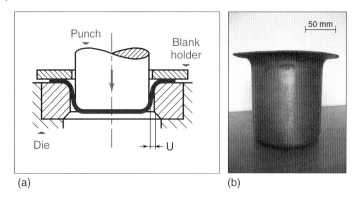

(a)   (b)

Figure 3.12 Deep drawing: (a) process principle and (b) a warm-drawn AZ31B cup (sheet thickness s_o = 1.2 mm, punch diameter d_p = 110 mm, drawing ratio β = 2.41, blank and die temperature $T_b = T_d$ = 250 °C, punch not heated).

formability. This requires the use of active tool heating and lubricants that are suitable for temperatures of typically 150–300 °C. For these warm-forming conditions, ductility of magnesium alloys is greatly enhanced and deep-drawability can be quite comparable to and even exceed that of steel and aluminum alloys in room-temperature deep drawing, as is obvious from the demonstration part in Figure 3.12. Technical solutions for warm sheet-metal forming are meanwhile state of the art as such processes were developed and industrially implemented for certain aluminum alloys in recent times. Notwithstanding such developments, still some challenges can be mentioned relating to the warm forming of magnesium sheet. Characterization of the sheet material is being addressed in such terms as constitutive equations and forming-limit curves in dependence on temperature and strain rate and also taking into account plastic anisotropy. The tailoring of process conditions involves the assessment of process windows for actual forming processes (e.g., of the limiting deep-drawing ratio as a function of temperature, blank holder force, etc.). A further issue is that elevated temperatures may induce microstructural effects that affect in-service properties. From a technical point of view, accurate temperature control and proper lubrication are becoming more critical as part size increases.

The following research example is about material characterization for sheet-metal forming. Figure 3.13 shows the forming-limit curves of some conventionally hot-rolled magnesium sheets in relation to the forming temperature [20]. Essentially, these curves record the combinations of in-plane plastic strains that mark the onset of necking and as such are representing the formability of the sheet. The forming-limit curves are measured by stretching sheet-metal samples over a hemispherical punch; by employing a variety of well-defined shapes (so-called Hasek-type samples), the stress state and hence the biaxiality of straining are varied, with each sample representing a single failure point. This point is assessed by providing a pattern on the sheet before testing, the deformation of which is quantified *in situ* by optical registration and image analysis. The graphs show that

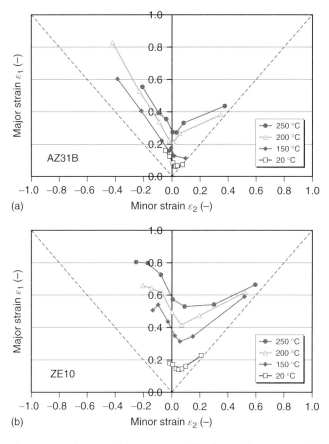

Figure 3.13 Sheet-metal forming: (a and b) forming-limit curves of two magnesium alloy sheet materials for ambient and elevated temperatures; sheet thickness s_o = 1.5 mm.

formability for both alloys increases with temperature within the explored limits. Moreover, differences between the alloys are substantial, with ZE10 in general featuring a better formability. The different plastic anisotropy of both alloys is also obvious from the positioning of the corresponding data points in the forming-limit diagram (for AZ31B tending more toward the left quadrant). For plane straining (ε_2 = 0), AZ31 in particular is quite susceptible to failure. This situation is technically relevant as it appears, for instance, in deep drawing at the transition from cup bottom to cup wall, as well as at the transition from cup wall to cup flange.

3.3
Cast Alloys and Associated Processes

Melt-metallurgical processing – which for magnesium alloys typically is done in the range 600–800 °C – is the main route for manufacturing magnesium (semi)finished products. Although casting is also an initial step in wrought products' manufacturing, this section is rather dedicated to shape casting, which, as pointed out before, accounts for the vast majority of all magnesium structural parts. Thus, magnesium technology has long since been focusing on casting; leitmotivs in this are alloys with better workability and in-service properties as well as processes yielding enhanced productivity and product quality. Meanwhile, a mature industrial supply chain for magnesium castings has been established.

One general aspect to be dealt with in the processing of a magnesium melt is its high affinity to oxygen, which implies both a safety hazard and the likelihood of excessive oxide formation and associated risk of melt contamination. While early solutions consisted of molten salts covering the melt, the use of sulfur hexafluoride gas (or more precisely, of mixtures of inert carrier gases such as nitrogen or argon with typically 0.2–0.5% SF_6) in closed furnaces has been the prevailing means of melt protection over the past decades. SF_6 is a very effective shielding agent as it forms a dense solid film on top of the melt in reaction with liquid magnesium. With SF_6 being an extremely potent greenhouse gas, however, its industrial use is presently being phased out under (European) legislation, which has triggered developments to come with more environmentally friendly alternatives. Mixtures in which SF_6 is substituted by SO_2 are now recommended practice, although this is regarded a temporary solution owing to their toxicity and corrosiveness. The ensuing generation of cover gases and their feeding systems is mainly based on fluorinated gases as known from cooling systems; examples are AM-cover (HFC-134a) by Advanced Magnesium Technologies, Australia, and HFE-7100 ($C_4F_9OCH_3$) and Novec 612 ($C_3F_7C(O)C_2F_5$) by 3M. Further developments are due though, with the overall trend being toward abandoning fluorinated gases completely. As an alternative approach, the addition of minor amounts of calcium compounds such as calcium oxide to the magnesium melt is being explored, which leads to a dense oxide film on top of the melt and effectively renders a cover gas redundant [21]. Here, however, the influence of calcium on the castability and properties of the product needs to be considered as well.

The following subsections outline the state of affairs in the field of cast alloys and their processes. These distinguish between the categories of die casting and gravity casting, each with their own subsets of processes. To start with, however, alloy development is discussed. Again, (research) examples are drawn from the immediate awareness spheres of the authors and are to be regarded as mere illustrations of the general trends.

3.3.1
Alloys

The mechanical properties of cast magnesium alloys are listed in Table 3.2, compiled from [12, 22–28] but noting occasional deviations in data between individual sources. Alloys of the AZ and AM series (with aluminum and zinc, and aluminum and manganese as the main alloying elements, respectively) can be

Table 3.2 Basic properties of magnesium alloys for cast applications (underlined: proprietary alloys).

Alloy (temper)[a]		Typical mechanical properties[b]						Corrosion rate[c] ($mg\,cm^{-2}\cdot d^{-1}$)
		At 20 °C			At 150 °C			
		TYS (MPa)	UTS (MPa)	El. (%)	TYS (MPa)	UTS (MPa)	El. (%)	
Die castings	AZ91D	160	260	6	105	160	18	0.05–0.10
	AM50	125	230	15	—	—	—	0.10
	AM60	130	240	13	—	—	—	0.09
	AS21	125	230	16	87	120	27	0.34
	AE42	135	240	12	100	160	22	0.06–0.21
	AJ62	142	239	8	108	163	19	0.04–0.08
	MRI-153M	170	250	6	135	190	17	0.09
	MRI-230D	180	235	5	150	205	16	0.10
	ACM522	158	200	4	138	175	7	—
Gravity castings	WE43 (T6)	180	260	6	178	210	7	—
	MRI-201S (T6)	170	260	6	170	245	11	0.10
	Elektron 21 (T6)	170	280	5	—	—	—	0.13–0.37
	AM-SC1 (T6)	130	206	4	128	190	11	—

[a]Temper condition affects mechanical properties.
[b]TYS, tensile yield stress; UTS, ultimate tensile strength; El., tensile elongation.
[c]ASTM B117 salt-spray test (for base metal).

regarded as the workhorses of the casting materials. Both alloy systems have been developed for die casting and give good mechanical properties at room temperature. Thus, most magnesium castings are made of AZ91 or AM50/AM60, with the former having more favorable castability and corrosion resistance (rendering this a true general-purpose alloy) and the latter having better ductility (which is of importance for crash-relevant parts). With a major application area for castings being in the drivetrain of vehicles with associated temperatures of typically 120 °C for oil-containing casings and up to 150 °C for actual engine components, creep-resistant magnesium alloys are another important class. Early alloys from the AS series (with aluminum and silicon) and especially AS21 and AS41 were used extensively for die casting of VW Beetle components. A successor is the AE series (with aluminum and rare earths), where AE42 has long been the reference, yet without extensive application. For gravity castings, the WE series (with yttrium and rare earths) is well established, with WE43 as its main representative being used in such niche products as helicopter transmission casings [22]. Patents for this alloy series recently ran out.

Alloy development for shape castings has been and is still strongly directed toward enhancing high-temperature properties. Improvements are being sought by modifying the AZ and AM series with a diversity of elements including calcium, rare earths, silicon, strontium, and tin. These additions target the introduction of stable precipitates (intermetallic compounds) within the grains and at the grain boundaries so as to hamper grain-boundary sliding at elevated temperatures and thus to limit creep of the thermally and mechanically loaded components. Especially, rare earth elements have been demonstrated to be effective in this respect.

Examples of alloy developments that recently made it to commercialization are as follows. As for die casting, Norsk Hydro, Norway, adopted the AS system for further modification. The original alloys showed limitations regarding castability, corrosion, and costs, but these were mainly overcome by the use of improved equipment for the upstream (impurity control) as well as downstream (die casting) processing. The modified AS31 alloy has been applied by Daimler Chrysler, USA, for the gearbox casing of the 7G-tronic transmission [29]. Noranda, Canada, went a different way and developed new alloys based on the AJ system (with aluminum and strontium) [23]. AJ62 is used for the BMW composite crankcase, which was introduced earlier on in this chapter. Creep-resistant alloys developed by the Magnesium Research Institute, Israel, are MRI-153M and MRI-230D [24]; these are basically modifications of AZ91 and AM60, respectively, obtained by the addition of calcium among others. As a further development, the ACM522 alloy (with aluminum, calcium, and rare earth misch metal) reviewed in [25] is to be mentioned. This alloy is used for the oil pan of the Honda Insight hybrid-car IMA engine. Finally, the AE system has seen further exploration within a Northern American automotive research project, which has led to the use of AE44 for the front engine cradle (subframe) of the 2006 Corvette Z06 [30]. As for gravity casting, developments have led to the introduction of such proprietary alloys as MRI-201S by Dead Sea Magnesium, Israel [26], Elektron 21 by Magnesium Elektron, England, as a successor for WE43 [27], and AM-SC1 by Advanced Magnesium Technologies, Australia [28]. These alloys are all rare earth based and can typically be employed to temperatures of up to 200–300 °C.

Figure 3.14 ranks a number of generic and proprietary alloys for die casting in terms of castability and creep resistance [31]. Two subsets can be distinguished, identified by different hatching: those alloys that can be used for long-term operation at temperatures of up to 150 °C (with a generally better castability) and those that can be used up to 180 °C. *Castability* in this respect is defined as the ease with which the alloy can be processed in view of shaping freedom (flow length, wall thickness, etc.) and the risk of defects (sticking to the mold, porosity, etc.).

As yet another means of comparison, Figure 3.15 represents the creep resistance of several alloys for die and gravity casting over a wide range of elevated temperatures (compilation from sources as cited earlier). Although the data sets are not fully complete and consistent in terms of experimental conditions, and occasional deviations in data occur between individual sources, differences in performance are apparent.

Notably, the selection of an alloy for any particular application goes well beyond the previously introduced characteristics and covers additional technical considerations regarding the processing (including recycling prospects) and product

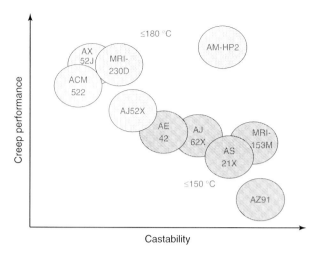

Figure 3.14 Comparison of workability and creep behavior of magnesium alloys for die casting.

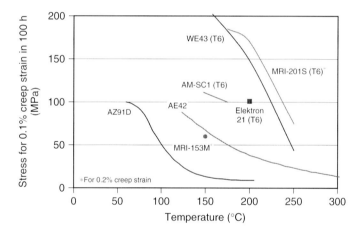

Figure 3.15 Comparison of creep behavior of magnesium alloys for die and gravity casting.

performance, and also nontechnical considerations such as costs and sourcing reliability.

3.3.2
Die Casting

The processes that are treated in this subsection are high-pressure die casting (HPDC), squeeze casting (SC), and semisolid casting.

In HPDC, the molten metal is injected into a mold cavity with filling speeds of up to 60 m s^{-1}. It is the process of choice when intricate components have to be manufactured in high volumes and with high geometrical accuracy, and is thus of particular relevance for transport (and for electronics) applications. Shot weights in the range 0.005–50 kg are common practice. Moreover, very large (up to 2 m wide) and thin-walled (down to 0.5 mm) components can be made because of the high fluidity of a magnesium melt. Solidification and cooling rates are typically high, which leads to short cycle times and fine microstructures with good (mechanical) properties. Owing to the low reactivity of the magnesium melt with the tool steel, molds are affected only by such phenomena as thermal fatigue and flow cavitation and may last up to three times longer than for aluminum HPDC. As a result of the entrapment of air during the turbulent mold filling, however, high-pressure die-cast components usually have a porosity of 2–4%. Implications of this are that static and dynamic mechanical properties are not that high as for fully dense counterparts (e.g., wrought products), that heat treatments to further tailor these properties cannot be applied, and that the castings may not be pressure-tight.

There are two basic types of die-casting machines: hot chamber and cold chamber, the working principles of which are illustrated in Figures 3.16 and 3.17, respectively. The former involves the use of a piston that traps a certain volume of molten metal within the furnace and forces it through a gooseneck and nozzle into the mold, while in the latter the melt is transferred from the furnace and poured into a shot chamber with a ladle, and is then forced into the mold by a piston. The main technical differences are that hot-chamber machines generally feature smaller closing forces, lower filling speeds, and lower (feeding) pressures. As a consequence, smaller parts are commonly cast on hot-chamber machines and larger and thin-walled parts on cold-chamber machines. Also, porosity of the parts is generally lower and surface quality is better for the latter. Hot-chamber machines are primarily used for AZ91 and AM50/AM60, while cold-chamber machines with their separate melting and casting zones are also suitable for the higher melting point alloys such as AS21 and AE42.

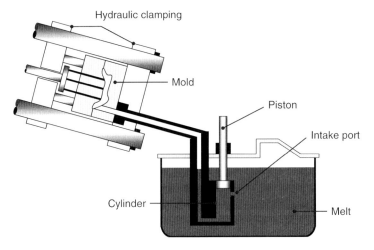

Figure 3.16 High-pressure die casting: the hot-chamber process.

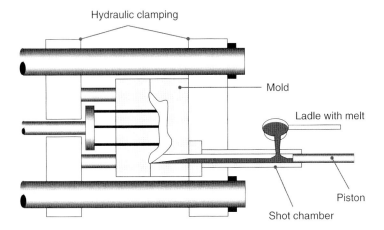

Figure 3.17 High-pressure die casting: the cold-chamber process.

HPDC equipment for magnesium is essentially the same as for aluminum and in that sense to be characterized as mature. Revolutionary developments are not anticipated anymore, yet there is still a trend toward higher performance (e.g., clamping forces), which expands production capabilities in terms of product size and quality as well as the alloys that can be processed. Meanwhile, the industrial knowledge base for HPDC on such issues as mold design has been and is still progressing, supported, among others, by process simulation tools. One particular development relates to vacuum-assisted casting, which was adopted from aluminum HPDC and is commercialized as the proprietary VACURAL process [32]. In this case, the mold parts are provided with an airtight sealing gasket. Just after mold closure and before injection, the air is evacuated from the mold through vent holes so that gas entrapment is limited, which effectively reduces porosity and renders the casting close to full density. This makes it an interesting option for safety-relevant components with high mechanical demands; however, the industrial application appears to be limited as yet, in view of the additional costs.

SC is in principle similar to HPDC. The main differences are that filling speed is much lower (in the range 1–2 m s^{-1}) to achieve laminar flow and that after mold filling a high pressure is maintained during solidification so as to avoid shrinkage porosity. This pressure is applied either by one part of the mold (direct SC) or through the feeding system (indirect SC). As a result, SC components are almost free of pores and usually have a fine microstructure on account of the high solidification and cooling rates, which translates into favorable mechanical properties as well as into the possibility of subjecting the components to heat treatment. Although SC of magnesium alloys has been the subject of study at several research organizations, this seems not to have resulted in any commercial applications as yet.

Semisolid casting covers a class of processes that revolve around the use of partially molten or partially solidified material as the feed for casting. The common concept is that globular instead of dendritic primary crystals are created in the material by the action of shearing in the semisolid state; this in turn leads to thixotropic rheological behavior with an associated laminar flow during mold filling, and consequently to reduced porosity and better mechanical performance of the component. The lower casting temperature as compared to liquid casting further brings advantages such as shorter cycle times and longer tool life.

In their actual implementation, however, these semisolid casting processes are distinctly different [33, 34]. In thixocasting, conditioned slugs are used that have a globular microstructure induced by electromagnetic stirring during their prior semicontinuous casting; in the cast-shop, these slugs are inductively reheated to the semisolid state (in which the globules remain solid) and then further processed in a conventional cold-chamber die-casting machine. In New Rheocasting (NRC), a melt is poured and thermally manipulated under well-controlled conditions so as to obtain a globular primary phase; the semisolid slugs are then fed into a conventional cold-chamber die-casting machine for further processing into components. In injection molding or thixomolding, feedstock in the form of granules or chips is introduced into the back end of a heated barrel and transported and compacted by means of a screw feeder while simultaneously being brought to the semisolid state; once enough material has accumulated at the front end of the barrel, the screw moves forward to inject this slurry into the mold. As such, this single-step process is similar to plastic injection molding. With oxidation of the feedstock being prevented by maintaining an argon atmosphere in the feeder to the barrel, operational (safety) requirements for injection molding are less strict than for other magnesium casting processes.

The following is a research example of the influence of the casting method on the structural quality of the product [35]. The MRI-153 alloy was processed by means of (cold-chamber) HPDC, injection molding, and NRC, using industrial equipment and customary operation conditions for each of these processes. Similar molds were used to obtain step-plate castings, representing a basic geometry with sections of different thicknesses. The light-optical images in Figure 3.18 show that the resulting microstructures are distinctly different. For the high-pressure die-cast state, the fine primary phase is dendritic with the eutectic phase solidified at the grain boundaries; further, gas entrapments and pores as typical for this process appear. For the injection-molded state, the primary phase is coarser and globular (with its fraction corresponding to the solid fraction on injection, which was 20–30%); again, some porosity is present. Finally, for the New-Rheocast state, the primary phase appears in the form of coarse rosettes (that are basically degenerated globules) with the eutectic phase surrounding these. That these differences have a bearing on the creep performance is apparent from the double-logarithmic plot. It shows the results of elevated-temperature tensile tests in which a constant stress is applied on the sample and the resulting creep rate measured. The high-pressure die-cast material has the poorest creep resistance, with the injection-molded material being marginally better, especially at the higher stress levels. The New-Rheocast material, however, shows a significantly better performance, with creep rates that are between 1 and 2 orders of magnitude lower than for the high-pressure die-cast

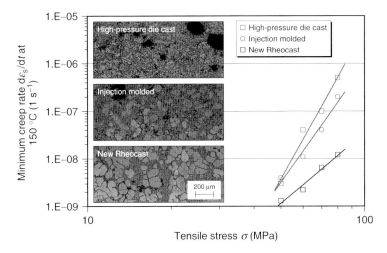

Figure 3.18 Microstructures and creep resistance of MRI-153 for different casting processes – same magnification for all light-optical micrographs.

material. Similar trends were observed for a somewhat lower exposure temperature, as well as for the AZ91D alloy.

Both thixocasting and NRC have as yet not seen sustained industrial application, supposedly because of a generally more demanding process control along with the need for auxiliary equipment and other provisions. In contrast, injection molding has established itself over the past few decades as a manufacturing technique for thin-walled components with several hundreds of manufacturing units up and running. Nevertheless, the vast majority of these are operating for the consumer electronics market and are accordingly based in Asia. The ease of adoption of the process by the plastic injection-molding industry and the practical assets in terms of operation and handling appear to be important reasons for this success. For these applications, component weight is generally less than 1 kg, limited by the capacity of the available manufacturing equipment. Efforts are undertaken now to also enter the automotive market.

3.3.3
Gravity Casting

The processes that are discussed in this subsection are sand casting and permanent-mold casting.

Sand casting is a traditional means of casting that basically consists of pouring a melt in a preformed sand mold. By its nature, the process is apt mainly for single-piece and low-volume production of highly complex and/or thick-walled and large shape castings, where surface quality is not a major concern. Solidification and cooling rates are typically low, implying high cycle times and relatively coarse microstructures. Partially assisted by the more advanced means of pressure or vacuum filling, mold-filling speed is generally low as is the resulting porosity,

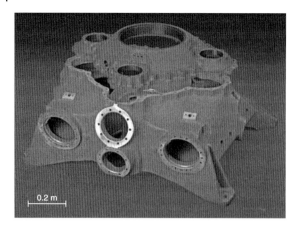

Figure 3.19 Sand casting: Lynx ZE41 main gearbox and WE43 top cover; overall dimensions 1.40 m × 0.85 m × 0.45 m, and weight ∼100 kg for the gearbox and ∼27 kg for the cover. (Courtesy of Stone Foundries Ltd, London, United Kingdom.)

rendering the casting heat-treatable. As such, the process is being used for elevated-temperature alloys in which mechanical properties are tuned for optimal strength and creep resistance by using the material in the solution-treated and artificially aged (T6) condition. A particular issue associated with this process is in the exothermic reactions of the magnesium melt with the oxides and humidity contained in the sand mold, which may induce shape aberrations and surface quality deterioration despite the permeability of the mold providing for effective degassing. In order to reduce these reactions, inhibitors to be added to the sand such as boric acid and ammonium fluorosilicate have been developed. In addition to that, diethylene glycol is commonly used as a binder of the sand to reduce water content of the mold. A typical aerospace sand casting example is shown in Figure 3.19.

Permanent-mold casting is similar to sand casting but uses steel molds with the advantage that better surface quality is obtained. As initial (mold) costs are higher, the process qualifies for low-to-medium volume production of otherwise comparable applications.

3.4
Other Aspects

Apart from the material and processing aspects as treated in the previous sections, there are some other issues that are to be briefly mentioned in order to place the subject of magnesium for (future) transport applications into proper perspective. These relate to corrosion and the means of corrosion control, safety hazards and procedures to cope with these, and sustainability including such angles as energy, recycling, and resources.

3.4.1
Corrosion and Coatings

Magnesium alloys are susceptible to corrosion mainly because of the following basic characteristics. Firstly, its low standard reduction potential implies that magnesium acts as the anode when in contact with most metals (e.g., iron and aluminum) in the presence of an electrolyte (such as salt water), which effectively leads to galvanic corrosion. Secondly, the natural passivation layer that forms on magnesium and that mainly consists of $Mg(OH)_2$ has a smaller specific volume than the metallic magnesium underneath; moreover, it is generally not resistant to aqueous electrolytes, so that the formation of a covering corrosion layer that shields the substrate is impeded. Nevertheless, contemporary magnesium alloys show (atmospheric) corrosion rates that are quite comparable to copper-containing aluminum alloys and mild steels. Although there are several types of corrosion, and corrosion rates depend highly on the particular circumstances, Tables 3.1 and 3.2 give an indication of the corrosion behavior of magnesium alloys.

The means of intrinsic corrosion control lie in alloy chemistry (composition) and in microstructural engineering (processing and heat treatment). Early alloys had relatively high impurity contents, and with such elements as copper, nickel, and iron leading to strong microgalvanic couples in the material, corrosion resistance was correspondingly low. The so-called high-purity or HP alloy variants with specific impurity limits (e.g., Cu $\leq$ 0.05 wt%, Ni $\leq$ 0.005 wt%, Fe $\leq$ 0.005 wt%) that are now common practice do perform substantially better in this respect. Further, to control impurities usually some manganese is added to aluminum-containing alloys: this forms Al–Mn compounds that tie up iron, and in conjunction with zinc offers a similar mechanism to neutralize copper. Rare earths as alloying elements appear to stabilize the corrosion layer and in that sense have a positive effect. Also, the intermetallic compounds that form in rare-earth-containing alloys are generally not that cathodic as other strengthening phases, and thus are not that detrimental to corrosion resistance. As for processing, a fine-grained microstructure generally yields a higher corrosion resistance than a coarse-grained microstructure; this is supported, for instance, by studies that have shown that a die-cast part outperforms a gravity-cast part in the same alloy. Finally, heat treatments for (mechanical) property modification aim to control the diffusion, dissolution, and precipitation of the alloying elements and their compounds; by their condition-dependent galvanic interaction with the magnesium matrix, this – often unintentionally – also affects the corrosion resistance.

The means of extrinsic corrosion control lie in coatings and in constructional measures. Mainstream coatings are chemical coatings, electrochemical coatings, and organic coatings. Chemically applied conversion coatings are appropriate for corrosion protection during storage and transport only, but provide good adhesion for a subsequent organic coating. Alternatives for chromate-based conversion coatings are meanwhile available, with further developments being anticipated (e.g., apt for multimaterial assemblies). Concerning electrochemical coatings, anodic oxidation of magnesium gives electrically insulating coatings with good wear resistance. State of the art are processes such as Anomag, Keronite, Tagnite, and Magoxid, which give hard ceramic coatings based on plasma-chemical reactions in electrolytes. Owing to the porosity of these coatings, they are often applied in conjunction with sealing or painting. Organic coatings such as painting systems are essentially applied as for other metals. These coatings basically prevent the access of electrolytes and oxygen to the metal substrate, and as such they are commonly used as a finishing treatment. Surface treatments are chosen so as to fit the aggressiveness of the environment. In general, the protection provided increases in the order acid pickling, conversion coating, anodizing, electroplating, and organic coating under the notion that combinations with a finishing organic coating as a sealant give the overall best performance. Constructional measures relate to those features provided in the design to avoid corrosion in general and galvanic (or contact) corrosion in particular. The latter is generally known to occur in assemblies with ferrous, aluminum, and other metal parts, and can principally be tackled by disturbing the electrical circuit between the magnesium, the electrolyte, and the mating metal. The most basic measure is to prevent the accumulation of an electrolyte. Further measures are the electrical insulation of the contact areas between the metal parts including the bolted joints by such means as nonmetallic washers; similarly, the metal parts may be coated.

With the durability of magnesium components at stake, corrosion (control) and coatings are still very actively investigated along the lines described earlier in order to enhance basic understanding and provide enabling solutions.

3.4.2
Safety

With substantial amounts of magnesium being incorporated in earlier as well as in present-day vehicles, it is generally a safe construction material – even in such events as in car fires. In view of the earlier outlined high affinity of magnesium to oxygen and its reactivity with water, however, care must be exercised in some specific manufacturing situations. Notably, this relates to casting, welding, machining, and hot forming. With respect to casting, it is crucial to avoid contact between a magnesium melt and water. The vigorous formation of steam and hydrogen gas may readily cause an explosion and fire, with catastrophes such as these having occurred in the past in cast-shops and being one of the main concerns in their operation. Similar remarks apply for welding, although this is commonly conducted on a smaller scale. With respect to machining, the generation of chips (with their high area–volume ratio) under the simultaneous generation of heat, as well as the creation of swarf and dust, may cause ignition. Coolants contribute to tackling this issue, but may require drums for storing the chips to be properly ventilated to remove any generated hydrogen gas. As a precaution for hot forming (extrusion and forging), furnaces for feedstock heating that use an open flame are to be avoided because of the ignition hazard of flashes and debris.

Such safety issues during the transport, storage, treatment, and removal of magnesium and its waste are to be dealt with by adhering to well-established codes of best practice (e.g., gathered in [36]). Safety management also includes the incorporation of these working procedures into quality systems, (awareness) training of personnel, and the acquiring and maintaining of the appropriate permits for the industrial operation of magnesium−related manufacturing activities.

3.4.3
Sustainability

Magnesium is often advertized as a "green material" on account of its weight-saving potential for transport applications, leading to increased fuel efficiency and reduced CO_2 and other emissions. For a proper assessment, however, the full life cycle has to be considered from primary magnesium production to postconsumer scrap recycling. Although the outcome of such reflections is highly dependent on the particular circumstances, there are some general notions that also direct further developments toward enhanced environmental compatibility.

Firstly, thermal reduction (e.g., the Pidgeon process) and electrolysis are both energy-intensive routes (45−80 vs 18−28 MWh t^{-1} of magnesium) for primary magnesium production [37]. Despite the former being associated with a higher environmental footprint, especially with coal being used as the source of energy, the reality is that this is currently the predominant means of production. Secondly, the high energy content of the metal highlights the importance of recycling. Although the recycling of manufacturing scrap is well established, it is to be pointed out here that substantial postconsumer scrap volumes are yet to come, considering the still relatively short period of intensified magnesium use in vehicles. Current practice is that such scrap is used for desulfurization in cast-iron production. The impending move toward closed-loop recycling, however, has meanwhile opened up a new area of development topics (e.g., impurity-tolerant recycling alloys). Thirdly, the aspect of the high global-warming potential of SF_6-based cover gas in casting and its substitution by less potent fluorinated and eventually nonfluorinated alternatives has already been mentioned. Regulation plays a prominent role in guiding these transitions. For the European situation, for instance, relevant EC legislation is in place with respect to chemicals and their safe use (the REACH directive, affecting magnesium imports), recycling quotas for vehicles (the end-of-life directive, affecting postconsumer scrap recycling), and the reduction of greenhouse gases (the F-gas regulation, affecting the means of melt protection).

Although magnesium in itself is not a scarce resource – being the sixth most abundant element in the Earth's crust and the third most abundant mineral in seawater – one further topical issue is that of critical raw materials and the dependency on imports. This primarily relates to the current (economically induced) limitation in Western magnesium production capacity and also extends to the availability of the alloying elements, especially of the rare earths.

3.5
Case Studies

The foregoing sections dealt with the state of affairs in magnesium alloys and their manufacturing processes. To make the recent advances in these areas more tangible, the following subsections present some concrete examples of demonstration components that were developed in the context of some of the underlying studies.

3.5.1
Case 1 Wrought Products: Forging

Although the density of magnesium is only two-thirds that of aluminum, the actual weight-saving potential for structural applications depends on the particular design criteria for the part under consideration as well as on the specific properties of the concerned materials. To illustrate this aspect, Figure 3.20 gives an overview of the anticipated weight reduction for some wrought magnesium alloys as related to the common aluminum forging alloy EN-AW 6082. The graph distinguishes between four distinct modes of loading a beam-shaped part, taking into account the relevant material properties (for other geometries and loading modes, other material indices apply). Although these data depend somewhat on the specific assumptions and on the actual property values, this basic comparison based on Ashby's material indices clearly demonstrates that benefits are anticipated for strength-related and especially for bending-relevant loading situations. For the former category, it also appears that substitution of this particular aluminum alloy only makes sense when high-strength magnesium alloys such as AZ80A and ZK60A are considered.

Figure 3.21 shows a number of sample parts that were developed in the context of a European Community FP6 research project on magnesium forging (MagForge;

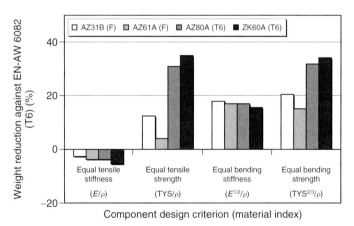

Figure 3.20 Weight-saving potential of magnesium over aluminum for some typical wrought alloys (tempers) and beam-loading situations; E = Young's modulus, TYS = tensile yield stress, and ρ = density.

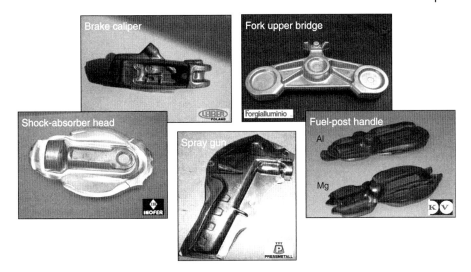

Figure 3.21 Forged AZ80A and ZK60A sample parts. (Courtesy of the MagForge project consortium.)

2006–2009). These were manufactured in an industrial setting to explore the forgeability and mechanical performance of some commercial-grade magnesium alloys; some parts are still with the flash, while for others it has been trimmed off [17]. All of these are small- or medium-sized forgings, with a length between 0.07 and 0.25 m, and were originally designed in aluminum. Apart from the different part shapes and complexities, the individual cases differed in terms of the used equipment (notably press type and slug/tool heating), forging sequence (single- vs multiple-stroke operation), and other forging parameters (temperatures and speeds). The forging trials with these sample parts confirmed that thermal control is essential for obtaining sound products and also that AZ80A is more critical in this respect than ZK60A.

3.5.2
Case 2 Cast Products: Injection Molding

Magnesium castings are now being used primarily for non-crash-relevant applications because of the porosity and associated toughness limitations of HPDCs. With one of the trends in car-vehicle development being toward large light-metal castings to be used in place of conventional multipiece stamped and welded body structures, this poses the question if advanced casting processes could be an enabling approach. In particular, the challenge is to achieve ductility levels and a consistency of mechanical properties for such ultralarge castings that are beyond the capabilities of conventional HPDC. The following case describes a successful effort in this direction on a prototyping level.

Figure 3.22 shows an inner front fender (also colloquially called a *shotgun*) that was developed in the context of a United States Automotive Materials Partnership

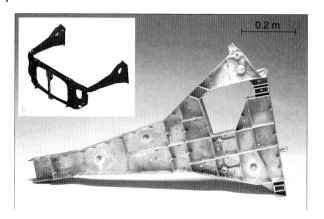

Figure 3.22 Injection-molded AM60B inner front fender with front-end structure (inset); overall dimensions 0.90 m × 0.35 m × 0.10 m, wall thickness ∼3–5 mm, and weight ∼2.9 kg.[1] (Courtesy of Ford Motor Company, Dearborn, MI, USA.)

(USAMP) / United States Department of Energy (DOE) research project on ultralarge castings for lightweight vehicle structures (ULC; 2004–2009). With a single one of these castings on either side, it is part of the three-piece front end of the vehicle where it serves to mount the radiator support to the passenger cell [38]. The component is manufactured with injection molding using the ductile alloy AM60B, which was especially demanding because of the sheer size of the casting and the generally restricted temperature range for semisolid processing of this alloy. The design and analysis of the fender were guided by functional testing of prototypes on a component as well as a system level, from which it appeared that it is possible to produce castings of acceptable quality with mechanical properties including ductility that meet the requirements for the primary vehicle structure. Moreover, it was demonstrated that the three ultralarge castings in the magnesium front end can replace a conventional multipiece steel and plastic front-end assembly for a 67% weight saving.

1) This material is based on the work supported by the Department of Energy National Energy Technology Laboratory under Award Number DE-FC26-02OR22910. This report was prepared as an account of work sponsored by an agency of the United States Government. Neither the United States Government nor any agency thereof, nor any of their employees, makes any warranty, express or implied, or assumes any legal liability or responsibility for the accuracy, completeness, or usefulness of any information, apparatus, product, or process disclosed, or represents that its use would not infringe privately owned rights. Reference herein to any specific commercial product, process, or service by trade name, trademark, manufacturer, or otherwise does not necessarily constitute or imply its endorsement, recommendation, or favoring by the United States Government or any agency thereof. The views and opinions of authors expressed herein do not necessarily state or reflect those of the United States Government or any agency thereof. Such support does not constitute an endorsement by the Department of Energy of the work or the views expressed herein.

3.6
Summary and Outlook

From the previous sections, it is clear that the use of magnesium for transport applications has seen a pronounced growth over the past decades, which is expected to persist for the foreseeable future. This expansion is primarily driven by the substitution of ferrous and aluminum components in an effort to reverse the upward trend in passenger car weight. In view of the differences in density and mechanical properties, an effective weight saving of 15–25% for magnesium structural parts as compared to their aluminum counterparts has been demonstrated for many actual redesigns.

Magnesium die castings have meanwhile become established practice and are currently even the preferred solution for certain interior automotive components. Their shaping possibilities (that provide options to reduce part count over conventional assembly designs) and cost-effectiveness definitely have added to this. Following the introduction of improved elevated-temperature alloys, die castings are meanwhile also established for drivetrain components. Besides automotive applications, castings are used to some extent in a diversity of other transport applications including sports and leisure, and military and (aero)space. Wrought magnesium applications are still rare despite their favorable mechanical properties and complementary shaping possibilities. At present, their costs generally are high as compared to designs in competing materials because of the limited size of scale, productivity, and industrial experience, which can only be offset for specific niche products.

Future developments will remain to be driven by cost-related issues in addition to product performance in a wide sense. This will be especially so for automotive applications. Some directions for further advancement are outlined next, distinguishing between alloys, processes, and products but noting that overlaps between these areas occur.

As for the alloys (including magnesium-based composites), Figure 3.23 presents a strategy for their medium- to long-term development. Depending on the demands that have to be met, distinct routes are anticipated that differ in alloying constitution and follow the general trend toward more complicated systems (from binary/ternary to quaternary and beyond). One particular direction relates to the commonly undesired directionality of mechanical properties in wrought products, to be decreased by the addition of alloying elements (such as rare earths) that randomize textures during thermomechanical treatments such as extrusion. Another direction relates to enhanced creep resistance for casting applications, which revolves around the exploration of alloying elements that retain stable precipitates in magnesium alloys at yet higher temperatures than current practice.

As for the processes, it is foreseen that industrial production will remain to be based on those that are implemented or considered today. Emphasis for further development will be on productivity increase, process window extension, and product quality improvement. For wrought products, conventional direct extrusion, forging, and (warm) deep drawing will be further tailored for magnesium, while some more exotic processes may find niche applications (e.g., hydrostatic extrusion for tube). A more revolutionary development has already started for sheet processing, where twin-roll casting may relieve current cost constraints for thin-gauge sheet and in that regard may serve as a truly enabling technology. For cast products, HPDC and sand casting will remain the backbone of production, although semisolid casting and especially injection molding may make an entrance for some automotive components as well. One particular aspect is the replacement of SF_6 for melt protection with (eventually nonfluorinated) alternatives.

As for the products, the use for mass-produced body and chassis components needs still to be conquered, while automotive applications for the interior and drivetrain can further be extended. Notably, in these first-mentioned areas, a substantial number of wrought products (sheet for the body and extrusions and forgings for the chassis) are foreseen. Corrosion protection is a critical factor for these products as well. Another particular aspect relates to recycling of postconsumer scrap, where the further tightening of legislation is triggering developments in (recycling) alloys and processes. Finally, new application areas may emerge such as for vehicle armor, based on magnesium's specific ballistic behavior (which further remained unaddressed in this chapter).

If and to what extent magnesium will evolve from its current modest role in the automotive material mix to a mainstream construction material will depend on the accomplishment of the outlined developments and also on the advances in competing materials (specifically aluminum and plastics). Apart from that, critical

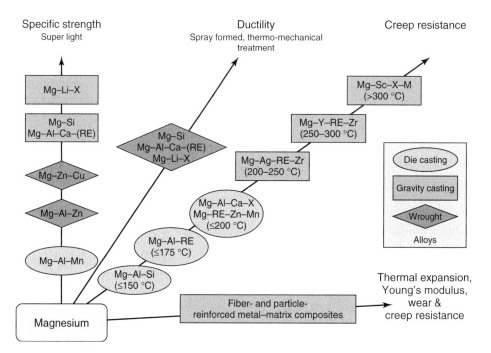

Figure 3.23 Vision on the further development of magnesium-based materials.

factors for success will be the augmentation of a reliable and competent supply base for magnesium alloys with environmentally compatible primary production and stable pricing, and the wider adoption of the material by the (supplying) industry. Geographical diversification would be in order with most primary magnesium and crucial rare earth alloying elements presently coming from China. The prospects for capitalizing on the assets of magnesium, however, will remain promising for the long term in the context of the societal issues of mobility, energy, and environmental concerns.

3.7
Further Reading

The successful utilization of magnesium as a construction material for lightweight transport components entails many different aspects, which go well beyond the descriptions in this chapter. This section gives some guidance to those readers who are after additional and/or in-depth information on the subject.

Probably the single most renowned handbook on magnesium was published by Beck [4] in 1939 with an English translation appearing shortly afterward [39], while a second printing in German was issued in 2001 [40]. Despite the entirely different spirit of the time and technological setting, many scientists and engineers still find inspiration in the detailed descriptions of, in particular, the metallurgy and the physical and mechanical properties, as well as in the processing and early application examples. Some further outstanding early works are those by Raynor [41] and Emley [42]. The handbook by Busk [43] that was published in 1987 takes the perspective of the product designer. It is a valuable and practical source of fabrication data (machining, joining, and forming) and especially the as-fabricated mechanical properties for the different categories of products (sand castings, permanent-mold castings, extrusions, sheet and plate, and forgings). Extensive practical information is also collected in the *ASM Specialty Handbook* on magnesium that appeared in 1999 [12]. This reference work deals with virtually all aspects associated with metallurgy and alloys, fabrication, finishing and inspection, engineering properties and performance, as well as with product design and alloy datasheets. An introductory book with an emphasis on materials and processes was edited by Kainer [44]; released in the German language in 2004, it also appeared in an English version.

During the past few years, some further books on magnesium technology were published that deal with the more recent progress. The contribution edited by Friedrich and Mordike [37] covers the full spectrum of magnesium-related topics ranging from primary production to recycling. As an update to the earlier mentioned handbooks, it primarily targets the metallurgy, design data, and applications and provides ample references of its contents to the underlying research papers. Die casting of both liquid and semisolid melts is dealt with in the book by Kaufmann and Uggowitzer [33], with its scope encompassing aluminum as well as magnesium alloys. The book by Czerwinski [34] specifically deals with magnesium injection molding; however, it also includes an elaborate introduction with a general overview of alloys and processes including surface protection. Most recently, a triptych of related book titles was published on welding and other joining methods [45], corrosion and its control [46], and wrought alloys, processes, and applications [47].

Promotion of the magnesium business is actively undertaken by the IMA. Dissemination channels include their web site (with general information on magnesium, highlighting of new applications, and a buyers' guide) [1] and regular events such as safety and end-user seminars. Leading-edge scientific and technological information can be gathered from the (proceedings of the) international conferences and symposia that are recurrently organized. Prominent examples of these are the three-yearly series of the "International Conference on Magnesium Alloys and Their Applications" (that in 2012 has seen its 9th edition) and the TMS Annual Meeting series of the "Magnesium Technology" symposia (in 2012 in its 13th edition). IMA organizes the industrially oriented "Annual World Magnesium Conference" (in 2012 in its 69th edition). Other conference series are "Light Metal Technology", "Magnesium Science, Technology and Applications", and "Magnesium in the Global Age". A comprehensive on-line listing of research papers on magnesium is compiled and kept up to date by Pacific Northwest National Laboratory, USA [48].

References

1. International Magnesium Association *www.intlmag.org* (accessed 17 February 2012).

2. Benedyk, J.C. (2008) Magnesium: Where have we been, where are we going? *Magnesium Article Archive: May 1943–June 2008*, Light Met. Age, 1–18.

3. Patzer, G. (2010) The magnesium industry today: The global perspective., *Magnesium Technology 2010*, TMS, pp. 85–90, ISBN: 978-0-87339-746-9.

4. Beck, A. (ed.) (1939) *Magnesium und seine Legierungen*, Verlag von Julius Springer, Berlin.

5. Friedrich, H. and Schumann, S. (2000) The second age of magnesium: Research strategies to bring the automotive industry's vision to reality, *Proceedings of the Second Israeli International Conference*, Magnesium Research Institute, pp. 9–18.

6. Luo, A. (2002) Magnesium: Current and potential automotive applications. *JOM*, **54**(2), 42–48.

7. Watarai, H. (2006) Trend of research and development for magnesium alloys: Reducing the weight of structural materials in motor vehicles. *J. Sci. Technol. Trends, Q. Rev.*, **18**, 84–97.

8. Deetz, J. (2005) The use of wrought magnesium in bicycles. *JOM*, **57**(5), 50–53.

9. Gwynne, B.A. (2010) Magnesium alloys in aerospace applications: Flammability testing and results, *Magnesium Technology 2010*, TMS, p. 13, ISBN: 978-0-87339-746-9.

10. Mathaudhu, S.N. and Nyberg, E.A. (2010) Magnesium alloys in U.S. military applications: Past, current and future solutions, *Magnesium Technology 2010*, TMS, pp. 27–32, ISBN: 978-0-87339-746-9.

11. N.N. (2006) Magnesium Vision 2020: A North American Automotive Strategic Vision for Magnesium, United States Automotive Materials Partnership (USAMP), *www.uscar.org* (accessed 17 February 2012).

12. Avedesian, M.M. and Baker, H. (eds) (1999) *Magnesium and Magnesium Alloys – ASM Specialty Handbook*, ASM

International, Materials Park, OH, ISBN: 0-87170-657-1.

13. Aghion, E., Bronfin, B., Friedrich, H., and Rubinovich, Z. (2004) The environmental impact of new magnesium alloys on the transportation industry, *Magnesium Technology 2004*, TMS, pp. 167–172, ISBN: 0-87339-568-9.

14. N.N. (2010) Elektron 675 Preliminary Data – Datasheet 102, Magnesium Elektron UK, Manchester, *www.magnesium-elektron.com* (accessed 17 February 2012).

15. Luo, A.A. and Sachdev, K. (2007) Development of a new wrought magnesium-aluminium-manganese alloy AM30. *Metall. Mater. Trans. A*, **38A**, 1184–1192.

16. Sillekens, W.H. and Van der Linden, D.C.W. (2007) Quality-affecting parameters in the direct extrusion of magnesium alloys, *Magnesium Technology 2007*, TMS, pp. 151–156, ISBN: 978-0-87339-663-9.

17. Sillekens, W.H., Kurz, G., and Werkhoven, R.J. (2010) Magnesium forging technology: State-of-the-art and development perspectives, *Proceedings of the 11th International Aluminium Conference, INALCO'2010 'New Frontiers in Light Metals'*, IOS Press, pp. 329–337, ISBN: 978-1-60750-585-3.

18. Park, S.S., Park, W.J., Kim, C.H., You, B.S., and Kim, N.J. (2009) The twin-roll casting of magnesium alloys. *JOM*, **61**(8), 14–18.

19. Krajewski, P.E., Friedman, P.A., and Singh, J. (2011) The warm forming performance of Mg sheet materials, *Magnesium Technology 2011*, Wiley–TMS, pp. 395–396, ISBN: 978-1-11802-936-7.

20. Stutz, L., Bohlen, J., Letzig, D., and Kainer, K.U. (2011) Formability of magnesium sheet ZE10 and AZ31 with respect to initial texture, *Magnesium Technology 2011*, Wiley–TMS, pp. 373–378, ISBN: 978-1-11802-936-7.

21. Kim, S.K. (2011) Proportional strength-ductility relationship of non-SF6 diecast AZ91D eco-Mg alloys, *Magnesium Technology 2011*, Wiley–TMS, pp. 131–136, ISBN: 978-1-11802-936-7.

22. N.N. (2006) Elektron WE43 – Datasheet 467, Magnesium Elektron UK, Manchester, *www.magnesium-elektron.com* (accessed 17 February 2012).

23. Pekguleryuz, M., Labelle, P., Argo, D., and Baril, E. (2003) Magnesium diecasting alloy AJ62X with superior creep resistance, ductility and diecastability, *Magnesium Technology 2003*, TMS, pp. 201–206, ISBN: 0-87339-533-6.

24. Dead Sea Magnesium Ltd MRI-153M & MRI-230D Alloys, Dead Sea Magnesium Ltd, Beer Sheva, *www.dsmag.co.il* (accessed 17 February 2012).

25. Luo, A.A. (2003) Recent magnesium alloy development for automotive powertrain applications. *Mater. Sci. Forum*, **419–422**, 57–66.

26. Dead Sea Magnesium Ltd MRI-201S & MRI-202S Alloys, Dead Sea Magnesium Ltd, Beer Sheva, *www.dsmag.co.il* (accessed 17 February 2012).

27. N.N. (2006) Elektron 21 – Datasheet 455, Magnesium Elektron UK, Manchester, *www.magnesium-elektron.com* (accessed 17 February 2012).

28. N.N. (2006) AM-SC1 Sand Casting Alloy for High Temperature Power Train – Technical Information Sheet, Advanced Magnesium Technologies, Milton, *www.am-technologies.com* (accessed 17 February 2012).

29. Barth, A. (2004) Mg application in the 7G Tronic gear, *Proceedings of the 61st Annual World Magnesium Conference*, IMA, pp. 81–86.

30. Carpenter, J.A. Jr., Jackman, J., Li, N., Osborne, R.J., Powell, B.R., and Sklad, P. (2007) Automotive Mg research and development in North America. *Mater. Sci. Forum*, **446–449**, 11–24.

31. Gibson, M.A., Bettles, C.J., Murray, M.T., and Dunlop, G.L. (2006) AM-HP2: A new magnesium high pressure diecasting alloy for automotive powertrain applications, *Magnesium Technology 2006*, TMS, pp. 327–331, ISBN: 978-0-87339-620-2.

32. FRECH *www.frech.com* (accessed 17 February 2012).

33. Kaufmann, H. and Uggowitzer, P.J. (2007) *Metallurgy and Processing of*

High-Integrity Light Metal Pressure Castings, Schiele & Schön, Berlin, ISBN: 978-3-7949-0754-0.

34. Czerwinski, F. (2008) *Magnesium Injection Molding*, Springer, New York, ISBN: 978-0-387-72399-0.

35. Frank, H., Hort, N., Dieringa, H., and Kainer, K.U. (2008) Influence of processing route on the properties of magnesium alloys. *Solid State Phenom.*, **141–143**, 43–48.

36. Landershammer, R. and Habelsberger, W. (2001) Leitfaden für einen sicheren Umgang mit Aluminium und Magnesium, Land Oberösterreich, Linz.

37. Friedrich, H.E. and Mordike, B.L. (eds) (2006) *Magnesium Technology: Metallurgy, Design Data, Applications*, Springer-Verlag, Berlin, Heidelberg, ISBN: 978-3-540-20599-9.

38. Sadayappan, K. and Vassos, M. (2010) Evaluation of a Thixomolded Magnesium Alloy Component for Automotive Application. SAE Technical Paper 2010-01-0403, SAE International.

39. Beck, A. (ed.) (1940) *The Technology of Magnesium and its Alloys*, F.A. Hughes & Co., Ltd., London.

40. Beck, A. (ed.) (2001, second printing) *Magnesium und seine Legierungen*, Klassiker der Technik, Springer, Berlin (in German), ISBN: 978-3-54041-675-3.

41. Raynor, G.V. (1959) *The Physical Metallurgy of Magnesium and its Alloys*, Pergamon Press, Ltd., London.

42. Emley, E.F. (1966) *Principles of Magnesium Technology*, Pergamon Press, Ltd., Oxford.

43. Busk, R.S. (1987) *Magnesium Products Design*, Marcel Dekker, New York, Basel, ISBN: 0-8247-7576-7.

44. Kainer, K.U. (ed.) (2004) *Magnesium: Alloys and Technology*, Wiley-VCH Verlag GmbH, Weinheim (originally issued in German as Magnesium: Eigenschaften, Anwendungen, Potenziale), ISBN: 3-527-29979-3.

45. Liu, L. (ed.) (2010) *Welding and Joining of Magnesium Alloys*, Woodhead Publishing, Ltd., Cambridge, ISBN: 978-1-84569-692-4.

46. Song, G.L. (ed.) (2011) *Corrosion of Magnesium Alloys*, Woodhead Publishing, Ltd., Cambridge, ISBN: 978-1-84569-708-2.

47. Bettles, C. and Barnett, M. (eds) (2012) *Advances in Wrought Magnesium Alloys: Fundamentals of Processing, Properties and Applications*, Woodhead Publishing, Ltd., Cambridge, ISBN: 978-1-84569-968-0.

48. Magnesium Research and Development Publication Database, *www.magnesium.pnl.gov* (accessed 17 February 2012).

4
Titanium and Titanium Alloys

Lothar Wagner and Manfred Wollmann

4.1
Introduction

Rutile TiO_2 and ilmenite $FeTiO_3$ are the two primary minerals that contain titanium and make up 24% of the Earth's crust. Hence, titanium is the ninth most abundant (0.6%) element on the planet and after iron, aluminum, and magnesium the fourth most abundant structural metal. Nevertheless, it was discovered as a metal in as early as 1791 by the British clergyman and amateur mineralogist William Gregor. It was independently discovered by the German chemist M.H. Klaproth in 1793. Klaproth named the new metal titanium after the titans of Greek mythology. He discovered four years later that his titanium was the same as Gregor's newly found element. However, the element was not successfully isolated until 1910. Many attempts were made to isolate the metal from the titanium ore using titanium tetrachloride ($TiCl_4$) as an intermediate step. While the Hunter process named after Matthew A. Hunter uses Na to reduce $TiCl_4$, the famous Kroll process named after William J. Kroll from Luxembourg uses Mg in an inert gas atmosphere. This process is the dominant process for titanium production today [1].

Titanium has a low density (4.5 g cm^{-3}), approximately 60% less than that of iron. It is nonferromagnetic. Its melting point of 1668 °C is higher than that of iron. It has a pronounced high passivity, which gives rise to its excellent corrosion resistance to most mineral acids and chlorides. The outstanding properties of titanium are its inherent corrosion resistance and high specific strength. Various alloys of titanium can be adjusted by specific thermomechanical treatments to almost any desired combination of properties.

Despite the excellent properties of titanium alloys, such as superior corrosion resistance in chloride-containing environments and high specific strength, the high production cost of titanium has limited its application in transport [1]. While titanium alloys have found wide application in aerospace, particularly in components of the jet engine, so far only limited usage is found in automobiles. Another well-known application of titanium is in the marine area, for example, in submarines. Although titanium alloys could be used in the powertrain, chassis, and the bodywork of automobiles, up to now, titanium is

Structural Materials and Processes in Transportation, First Edition.
Edited by Dirk Lehmhus, Matthias Busse, Axel S. Herrmann, and Kambiz Kayvantash.
© 2013 Wiley-VCH Verlag GmbH & Co. KGaA. Published 2013 by Wiley-VCH Verlag GmbH & Co. KGaA.

incorporated only in luxury or sports cars. A wider distribution of titanium in mass production automotive applications would require drastic reduction in its price. However, a cost-effective manufacturing process for titanium is not yet in sight [2, 3].

4.2
Fundamental Aspects

4.2.1
Phase Diagrams and Alloy Classes

Depending on the temperature, the crystal structure of pure titanium is either hexagonal (α-phase) or body-centered cubic (β-phase). The transition temperature from the α-phase to the β-phase at $T = 882\,°C$ is called β-*transus temperature* (Figure 4.1).

The β-transus temperature is increased by α-stabilizing elements such as Al, O, N, and C or decreased by β-stabilizing elements such as V, Mo, and Fe (Figure 4.2). Furthermore, elements such as Zr and Sn are called *neutral* because they do not change the β-transus temperature.

The various titanium alloy classes are usually illustrated in a pseudobinary phase diagram where the temperature is plotted versus the β-stabilizer content at a fixed Al-concentration (Figure 4.3). Above a certain amount of β-stabilizing elements, a two-phase (α + β) region exists. As illustrated in Figure 4.3, a martensite's start temperature (M_s) runs through this (α + β) phase region indicating that the β-phase will martensitically transform at sufficiently high cooling rates. Alloys in the (α + β) phase region that do not undergo a martensitic transformation because they are

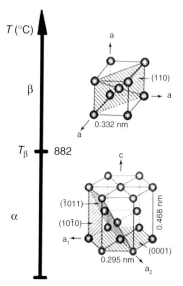

Figure 4.1 Allotropic phase transformation in pure titanium.

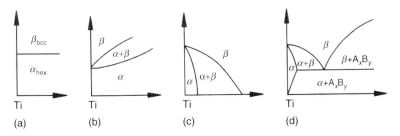

Figure 4.2 Typical binary phase diagrams of titanium alloys (schematically): (a) neutral, (b) α-stabilizing, (c) β-isomorphous, and (d) β-eutectoid [4].

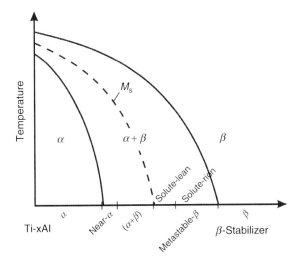

Figure 4.3 Pseudobinary section through a β-isomorphous phase diagram (schematically).

located on the right side of the position where M_s approaches room temperature are called *metastable β-alloys*. Metastable β-alloys are divided into solute-lean alloys if they can precipitate out primary α-phase similar to (α + β) or near-α alloys. As opposed to solute-lean alloys, solute-rich metastable β-alloys do not contain primary α. Stable β-alloys are located even more to the right in the single β-phase region. However, these alloys do not exist as commercial materials. Therefore, the metastable β-alloys are often simply called *β-alloys*.

Selected commercial titanium alloys of the various alloy classes are listed in Table 4.1.

4.2.2
Hardening Mechanisms

Depending on the alloy class, one or more of the basic strengthening mechanisms in metallic materials can be used to improve the strength of the particular material.

Table 4.1   Titanium alloy classes and selected commercial alloys.

(a)

cp-Ti and α-alloys	Composition
Grade 1	Ti–0.2Fe–0.18O
Grade 2	Ti–0.3Fe–0.25O
Grade 3	Ti–0.3Fe–0.35O
Grade 4	Ti–0.5Fe–0.40O
Ti–2.5Cu	Ti–2.5Cu
Ti-3-2.5	Ti–3Al–2.5V

(b)

Near-α and α + β alloys	Composition
Ti-6242	Ti–6Al–2Sn–4Zr–2Mo–0.1Si
Ti-834	Ti–5.8Al–4Sn–3.5Zr–0.5Mo–0.7Nb–0.35Si–0.06C
Ti–6Al–4V	Ti–6Al–4V–0.2O
Ti–6Al–4V ELI	Ti–6Al–4V–0.1O
Ti-662	Ti–6Al–6V–2Sn
Ti-550	Ti–4Al–2Sn–4Mo–0.5Si

(c)

Metastable β-alloys	Composition
Ti-6246	Ti–6Al–2Sn–4Zr–6Mo
Ti-10-2-3	Ti–10V–2Fe–3Al
Ti-LCB	Ti–4.5Fe–6.8Mo–1.5Al
Ti-15-3	Ti–15V–3Cr–3Al–3Sn
Beta C	Ti–3Al–8V–6Cr–4Mo–4Zr
B120VCA	Ti–13V–11Cr–3Al

4.2.3
cp-Ti and α-Alloys

Grain refinement, solid solution hardening by oxygen and dislocation strengthening by cold work can be used to strengthen the materials of this alloy group. The effect of swaging on the reduction in grain size of cp-Ti grades 1 and 2 is illustrated in Figure 4.4 and Figure 4.5, respectively.

The effect of the deformation degree on yield stress of cp-Ti grades 1 and 2 (RT 15) in swaging is illustrated in Figure 4.6. The oxygen level in grade 2 being higher than in grade 1 results in a shift of the yield stress–deformation curve to stresses roughly 100 MPa higher (Figure 4.6). The very marked increase in yield stress with deformation degree in swaging is due to the observed pronounced grain refinement through severe plastic deformation (Figures 4.4b and 4.5b) and dislocation strengthening as well.

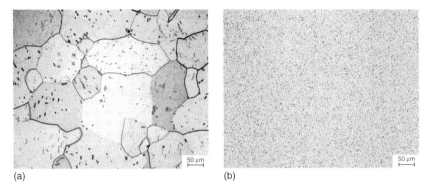

(a)   (b)

Figure 4.4 Effect of rotary swaging ($\varphi = 3.2$) at room temperature on grain refinement in cp-Ti (grade 1): (a) before and (b) after.

(a)   (b)

Figure 4.5 Effect of rotary swaging ($\varphi = 2.5$) on grain refinement in cp-Ti (grade 2): (a) before and (b) after. (Source: R. Wan and L. Wagner, TU Clausthal, unpublished results.)

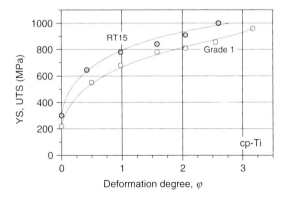

Figure 4.6 Yield stress (YS) versus deformation degree in rotary swaging of cp-Ti grades 1 and 2 (RT 15). (Source: R. Wan and L. Wagner, TU Clausthal, unpublished results.)

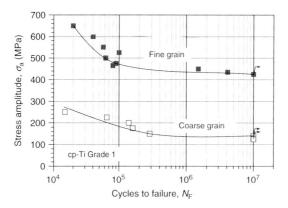

Figure 4.7 *S–N* curves in rotating beam loading (*R* = −1) of cp-Ti grade 1. (Source: R. Wan and L. Wagner, TU Clausthal, unpublished results.)

The high cycle fatigue (HCF) performance of the severe plastically deformed (φ = 3.2) condition is compared with the coarse grained reference of cp-Ti grade 1 in Figure 4.7. As seen, there is a drastic increase in the fatigue strength at 107 cycles from about 140 to 425 MPa [4].

4.2.4
Near-α and α + β Alloys

In addition to the aforementioned hardening mechanisms, precipitation hardening by Ti_3Al coherent particles in the α-phase and by fine secondary α particles in the β-phase can be achieved. Basically, three different microstructures can be generated, namely, fully lamellar, fully equiaxed, and duplex (or bimodal) [5, 6]. While fully lamellar structures are produced simply by cooling from the β-phase field, both fully equiaxed and bimodal microstructures need to be deformed first in the α + β phase field because they require recrystallization.

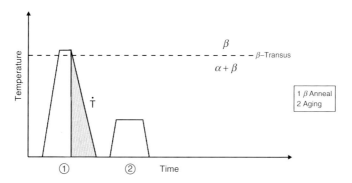

Figure 4.8 Thermal processing sequence for producing fully lamellar microstructures (schematically).

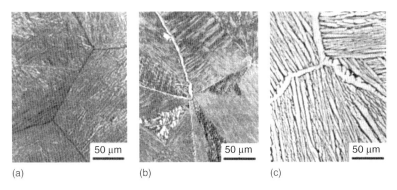

(a)   (b)   (c)

Figure 4.9 Resulting fully lamellar microstructures in Ti–6Al–4V: (a) water quenched (WQ), (b) air cooled (AC), and (c) furnace cooled (FC).

The thermal processing route for generating fully lamellar microstructures in Ti–6Al–4V is illustrated in Figure 4.8.

Depending on the rate of cooling from the β-phase field, fine lamellar (martensitic), lamellar, or coarse lamellar microstructures can be achieved (Figure 4.9).

While the yield stress strongly depends on the width of the α lamellae, their length determines tensile ductility (Table 4.2).

The HCF performance of the various lamellar microstructures is illustrated in Figure 4.10. As seen, the HCF strength is very much affected by the width of the alpha plate thickness.

The processing route for generating fully equiaxed microstructures is illustrated in Figure 4.11.

Depending on the holding time at the recrystallization temperature (Figure 4.12), fine equiaxed or coarse equiaxed microstructures with different mechanical properties can be achieved (Table 4.3).

Table 4.2 Tensile properties of fully lamellar microstructure.

Cooling rate ($^\circ$C min^{-1})	Yield stress (MPa)	ε_F
8000	1040	0.20
400	980	0.25
1	935	0.15

Table 4.3 Tensile properties of fully equiaxed microstructures.

Microstructures	Annealing time (h)	Yield stress (MPa)	ε_F
Fine equiaxed	1	1170	0.55
Coarse equiaxed	100	1075	0.38

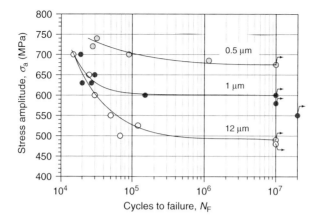

Figure 4.10 *S–N* curves of fully lamellar microstructures in Ti–6Al–4V in rotating beam loading (*R* = −1) [6].

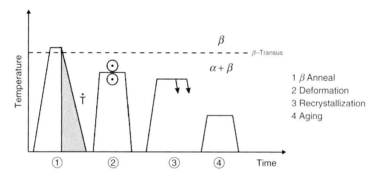

Figure 4.11 Thermomechanical processing sequence for producing fully equiaxed microstructures (schematically).

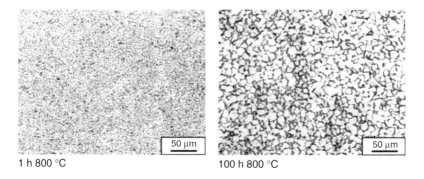

1 h 800 °C   100 h 800 °C

Figure 4.12 Resulting fully equiaxed microstructures in Ti–6Al–4V.

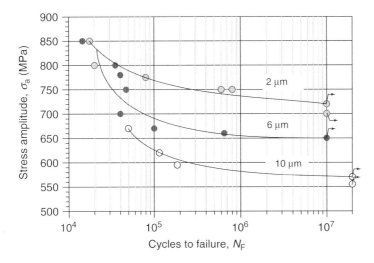

Figure 4.13 S–N curves of fully equiaxed microstructures in Ti–6Al–4V in rotating beam loading ($R = -1$) [6].

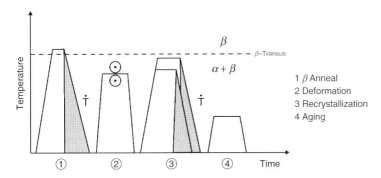

Figure 4.14 Thermomechanical processing sequence for producing duplex microstructures (schematically).

Decreasing the α grain size from about 12 to 2 μm not only increases the yield stress by about 100 MPa, but also markedly enhances tensile ductility (Table 4.3) as well as HCF strength (Figure 4.13).

The processing route for generating duplex microstructures is illustrated in Figure 4.14.

The main difference to the thermomechanical processing sequence for equiaxed microstructures is that the recrystallization annealing for duplex microstructures is performed at temperatures much closer to the β-transus temperature, where higher volume fractions of β-phase are present (Figure 4.14).

Similarly, as observed in the fully lamellar microstructures (Figure 4.9), increasing the cooling rate markedly increases the yield stress of the duplex structures

Table 4.4 Process parameters and tensile properties of duplex microstructures in Ti–6Al–4V.

Microstructure	Cooling rate ($^\circ$C min^{-1})	Yield stress (MPa)	ε_F
D20/WQ	8000	1050	0.40
D40/WQ	8000	1045	0.50
D40/AC	400	975	0.42

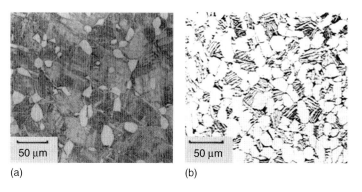

(a)   (b)

Figure 4.15 Resulting duplex microstructures in Ti–6Al–4V: (a) D20/WQ and (b) D40/WQ.

(Table 4.4). On the other hand, the tensile ductility increases with the volume fraction of primary α-phase. This is because the length of the α lamellae and the packet size decrease with the increasing volume fraction of the primary α-phase (Figure 4.15a,b).

Figure 4.16 indicates the effect of the width of the α lamellae in the lamellar portion of the duplex microstructure on HCF strength.

4.2.5
Metastable β-Alloys

Age-hardening effects are much more pronounced in this alloy class than in α or $\alpha + \beta$ alloys. These types of alloys can be categorized as either solute-rich or solute-lean. Classification criterion is the content of alloying elements. The thermomechanical processing window for the solute-lean alloys is critically related to the development of the necessary microstructure. Solute-rich alloys are too stable to decompose isothermally in different phases (β- and Ω-phase mixtures). The metastable Ω-phase may form in the solute-lean β-alloys.

A typical processing sequence for generating various microstructures in a solute-lean alloy is illustrated in Figure 4.17.

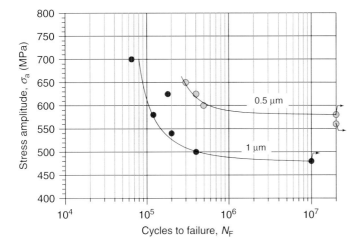

Figure 4.16 *S–N* curves of duplex microstructures (40% α_p volume fraction) in Ti–6Al–4V [6].

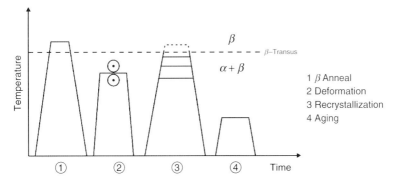

Figure 4.17 Processing sequence for generating various microstructures in Ti-10-2-3.

Characteristic microstructures of the alloy Ti–10V–2Fe–3Al (Ti-10-2-3) are shown in Figure 4.18.

Ti-10-2-3 was developed primarily for high-strength applications. This alloy provides weight savings over steels in airframe forging applications. It should be emphasized that this material possesses the best hot-die forgeability compared to any commercial titanium alloy. Ti-10-2-3 also offers pronounced strength/toughness combinations and is deep-hardenable.

Table 4.5 demonstrates the yield stress/primary α-phase relationship.

The influence of primary α content in Ti-10-2-3 on the fatigue performance is illustrated in Figure 4.19. The poor HCF strength of the condition with 0% primary α content is caused by a soft and very thin α_p layer along the β-grain boundaries.

A typical processing sequence for generating various microstructures in a solute-rich alloy is illustrated in Figure 4.20.

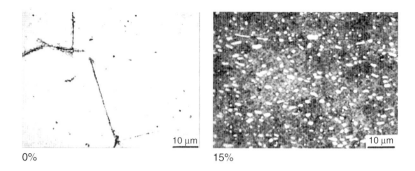

0%   15%

Figure 4.18 Typical microstructures of Ti-10-2-3: (a) 0% α_p content and (b) 15% α_p content.

Table 4.5 Process parameters and tensile properties of Ti-10-2-3.

α_p (%)	Aging treatment	Yield stress (MPa)	ε_F
0	8 h 480 °C	1555	0.02
5	8 h 480 °C	1370	0.09
15	8 h 480 °C	1330	0.11
30	8 h 480 °C	1195	0.25

Figure 4.21 demonstrates the way in which the microstructure is being affected by means of a modified aging treatment (compare Figure 4.21a,b). Simplex aging leaves soft precipitate free zones inside the material. These zones are prefered sites for crack nucleation [20].

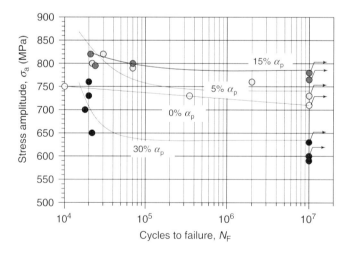

Figure 4.19 Fatigue performance of Ti–10V–2Fe–3Al with various primary α contents [6].

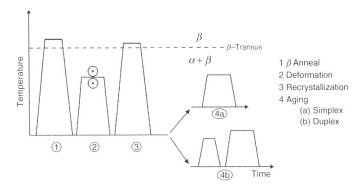

Figure 4.20 Processing sequence for generating various microstructures in Ti–3Al–8V–6Cr–4Mo–4Zr (Beta C).

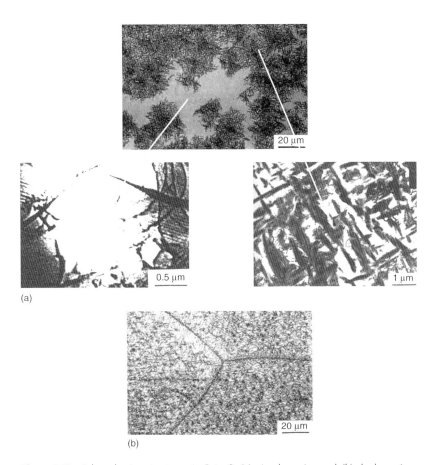

Figure 4.21 Selected microstructures in Beta C: (a) simplex aging and (b) duplex aging.

Table 4.6 Heat treatment and tensile properties of Beta C.

Aging	Yield stress (MPa)	ε_F
16 h 540 °C	1085	0.26
4 h 440 °C + 16 h 560 °C	1085	0.27

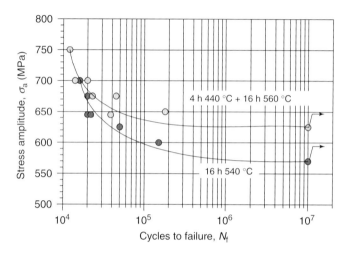

Figure 4.22 *S–N* curves of Ti–3Al–8V–6Cr–4Zr–4Mo (Beta C) [6].

While yield stress and ductility are hardly affected by using duplex instead of simplex aging (Table 4.6), the HCF strength is clearly improved (Figure 4.22).

4.2.6
Conclusions Regarding Fatigue Performance

In general, the HCF strength as a measure of the resistance to fatigue crack nucleation can be correlated quite well with the yield stress or the free slip length of dislocations in a given alloy [5]. Therefore, increasing the yield stress by the various hardening mechanisms usually also increases HCF strength. However, fatigue cracks in two-phase or heterogeneous materials tend to nucleate in microstructural areas, which are weaker than other areas. If the volume fraction of the weak microstructural constituent is low, the fraction of the HCF strength and yield stress will also be low. Typical examples can be found in duplex microstructures of Ti–6Al–4V (Figure 4.16) as well as in solute-lean and solute-rich alloys of the metastable β-alloy classes (Figure 4.19 and Figure 4.22). In addition, cyclic softening during fatigue loading as observed in metastable β-alloys may explain the noncorrelation of HCF strength in high-strength conditions with the yield stress.

4.3
Applications in Automobiles, Aerospace, and ShipBuilding

4.3.1
Automobiles

Depending on the particular requirements, commercially pure titanium or titanium alloys can be used. There are enough reasons for a potential application of titanium in the automotive industry [7]. All the technical property requirements for use in automobiles are met by titanium [8, 9]. Some examples of the use in engines are connecting rods, components for turbochargers, piston pins, and valve rods.

Every gram of weight saving is especially useful for fast-moving mass. This applies particularly to engine components such as connecting rods and piston pins.

Titanium exhaust systems is another interesting application. Not only the weight saving is a convincing reason for the usage of titanium in cars. The excellent corrosion resistance is another compelling reason to use titanium. Suspension springs made of titanium materials can be cited as another example. A car consists of three main modules: powertrain, chassis, and car body. Each module is subject to a specific stress scenario. Accordingly, the selection will be made for a specific titanium-based material.

A selection of potential applications for titanium alloys in an automobile is illustrated in Figure 4.23. In addition to the above-mentioned examples, the crankshaft and camshaft are listed. Furthermore, crash elements and reinforcements are highlighted.

These components contribute significantly to improve the safety of a modern motor vehicle.

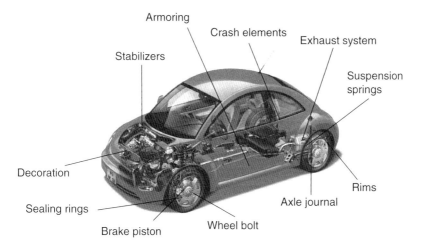

Figure 4.23 Potential applications of titanium in automotive. (Courtesy of Volkswagen.)

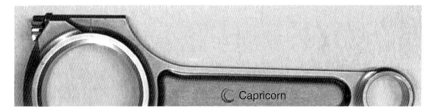

Figure 4.24 Connecting rod of Ti–6Al–4V.

4.3.1.1 **Powertrain**

Examples of components in the powertrain are connecting rods, piston pins, valves, valve springs, spring cups, camshafts, and crankshafts [3]. Many components are subjected to cyclic loading with high numbers of cycles. The alloys used must have superior HCF strength. This is precisely the case with connecting rods [8]. The required properties are met by the two-phase titanium alloy Ti–6Al–4V without problems. However, the low Young modulus of the titanium alloy needs to be compensated by a suitable design. Figure 4.24 shows a connecting rod of Ti–6Al–4V.

Gamma titanium aluminides have a higher Young's modulus and an even lower density than Ti–6Al–4V. Thus, titanium aluminides would be excellent alloys for innovative applications in this scope [2, 10]. A historical overview of the use of connecting rods made of titanium alloys by various manufacturers is illustrated in Table 4.7.

The operating temperatures in the combustion chamber of direct-injection turbo chargers have increased as a consequence of technical developments. Thus, the scope of aluminum-based piston pins is limited. These are very small and lightweight components (Figure 4.25). Yet, even here, a weight saving makes sense because this mass has to be moved at high speeds [9]. Moreover, the temperature resistance compared to aluminum is a convincing reason for the use of titanium alloys.

Turbocharger wheels (Figure 4.26) as components of the powertrain are subjected to various stresses in service. The blades have to demonstrate significant endurance

Table 4.7 Connecting rods made of titanium alloys.

Year	Material	Manufacturer	Model
1992	Ti–3Al–2V – RE	Honda	NSX
1994	Ti–6Al–4V	Ferrari	All 12 cylinder
1999	Ti–6Al–4V	Porsche	GT3
2002	Ti–6Al–4V	GM	Corvette Z06
2005	Ti–6Al–4V	Bugatti	Veyron 16.4
2006	Ti–6Al–4V	Koenigsegg	Koenigsegg
2011	Ti–6Al–4V	Porsche	GT3 RS 4.0

Figure 4.25 Piston pins made of titanium. (Courtesy of Nicoll Racing.)

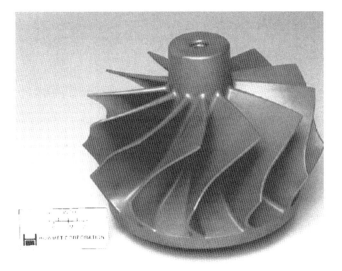

Figure 4.26 Turbocharger wheel. (Courtesy of Howmet Corporation.)

limits at very high numbers of cycles. For this application, the use of titanium aluminides is particularly suitable [10].

The transition region to the disk must have a high creep resistance.

The operating temperature is 750 °C for diesel engines and 950 °C for gasoline engines. Sufficient oxidation stability is also required. Additionally, resistance against erosion has to be considered. Table 4.8 lists early examples of turbo charger applications made of titanium alloys.

Valve rods, valve springs, and retainers may be manufactured from titanium alloys [2]. If the complete valve rod is made of titanium alloys, a weight saving of about 70% is possible, which gives rise to an increase of rpm by about 10%. Again, high fatigue strength, creep resistance, oxidation resistance, and high resistance to wear are required. For standard applications valves can be made of Ti–6Al–4V. Additionally, near-α or particle-reinforced alloys ($\alpha + \beta$) may be used in this area. Table 4.9 shows some manufacturers that use titanium alloys for this application.

Table 4.8 Turbocharger wheels of various Ti alloys and manufacturers.

Year	Material	Manufacturer	Model
1999	Ti–6Al–4V	Daimler-Benz	Truck Diesel
2000	γ-TiAl	Mitsubishi	Lancer RS

Table 4.9 Titanium alloys for valves and retainers.

Year	Components	Material	Manufacturer	Model
1998	Valves	Ti–6Al–4V	Toyota	Altezza 6-cylinder
2002	Valves	Ti–6Al–4V	Nissan	Infinity Q45
2003	Spring retainer	β-Alloy	Mitsubishi	All 1.8 l – 4- cylinder
2005	Valves	Ti–6Al–4V	GM	Corvette Z06

β-Alloys are also used in valve systems [7]. The creep resistance is lower compared to α-alloys, but is sufficient for valve spring retainers. An example can be seen in Figure 4.27.

In the luxury car segment and in motorcycles [11], cp-titanium is often used for exhaust systems. The excellent electrochemical corrosion resistance, oxidation resistance, and the technological properties such as high ductility, cold drawability, and weldability are essential for this application. In addition, the cyclic loading requires high fatigue strength. The usage of titanium in exhaust systems for motorcycles is increasing in recent years. A selection of manufacturers that use cp-titanium for their exhaust systems are listed in Table 4.10. However, such applications are optional and are not the standard equipment. For example, the car manufacturer Audi offers the purchase of titanium components from a supplier company.

It was demonstrated that the substitution of steel with titanium in the exhaust system of the Volkswagen Phaeton results in a weight saving of more than 12 kg. The use of titanium in the Golf model would give rise to a mass reduction of at

Figure 4.27 Valve spring retainer. (Courtesy of Neff Prazision.)

Table 4.10   Titanium in exhaust systems.

Year	Material	Manufacturer	Model
2001	cp-Ti grade 2	GM	Corvette Z06
2003	cp-Ti grade 2	Nissan	Fair Lady Z
2004	cp-Ti grade 2	Koenigsegg	Koenigsegg CCR
2005	cp-Ti grade 1 Al-plated	Bugatti	Veyron 16.4
2007	TIMETAL Exhaust XT	Porsche	GT 2
2008	cp-Ti grade 2	Audi	TTS
2011	cp-Ti grade 2	Porsche	Cayenne II S V8
2011	cp-Ti grade 2	Porsche	GT3 RS 4.0

Figure 4.28   Exhaust system of the Porsche GT 2: Ti–0.45Si–0.25Fe (TIMETAL Exhaust XT).

least 8 kg. Comparable mass reductions were also demonstrated by Porsche sports cars. The titanium muffler weighs about 9 kg. This is around 50% less than a similar component made of stainless steel. The advantage of weight savings can be integrated with another interesting aspect. The smaller footprint of the titanium silencer increases the distance to surrounding components and improves heat dissipation.

The exhaust system of the Porsche GT 2 made of cp-titanium with small amounts of silicon and iron is illustrated in Figure 4.28. At low operating temperatures, use of cp-titanium would be sufficient. However, for very high gas temperatures, the oxidation resistance has to be improved. At temperatures in excess of $800\,^\circ$C, severe embrittlement and a significant loss of material will occur [3]. Figure 4.29 shows the influence of temperature, reaction time, and Al-plating on the microstructure and material removal.

As shown in Figure 4.29d, Al-plating ensures effective protection against oxidation processes and the concomitant loss of material. It is feasible to impede the coarse grain growth by means of alloying measures (cerium mixed metal).

Figure 4.30 shows the influence of cerium mischmetal on the microstructure of Al-plated titanium, before (Figure 4.30a) and after exposure at high temperature

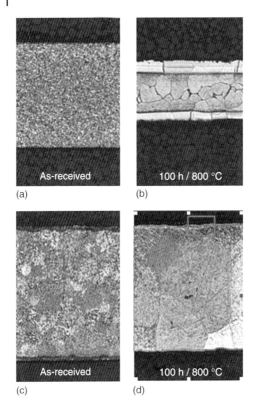

Figure 4.29 Effect of temperature on the microstructure and material loss: (a) before exposure at high temperature, (b) grain growth and material loss after exposure at 800 °C for 100 h, (c) Al-plated, before exposure at high temperature, and (d) Al-plated, coarse grain formation but no material loss after exposure at 800 °C for 100 h.

(Figure 4.30b). Despite the long time exposure at 800 °C, no marked grain growth is observed (Figure 4.30b).

4.3.1.2 Suspension

Compared to steel, titanium possesses material properties that are superior for spring production. First of all, the use of titanium gives rise to lighter springs. One important advantage of titanium alloys is their low Young's modulus, which is roughly 50% of that of steel. Thus, the stored elastic energy of titanium alloys can be much higher, provided that the alloy is tailored to high strength, particularly to high endurance limits. Because of their spring characteristics, titanium alloys can offer designers smaller suspension heights [7]. This results in lower hoodlines, better aerodynamics, and more cargo space. Metastable β-titanium alloys such as TIMETAL LCB can be formed into springs at room temperature in the low strength solution heat-treated condition. Later, aging leads to the desired high strength. A list of manufacturers that produce cars and motorcycles having titanium springs is presented in Table 4.11.

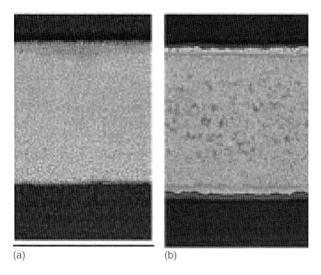

(a)   (b)

Figure 4.30 (a,b) Effect of addition of cerium mischmetal on the thermal stability of Ti–0.1Fe–0.35Si.

Table 4.11 Titanium suspension springs.

Year	Material	Manufacturer	Model
2000	TIMETAL	Volkswagen	Lupo FSI
2003	Low-cost beta titanium (LCB)	Ferrari	360 Stradale
2005		Bugatti	Veyron 16.4
2006		Yamaha	YZ450F

LCB was developed specifically for suspension springs in auto applications. A key to LCB's lower cost is the use of an iron–molybdenum master alloy during alloy production instead of adding Fe and Mo in elemental form. High strength can be realized by means of precipitation hardening. The yield stress in the age-hardened condition can be as high as 1600 MPa at a Young's modulus of 115 GPa. Figure 4.31 shows the microstructure of LCB. Primary α-grains are embedded in the β-phase that is age-hardened by fine secondary alpha (α_s) precipitates.

In summary, research and development activities for potential applications of titanium and titanium alloys in automotive engineering are making good progress. New applications have been introduced and existing applications have been improved through new compositions and new manufacturing technologies. Titanium is now established for many interesting but expensive applications in automobiles. Hence, applications are restricted. However, for mass production, more cost-efficient titanium production processes are highly needed. Unfortunately, a more economical method of production of titanium is currently not in sight.

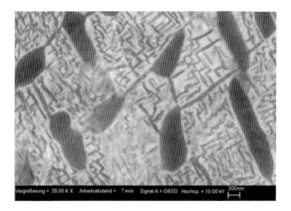

Figure 4.31 TEM microstructure of TIMETAL LCB after precipitation hardening [12].

4.3.2
Aerospace Applications

4.3.2.1 **Engines**
In gas turbine engines, Ti–6Al–4V is applied for static and rotating components. Rotating components can be made of forgings. Ti–6Al–4V is used as material for engine components up to temperatures of 350 °C. Well-known examples are fan disks (Figure 4.32) and blades.

Compared to Ti–6Al–4V, the alloy Ti–6Al–2Sn–4Zr–6Mo (Ti-6246) can be used at moderately higher temperatures up to 450 °C [5]. Ti-6246 was developed by Pratt and Whitney (P&W) [13] and provides higher strength than Ti–6Al–4V because of the additions of Sn and Zr as solid solution strengtheners. The strength is also increased because of the addition of the β stabilizer Mo, which increases the amount of the β-phase present in the alloy. In general, the possible section sizes of high strength are larger in Ti-6246 than in Ti–6Al–4V. Furthermore, Mo improves the corrosion resistance compared to Ti–6Al–4V.

Figure 4.33 illustrates a Rolls-Royce engine with components made of Ti-6246.

Figure 4.32 Fan disk made of Ti-6-4. (Courtesy of Otto Fuchs.)

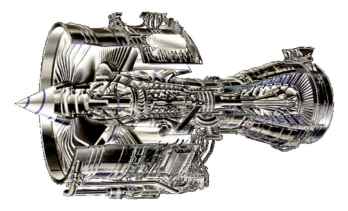

Figure 4.33 Rolls-Royce jet engine with Ti-6246 compressor disks in the intermediate and high compressor stages.

Applications of Ti-6246 are, for example, intermediate compressor stages of turbine engines for disks, blades, seals, and airframe parts. Ti-6246 can be tailored by thermomechanical treatments to a wide variation in microstructure and mechanical properties. The alloy Ti–5Al–2Sn–2Zr–4Cr–4Mo (Ti-17) was developed by General Electric as a high-strength, deep-hardening alloy for fan and compressor disks and for other voluminous components [14, 15]. Ti-17 may be processed in the β region or in the two-phase ($\alpha + \beta$) region. The tensile and creep strengths are higher than in Ti-6Al-4V. The minimum strength is 1125 MPa. Upper temperature limit is 400 °C. While its hardenability is similar to metastable β-alloys, Ti-17 has a lower density and a higher Young's modulus than most β titanium alloys. The temperature limitation for Ti-17 is the result of its rather poor creep resistance.

Near-α alloys are applied at elevated temperatures where the conventional ($\alpha + \beta$) alloys cannot meet the required properties [16–19]. Superior creep resistance and good elevated temperature LCF and HCF strengths are characteristic for this group of titanium alloys. One example is Ti–6Al–2Sn–4Zr–2Mo–0.08Si (Ti-6242). The alloy possesses an excellent combination of tensile strength, creep strength, fracture toughness, and reliable high-temperature stability for long-term application at temperatures up to 540 °C. The forging and machining characteristics are comparable to Ti–6Al–4V. Ti-6242 is used for gas turbine compressor components (blades, disks, and impellers). This alloy is also used in sheet metal form for engine afterburner structure and for hot airframe skin applications. The near-α alloy Ti–5.8Al–4Sn–3.5Zr–0.7Nb–0.5Mo–0.3Si–0.06C (TIMETAL 834) was developed to be capable of being used at even higher temperatures up to 590 °C [5]. Ti-834 is used for compressor disks in the last two stages of the intermediate-pressure compressor and the first four stages of the high-pressure compressor. Application is realized in various sections of the Trent 800 jet engine produced by Rolls-Royce (Figure 4.34). In 2011, Trent 800 became the market leader with a share of 41%. For example, this engine is installed on the Boeing 777.

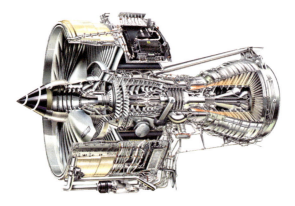

Figure 4.34 Trent 800 with the application of TIMETAL 834 as compressor disk material. (Courtesy of Rolls-Royce.)

4.3.2.2 Nonengine Applications

Ti–6Al–6V–2Sn (Ti-662) was used in components, where higher strengths up to 1170 MPa are required. The alloy was used in the solution-treated and aged conditions and offered higher weight savings, although both fracture toughness and stress corrosion resistance were reduced relative to Ti–6Al–4V. Ti-662 was used in airframe structures and in the landing gear support structure of the 747, but has not been used on later Boeing aircraft. Ti-662 has been replaced by the solute-lean metastable β-alloy Ti–10V–2Fe–3Al (Ti-10-2-3) [19] that offers better weight savings, better fracture toughness, and more resistance against stress corrosion cracking compared to Ti-662.

The main landing gear of the Boeing 777 (Figure 4.35) is almost entirely produced from Ti-10-2-3, except the inner and outer cylinders and the axles. Ti-10-2-3 applications in the main landing gears give rise to a weight savings of more than 250 kg per aircraft. The potential of stress corrosion cracking associated with components made of steel is eliminated.

In retrospect, the first commercially important β-alloy was Ti–13V–11Cr–3Al (Ti-13-11-3), and it is well known from its use in the SR-71 "Blackbird" reconnaissance aircraft. Probably, about 95% of its low structural weight was caused by using

Figure 4.35 Landing gear of the Boeing 777. (Courtesy of Boeing.)

Figure 4.36 The Lockheed SR-71 "Blackbird," a long-range, Mach 3+ strategic reconnaissance aircraft.

titanium. The most of it was Ti-13-11-3. Among other applications, Ti-13-11-3 was used for the main and nose landing gears.

One of the most important reasons for the use of T-13-11-3 was the thermal stability of this alloy. Some of the skins of the SR71 (Figure 4.36) were spot welded on assembly. The temperature caused effects, which were comparable with a solution heat treatment. After the welding process the weld nuggets stayed in a solution-treated condition. In the 1960s, Ti-13-11-3 was the only heat treatable alloy that could be used at the operating temperatures of up to 315 °C without causing embrittlement in the critical weld regions. Later, this alloy was partially replaced by Ti–3Al–8V–6Cr–4Mo–4Zr (Beta C), in particular, for spring applications because of the strength levels of 1200–1400 MPa could be easily achieved [20].

For applications where mechanical loads are not as high as discussed before, Ti–6Al–4V is sufficiently strong. The use is limited to moderate temperatures. Examples for typical applications are airframe components, landing gear components, wing components, and empennage. Ti–6Al–4V has shown outstanding resistance against stress corrosion cracking and corrosion fatigue. Windshield frames in the Boeing 757, 767, and 777 are made from β-annealed Ti–6Al–4V forgings. For the fin deck on the 777, where the composite vertical fin attaches to the fuselage, Ti–6Al–4V hot-formed plate is used. The main reason for this application is the coefficient of thermal expansion (CTE) of titanium. Compared to other structural materials, titanium very well meets the coefficient of the carbon fiber in the polymer matrix composite (PMC). Ti–6Al–4V β-annealed material for forgings is also used for the critical fittings that attach the composite horizontal and vertical fins to the fuselage of the 777. Again, the main reason for this application of Ti–6Al–4V is its excellent corrosion resistance.

Ti–6Al–4V sheets are also used in the floor support structure in areas of the galleys and lavatories. In this part of the aircraft, severe corrosion problems occur. Corrosion problems can be due to aggressive fluids that are ubiquitous in this area. The potential for corrosion was aggravated on the 777 because PMC floor beams were installed. In galvanic contact with aluminum structural elements, these systems may suffer from contact corrosion.

Figure 4.37 Structural beam from Ti–6Al–4V investment casting for the Airbus A400. (Courtesy of Tital.)

The 777 tail cone and auxiliary power unit (APU) exhaust duct are made from Ti–6Al–4V. The cone is produced from sheet. In contrast, the exhaust duct is manufactured from a casting. In the case of more complex geometries, casting is the most cost-effective means of producing components. Figure 4.37 shows a rather complex structural beam of Ti–6Al–4V made by investment casting.

It should be noted that unlike the situation in aluminum and magnesium alloys, there are no special cast alloys made from titanium. Generally, cast products have the same chemical composition as wrought products.

4.3.3
Shipbuilding

The applications of titanium materials in standard and specialized ships are limited to areas with extreme requirements [21]. These are, for example, in pipes and heat exchangers. Figure 4.38 shows the components for shipbuilding; the material used is cp-titanium (3.7025), ASTM grades 1–4.

Additional examples of the use of titanium in ships are pressure vessels, tanks, pipes, vessels, containers, and exhaust systems. Other applications are components for submarines. An actual development is a submersible that is being manufactured. Responsible for fabrication is the Southwest Research Institute (SWRI) in San Antonio, Texas. The inner diameter of this submersible personal sphere is 2.1 m. The maximal operating depth is 6500 m in seawater. The material for the hull construction is Ti–6Al–4V ELI (extra low interstitials). The working

Figure 4.38 Tubes made from titanium for application in ships. (Courtesy of Knaack & Jahn.)

space is conceived for three people. At the time of this writing (2012), the two hemispheres have been joined by electron beam welding [22].

4.4
Future Trends

4.4.1
New Alloy Developments

Since 2007, only two new conventional wrought alloys of the US production have achieved commercial status: Ti–5Al–3V–0.6Fe–0.17O (TIMET) and Ti–4Al–2.5V–1.5Fe–0.25O (ATI 425, Allegheny Technologies) [21].

The ($\alpha + \beta$) alloy Ti–5Al–3V–0.6Fe–0.17O was developed by TIMET for armor applications. The ballistic and mechanical properties are similar to Ti–6Al–4V. The formability seems to be slightly better. A scale-up for industrial production is underway. For example, the material can be used for components of ground combat vehicles. This material offers producibility benefits and lower manufacturing cost compared to Ti–6Al–4V. Cost reduction can be achieved by the use of Ti–6Al–4V turning, which is enriched with elemental Fe and Al. This leads to reduced requirements regarding master alloy and sponge qualities.

ATI 425 titanium was designed in 2006 and designated by the ASTM as grade 38 titanium. Since that time, the alloy has gained Aerospace Material Specification (AMS) approval for applications in aerospace. It was a key material for the Phoenix Mars Lander Analytical Componentry. ATI425 is stronger than CP grades and combines strength comparable to Ti–6Al–4V, improved workability, and higher ductility. A pronounced characteristic of ATI 425 is the excellent cold ability it exhibits. ATI 425 realizes a good combination among strength, finish, and workability for a broad range of applications. All the properties make it in transports a potential candidate for a wide variety of aerospace, shiphold structural materials, and ground vehicles. The availability of ATI 425 cold-rolled sheet and coil in long lengths facilitates its use in manufacturing methods such as roll forming and may allow structures to be designed with fewer joints and fasteners [21].

Ti–5Al–4V–0.75Mo–0.5Fe (TIMETAL-54M) was developed by TIMET to offer improved machinability and formability over Ti–6Al–4V. Cost savings are expected for structural parts that require extensive machining. Mechanical properties were shown to be slightly superior to Ti–6Al–4V, presumably due to solid solution strengthening effects of Mo and Fe [23].

4.4.2
Developments in Powder Metallurgy (P/M)

Powder metallurgy offers the potential to reduce the cost of manufacture of titanium components. The Materials and Electrochemical Research Corporation (MER) has produced various shapes with a wide variety of complexity, for example, blisks. The costs are much lower compared to those of conventional production. MER estimated the costs to about $22 kg^{-1}$ (2007). The P/M manufacturing of Ti–6Al–4V components that have lesser demanding requirements was studied, for example, by Boeing. P/M billets were subsequently extruded. Data were encouraging, although the oxygen content exceeded the upper limit for aerospace specifications. Presumably, the observed high oxygen content is responsible for the measured yield and tensile strengths being higher than in extrusions of wrought materials. No significant differences in fatigue performance were observed comparing wrought and P/M products. However, the fracture toughness in the P/M material was lower and macrocrack growth rates higher than in the wrought counterpart. Probably, this is because of the high oxygen content and the finer grain sizes of P/M products. Material with low oxygen contents (O $\leq$ 0.2%), which meets the AMS specification, is currently being evaluated.

Advanced titanium alloys and titanium metal matrix composites (MMCs) for automotive power components are being developed by Dynamet Incorporated in cooperation with the Ford Motor Co. Increased fuel efficiency, reduced emissions, and economic advantage would be the benefits. The goal is to develop titanium components that are more economic than the conventionally produced wrought products.

4.4.3
Protection against Galvanic Corrosion

An aircraft such as the Dreamliner (Boeing 787) contains approximately 35 short tons (32 000 kg) of carbon fiber-reinforced plastic (CFRP), made with 23 tons of carbon fiber. The use of carbon fibers in aviation technologies is a new trend because it increases the fuel efficiency significantly. Traditionally, a lot of aircraft components are made of aluminum. However, unlike titanium, aluminum in contact with carbon fibers is susceptible to corrosion. Another example of the utilization of titanium on the Boeing 787 is for landing gear application where Ti–5Al–5V–5Mo–3Cr (Ti-5-5-5-3) replaces Ti-10-2-3 that is used on the Boeing 777 and AirbusA380.

4.4.4
Titanium Aluminides

For several years, Ti-aluminides were promising candidates for the use in aircraft engineering. The Gamma Ti Aluminide Ti–48Al–2Nb–2Cr for low-pressure turbine (LPT) blades in the GEnx engine for the Boeing 787 and 747-8 represents the first production example of this special class of titanium materials that are used for aircraft engines. γ-TiAl for LPT blades are produced by means of investment casting. Obvious is the weight reduction compared to Ni-based alloys. In total, the weight reduction per aircraft amounts to roughly 182 kg.

4.5
Further Reading

A fascinating and exciting overview of the history of titanium from the very beginning to the end of the Cold War is presented in the book entitled *Black Sand – The History of Titanium* – Metal Management Aerospace, Inc. (2007), written by K.L. Housley.

The book *Titanium*, written by G. Lütjering and J.C. Williams, Springer-Verlag Berlin, Heidelberg, New York (2007), is a comprehensive summary of the current state of the art of titanium and its alloys. The book covers physical metallurgy, extractive metallurgy, the various production processes, and, in particular, the correlations among processing, microstructures, and properties. Additionally, applications and economic aspects are presented.

The book *Surface Performance of Titanium*, edited by J.K. Gregory, H.J. Rack, and D. Eylon (1996), is a valuable resource for those who are both practiced in the art and just entering the field of shot peening, residual stresses, fatigue, surface modification, biomedical applications, wear, and oxidation of titanium and titanium alloys.

The interested reader can find plenty of information about all aspects of titanium in the various proceedings of the World Conferences on Titanium, which have been published every four years since 1968. The current proceedings (three volumes, edited by Lian Zhou, Hui Zhang, Yafeng Lu, and Dongsheng Xu) are from the 12th World Conference that took place in Beijing in 2011.

References

1. Housley, K.L. (2007) *Black Sand-The History of Titanium*, Metal Management Aerospace Inc.
2. Friedrich, H., Kiese, J., Haldenwanger, H.-G., and Stich, A. (2003) in *Titanium in Automotive Applications – Nightmare, Vision or Reality* (eds G. Luetjering and J. Albrecht) Ti-2003, Wiley-VCH Verlag GmbH, Weinheim, p. 3393.
3. Wagner, L. and Schauerte, O. (2007) in *Status of Titanium and Titanium Alloys in Automotive Applications* (eds M. Niinomi, S. Akiyama, M. Hagiwara, M. Ikeda, and K. Maruyama) Ti-2007, JIMIC 5, p. 1371.
4. Peters, M. *et al.* (2003) in *Titanium and Titanium Alloys* (eds M. Peters and C. Leyens), Wiley-VCH Verlag GmbH, Weinheim, p. 1.
5. Lütjering, G. and Williams, J.C. (2007) *Titanium*, Springer-Verlag, Berlin, Heidelberg, New York.
6. Wagner, L. and Bigoney, J.K. (2002) in *Titanium and Titanium Alloys* (eds M. Peters and C. Leyens), Wiley-VCH Verlag GmbH, Weinheim, p. 153.
7. Faller, K. and Froes, F.H. (2001) *JOM*, 53(4), 27.
8. Sherman, A.M., Sommer, C.J., and Froes, F.H. (1997) *JOM*, 38.
9. Froes, F.H., Suryanarayana, C., and Eliezer, D. (1991) *ISIJ Int.*, 31, 1235.
10. Liu, C.T., Stiegler, J.O., and Froes, F.H. (1990) *"Ordered Intermetallics", Properties and Selection: Nonferrous Alloys and Special Purpose Materials*, Metals Handbook, 10th edn, Vol. 2, ASM International, Materials Park, OH, p. 913.
11. Froes, F.H., Friedrich, H., Kiese, J. and Bergoint, D. (2004) Titanium in the family automobile: The cost challange, *JOM*, 40.

12. J. Kiese, W. Walz, B. Skrotzki, in *Influence of Heat Treatment and Shot Peening on Fatigue Behavior of Suspension Springs Made of TIMETAL LCB*, Ti-2003 (G. Luetjering and J. Albrecht (eds) Wiley-VCH Verlag GmbH, Weinheim (2003) p. 3043.

13. Boyer, R.R. (1996) An overview on the use of titanium in the aerospace industry. *Mater. Sci. Eng.*, **A213**, 103.

14. Redden, T.K. (1984) in *Beta Titanium Alloys in the 1980s* (eds R.R. Boyer and H.W. Rosenberg), TMS, Warrendale, PA, p. 239.

15. Rosenberg, H.W. (1984) in *Beta Titanium Alloys in the 1980'* (eds R.R. Boyer and H.W. Rosenberg), TMS, Warrendale, PA, p. 433.

16. Thiehsen, K.E. *et al.* (1993) The effect of nickel, chromium, and primary alpha phase on the creep behavior of Ti-6242Si. *Metall. Trans. A*, **24A**, 1819.

17. Seagle, S.R., Hall, G.S., and Bomberger, H.B. (1972) *Met. Eng. Q*, 48.

18. S.R. Seagle and H.B. Bomberger, in (R.I. Jaffee and N.E. Promisel (eds) *The Science Technology and Application of Titanium*, Pergamon, New York, 1970, pp. 1001–1008.

19. Boyer, R.R. (1980) Design properties of a high strength titanium alloy, Ti-10V-2Fe-3Al. *J. Organomet. Chem.*, **32**, 61.

20. Wagner, L. and Gregory, J.K. (1993) Improvement of mechanical behavior in Ti-3Al-8V-6Cr-4Mo-4Zr by duplex aging, in *Beta Titanium Alloys for the 1990s*, TMS-AIME.

21. Boyer, R.R. and Williams, J.C. (2012) in *Developments in Research and Applications in the Titanium Industry in the USA* (eds L. Zhou, H. Chang, Y. Lu, and D. Xu) Ti-2011, Science Press, Beijing, p. 10.

22. Malletschek, A. (2011) Einfluss von Titan auf den Entwurf von Unterwasserfahrzeugen. Dr-Ing. thesis. TU Hamburg-Harburg.

23. Yang, S., Wu, Z., Zay, K., Kosaka, Y., and Wagner, L. (2012) in *Effects of Thermomechanical Treatments on Microstructures and Mechanical Properties: TIMETAL-54M vs. Ti-6Al-4V* (eds L. Zhou, H. Chang, Y. Lu, and D. Xu) Ti-2011, Science Press, Beijing, p. 984.

Part II
Polymers

James Njuguna

Polymer materials are widely used for many transport structure applications owing to their many engineering designable advantages such as specific strength properties with weight saving, potential for rapid process cycles, ability to meet stringent dimensional stability, lower thermal expansion properties, and excellent fatigue and fracture resistance over other materials such as metals and ceramics. The need for high performance and capability to design material characteristics to meet specific requirements has made polymeric materials a first choice for many transport applications. Such materials can be tailored to give high strength coupled with relatively low weight and corrosion resistance to most chemicals, and offer long-term durability under most environmental severe conditions. Polymer materials have key advantages over other conventional metallic materials owing to their specific strength properties with weight saving of 20–40%, potential for rapid process cycles, ability to meet stringent dimensional stability, recyclability, low cost, manufacturability, lower thermal expansion properties, and excellent fatigue and fracture resistance.

Polymers are substances whose molecules have high molar masses and are composed of a large number of repeating units. There are both naturally occurring and synthetic polymers. Naturally occurring polymers exist in plants and animals in the form of proteins, starches, cellulose, and latex. Synthetic polymers are derived mainly from oil-based products, and include engineering polymers such as epoxies, polyamides, polypropyrenes, polyurethanes styrene–butadiene rubber, aromatic polyamides, and poly(ether ether ketone)/polybenzimidazoles, among others. The materials commonly called *engineering polymers* are all synthetic polymers.

The polymers are formed by chemical reactions in which a large number of molecules called *monomers* are joined sequentially, forming a chain. In many polymers, only one monomer is used. In others, two or three different monomers may be combined. Polymers are classified by the characteristics of the reactions by which they are formed. If all atoms in the monomers are incorporated into the polymer, the polymer is called an *addition polymer*. If some of the atoms of the monomers are released into small molecules, such as water, the polymer is called a *condensation polymer*. Most addition polymers are made from monomers containing a double bond between carbon atoms. Such monomers are called *olefins*,

Structural Materials and Processes in Transportation, First Edition.
Edited by Dirk Lehmhus, Matthias Busse, Axel S. Herrmann, and Kambiz Kayvantash.
© 2013 Wiley-VCH Verlag GmbH & Co. KGaA. Published 2013 by Wiley-VCH Verlag GmbH & Co. KGaA.

and most commercial addition polymers are *polyolefins*. Condensation polymers are made from monomers that have two different groups of atoms, which can join together to form, for example, ester or amide links.

In general, polyolefins are some of the largest volume commodity that the organic chemicals industry produces, and are produced with a variety of processes. This results in a wide range of polyolefin grades, differing in tacticity, morphology, degree of branching, molecular weight distribution, and other properties (such as thermal stability) that can require significantly different stabilizer formulations. Examples of polyolefins group are polyetheylenes, polypropylene (PP), polyisobutylene (PIB), cyclic olefin copolymers (COCs), and diene polymers.

In recent years, there has been increased use of inorganic polymers. Inorganic polymers have macromolecular substances whose principal structural features are made up of homopolar interlinkages between multivalent elements other than carbon. Inorganic polymers do not preclude the presence of carbon-containing groups in side branches, or as interlinkages between principal structural members and are mainly found in nature, for example, mica, clays, and talc. Polysiloxanes, polyphosphazenes, polysilazanes, polygermanes, and polystannanes are the most important class of inorganic polymers from the applications point of view. High molecular weight polymers with inorganic elements in their backbone are attractive and challenging, because of their physical and chemical differences with their organic counterparts. These polymers offer a unique combination of high temperature stability and excellent low temperature elastomeric properties.

Another important category of polymers that has seen major developments are high temperature polymers. The main goal of the research in this area is the preparation of polymers possessing good thermal and oxidative stability, toughness, stiffness, and retention of physical properties at high temperature. Most materials of this type are based on aromatic rings incorporated into the main chain. For such polymer systems, a balance must be achieved between thermal stability and processability. On the other hand, extensive research on aromatic thermoplastic polymers, polyphenylenes, polyether ether ketones (PEEKs), polybenzimidazoles (PBIs), and polybismaleimides among others potentially offer the favorable properties that make them very suitable for such applications.

In this part, three important classes of polymers (i.e., thermosets, thermoplastics, and elastomers) that are widely used in transport structures and related applications are covered.

5
Thermoplastics

Aravind Dasari

5.1
Introduction

Thermoplastic polymeric materials have several attractive features such as lightweight, cost, performance, and recyclability. In fact, these are the main drivers for their increasing usage in various structural applications. As a majority of the fuel consumption is directly related to the vehicle weight, usage of lightweight materials in car and other mass transit manufacture has grown during the last decade, and as of 2009, about 9.3% of thermoplastics are used in cars [1]. These materials are mostly used for structural and safety parts such as bumpers, passenger protection, floor segments, side body panels, structural frame segments, battery-box access doors, air-conditioning roof-covers, and pedestrian impacts. Even in 1986, Ford Taurus series featured a thermoplastic bumper utilizing a fascia and a reinforcement beam that were glued together with toughened methacrylate adhesive. Another example is the CompoBus® that uses glass fiber-reinforced vinyl ester skins and wood core sandwich structure as the primary structure. This bus is ~30% lighter than a conventional metallic bus, and requires 60% less power to run [2, 3].

It is also estimated that a 20% weight reduction could yield 12–14% fuel efficiency [4–6]. As evident from Figure 5.1a, fuel efficiency in cars and trucks increased significantly during the period from 1975 to 1985, and then leveled off [1]. During that period, the weight of an automobile decreased by almost 1000 lb (~453 kg); but interestingly, recent trends suggest a reversion of this trend despite the increased usage of polymeric materials, which was attributed to the demand for performance and acceleration time (Figure 5.1b). This resulted in the requirement and evolution of multifunctional materials.

In the following sections, the fundamentals and recent advancements in the field of thermoplastics are given, with particular emphasis on high-performance materials for structural applications; this is followed by discussions on the processing and evolution of structure in thermoplastic materials, and the resultant properties.

Structural Materials and Processes in Transportation, First Edition.
Edited by Dirk Lehmhus, Matthias Busse, Axel S. Herrmann, and Kambiz Kayvantash.
© 2013 Wiley-VCH Verlag GmbH & Co. KGaA. Published 2013 by Wiley-VCH Verlag GmbH & Co. KGaA.

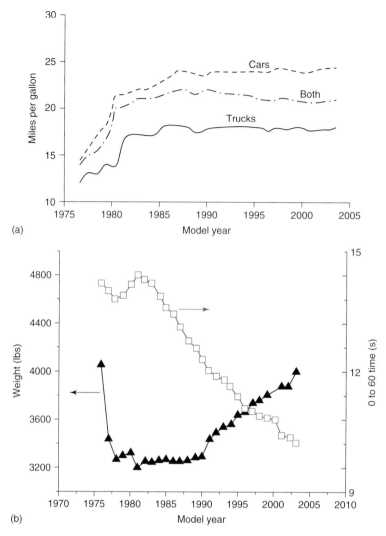

Figure 5.1 Characteristics and performance by model year (three-year moving average) for light vehicles during 1975–2004: (a) adjusted fuel economy and (b) weight and acceleration. (Source: Adapted from Ref. [1].)

5.2
Fundamentals and Recent Advancements in Thermoplastics

Generally, some thermoplastics possess good mechanical properties, particularly better toughness than many thermosetting materials. The cross-linked thermosets, expectedly, have limited chain mobility, while the relatively weak secondary bonds between chains and lack of three-dimensional cross-linking in thermoplastics impart a high level of plastic flow via chain stretching and pulling, as well as

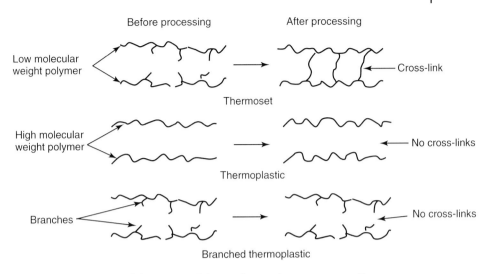

Figure 5.2 Comparison of thermoset and thermoplastic polymer structures. (Source: Adapted from Ref. [8].)

providing sufficient strength (Figure 5.2 for a schematic showing structural differences between thermosets and thermoplastics). Another advantage of many high-performance thermoplastics is their low moisture absorption compared to thermosets [7]. This issue is important for aerospace applications, as the absorption of moisture lowers their elevated temperature performance. Also, thermoplastic polymers can be recycled, which is important, considering the life cycle of the product. Another advantage of thermoplastic materials is their superior impact resistance property. This property in turn makes these materials excellent candidates for structural materials for mass transit systems. Furthermore, rapid processing cycles and long shelf-life are the other advantages of these materials.

On the contrary, there are many disadvantages of thermoplastics. As they are high molecular weight resins, the viscosities of thermoplastics during processing are orders of magnitude higher than those of thermosets. This in turn results in difficulties in wetting the fibers during prepreg manufacturing process, and the requirement of high pressures during processing [8–10].

Thermoplastic polymers are generally divided into two classes based on the maximum service temperature: high-temperature thermoplastics and engineering thermoplastics. Examples of engineering thermoplastic polymers include polypropylene (PP), acrylonitrile butadiene styrene (ABS), polyvinyl chloride (PVC), polyamides (PAs), polyethylene (PE), and poly(ethylene terephthalate) (PET) and high-temperature/high-performance thermoplastics include semicrystalline polyether ether ketone (PEEK), polyether ketone ketone (PEKK), polyphenylene sulfide (PPS), polyethersulfone (PES), and amorphous thermoplastic polyethylenimine (PEI). As evident, these high-performance polymers are aromatic and

Table 5.1 Selected primary bond dissociation energies [11, 12].

Bond	Dissociation energy (kcal mol^{-1})
C–C	83
C=C	145
C–H	99
C–N	70
C=N	147
C=O	84
C–F	123
N–N	38
C–S	62
C–Si	69
Si–O	88
Ti–O	160
B–O	185

therefore, have higher T_g's. The presence of rigid aromatic rings enhances the intermolecular forces and restricts the mobility of the backbone chain [9], which results in greatly improved mechanical properties, thermal stability, and solvent resistance. Also, the differences in the primary bond strength or the bond dissociation energy between C–C with C=C and others explain this behavior (Table 5.1); however, the mechanism of bond cleavage should also be considered, as recombination or unzipping can propagate rapidly and the ends of the molecules continuously degrade to release small monomer fragments [11].

The structures, processing, and heat distortion temperatures of some of these commonly used thermoplastic polymers are given in Table 5.2 [11, 12]. According to [9, 13], service temperature of a composite should be 25–30 °C below the T_g of the matrix, but high T_g's of thermoplastics are generally associated with higher melt viscosities (e.g., Figure 5.3) and higher processing temperatures. As shown in Table 5.2, processing temperatures for thermoplastics vary from 240 to 470 °C compared to 120–180 °C for epoxies. For amorphous thermoplastics, processing temperatures should exceed T_g and for semicrystalline thermoplastics, T_m, to achieve a low melt viscosity and expected morphology.

Before proceeding further, it is interesting to note some of the specific definitions used to categorize polymers into high-temperature/high-performance group. According to Hergenrother [11], they are (i) long-term durability (>10 000 h) at 177 °C; (ii) 5% weight loss thermal decomposition temperatures >450 °C; (iii) high heat deflection temperatures >177 °C (temperature where 10% deflection occurs under a load of 1.52 MPa); and (iv) high aromatic content.

Recently, much focus was also diverted to the aspect of self-healing in polymeric materials. This is important not just from the viewpoint of damage tolerance

Table 5.2 Chemical structures, processing, and heat distortion temperatures of some selected high-performance thermoplastics.

Polymer	Chemical structure	Processing temperature range (°C)	Heat deflection temperature[a] (°C)
PEEK		350–400	152–160
Polyether-ketone (PEK)		385–410	165–180
Polyimide		350–400	—
PEKK		335–360	160
PPS		315–340	135
Torlon® Polyamide-imide (PAI)		330–400	274

[a]At 1.82 MPa.
Processing and distortion temperatures will vary depending on the grade. The shown values are for guidance only. PEEK, PEK, and PEKK are based on Victrex products.

(from impacts or punctures) but also from maintenance and related economics. The concept of healing in these materials is based on the viscoelastic response to the energy input; that is, after the development of a local melt state in the polymer material during an event, the melt elasticity of the material should be high enough to snap back and heal the damage. For example, Gordon *et al.* [15] compared several materials for puncture healing from ballistic tests and revealed that commercial Surlyn® poly(ethylene-*co*-methacrylic acid) and Affinity EG8200 (a polyolefin elastomer) materials self-healed when tested at ambient temperature,

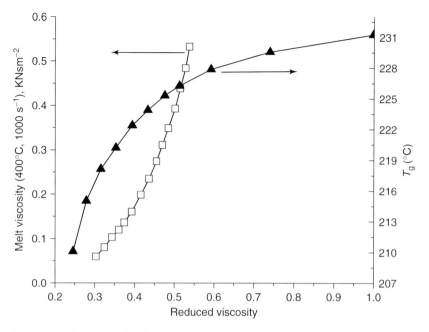

Figure 5.3 Relationship of reduced viscosity versus melt viscosity and T_g of PES. (Source: Adapted from Ref. [14].)

whereas poly(butylene terephthalate) (PBT), Lexan (a thermoplastic polycarbonate), and poly(butylene terephthalate)-co-poly(alkylene glycol terephthalate) (PBT-co-PAGT) did not display any healing. However, as viscoelastic response is critical, T_g's of these materials will play an important role. Considering that T_g of Surlyn and Affinity EG8200 are −100 and −68 °C, respectively, this behavior is expected and the very low T_g's are obviously, a compromise for structural applications.

5.3
Processing and Evolution of Structure – Basics and Recent Developments

5.3.1
Synthesis of Thermoplastics

Thermoplastics are synthesized via polymerization, that is, linking of monomers to form long chains with repeat units. Polymerization can in turn be addition or condensation. Addition polymerization involves the linkage of reactive monomers to form a linear macromolecule without any other by-products. The process consists of the initiation, propagation, and termination steps. Condensation polymerization involves stepwise intermolecular chemical reactions that may involve more than one monomer and the elimination of a small molecular weight by-product. Most of the engineering and high-performance polymers are formed by condensation reactions. For example, polyimides are generally synthesized in a chain-extendable manner via condensation route. This has obvious advantages when incorporating fibers; that is, during the prepreg stage, only polymer precursors are used, which can wet the fibers more uniformly. But on the contrary, it involves the removal of large number of volatiles formed during the condensation step. Other methods such as impregnation from monomer and subsequent polymerization were also exploited, particularly for acrylic and styrenic polymers [9, 16].

5.3.2
Processing of Thermoplastic Structural Components

For the preparation of thermoplastic prepregs, solution or melt coating of the fiber reinforcement is generally carried out. Apart from the viscosity, the choice of solvents is limited for solution processing; melt processing requires higher temperatures/pressures; and there are issues with formation of double curvature structures from inextensible materials [8]. Incomplete removal of solvent before final laminate consolidation is often a major problem in solution coating with thermoplastic resins. If the individual fibers are not homogeneously coated, this can result in uneven distribution of fibers in the matrix affecting the final properties of the material. For this purpose, various efforts were made to tackle this problem including the application of thermoplastic polymer in powder form to fiber tows (using a fluidized bed) followed by direct electrical heating of fiber tows to temperatures above the melt flow temperatures of the polymer [8].

For the processing of thermoplastic materials for structural components, many technologies are used and include: injection molding, rotational molding, consolidation, thermoforming, stamping, machining, and joining. Even fiber placement technologies and autoclave consolidation processes that are used for thermoset materials have been used for processing thermoplastic structural materials, in addition to the continuous forming processes such as pultrusion and roll forming. The use of these techniques depends on the preform configuration and the degree of shaping required for producing parts of various geometries. But the main advantage of high-performance thermoplastics is their rapid processing. Figure 5.4 shows the differences between the production cycles of high-performance thermosets and thermoplastic composites [17]. Irrespective of the various processing techniques, the basic principle during processing of thermoplastics involves heating the polymer to an elevated temperature and forcing it into a mold. The principle stages of pressure and temperature in processing of preimpregnated tape/tow is schematically shown in Figure 5.5. Moreover, at these flow temperatures, thermoplastics are more prone to oxygen attack resulting in degradation. Antioxidants (hindered amines) are generally added during processing to tackle this purpose.

On the contrary, highly viscous thermoplastic melt is a major problem to achieve a complete and homogeneous impregnation of fibers. To overcome the problems associated with wetting of the fibers during processing of thermoplastic composites, several impregnation processes were developed. They include

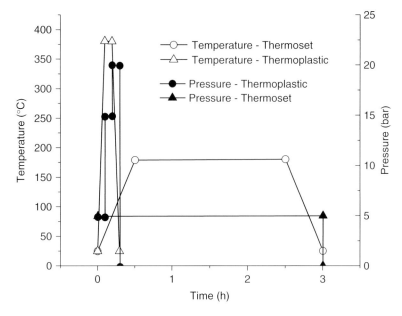

Figure 5.4 Typical processing cycles of advanced composites with thermoset and thermo-plastic matrices. (Source: Adapted from Ref. [17].)

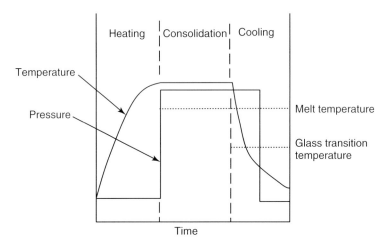

Figure 5.5 Typical processing cycles of advanced thermoplastic and thermoset composites. (Source: Adapted from Ref. [8].)

comingled thermoplastic fabric, powder impregnated fabric, powder/sheath-fiber bundle process, direct reinforcement fabrication technology (DRIFT), filament winding, wet method, and film stacking [3, 18–22]. Figure 5.6 shows an overview of the common impregnation techniques. In filament winding, the process starts with fully or partially impregnated prepreg similar to tape lay down. But precise

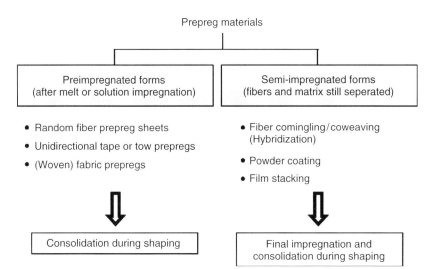

Figure 5.6 An overview of impregnation techniques. (Source: Adapted from Ref. [23].)

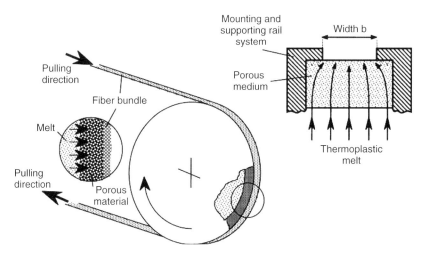

Figure 5.7 Schematic of impregnation wheel technique. The wheel consists of a porous ring through which the molten polymer is squeezed under pressure. As the fiber bundle is in tight physical contact with the outer surface of the ring, it ensures good wetting. (Source: After Ref. [23], with permission from Elsevier.)

control over the local tape placement geometry, pivoting and tilting movements, and winding speed is required to maintain the quality of impregnated tows.

Lutz *et al.* [24] developed an innovative melt impregnation technique, termed as *impregnation wheel* to produce unidirectional reinforced tapes at considerable speeds. The concept behind this technique is schematically illustrated in Figure 5.7. Henninger and Friedrich [23] used this impregnation wheel technology and developed a process combination system to realize fiber bundle impregnation and

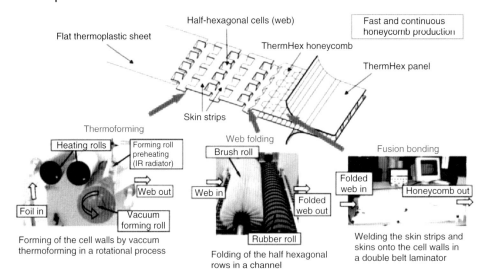

Figure 5.8 Schematic of the process of ThermHex production. (Source: After Ref. [27], with permission from Sage Publications.)

filament winding. In another approach, fiber bundles were pulled into a film of thermoplastic melt, which was in fact, extruded onto the mandrel beforehand [25]. Other processing technologies such as liquid monomer processing were also established to exploit the properties of thermoplastics. Here, liquid monomer materials (such as PBT) can be processed isothermally: injected-polymerized-crystallized (controlling the cycle time) and de-molded at the same temperature [26]. Even advancements were made in continuous processing technologies to produce thermoplastic honeycombs. For example, a schematic of continuous production of thermoplastic honeycomb from "ThermHex" based on PP with 6.4 mm cell size and 0.2 mm cell wall thickness, is shown in Figure 5.8 [27]. It was stated that the production speed of ThermHex is 12.5 m min^{-1} and the resulting price of the final honeycomb is $<\$2$ m^{-2}.

Moreover, as these high-performance materials have to be joined to the rest of the structure, it is critical to have a good joint that can transfer the load from these polymeric materials to metallic or other substrates. Although the current technologies such as mechanical fastening and adhesive bonding are applicable to both thermosets and thermoplastics, the repeat processability of thermoplastics was exploited to provide an effective "fusion bonding" [8, 14, 28]. Melt fusion essentially produces joints as strong as those of the parent resin. Apart from fusion bonding, thermoplastics can be joined by dual resin bonding, resistance welding, ultrasonic welding, or induction welding. There is always an option of using extra layers of neat thermoplastic polymer to fill gaps along the bond-line and facilitate a good bond. In the dual resin bonding approach, which is also termed as *amorphous bonding* or the "Thermabond" process, a layer of amorphous polymer is generally used to bond two thermoplastics. In fact, the amorphous polymer is fused to both

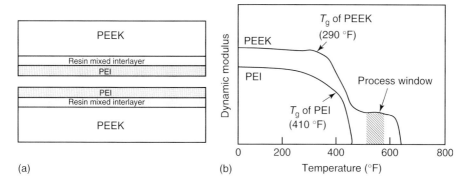

Figure 5.9 Amorphous bonding of PEEK with PEI (a) schematic and (b) dynamic modulus curves illustrating the feasibility of this process. (Source: Adapted from Ref. [8].)

surfaces of thermoplastic before bonding to enhance the resin mixing process. For example, Figure 5.9 shows the bonding of two PEEK composite laminates with amorphous polyetherimide (PEI). As the processing temperature for PEI is below the melt temperature of PEEK, there is no evident danger of delamination within the PEEK substrates [8].

When using adhesive bonding, surface treatment or surface preparation is important. Some of the preparatory techniques for joining two similar high-performance thermoplastics include: mechanical abrasion, corona discharge, plasma treatment, and chemical etching. With mechanical abrasion, results (as indicated by lap shear strength) were found to vary quite considerably depending on the preparation of surface, that is, the magnitude of roughening and removal of surface layer [28]. Figure 5.10a illustrates this point for carbon fiber/PEEK joints bonded by curing modified epoxy film adhesive at 177 °C. The objective of other chemical and discharge methods is to increase the reactivity of thermoplastics as they generally show low surface energies. Furthermore, plasma treatments were shown to produce an etching effect so that both roughening and chemical activation occur. An example of oxygen plasma treatment effect on lap shear strength of carbon fiber/PEEK composites is shown in Figure 5.10b, and the etched surfaces are shown in Figure 5.10c,d.

5.3.3
Evolution of Structure in Thermoplastic Polymers

Changing the processing conditions, for example, combination of shear and extensional flow, and/or addition of fillers, generally influences the spatial organization and alignment of crystallites in semicrystalline polymers, and may thus affect the material properties such as strength, hardness, toughness, and transparency [29–34]. In neat polymers, the negligible shear in the center of a (injection) molded bar and the presence of intense shear near the surface are shown to produce different spatial organizations of the crystalline lamellae [34–36]. Even processing cycles affect the spherulite size, as well as the degree of crystallinity. A low cooling

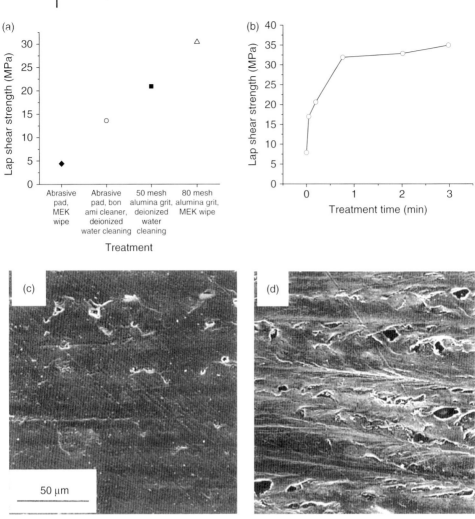

Figure 5.10 Influence of (a) mechanical and (b) oxygen plasma pretreatments on lap shear strengths of carbon fiber/PEEK joints. (c,d) SEM images of oxygen plasma treated surfaces for (c) 30 s and (d) 7.5 min. (Source: (a,b) are adapted from Ref. [28]; (c,d) are after Ref. [28], with permission from Carl Hanser Verlag Publishers.)

rate results in the formation of large spherulites, while smaller spherulites can be formed by fast cooling. Crystallinity is also influenced by other processing parameters, such as the temperature to which the polymer is heated, and its holding time. As well known, for semicrystalline thermoplastics, a temperature high enough to melt all the crystalline material formed during previous thermal treatments has to be chosen to assure complete removal of thermal history. Generally, crystallinity increases strength, stiffness, and temperature resistance, but has a negative effect on toughness/ductility. In the case of amorphous thermoplastic polymers, they contain a random array of entangled molecular chains, which are held together by strong covalent bonds. But the bonds between the chains are weaker secondary bonds. These secondary bonds are easily broken on mechanical stress and thus allow the chains to move and slide past one another. This leads to good elongation, toughness, and impact resistance.

Many studies have shown that fillers, particularly fibers, act as heterogeneous nucleating agents in semicrystalline thermoplastic materials and that they initiate crystallization along the interface [37–41]. However, the essential prerequisite for this to happen is the presence of a high density of active nuclei on the filler surface. These nuclei hinder the lateral extension of spherulites and force their growth normal to the fiber surfaces resulting in columnar crystalline layers, called *transcrystalline layers*, with some finite thickness [42, 43]. An example showing the formation of transcrystalline layers in the vicinity of a carbon fiber in PP is shown in Figure 5.11 [44].

Generally, transcrystallization is believed to occur only on solid, crystalline surfaces and is governed by epitaxy [45–47]. Epitaxy involves the oriented overgrowth of one crystal on another in one or more strictly defined crystallographic orientation (lattice matching). In most cases, it is noticed that the polymer chains lie with their chain axis parallel to the substrate surface, that is, lamellae that are developed on further growth stand edge-on (normal to that surface). Wittmann and Lotz [48] have described that the amount of disregistry, Δ, between lattice spacings can be calculated from the following equation:

$$\Delta = 100 \times \frac{(d - d_0)}{d_0}\% \tag{5.1}$$

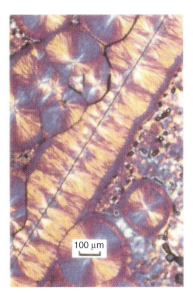

100 µm

Figure 5.11 Polarized optical photograph showing the formation of transcrystalline layers around carbon fiber in polypropylene crystallized at 135 °C [44].

where d is the lattice spacing of deposit and d_0 the lattice spacing of substrate. Normally, 10–15% disregistries between the matching lattice spacings of host and guest crystals are considered as an upper limit. But, it is also important to note that epitaxy is not confined to cases where there is such a small misfit. For example, Koutsky *et al.* [49] demonstrated that this is not a requirement as epitaxial overgrowth of PE (unit cell parameters in nanometers – *a*: 0.736; *b*: 0.492; and *c*: 0.253) was noted on substrates with a wide range of cell parameters (from 0.4 to 0.7 nm) including halide surfaces.

Despite the general belief that transcrystallization is only governed by epitaxy, it is important to note that there are other means through which transcrystallization can occur. These include differences in thermal conductivity, surface energy, chemical composition, thermal expansion coefficients between filler and matrix, and stress-induced crystallization by local flow [40, 50–53]. For example, if a temperature gradient is generated because of the thermal conductivity mismatch, filler surface can be cooler than the matrix. The lower surface temperature can result in a larger supercooling at the filler–matrix interface and subsequently increases the nucleation rate and results in transcrystallization [40]. This phenomenon was used to explain the formation of transcrystallized layers in polyacrylonitrile (PAN)-based carbon fiber/PEKK and amorphous E-glass fiber/PEKK systems [40]. Even transcrystallization is reported in polymers that are crystallized on planar substrates (rubber and calcite) suggesting that the crystallization behavior near an incoherent (or dissimilar) interface is different from that occurring in the bulk and is independent of the nature of the substrate, whether it be amorphous or crystalline [54].

Although the exact mechanisms of the transcrystalline growth are still not clear, it is commonly agreed that extensive heterogeneous nucleation of polymer melts at high energy surfaces generates transcrystalline layers in the interfacial region if sufficient time is allowed for the polymer melt to achieve extensive and intimate contact with the filler. Therefore, the spherulites disappear and the transcrystals appear at the filler surface and become dominant when the filler surface energy increases. Cho *et al.* [53] compared the effect of substrate (silicon) surface energy on the transcrystalline growth at the interface of semicrystalline isotactic PP by treating

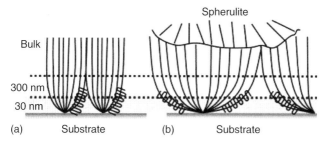

(a)   Substrate   (b)   Substrate

Figure 5.12 Schematic diagrams of the crystalline microstructure both near the interface and in the bulk for (a) a high-surface-energy substrate and (b) a low-surface-energy substrate. (Source: After Ref. [53], with permission from the American Chemical Society.)

the substrate with various silane coupling agents so as to change the surface energy. They showed that with a high-surface-energy substrate, the transcrystalline region is very thick when compared to that with a low-surface-energy substrate (Figure 5.12); however, it is still unclear whether the transcrystalline layers are beneficial for shear stress transfer at the interface, thereby improving the mechanical properties, or whether they have a negative or negligible effect on the ultimate mechanical properties. Nevertheless, this behavior shows the molecular scale proximity or the interaction of polymer and fiber, which in turn indicates the quality of impregnation process.

5.4
Properties

This section is mainly intended to give a short general overview of the physical and mechanical performance of high-performance thermoplastics without considering any specific applications.

Fundamentally, properties of high-performance thermoplastic polymers depend on the following factors: rigid aromatic units, flexible and stiff spacers, number of times the monomer sequence is repeated, and the end groups [9]. Flexible linkages, as the name indicates, provide some degree of freedom for the chain to rotate, which is critical for processing. Ether link is a good example of flexible linkage. Examples of stiffer linkages include sulfone group or ketone group. Besides, as mentioned earlier, molecular weight control as well as end-capping are important parameters to be controlled [11]. End-capping offers advantages, such as thermo-oxidative stability (durability) and melt stability. At a different scale, as mentioned before, mechanical properties of semicrystalline thermoplastics depend on their morphology and the degree of crystallinity (including the number and size of spherulites, the crystalline structure, and the crystalline orientation). Orientation can be responsible for anisotropy in the mechanical properties. Even thermal treatments such as annealing of polymers above the T_g, can enhance molecular packing and crystallinity. As in cross-linking, crystallization can enhance stiffness, strength, and T_g by constraining molecular mobility [55]. Figure 5.13 shows representative T_g for different thermoplastic polymers as well as an epoxy matrix [14]. It is clear that except for PPS, all other polymers have T_g's over 120 °C. Unfortunately, as mentioned before, higher T_g's of high-performance thermoplastics are generally associated with a higher melt viscosity and a higher processing temperature.

However, on the flip side, a direct consequence of constraining molecular mobility (or increasing crystallinity) is reduction in fracture toughness. An example of this behavior is shown for PEEK in Figure 5.14 [56]. Nevertheless, in general , high-performance thermoplastics exhibit a good balance between modulus and fracture toughness when compared to thermosets (Figure 5.15) [14]. Except for PPS, the G_{IC} values for other thermoplastics far exceed generally accepted values of

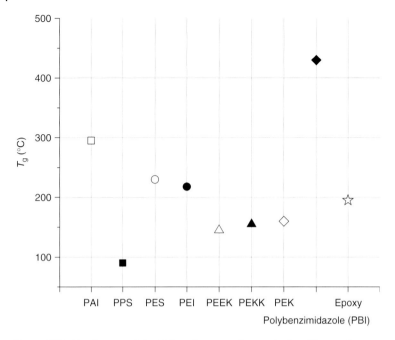

Figure 5.13 T_g of some of the high-performance thermoplastics. These values are only representative and might vary depending on the processing conditions and nature of other groups attached. (Source: Adapted from Ref. [14].)

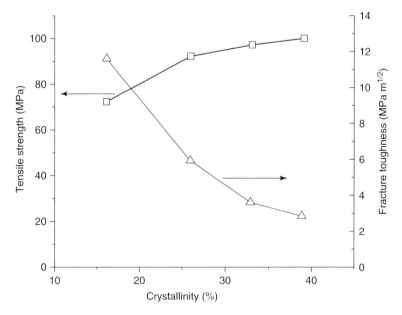

Figure 5.14 Variations in tensile strength and fracture toughness of PEEK as a function of crystallinity. (Source: Adapted from Ref. [56].)

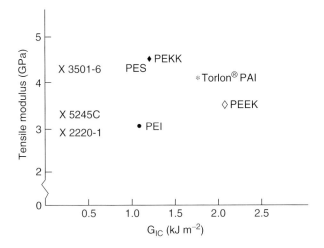

Figure 5.15 Tensile modulus versus fracture toughness of some of the selected thermo-plastics and thermosets. As before, these values are only representative and vary depending on the processing conditions and nature of other groups attached. (Source: Adapted from Ref. [14].)

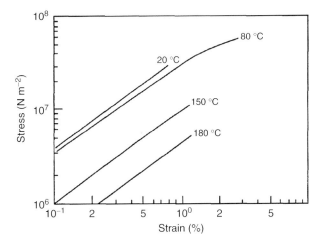

Figure 5.16 100s isochronous data for PEEK. (Source: After Ref. [57].)

$0.7–1$ kJ m^{-2} for second generation thermosets in structural applications involving civilian transportation systems.

The viscoelastic behavior of thermoplastics makes them more complex as it is primarily determined by molecular configuration. An example of this behavior for PEEK at temperatures in the range $20–180\,°C$ is shown in Figure 5.16 [57]. As the T_g for PEEK is $\sim143\,°C$, a significant shift in the stress–strain curves is evident between 80 and $150\,°C$. In thermosets, no considerable variations can be observed in their nonlinear stress–strain behavior and time-dependent response.

Table 5.3 Tensile strengths and flexural properties of some selected thermoplastic polymers at different temperatures.

Polymer		Tensile strength (MPa)		Flexural properties at 23 °C	
	At 23 °C	Between 100 and 140 °C	>140 °C	Flexural strength (MPa)	Flexural modulus (GPa)
PEEK	100	66 at 100 °C	35 at 150 °C	110	3.9
PEK	105	No data available	20 at 250 °C	—	3.7
PEI	105	90 at 100 °C	76 at 150 °C	145	3.3
PES	84	69 at 100 °C	55 at 150 °C	129	2.6
PBI	159	165 at 93 °C	138 at 205 °C	220	6.5

These values are only representative and vary depending on the processing conditions and nature of the other groups attached.
Source: After Ref. [14].

Depending on the temperature of operating conditions, the extent of degradation in some of the mechanical properties of high-performance thermoplastics is given Table 5.3. Crystallinity is also a determining factor when considering dimensional stability, creep, and environmental resistance. Oxidation of thermoplastics such as PEEK, should also be considered as it accelerates aging. Even the high-performance thermoplastics are notch sensitive, and the notched Izod impact energies are much lower than unnotched Izod impact energies [14]. Moreover, thermal stabilities of thermoplastics, such as PEEK, PI, and PPS are generally higher than epoxies.

There are many other properties that are not mentioned here, but are important in commercial applications. These include electrical conductivity, optical properties, noise dampening, dimensional stability, corrosion, photo-degradation, and scratch and wear resistance. Flame retardant properties are also worth mentioning. Significant amount of research in the recent past has been diverted to this facet because of the increasing usage of composites in transportation technologies.

5.5
Summary

The last few decades have seen significant growth in the development of new high-performance/high-temperature thermoplastic polymers. Depending on the application and its requirement, a broad choice of materials is available, be it for supersonic aircraft or screws/bushes. Generally, strength/stiffness of these materials is comparable and is often better than those of thermosets. But their huge advantages include outstanding toughness, ductility, low moisture absorption, and high T_g's. On the contrary, requirement of higher processing temperatures and pressures is a drawback, on account of their inherently high viscosities.

Acknowledgment

A.D. acknowledges his Start-up Grant provided by Nanyang Technological University.

References

1. Vaidya, U. (2010) *Composites for Automotive, Truck and Mass Transit: Materials, Design, Manufacturing*, DEStech Publications, Inc., Lancaster, PA.

2. Vaidya, U.K., Samalot, F., Pillay, S., Janowski, G.M., Husman, G., and Gleich, K. (2004) Design and manufacture of woven reinforced glass/polypropylene composites for mass transit floor structure. *J. Compos. Mater.*, **38**, 1949–1972.

3. Ning, H., Janowski, G.M., Vaidya, U.K., and Husman, G. (2007) Thermoplastic sandwich structure design and manufacturing for the body panel of mass transit vehicle. *Compos. Struct.*, **80**, 82–91.

4. Ning, H., Pillay, S., and Vaidya, U. (2009) Design and development of thermoplastic composite roof door for mass transit bus. *Mater. Des.*, **30**, 983–991.

5. Plotkin, S.E. (2009) Examining fuel economy and carbon standards for light vehicles. *Energy Policy*, **37**, 3843–3853.

6. Cheah, L. and Heywood, J. (2011) Meeting U.S. passenger vehicle fuel economy standards in 2016 and beyond. *Energy Policy*, **39**, 454–466.

7. Mascia, L. (1989) *Thermoplastics: Materials Engineering*, 2nd edn, Elsevier Science Publishers Ltd., Barking, Essex.

8. Campbell, F.C. (2010) *Structural Composite Materials*, ASM International, Materials Park, OH.

9. Cogswell, F.N. (1992) *Thermoplastic Aromatic Polymer Composites*, Butterworth-Heinemann, Oxford.

10. Davies, P. and Plummer, C.J.G. (1993) in *Advanced Thermoplastic Composites: Characterization and Processing* (ed. H.H. Kausch), Carl Hanser Verlag, Munich, pp. 141–170.

11. Hergenrother, P.M. (2003) The use, design, synthesis, and properties of high performance/high temperature polymers: an overview. *High Perform. Polym.*, **15**, 3–45.

12. Cottrell, T.L. (1958) *The Strength of Chemical Bonds*, 2nd edn, Butterworth-Heinemann, London.

13. Leach, D.C., Cogswell, F.N., and Nield, E. (1986) High temperature performance of thermoplastic aromatic polymer composites. 31st International SAMPE Symposium, 1986, pp. 434–448.

14. Beland, S. (1990) *High Performance Thermoplastic Resins and their Composites*, National Research Council of Canada and Noyes Publications, Park Ridge, NJ.

15. Gordon, K.L., Working, D.C., Wise, K.E., Bogert, P.B., Britton, S.M., Topping, C.C. et al. (2009) Recent Advances in Thermoplastic Puncture-Healing Polymers. NASA-Langley Research Center, Hampton, pp. 1–2.

16. Lubin, G. and Dastin, S.J. (1982) in *Handbook of Composites* (ed. G. Lubin), Van Nostrand Reinhold, New York, p. 740.

17. Muller, J. (1993) in *Advanced Thermoplastic Composites: Characterization and Processing* (ed. H.H. Kausch), Carl Hanser Verlag, Munich, pp. 303–336.

18. Van West, B.P., Pipes, R.B., and Advani, S.G. (1991) The consolidation of commingled thermoplastic fabrics. *Polym. Compos.*, **12**, 417–427.

19. Ye, L. and Friedrich, K. (1995) Processing of thermoplastic composites from powder/sheath-fibre bundles. *J. Mater. Process. Tech.*, **48**, 317–324.

20. Hartness, T., Husman, G., Koenig, J., and Dyksterhouse, J. (2001) The characterization of low cost fiber reinforced thermoplastic composites produced by the DRIFT™ process. *Composites Part A*, **32**, 1155–1160.

21. Henninger, F., Hoffmann, J., and Friedrich, K. (2002) Thermoplastic filament winding with online-impregnation.

Part B. Experimental study of processing parameters. *Composites Part A*, **33**, 1677–1688.

22. Gamstedt, E.K., Berglund, L.A., and Peijs, T. (1999) Fatigue mechanisms in unidirectional glass-fibre-reinforced polypropylene. *Compos. Sci. Technol.*, **59**, 759–768.

23. Henninger, F. and Friedrich, K. (2002) Thermoplastic filament winding with online-impregnation. Part A: process technology and operating efficiency. *Composites Part A*, **33**, 1479–1486.

24. Lutz, A, Velten, K, and Evstatiev, M. (1998) A new tool for fibre bundle impregnation: experiments and process analysis. Proceedings of ECCM 8, Naples, Florida, 1998, pp. 471–477.

25. Christen, O. (1999) Development of a Novel Thermoplastic Direct Impregnation Method for the Winding-Scale Production of Pressure Vessels. Dusseldorf, Kaiserslautern.

26. Eire Composites (2011) Thermoplastic Composites Explained, *http://www.eirecomposites.com/services/thermoplastic-composites-explained.html*.

27. Fan, X., Verpoest, I., Pflug, J., Vandepitte, D., and Bratfisch, P. (2009) Investigation of continuously produced thermoplastic honeycomb processing–part I: thermoforming. *J. Sandwich Struct. Mater.*, **11**, 151–178.

28. Davies, P. and Cantwell, W.J. (1993) in *Advanced Thermoplastic Composites: Characterization and Processing* (ed. H.H. Kausch), Carl Hanser Verlag, Munich, pp. 337–366.

29. Wang, K.H., Chung, I.J., Jang, M.C., Keum, J.K., and Song, H.H. (2002) Deformation behavior of polyethylene/silicate nanocomposites as studied by real-time wide-angle X-ray scattering. *Macromolecules*, **35**, 5529–5535.

30. Kojima, Y., Usuki, A., Kawasumi, M., Okada, A., Kurauchi, T., Kamigaito, O. et al. (1995) Novel preferred orientation in injection-molded nylon 6-clay hybrid. *J. Polym. Sci., Part B: Polym Phys.*, **33**, 1039–1045.

31. Kim, G.M., Lee, D.H., Hoffmann, B., Kressler, J., and Stoppelmann, G. (2001) Influence of nanofillers on the deformation process in layered silicate/polyamide-12 nanocomposites. *Polymer*, **42**, 1095–1100.

32. Corte, L., Beaume, F., and Leibler, L. (2005) Crystalline organization and toughening: example of polyamide-12. *Polymer*, **46**, 2748–2757.

33. Muratoglu, O.K., Argon, A.S., Cohen, R.E., and Weinberg, M. (1995) Toughening mechanism of rubber-modified polyamides. *Polymer*, **36**, 921–930.

34. Kornfield, J.A., Kumaraswamy, G., and Issaian, A.M. (2002) Recent advances in understanding flow effects on polymer crystallization. *Ind. Eng. Chem. Res.*, **41**, 6383–6392.

35. Pearson, J.R.A. (1988) *Mechanics of Polymer Processing*, Elsevier, New York.

36. Schultz, J.M. (2001) *Polymer Crystallization: The Development of Crystalline Order in Thermoplastics*, American Chemical Society, New York.

37. Zafeiropoulos, N.E., Baillie, C.A., and Matthews, F.L. (2001) A study of transcrystallinity and its effect on the interface in flax fibre reinforced composite materials. *Composites Part A*, **32**, 525–543.

38. Gassan, J., Mildner, I., and Bledzki, A.K. (2001) Transcrystallization of polypropylene on different modified jute fibers. *Compos. Interfaces*, **8**, 443–452.

39. Arbelaiz, A., Fernandez, B., Ramos, J.A., and Mondragon, I. (2006) Thermal and crystallization studies of short flax fibre reinforced polypropylene matrix composites: Effect of treatments. *Thermochim Acta.*, **440**, 111–121.

40. Chen, E.J.H. and Hsiao, B.S. (1992) The effects of transcrystalline interphase in advanced polymer composites. *Polym. Eng. Sci.*, **32**, 280–286.

41. Thomason, J.L. and Vanrooyen, A.A. (1992) Transcrystallized interphase in thermoplastic composites. Part II. Influence of interfacial stress, cooling rate, fiber properties and polymer molecular-weight. *J. Mater. Sci.*, **27**, 897–907.

42. Varga, J. and Kargerkocsis, J. (1995) Interfacial morphologies in carbon fiber reinforced polypropylene microcomposites. *Polymer*, **36**, 4877–4881.

43. Quan, H., Li, Z.M., Yang, M.B., and Huang, R. (2005) On transcrystallinity

in semi-crystalline polymer composites. *Compos. Sci. Technol.*, **65**, 999–1021.

44. Thomason, J.L. and Vanrooyen, A.A. (1992) Transcrystallized interphase in thermoplastic composites 1. Influence of fiber type and crystallization temperature. *J. Mater. Sci.*, **27**, 889–896.

45. Campbell, D. and Qayyum, M.M. (1980) Melt crystallization of polypropylene - effect of contact with fiber substrates. *J. Polym. Sci., Part B: Polym. Phys.*, **18**, 83–93.

46. Varga, J. and Kargerkocsis, J. (1994) The difference between transcrystallization and shear-induced cylindritic crystallization in fiber-reinforced polypropylene composites. *J. Mater. Sci. Lett.*, **13**, 1069–1071.

47. Varga, J. and KargerKocsis, J. (1996) Rules of supermolecular structure formation in sheared isotactic polypropylene melts. *J. Polym. Sci., Part B: Polym. Phys.*, **34**, 657–670.

48. Wittmann, J.C. and Lotz, B. (1990) Epitaxial crystallization of polymers on organic and polymeric substrates. *Prog. Polym. Sci.*, **15**, 909.

49. Koutsky, J.A., Walton, A.G., and Baer, E. (1966) Epitaxial crystallization of homopolymers on single crystals of alkali halides. *J Polym. Sci. Part A2*, **4**, 611.

50. Chatterjee, A.M., Price, F.P., and Newman, S. (1975) Heterogeneous nucleation of crystallization of high polymers from melt - II - Aspects of transcrystallinity and nucleation density. *J. Polym. Sci., Part B: Polym. Phys.*, **13**, 2385–2390.

51. Chatterjee, A.M., Price, F.P., and Newman, S. (1975) Heterogeneous nucleation of polymer crystallization from melt – I – Substrate induced morphologies. *Bull. Am. Phys. Soc.*, **20**, 341.

52. Goldfarb, L. (1980) Transcrystallization of isotactic polypropylene. *Makromol. Chem., Macromol. Chem. Phys.*, **181**, 1757–1762.

53. Cho, K.W., Kim, D.W., and Yoon, S. (2003) Effect of substrate surface energy on transcrystalline growth and its effect on interfacial adhesion of semicrystalline polymers. *Macromolecules*, **36**, 7652–7660.

54. Bartczak, Z., Argon, A.S., Cohen, R.E., and Kowalewski, T. (1999) The morphology and orientation of polyethylene in films of sub-micron thickness crystallized in contact with calcite and rubber substrates. *Polymer*, **40**, 2367–2380.

55. Cogswell, F.N., Leach, D.C., McGrail, P.T., Colquhoun, H.M., MacKenzie, P., and Turner, R.M. (1987) Semi-crystalline thermoplastic matrix composites for service at 350° F. 32nd International SAMPE Symposium, 1987, pp. 382–395.

56. Talbott, M.F., Springer, G.S., and Berglund, L.A. (1987) The effects of crystallinity on the mechanical properties of PEEK polymer and graphite fiber reinforced PEEK. *J. Compos. Mater.*, **21**, 1056–1081.

57. Jones, D.P., Leach, D.C., and Moore, D.R. (1985) Mechanical properties of poly(ether-etherketone) for engineering applications. *Polymer*, **26**, 1385–1393.

6
Thermosets

Sérgio Henrique Pezzin, Luiz Antonio Ferreira Coelho, Sandro Campos Amico, and Luiz Cláudio Pardini

6.1
Introduction

Thermosets are highly cross-linked polymers that are cured (or set) using heat, or heat and pressure, and/or light irradiation. Thermosetting reaction is not reversible under heat; that is, cross-linked (cured) thermoset resins cannot be reheated and remolded.

These materials have an important role in industry because of their distinctive flexibility for tailoring desired properties. Their high cross-linking densities lead to high modulus, strength, durability, and thermal and chemical resistances; however, they have inherently low-impact resistance [1, 2].

The market for thermosetting materials, especially as polymer composite matrices, has grown constantly in the last three decades, including numerous applications in transportation (aerospace, automobile, marine, and rail industries) and civil infrastructure. Thus, many professionals are in charge to design new products using these materials.

The different thermosetting materials present diverse characteristics, and their uses in different applications depend on a number of factors, such as structural performance, prize and availability of raw materials, and suitability of the fabrication process, among other parameters.

There is also a growing concern about thermosets, because of the necessity to preserve the environment from the many aggressions from industrial processes and to minimize the dependence on non-renewable resources, as petroleum. Nevertheless, thermosetting materials are known to be very hard to recycle and contribute to increase urban waste. Aiming at sustainable development, the use of raw materials from renewable sources has been increasing during the past years.

In this chapter, we present a survey of the most important thermosetting materials, as well as advanced thermosets, with an application-oriented approach to professionals working in the transportation field.

Structural Materials and Processes in Transportation, First Edition.
Edited by Dirk Lehmhus, Matthias Busse, Axel S. Herrmann, and Kambiz Kayvantash.
© 2013 Wiley-VCH Verlag GmbH & Co. KGaA. Published 2013 by Wiley-VCH Verlag GmbH & Co. KGaA.

6.1.1
Historical Development

The beginning of thermosets dates back to the development of phenolic resins by Leo Baekeland in 1909, although the discovery of vulcanization in 1839, by Goodyear, is considered as the first step in the production of thermosets [1]. Baekeland not only produced the first synthetic cross-linked polymer (Bakelite), but also developed the molding process to produce commercial articles. Another milestone in the thermoset's history was the appearance of alkyd resins (polyesters modified by the addition of fatty acids) in 1926. Unsaturated polyester resins (UPRs) were patented by Ellis in 1933, while polyurethanes (PUs) were first produced in 1937, and epoxy resins were commercially introduced in 1947. Silicones were first produced industrially by Dow Corning in 1942 and polyimides (PIs) after 1964 [1].

Further modified polymers and composites followed for high-performance applications. The advent of synthetic fibers, for example, glass, carbon, and polyaramid, permitted the development of reinforced polymer matrix systems of outstanding performance levels. Fine tuning of coatings and adhesives have also been very important developments in the field.

6.1.2
Current Use and Global Supply Base

The worldwide current consumption of thermoset resins across the whole industrial spectrum probably exceeds 24 million tons, based on previous market reports [3]. Markets are highly dynamic nowadays. Growth rates may vary considerably, depending on the areas of component application, the production/processing techniques, and the company size. Major markets for thermoset composites are transportation, construction, and marine, representing 70% of the total consumption [4].

The use of fiber-reinforced composites was estimated as 2 million tons in 2008. UPR (33%), epoxy resin (5%), and PU (2%) account for the majority of the resins. Reinforcements (50%) and fillers (10%) account for the rest. The total amount of resin in reinforced thermoset composites is 0.8 million ton. Another 200 thousand tons of UPRs are used in color sensitive applications (gel coats, etc.).

Thinking "green," and assuming that the thermoset resins could be totally replaced with soy-based ingredients, the total amount would be 0.5 million ton (in 2008), which would represent the quantity of soy oil from 100 million bushels of soybeans [5].

6.1.3
Basic Thermoset Classes: General Aspects

6.1.3.1 Polyester Resins
These resins are formed by the reaction of polyols and carboxylic acids or anhydrides. If one or both precursors are unsaturated, that is, contain a reactive

carbon–carbon double bond, the resulting resin is called *unsaturated* polyester resins [6].

Polyester resins are supplied as a viscous liquid and become a rigid, infusible solid, a thermoset, through an exothermal cure reaction. However, the cure of these resins is normally very slow, as the constituting molecules have low mobility. This problem is solved by the addition of unsaturated monomers of low molecular weight to the polyester resin, increasing the reactions between broken carbon–carbon double bonds. In the beginning of the reaction, the double bonds need to break to create cross-linkings between polymer chains [7]. This is achieved by heating the resin, through electromagnetic radiation, or by the addition of catalysts and reaction accelerators. The free radicals from the catalyst attack the unsaturations on the polyester or on the monomer styrene, for example, to start a polymerization chain reaction, giving a three-dimensional network styrene-polyester copolymer thermoset [6]. For curing at room temperature, the often used catalyst is methyl-ethyl-ketone peroxide (MEKP), together with cobalt naphthenate (CoNap) or dimethylaniline as accelerators. MEKP is in fact, a mixture of peroxides, permitting changes in reactivity by modifying the proportions of each peroxide. The amounts of catalyst and accelerator control the reaction rate and, therefore, the gel-time and the peak exothermic temperature of the curing reaction. Generally, curing systems at room temperature does not achieve a total cure and a post-curing at a higher temperature is required to complete the reaction [8].

Resins formulated for curing at room conditions are slightly unstable and curing inhibitors are required to prolong the shelf-time during resin storage at room temperature. For hot pressing molding, the catalysts are used without accelerators and the mixture is stable for a relatively long time at low temperatures. Once the reaction is started, the exothermal curing takes place usually in a short period of time by increasing the temperature.

As there are a number of available acids and polyols, it is possible to obtain a large number of resin variations. However, the cost of raw materials and the ease of processing limit the number of variations. To avoid an excessively fragile and brittle material, saturated diacids are included in the resin formulation as chain extensors. The higher the saturated acid content, higher the toughness, and lower the cure shrinkage of the polyester. The most used saturated acids are orthophthalic acid (in anhydride form) and its isomer, isophthalic acid [9]. Orthophthalic polyester resins are stiffer, have longer gel time, lower chemical resistance, lower tensile and impact resistances, and lower viscosity than isophthalic resins.

The most used unsaturated acids in the synthesis of polyester resins are maleic acid and its anhydride. Polyesterification in the presence of propylene glycol converts maleic acid to its isomer, fumaric acid. Chlorinated acids are also used when flame resistance and self-extinguishing resins are required, but the use of chlorinated resins has been discontinued owing to emission of toxic fumes during burning [8].

The choice of an adequate glycol or polyol for the polyester synthesis affects the toughness and the physical-chemistry properties of the cured resin. Glycols with longer molecular chains give tougher polyesters, while branched propylene glycol polyethers gives resins with higher impact resistance. UPRs can also be synthesized by the reaction of an unsaturated carboxylic acid, usually methacrylic acid, and bisphenol-A diglycidyl ether (BADGE). These resins, named as ester vinylic, differ from conventional unsaturated polyesters in the following: they present only one terminal unsaturation; possess pendant hydroxyl groups; terminal carboxylic and hydroxyl groups are absent; and they are less susceptible to chemical attacks [8].

UPRs have tensile strength in the range of 40–90 MPa, tensile modulus from 2 to 4.5 GPa and elongation at break between 3 and 5%. Compression strength can vary from 90 to 250 MPa [10]. UPRs are the most widely used resin type for composites, comprising more than 80% of all thermoset resins [11]. The possibility of curing at room temperature and under atmospheric conditions makes these resins versatile materials for fabrication of large structures (as composites) such as boat hulls, automobile bodies, aircraft radomes, and storage tanks. These UPR applications in transportation industry are largely due to the development of bulk molding compound (BMCs) and sheet molding compounds (SMC) using glass fibers [3].

One of the biggest problems with these resins is their high shrinkage during curing reaction (>7%), resulting in warpage and poor surface finish, as the glass fibers can raise from the surface of the molded part [12]. In order to meet different performance requirements, this drawback can be reduced or eliminated, by an adequate selection of different acid/anhydrides and glycols and changes in the ratio of saturated/unsaturated components and the reactive diluent. Addition of thermoplastics in the resin formulation can minimize the thermal and polymerization shrinkage of UPR systems. This low shrink technology has allowed molding compounds to compete successfully with steel in exterior automotive applications [12].

6.1.3.2 Epoxy Resins

Epoxy resins are used to produce high-performance thermosets in a number of industrial sectors, as electro-electronic, packaging, civil construction, and transportation. The main applications in transportation include protective coatings, adhesives, and structural composites. The most important epoxy resin producers are Shell, Dow Chemical, and Huntsman, accounting for approximately 70% of the world's production.

The most used epoxy resins (almost 90% of the world production [2]) are produced from BADGE, Figure 6.1, which is synthesized from epichlorohydrin and bisphenol A [13]. The epichlorohydrin/bisphenol A molar ratio can vary from

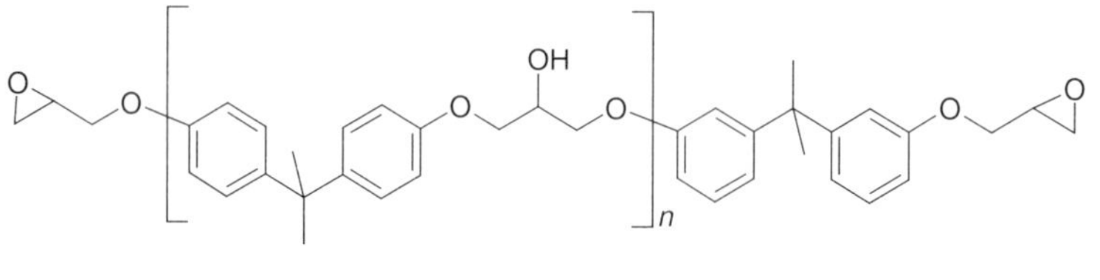

Figure 6.1 Chemical structure of bisphenol-A diglycidyl ether (BADGE).

10:1 to 1.2:1, giving liquid or solid resins. These resins are, in fact, mixtures of oligomers and the average number of repeating units (n) in a molecule can vary from 0 to 25. Low viscosity liquid resins have $0 < n < 1$, while solid resins have $n > 1$.

Multifunctional epoxy resins, as novolac glycidyl ether and tetraglycidyl methylene dianiline present high viscosity at room temperature ($\eta > 50$ Pa·s) and give materials with higher cross-linking degree, as compared with DGEBA-type resins. These factors enhance the performance at higher temperatures, and thus, these resins are used in the manufacture of preimpregnated materials for aeronautic and space industries [14].

A huge number of curing agents (hardeners) are used in the processing of epoxy resins. The curing agent determines the cure reaction, influencing the kinetics, gelation, and the processing cycle. Amine curing agents are divided into aliphatic and aromatic amines. Aliphatic amines are highly reactive, and are liquid and volatile at room temperature, providing relatively short gel-times in the production of epoxy thermosets. Aromatic amines are commercialized as powder or pellets, have lower reactivity and require curing at high temperatures (150–180 °C). These amines can provide a partial curing stage with DGEBA, being adequate in the manufacture of preimpregnated materials (prepregs). Anhydride curing agents have lower reactivity than aromatic amines, permit longer processing times, have low exothermic temperature and have relatively long curing cycles. Generally, they are used with benzyldimethylamine (BDMA) accelerators [15].

The stoichiometry of the epoxy/hardener system is of fundamental importance for the thermoset properties. In principle, all epoxy groups and the curing agent must have reacted to achieve optimized properties. After cure, the epoxy/hardener ratio affects the glass transition temperature, the tensile modulus, and the mechanical strength. The tensile strength of cured epoxy systems varies from 40 to 90 MPa, while the tensile modulus goes from 2.5 to 6.0 GPa, and the elongation at break varies from 1 to 6%. Compression strength varies from 100 to 220 MPa [10].

Epoxy-based prepregs have been used in numerous aircraft components such as rudders, stabilizers, elevators, wing tips, landing gear doors, radomes, and ailerons. The epoxy composite materials constitute 3–9% of the total structural weight of commercial aircrafts such as Boeing 767 or 777 [16]. Composite and laminate industry uses about 28% of epoxy resins produced, including processing by filament winding (tank and rocket motor housings, pressure vessels), pultrusion, casting, and molding (graphite composites for aerospace applications). Besides these applications, the other major user of epoxy is the coating industry [12]. Epoxy resins are also used as adhesives in automotive and aircraft industries to bond metals and composites.

6.1.3.3 Phenolic Resins

Phenolic resins are prepared from phenol and formaldehyde, and even after a century of their first commercial product (Bakelite) and with the advent of new high-performance thermoset polymers, these resins still remain as important commodity and engineering materials in the high-technology transportation industry [17].

This importance is related to their desirable characteristics, such as superior mechanical strength, heat, and flame resistance (low smoke emission on incineration) and dimensional stability, as well as high resistance against various solvents, acids, and water [2].

Phenolic resins can be synthesized under both acidic and alkaline conditions [18]. Novolacs are acid-catalyzed resins prepared at a formaldehyde/phenol ratio between 0.75 and 0.85, and thus, only partially cross-linked. A subsequent curing step, with the addition of hexamethylenetetramine as a hardener, is necessary to achieve a fully cross-linked material. On the other hand, *Resoles* are alkaline-catalyzed resins prepared in formaldehyde/phenol ratios higher than 1, giving a thermosetting material in just one step [17].

Although phenolic resin composites are still in the market for thermo-structural applications in the aerospace and railway industries, because of their good heat and flame resistance, excellent ablative properties, and low cost, phenolics cannot compete with epoxies and PIs for superior engineering applications [2]. New innovative phenolic products and applications continue to emerge, showing their capability to fulfill the requirements and challenges of advanced technologies [19, 20].

6.1.3.4 Polyurethane Resins

PUs are produced from polyols and isocyanates in a reaction usually catalyzed with tin derivatives [21, 22]. These resins represent an important class of thermoplastics and thermosets as their mechanical, thermal, and chemical properties can be tailored by reactions with different polyols and isocyanates. PU chains are linked by urethane linkages ($-NH-COO-$) and often a cross-linking agent, such as melamine, is required to obtain a thermoset polymer.

Nowadays, PUs are one of the most versatile materials, with applications in the manufacture of parts for cars, trucks, trains, planes, tractors, tractor trailers, and recreational vehicles; materials for ships, boats, and flotation devices; automotive cushioning, insulation, and sealing; and adhesives and protective coatings.

The automotive market utilizes both rigid and flexible PU foams. Soy-based, flexible PU foams are being used, for example, in Ford Mustangs, F-150s, and Expeditions. Toyota is also using similar technology in Corollas and the Lexus RX. PU composites fabricated by the pultrusion process are also being evaluated in the transportation industry [5].

6.1.3.5 Typical Processing, Safety, and Handling

To extend the use of thermosets to more high-performance industrial applications, such as transportation [23, 24], a variety of fillers such as continuous or chopped fibers (glass, carbon, aramid, or natural fibers) and/or (nano) particles (clays, silica, and carbon nanotubes) are often added to the resin to form composite materials. Most of these thermosetting-based composites are currently prepared using manual lay-up and spray-up as techniques. The manual lay-up involves the setting of a fibrous fabric on an one-sided open mold, and then spreading the resin over the mold with a manual roller across the fabric [23]. In spray-up, resin and chopped fibers (a few centimeters thick) are spread with an air-assisted gun on an one-sided mold. As an open mold is used in both processes, controlling the final shape can be a difficult task. Another important drawback is an undesirable release of volatile organic compounds (VOCs) during the curing process.

For better dimensional accuracy and safer work conditions, liquid composite molding process, such as resin transfer molding (RTM) is thus preferred. In RTM, a liquid resin is injected into a closed mold containing a fibrous preform. The resin then cures/polymerizes into a cross-linked network, entrapping the fibers [25]. Alternatively, preimpregnation of fabrics with a prepolymeric resin ("prepregs") may be carried out, and after compressing the mixture into a closed mold, curing/polymerization is normally activated by heating in an autoclave.

Hot-press compression molding is also used for applications requiring mechanical performance, complex shapes, dimensional stability, and chemical resistance (e.g., liquid storage tanks and automotive bodyworks). There are two main techniques widely used in industry, generally based on UPR materials: BMC/DMC (dough molding compound) and SMC. BMC/DMC are used to process short fiber (4–8 mm) composites and deep drawn parts can be easily molded. BMC moldings are well known for their creep resistance at elevated temperatures, fire retardancy, and dielectric strength. SMC is used for "prepregs" with much longer (40–50 mm) reinforcements (generally glass fibers). This technique provides better mechanical properties as compared with BMC/DMC, and eliminates the need for post-molding painting on structural and exterior applications [3].

Thermoset composite parts (profiles) with a constant cross-sectional shape may be manufactured through a continuous technique, called pultrusion [3]. This process consists of impregnating the reinforcement fiber with resin by dipping into a bath, and then passing the impregnated fiber continuously toward a heated, shaping die. The resin is cured as it travels through the heated die and the composite is pulled out from the back of the die. Pultrusion has arguably, been a burgeoning sector of the composite industry and every manufacturer of UPR, epoxy or phenolics is able to offer special grades for pultrusion. Filament winding is another way to continuously manufacture hollow composite parts (such as pipes and pressure vessels). After impregnation in resin bath, fiber yarns are continuously conveyed to a rotating mandrel [26]. This technique is normally used for specialized military and space applications, such as rocket launchers and rocket engine components.

6.1.3.6 Ecological Aspects

Owing to environmental issues, waste disposal, and depletion of non-renewable resources, there is presently an increasing demand for natural products in industrial applications [27, 28]. Thus, there is a growing interest in the use of raw materials from renewable resources, for example, soy-derived resins (especially UPR), in areas such as transportation.

Renewable resources can substitute partially, and to some extent totally, petroleum-based polymers. Nevertheless, bio-based polymers can compete or even surpass the existing petroleum-based materials on a cost-performance basis [29, 30].

Solvay and Dow Chemical have both recently announced making epichlorohydrin for thermoset epoxy from glycerin, and Ashland, Inc. and Cargill have announced a joint venture for making propylene glycol used in thermoset UPR [5]. Glycerin is also of interest as a low-cost raw material from biodiesel manufacture.

Interest in renewable resources compels the development, but cost parity is a must. Thus, the conversion of glycerin for materials to be used in thermoset composites is driven by the economics of glycerin feedstock and the technology to find economical conversion processes [2]. Today, soy polyols are priced competitively with petroleum-based polyols.

There is also a great concern to reduce VOC emissions, by reducing the use of solvents and bringing in ultraviolet curing formulations. This is particularly important in the coatings industry. Important efforts are also being taken to replace manual lay-up and spray-up processing techniques with other more environment-friendly techniques, such as RTM.

Recycling is another key environmental issue that is being tackled by manufacturers. Thermoset materials are designed for a long lifetime and can therefore be difficult to break down. However, methods of reuse and recycling are available. Innovations in materials include the use of natural fiber reinforcement in composites and the application of biopolymers as the matrix materials [2, 29, 31].

6.2
Advanced Thermosets and Associated Processes

6.2.1
Polyimides

PIs, that exist both in thermoset and thermoplastic forms, are usually produced by polycondensation of a dianhydride and an aromatic diamine, which gives an acid that is soluble in polar solvents. Usually, heating above 250 °C to remove water results in the formation of the PI cycle. A variety of routes exist for making PIs, and polymerization by addition is also possible. The curing chemistry is difficult to control, and there are problems related to the stability of the reactants, the molecular weight distribution, and porosity, due to volatile entrapment, especially in thick parts [32, 33].

Molding is carried out at around 300 °C with a post-cure at 400 °C to obtain full thermal stability. The elimination of water during cure is another added complication, [34] and when cured, the material is prone to microcracking [35]. Thus, PI is more difficult to work with than epoxies or cyanate esters (CEs), although special formulations and processing techniques have been developed to reduce voids [36].

Polymides are more expensive and less widely used than polyesters or epoxies but comprise some of the best thermally stable organic resins [34]. They have perhaps the best upper working temperature attainable in a readily available polymer system (≈300 °C) [35], and can withstand relatively high service temperatures, that is, 425 °C for several hours or even over 500 °C for a few minutes. This characteristic, along with its rigidity, is a consequence of the presence of ring structures on the PI chain [37].

Distinct classes of PIs have been developed to meet various service temperature regimes [33] and their development has given polymer composites new opportunities for applications requiring moderately high service temperatures (≈250–300 °C) [38], for example, electrical insulation [32], laminating adhesives, wire enamels, gears, covers, piston rings, and valve seats [39]. Some uses in structural composites include missiles and turbine engine (e.g., engine flaps) components, which must function in temperature ranges beyond the capability of epoxy-resin matrices [33, 40].

6.2.2
Bismaleimide

Bismaleimides (BMIs) are a class of resins of somewhat complex chemistry related to PIs [35]. They are addition-PIs made by addition curing of maleic anhydride with MDA, but the latter is being increasingly replaced by other aromatic diamines because of health risks [32]. The imide monomer is terminated by reactive maleimides. Free radical thermally initiated cross-linking occurs across the terminal double bonds. These reactive sites can react with themselves, when high cross-linking density and good thermal stability are achieved, or with other co-reactants such as vinyl, allyl, or amine functionalities [38], which decrease cross-linking density. The homopolymerized resin is inherently very brittle, but this may be countered using copolymerization [32]. Alternatively, the introduction of epoxies, reactive rubbers, or a thermoplastic phase (e.g., vinyls, polysulfone, and polyetherimide) into the resin [38] may improve fracture toughness and processability [41].

Bismaleimides are generally characterized as having relatively low viscosity and long gel times at high temperatures [41]. The resins are usually solid at room temperature, usually requiring preheating (≈125 °C) during processing to melt and reduce viscosity further. Pot life may vary from 30 min to several hours and curing requires temperatures as high as 150–200 °C [41], usually above 175 °C. These resins may reach an upper working temperature of around 220 °C, better than average epoxies [32].

BMIs bridge the gap between epoxies, which are relatively easy to process but whose properties suffer under hot wet conditions, and PIs, which have excellent properties at high temperatures but are generally very difficult to process [41]. Indeed, BMIs can be processed almost as easily as epoxies, requiring only low pressures for processing, and are inexpensive. Compared to epoxies, BMIs are usually more brittle and prone to microcracking [32], and may also show fabrication difficulties with the material in prepreg form due to limited tack [35].

BMIs and PIs are used in high-temperature applications in aircraft and missiles (e.g., for jet engine nacelle components and external hot spots in aircrafts) [42]. BMIs offer hot/wet service temperatures (up to 232 °C), while some PIs can be used up to 371 °C for short periods of time. BMIs exhibit long-term high-temperature stability and are used to make lightweight composite tooling. BMIs and PIs exhibit higher moisture absorption and lower toughness compared to CEs and epoxies, even though significant progress has been made recently to create tougher formulations [36].

6.2.3
PMR (*In situ* Polymerization with Monomeric Reactants)

The basic chemistry of this resin was originally invented and developed at NASA Lewis Research Center [43]. It differs from other resins, because it is used as monomers dissolved in a low boiling point alcohol that is stable at low temperatures, but react *in situ* at elevated temperatures to form a stable PI resin [35].

In a typical use of PMR-15 to produce composites, reinforcements are impregnated with a methanol solution containing an aromatic diamine, an aromatic diester diacid, and a monoalkyl ester of 5-norbornene-2,3-dicarbonylic acid that is used as an endcapper. The monomers undergo cyclodehydration to form a norbornene-endcapped, low molecular weight imide prepolymer in the temperature range of 100–250 °C. Volatile by-products produced at this stage are easily removed. Additional polymerization of the norbornene endcaps occurs at high temperatures without volatile by-products, and yields a network structure with a T_g of about 300 °C [38]. Composites with void contents <1% can be prepared [44].

PMR-15 resins have an average molecular weight of 1500 g mol^{-1}, hence the name, and are considered the most widely used resin group with high in-service temperature capabilities, up to 316 °C for at least 1500 h in air. They are brittle and of primary interest to the aerospace industry for high temperature uses, including aircraft engine applications [45, 46]. PMR-15 is likely to be displaced with modified PIs, which are considered tougher and less toxic during processing [32].

The second generation of PMR-15, developed to overcome difficulties related to thermal-oxidative instability and high brittleness [43], was called *PMR-II*. It is synthesized from the perfluorinated isopropylidene known as *6FDE* (diethyl ester of 4,4′-(hexafluoroisopropylidene)diphthalic acid) and phenylenediamine, reaching a molecular weight in the 3000–5000 g mol^{-1} range [1].

6.2.4
Polystyryl Pyridine

Polystyryl pyridine (PSP) is synthesized by condensation of a pyridine containing at least three methyl groups with an aromatic dialdehyde [47]. The PSP chain can be terminated with unsaturated methyl vinyl pyridine, and the resulting unsaturated prepolymer can then be cured by an addition reaction. The addition reaction does not generate water and consequently the prepolymer can be cured without the problems created by volatile water [48]. Fabrication is difficult because solvents must be used in a high-temperature curing cycle [35].

PSP was originally developed as an ablative system and hence has excellent high-temperature properties, even in the presence of moisture, as well as high char yield (about 70%) in nitrogen, with a glass transition temperature around $280\,^{\circ}C$ [35]. These properties are particularly desirable for the matrix of graphite fiber composites, which may be subjected to fire. PSP resin can also be used to prepare marine antifouling painting [49].

6.2.5
Silicone

Silicone resins, also called organopolysiloxanes, are three-dimensional cross-linked compounds that are based on a silicon/oxygen skeleton. Silicone resins are formed by polycondensation of silanes, yielding viscous and solid polymeric siloxanes, which are soluble in organic solvents. These silicone resins – almost exclusively methyl silicone resins – dry from organic solution or emulsion to form a non-tacky water-repellent film. In a process referred to as *chemisorption*, the remaining alcohol groups undergo a condensation reaction with available substrate groups (usually OH–), forming silanol. The alcohol groups also react with one another, thus further cross-linking the silicone resin into a silicone resin network [50]. Cross-linking by heating occurs in the presence of a catalyst such as CoNap, zinc octoate, or amines such as triethanolamine, producing complex, heavily branched oligosiloxanes, which are available in the form of flake resins and solvent-based or solvent-free liquids.

Silicone resins are among the most heat resistant resins used for composites, exhibiting good mechanical and electrical insulation properties, water repelling characteristics, good moisture, and chemical resistance, and they also resist weathering and UV radiation. Owing to their high cost, silicone resins are primarily used in coating applications requiring outstanding thermal stability, in the $300{-}500\,^{\circ}C$ range, being mineral-filled for maximum heat resistance [51]. They are suitable for many electrical applications, particularly in electrical insulation exposed to high temperatures, and as encapsulating materials for electronic devices. Some examples include: binders for high-temperature paints, protective coatings, automotive clear coats, antifouling paints, varnishes, impregnating and protecting electrical coils and resistors, protective conformal coatings on rigid and flexible circuit boards, and coatings to protect transformers and solenoids from moisture. Silicone resins are further used in the formulation of pressure sensitive adhesives, release coatings, specialty elastomers, antifoams, and other surfactants [52], and also applied to prepreg manufacture [34].

6.2.6
Cyanate Esters

CE resins are a family of high-temperature thermoset resins – more accurately named polycyanurates – that bridge the gap in thermal performance between engineering epoxy and high-temperature PIs. CEs are based on a bisphenol or novolac derivative in which the hydrogen atom of the phenolic OH group has been substituted by a cyanide group, that is, they have a CE-OCN functional end group. The chemistry of the cure reaction regards cyclotrimerization of CN groups producing triazine ring systems. CEs can be cured and post-cured either alone at elevated temperatures or at lower temperatures (150–200 °C) in the presence of a suitable catalyst in an addition polymerization process. The most common catalysts are transition metal complexes of cobalt, copper, manganese, and zinc [38, 53].

Product properties can be fine tuned by the choice of substituents in the bisphenolic compound. Bisphenol-A and novolac-based CEs are the major products, but bisphenol-F and bisphenol-E are also used. The aromatic ring of the bisphenol can be substituted with an allylic group for improved toughness of the material. CEs can also be mixed with BMIs, to form bismaleimide-triazine (BT) resins, or epoxy resins for optimizing the end use properties and reducing the cost. Besides, CEs are usually toughened with thermoplastics or spherical rubber particles [36].

CEs process is similar to epoxies, but their curing process is simpler due to lower viscosity and low outgassing of volatiles – for instance, aryl CEs such as bisphenol A dicyanate are used for carbon fiber prepregs. The unique trimerization of CEs contributes to their high T_g (250–290 °C) and their long-term performance in hot and wet environments (up to 149 °C) ranges between BMIs and epoxies [38]. They are versatile matrices that provide excellent toughness and strength, low shrinkage, allow very low moisture absorption and possess superior electrical properties (low dielectric constant and dielectric loss), although at a higher cost than other matrices [36]. Other characteristics include very good fire, smoke, and toxicity performance.

CEs can be formulated for use as high-performance adhesives, syntactic foams, honeycomb for sandwich construction, fiber-reinforced composites, and blends. Applications range from radomes, antennae, nose cone, missiles, damage-resistant aircraft composites, and ablatives to microelectronics packaging (e.g., conductive adhesives and encapsulants), printed multilayer circuit boards, multichip modules, optoelectronics (e.g., optical waveguides and nonlinear optical devices), and microwave products. CEs have been qualified in the areas of satellite, radar, air frame/missile, and dielectric structures [36], but because of their high cost and lack of a comprehensive database, they did not penetrate into the large commercial aircraft and structural composite industry, even though lower-cost CE resins and CE blends were developed [53, 54].

6.2.7
Furane

Furane is a liquid thermoset resin in which the furan ring is an integral part of the polymer chain. Furan resins, produced by the reaction between phenols and furans such as furfurals (aldehyde) and furfural derivatives, are relatively new. Furfuryl alcohol-based resins are the most important industrial furan materials in terms of usage and volume.

Furfural replaces formaldehyde in the conventional production of phenolic resins. It reacts easily with phenol in the presence of an alkaline catalyst to form a novolac phenol-furfural resin. Furfuryl alcohol readily homopolymerizes in the presence of an acid catalyst to produce liquid linear chains (oligomers) in a highly exothermic process. The resulting liquid resin has a shelf-life of more than six months. Furfuryl alcohol also undergoes copolymerization with aldehydes (e.g., formaldehyde and furfural) and with phenols and urea in the presence of an aldehyde. In the presence of a strong acid catalyst, especially aromatic sulfo acids and mineral acids, the furfuryl alcohol resins cross-link via condensation [55]. As with phenolic resins, the need for acidic catalysts leads to considerable fabrication problems [34].

The furan resins were originally introduced to complement phenolic resins, and as such, have comparable, but sometimes better, properties than the latter. The cured products are known for their high resistance to heat, acids, and alkalis, as well as high coking values. They are primarily used for chemical resistant applications, showing perhaps the best chemical resistance of any thermoset resin in nonoxidizing conditions. They have excellent solvent resistance, unlike many of the commonly used resins in chemical plant applications [34]. Their characteristic light colors imply that they can be used in various colored products. The main market for furan resins is adhesive and bonding, but they are also used as binders in core moldings and friction materials [55].

6.2.8
Benzoxazine

Benzoxazine can be considered a subclass of phenolic resins. The chemistry involved, comprising the reaction of a phenol with an aldehyde and an aromatic amine has been known since the 1940s. The benzoxazine ring is stable at room temperature for a long period of time and starts to homopolymerize to high molecular weight oligomer or polymer at elevated temperature. They are halogen-free systems and some require no refrigeration during storage [36].

The ring opening polymerization of benzoxazine does not generate any volatiles due to addition reaction even though it requires high temperature for gelation and curing. They may be combined with epoxy resins under heat to produce new thermoset materials with very high glass transition temperatures, and can also be toughened by adding traditional agents such as thermoplastic rubber and nanoparticles [56].

Depending on the backbone chemistry, benzoxazine systems can provide polymers with high-temperature resistance and thermal mechanical properties along with excellent flame/smoke/toxicity performance. The flammability characteristics of benzoxazine are very similar to the phenolic resin owing to the formation of very similar network after polymerization. Other characteristics include: low moisture absorption, dimensional stability (i.e., low shrinkage), low dielectric constant, very high hot/wet property retention, and high strength [36, 56].

The rigid polymer may be used in producing composites, coatings, adhesives, and encapsulants, for example, for electronic packaging applications [56, 57].

Benzoxazines may be used to replace phenolic resins, because of toxicological issues in the workshops before or on curing, along with very good fire retardant properties. Aerospace industry uses huge quantities of phenolic prepregs to manufacture semi-structural interior parts, which must satisfy the more stringent fire smoke and toxicity (FST) regulations [58].

6.2.9
Urea-Formaldehyde and Melamine-Formaldehyde

Urea formaldehyde (UF) thermosets are strong, glossy, and durable resins. They are not affected by fats, oils esters, ether, petrol, alcohol, or acetone, nor by detergents or weak acids, and they exhibit good resistance to weak alkalis. Their high mechanical strength, heat, and fire resistance, and good electrical arc and tracking resistance make them useful for numerous industrial and household applications, including electrical components, cosmetic enclosures [59], adhesives, finishes, medium-density fiberboard (MDF), and various molded objects (e.g., doorknobs and toilet seats).

Melamine formaldehyde (MF) resins are similar to UF resins, but have even better resistance to heat, chemicals, moisture, electricity, and scratching. MFs are ideal for dinnerware, kitchen utensils, bathroom accessories, and electrical components, and some of the uses include: electrical breakers, receptacles, closures, knobs and handles, appliance components, adhesives, coatings, and laminates [59].

6.2.10
Phthalonitriles

Phthalonitrile is a lesser known resin class originally developed by the US Naval research Laboratory for service temperatures approaching 370 °C [36]. Phthalonitrile-based polymers are high-temperature polymers with a wide range of potential applications including adhesives, coatings, electrical conductors, and composites (e.g., high-temperature engine parts [36]). In addition, bisphthalonitrile monomers have been established as candidates for matrix materials in advanced composites [60].

Phthalonitrile resins are obtained by heating a diphthalonitrile monomer with a primary amine at a temperature from about the melting point of the monomer to about the decomposition temperature of the resin. The resins can also be prepared by dissolving a diphthalonitrile monomer in a solvent, for example, acetonitrile, adding a primary amine to form a suspension, heating the suspension to the B-stage, quenching the suspension, removing the solvent, and heating the residue at a temperature from about its melting point to about the decomposition temperature of the resin [61]. On curing, the cyano groups of the phthalonitrile react to form triazine and other heterocyclic ring structures. The formation of heterocyclic cross-linked structures promotes high thermo-oxidative stability, as well as good mechanical properties, for example, toughness. However, extensive curing times and temperatures are necessary to achieve these properties. Furthermore, the processability of these resins is limited as the crystallinity of the phthalonitrile monomers increases [60].

It is important to bear in mind that the processing of high-temperature resin systems in general is not straightforward as they inevitably involve curing at high temperatures for extended periods of time. The processing cycle can be further complicated by the need to remove the solvent used to ease processing or the requirement to heat the material to reduce viscosity. Also, increased temperature capability most likely results in loss of toughness, which can affect performance or even give rise to cracking because of the residual stresses generated at high-temperature curing [35]. Furthermore, most resins which are said to retain their properties to above 300 °C, however, suffer a decrease in T_g, when subjected to temperature cycles in a humid atmosphere [32].

6.3
Thermosets for Coatings and Adhesives

Generally, parts are coated to provide color, texture, and protection. In more specific cases, coatings might attenuate electromagnetic interference/radio frequency interference.

In the transportation area the importance of coatings can be estimated. As around 10% of a car (in mass) is composed of polymers, and 40 types of different polymers have been used lately, a reduction in 10% of a car's mass reduces gas mileage by 7% approximately. In addition, 40% of polymers used in cars are molded in color and unpainted, although manufacturers do decide to paint for several reasons such as: esthetics, ultraviolet resistance, chemical resistance, and scratch resistance. In case of etching resistance coatings, thermosets play a very important role in avoiding diffusion of acids caused by acid rain [62].

Coatings are in general applied to metals or polymer surfaces. In general, an anticorrosive coating system usually consists of several different layers of different coating with different properties and purposes. According to Schweitzer [63] there are four types of coatings: organic (acrylics, urethanes, epoxies, alkyds, etc.), inorganic (silicates, ceramics, and glasses), conversion (anodizing, phosphating, chromate, and molybdate), and metallic (galvanizing, vacuum vapor deposition, electroplating, and diffusion). The most common anticorrosive system may consist of a primer, one or several intermediate coats and a topcoat. The function of a primer is to protect the substrate from corrosion and ensure strong adhesion to the substrate [64].

As very well pointed out in the literature, one of the main reasons for the limited number of high performance coating systems is the complexity of the coating-substrate system and the number of factors affecting the performance and service life of anticorrosive coatings. Recently, another challenge is to develop anticorrosive coatings with low amount of volatile organic solvents [64, 65].

Bierwagen [66] developed a coating based on PU automotive clearcoat-type of coatings by modifying the polyol portion of the reactive polymer system to introduce water sensitivity/removability at high pH.

6.3.1
Corrosion Protection

The essential role of an organic coating is to delay the transport of ions to the substrate inhibiting the formation of corrosion cells [63]. It should be pointed out that most of the organic coatings without pigments and nanoparticles are not able to avoid the transport of water, oxygen, and ions to the metal surface.

In this scenario nanocomposites can play a major role. Allie *et al.* [67] studied the corrosion protection of two nanocomposites of epoxy and poly(vinyl chloride-co-vinyl acetate) with clays. Steel substrate were prepared, coated, and submerged in a tank with 5% NaCl solution. Samples were tested using electrochemical impedance spectroscopy to study the effect of nanoclay on the corrosion protection of the coatings. Tests showed that poly(vinyl chloride-co-vinyl acetate) with 0.5% nanoclay provided superior barrier protection.

Yeh *et al.* [68] showed that polymer clay nanocomposites consisting of a siloxane-modified epoxy resin with nanolayers of a montmorillonite used as coating on cold-rolled steel (CRS) were found to be much superior in corrosion protection over those of pure epoxy resin when tested for performance in a series of electrochemical measurements of corrosion potential, polarization resistance, corrosion current, and impedance spectroscopy in 5 wt% aqueous NaCl electrolyte. The permeability of oxygen, nitrogen, and water was lower than in those materials with low content of nanoclays.

PU combined with nanoclays can also present an interesting option for anticorrosive protection. By using electrochemical impedance spectroscopy it is shown that nanocomposites of PU are very superior than the neat PU in terms of anticorrosive characteristics [69, 70].

It can be seen in the literature that a majority of nanocomposites coatings are based on clays and silica. Nevertheless, recent investigation on the role of other inorganic particles, such as TiO_2, have been also carried out [71]. Another interesting approach is related to the use of self-healing coatings, and this concept is well presented by Sauvant-Moynot *et al.* [72] and Samadzadeh *et al.* [73].

6.3.2
Adhesive Thermosets

Adhesives are designed to hold materials together as alternatives to mechanical fastening systems. They can come in several forms and in a wide range of strengths and can be used from holding papers to bonding parts of cars and aircrafts. It can be seen as a competition between adhesives and mechanical fastening systems in different industries [74].

An important difference between a coating and an adhesive is that the adhesive must adhere to two substrates. An adhesive is generally sandwiched between two substrates and it is expected to have a bond strength high enough to fracture or tear at least one of the substrates. In general, it is expected that bond strength is not affected by heat, humidity, and solvents.

Epoxy and PU adhesives have long been associated with quality, reliability, and innovative chemistry. In general, most of these systems are qualified to comply with aircraft manufacturer's specifications for repairing parts [75].

It is now common for modern vehicles to employ extensive use of both structural and semi-structural adhesives. In 2004, according to Montimer [76], a DB9 sports car used the first UK applications of precision robot-dispensed two-part adhesives for bonding aluminum body-in-white (BIW) structures.

In the automotive industry, as very well pointed out by Hutchinson *et al.* [77], disassembly strategies for repair and end-of-life are increasingly important. Nowadays, dismantling is labor intensive and inefficient, ranging from mechanical cutting to thermal degradation. As such there is a need for reversible adhesive, since currently the only accepted solution for the disassembly of structurally bonded joints in an automotive application is mechanical destruction.

Schmid *et al.* [78] developed a polypyrrole-coated thermally expandable microspheres, which according to authors is suitable for niche applications such as the reversible adhesion of car glazing and panels. In addition these authors pointed out an interesting direction for near future where metal panels and welded joints are being gradually replaced by lightweight composite materials prepared using adhesive bonding technologies.

According to Loureiro *et al.* [79], adhesive bonding is increasingly being used in structural applications such as in automotive joints. Elastomeric adhesives such as PUs have been used in windshield bonding, representing the important advantages in terms of damping, impact, fatigue, and safety, which are critical factors in the automotive industry. In addition, there are other structural applications in the main body where PUs could also be used.

6.4
Case Studies – Thermoset Composites

Use of composite materials in the transportation sector has been the object of many studies [80–82]. The commercial use of thermoset composite materials has an extensive history in the transport industries. Typically, thermoset resins can be used for molding composites by prepreg/autoclave, liquid molding techniques (RTM, vacuum assisted RTM, hand lay-up), BMC and SMC, and by automated processes, such as tape laying and filament winding [83–85].

The prepreg method uses a preimpregnated reinforcement, unidirectional or fabric, with an amount of resin partially cured, called *B-stage*. The prepreg is stored at $-18\,^{\circ}$C. High temperature and pressure are required for final curing. The fiber lay-ups are placed on a mold and vacuum bagged according to design guidelines.

Pressure is applied to the lay-up in the autoclave. Autoclaving is a clean process and the drawback is the prepreg limited out-life and limitation on the size and complexity of the part [86].

The liquid molding techniques are varied represented mainly by RTM and resin infusion molding. The RTM is a closed molding process that requires a quantity of dry reinforcement in the correct size, shape, and orientation in the form of preform is placed in the specially designed and built closed mold. Further, a required quantity of resin is injected into the mold under pressure. Curing of the resin takes place with or without heat. The fiber loading in this process is limited to <50%. The resin infusion technique is a hybrid of the autoclave method and RTM. The tooling is similar to autoclave method where the flow medium is the fiber preform. A low viscosity resin is appropriately sucked into the flow medium. Using this method the fiber loading can be as high as 60%. This is very close to prepreg technology and there is no limitation due to out-life. Large parts can be molded as also co-cured parts.

Another popular method of molding composites, mainly for the rail and automotive sectors, are the SMC and BMC. SMC/BMC are designed for compression and injection molding. The compound consists of a blend of resin (the most important part), usually unsaturated polyester or vinyl ester thermoset resins, reinforcement, catalyst, filler, mold release agents, and other additives. Both SMC and BMC are flowable compounds that are transformed under heat and pressure. SMC is a leather-like sheet material supplied in rolls, which has a longer fiber length compared to BMC. BMC comes in dough form and can be transformed by either compression or injection molding. On the other hand, fiber loading is low (<35%).

More recently, automated and robotic processes are being fast introduced in the aerospace industry. The tape laying is a continuous process, similar to filament winding, replacing it in aeronautical construction, mainly for surfaces of revolution, such as fuselages.

All the processes mentioned earlier have taken advantage of the easy processing characteristics of the thermosetting resins (Table 6.1). Among thermosets the epoxy family, phenolics, polyester, vinyl ester, and more recently, the benzoxazine family are the most used [87].

The selection of the manufacturing process for a composite component is a function of the nature of the part and the required production volumes. Hand lay-up technique, results in very low production volumes (<100 parts per year) and low quality parts. Vacuum infusion (VI) are suitable for low-medium production volumes (<500 parts per year), large parts (>1 m) with intermediate fiber content (<35% by volume). RTM is suitable for medium production volumes (<30 000 parts per year, small-medium sized parts (<2 m) with high fiber content (30–55% by volume) [88].

In the transport sector, lightweighting is a key issue. The estimated cost benefit for a kilogram saved in the automotive, rail, and aeronautical sectors are shown in Figure 6.2 [89].

Table 6.1 Composite quality according to different manufacturing processes.

Quality	VARTM	RTM	Autoclaving	SMC/BMC
Fiber content	Changeable	Adjustable	Fixed	Adjustable
Surface quality	Adequate	Good	Good	Good
Part thickness	Irregular	Uniform	Uniform	Uniform
Void content	Low	Less	Less to least	Low
Racetracking	No	Yes	No	No
Curing control	Normal	Simple	Simple	Simple
Dimensional control	Adequate	Good	Good	Good

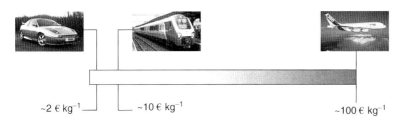

~2 € kg^{-1} ~10 € kg^{-1} ~100 € kg^{-1}

Figure 6.2 Cost benefits of lightweighting in the transport sector [89].

6.4.1
Composites in Automotive Vehicles

Traditionally, the automotive sector has worked with isotropic sheets of metal that are joined by welding processes. However, the opportunities for composites in the automotive industry are significant, especially if one considers their potential advantages compared to metals. Lightweighting is one of the major drivers for the use of composites in the automotive industry and are an enabling technology to this. In sport cars weight reduction leads to increased performance in terms of acceleration and top speed. In mass production vehicles, the most important driver for lightweighting is the reduction of fuel consumption and the associated reduction of CO_2 and other emissions [81, 90].

Composites require specialized knowledge of both the materials and the related processes, if the opportunities they present in terms of functional integration, such as lightweighting, orthotropic behavior, and styling freedom are to be properly exploited. Moreover, composites can offer benefits in terms of weight reduction, tooling cost savings, and design and styling freedom. Niche production cars have already demonstrated the feasibility of using composites and automotive designers are now looking to widen their application to passenger cars and heavy duty trucks [91].

Historically, the use of composites have been limited to secondary automotive structures such as appearance panels and dash boards, and as semi-structural application mainly in sport cars. The major obstacles to implementation of polymer-based composites the in automotive industry stem from a variety of factors, and the important ones are inexperience with these materials, undeveloped high production rate processes, joining techniques, lack of knowledge about material responses to automotive environments, lack of crash models, immature recycling technologies, and a small supplier base [92]. Composites can locally vary in types of reinforcements, which significantly offer advantages over metal tailored blanks. For instance, different thicknesses of sheet metal that are laser welded before stamping, could apply for car doors. The use of composites in the aerospace and military applications has brought a pack of knowledge about durability and fatigue life, and nowadays it is a consensus that this is not a concern.

The debut of composite materials in the automotive industry was in 1953, when GM first used it for the body panel in the Corvette [93]. At that time, the Corvette body was made from hand lay-up polyester resin and fiberglass. Although plastics had been used in many car components since the beginning of the forties, such as in distributor caps and steering wheels, the Corvette was a significant milestone.

However, there has been no wide switch from metals to composites in the automotive sector. This is because there are a number of technical issues relating to the use of composite materials that still have to be resolved, such as accurate material characterization, manufacturing, painting, and coupling with metals.

Thermosetting polymers (including those reinforced with fibers) are less widely used, and current forecasts do not show big opportunities. In fact, only two processing methods for thermoset composites have been sufficiently automated to make them accessible for medium/high volume automotive applications – SMC and RTM, mainly for non-structural applications.

A good example of a high-volume vehicle that has pioneered the use of these technologies is the Renault Espace. The 1984–1996 versions featured polyester SMC body panels and an RTM tailgate. At production levels of 70 000 vehicles per year, the Espace represents one of the most significant automotive applications of SMC [81].

Other vehicles to feature extensive use of SMC include the Renault Megane II (tailgate and wings), the Fiat Coupé and Alfa Romeo GTV (integrated bonnet/wing), the Mercedes-Benz CL-500 (boot lid), the 2002 Ford Thunderbird (body panels and hard top roof), and the Volvo V70 (tailgate). In many of these examples, it would have been difficult, if not impossible, to achieve the desired styling with metals [81] (Figures 6.3 and 6.4).

Research priorities for composites used in the automotive sector are concerned with processes suitable for high-volume production and process automation and recycling [94]. Also, high-volume production using out-of-autoclave processing need fast-curing thermosets, for reducing manufacturing costs and cycle times [95]. Specific test methods and design procedures for the automotive sector are also needed. Composite materials failure criteria and new numerical models for automotive composite materials are also an issue for research. Current automotive fire safety regulations are largely based on metals, and more stringent fire safety requirements can be expected as the introduction of resin-based materials might

Figure 6.3 Renault Espace (1984–1996) – polyester SMC body panels and a resin transfer molded tailgate.

Figure 6.4 Volvo V70 XC AWD support lid.

lead to new regulations. The development of integrated product/process analysis tools can reduce the number of experimental tests required during the development of composite parts [92, 96].

6.4.2
Composites in the Rail Vehicles

Composite materials can be used in many applications in the railway industry: infrastructure, interior, and exterior parts in rolling stocks. The main advantage in this sector is the better corrosion resistance as compared to metal [96, 97]. In the case of composite poles or signaling equipment, for instance, the advantages are lower maintenance costs and longer lifetime than wood or steel. The composite sleepers are another application that last more and have lower weight than concrete, enabling faster and easier installation, which are key parameters for water drains (on the side of the tracks). In terms of rolling stocks, the main advantage is the lightweight that results in saving energy. Compared to metals, a 30% in weight reduction can be easily achievable. Also, freedom of design leads to optimization of internal and external parts. It also allows to make aerodynamic exterior parts having complex geometries, to reduce the energy consumption of the train. Another advantage would be the low tooling cost so that low series can be done, and very good fire resistance with new resins such as the Norsodyne H 81269 TF that have

Table 6.2 Typical properties of fiberglass composite, steel, and aluminum.

	Glass fiber composite[a]	Steel	Aluminum
Density (g cm^{-3})	1.6–2.1	7.85	2.70
Tensile strength (MPa)	100–1400	200–1300	100–400
Tensile modulus (GPa)	12–40	210	70
Thermal conductivity (W m K^{-1})	0.50	26–46	170–237
Coefficient of linear expansion ($0.10^{-6}/^{\circ}$C)	5–10	10–18	27

[a]Depending on glass content.

been specifically developed by Cray Valley for the railway industry. A comparison of typical properties of fiberglass composite against metals (steel and aluminum) is shown in Table 6.2. Train cabins and rail seats are another example of replacing steel components with fiberglass composites, as shown in Figures 6.5 and 6.6.

Trains increased their weight since the beginning of the 1970s [89]. The estimated increase in weight is about 40% heaving per seat. The reasons for that are the performance (e.g., tilting mechanisms), crashworthiness (e.g., crumple zones), accessibility (e.g., disabled lavatories and seating rules), and passenger comfort and convenience (e.g., air conditioning and power points for laptops). In the 1970s the hardware of a train was made almost entirely of metals. Nowadays, composite materials can provide an enabling technology to assist in meeting the capacity increases targeted by the railways, by using double-decks and longer trains [82, 98].

Figure 6.5 Composite high speed train cab produced by vacuum infusion. Source: Courtesy of Alcan Airex Composites.

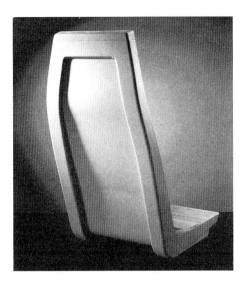

Figure 6.6 Composite rail seat produced by resin transfer molding (RTM). Source: Courtesy of Ashland Composite Polymers.

Research priorities related to the use of composites in rail sector are mainly related to modeling methodologies (structural analysis, long-term behavior, damage mechanisms and failure modes), understanding the behavior of composites at elevated strain rates, and characterizing composites under dynamic loading [82]. Also, important issues are related to process simulation to improve overall efficiency, process consistency and optimization. Accurate life cycle analysis is also an important issue to quantify their environmental, financial benefits, and repair procedures. The development of new resin systems that provide good all-round performance in terms of fire, smoke and toxicity, processability, mechanical properties, and surface finish, as well as, the implementation of low-cost fire test protocols to assist the research and development of new fire-safe composites [99].

The development of rail gear-cases in fiberglass composite is one example of replacing steel parts in the rail sector. A work on this subject was reported by M/s Permali Wallace Ltd, India, as shown in Figure 6.7 [98]. The gear-cases are used in the traction motors for diesel and electric locomotives and the aim is replacement of existing steel gear-cases. Weight reduction, improved assembly and maintainability, optimization of properties, component testing, standardization, were the main goals to make this component technologically and commercially attractive. The heavy weight of steel gear-case sometimes causes its detachment from the traction motor due to violent jerks or impact, which is very common in locomotives. The lightweight of the FRP gear-case prevents such occurrences. These gear-cases are expected to last for six years in service as against two to three years of conventional steel gear-case. Various design features of the gear-case were tried to bring down the cost by trying different resin systems. It was observed that unlike steel gear-case, fiberglass composite gear-case prevented the leakage

Figure 6.7 Rail gear-case in fiberglass composite [98].

of lubricating medium. They were also safe from damages by stone ballast. Also, they are easy handled as mounting and demounting are quick, and consequently requires less man-hours and engine downtime for any maintenance. In addition, weight saving of 372 kg per six-axle system has also been achieved in relation to the steel counterpart.

The work done by M/s Permali Wallace Ltd, showed that experimental trials with fiberglass composite gear-cases have broken the myth of using only steel/metals for such stringent functional requirements in locomotives. The fiberglass composite gear-cases have glass content of 64%, a density of 1.94 g cm^{-3}, a tensile strength of 326 MPa, a flexural strength of 463 MPa, and an Izod impact strength of 133 kg cm. The water absorption at 27 °C for 24 h was 0.118%.

6.4.3
Composites in the Aeronautical/Aerospace Industry

The aeronautical and aerospace industry has plenty of examples of composite usage and nowadays the development of this sector without composites [100, 101] is unthinkable. The versatile nature of composites attracted the designers to use these materials for several critical aircraft structural applications. As a result, a large number of materials (fibers and resins) are developed for use. Added to this, a large number of fabrication processes are also developed. To realize a cost-effective civil aircraft structure, what material and process need to be used is not an easy matter to decide. An attempt is made in this chapter to logically select the material and the corresponding fabrication process for various components of an aircraft for civil application [102].

The Boeing 787, scheduled to start flight testing in 2007/2008, was a smaller aircraft than the A380 and was designed to carry approximately 300 passengers. The B787 materials design was based substantially on carbon fibers in an advanced toughened epoxy resin matrix systems for the fuselage and a composite wing. Tape laying techniques had been developed to produce the composite fuselage,

Figure 6.8 A cured 787 fuselage barrel section [103].

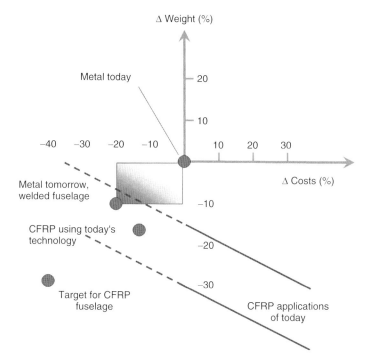

Figure 6.9 The goals for weight and cost trade on airbus planes.

including composite fuselage sections 6.7 m long and nearly 6 m wide, as shown in Figure 6.8.

The challenge for all developments in aerospace is always to find a good compromise between performance and cost [104]. Depending on the nature of the mission and the market, one or the other will dominate. As an example, the Figure 6.9 shows the goals for the next generation of Airbus planes. Clearly, a 40% cost saving and a 30% weight saving compared to the state of the art cannot be reached in small steps.

An integrated approach that takes all disciplines into account is necessary. With this in mind, recommendations for future research priorities in the aeronautical

Figure 6.10 Airbus begins A350 XWB fuselage manufacturing by automated fiber placement.

sector rely mainly on improved manufacturing technologies for better affordability and quality. The main goals are higher degrees of automation, better quality control, reduced tooling costs, and shorter cycle times to be consistently attained [105]. Recycling of aircraft is also an issue of priority because of political pressure and regulations.

When optimizing manufacturing technologies, such as non-autoclave injection technologies and microwave curing, a multi-disciplinary approach involving the views and technologies from across the transport sectors (automotive, rail, marine, and aerospace) can be very useful. Continuous improvement in materials is also necessary in the fields of carbon fibers (e.g., cost reduction and improved textile processability), binders, and matrix systems (e.g., fast curing, high-temperature capability, and high toughness). Significant advances can also be expected in nanotechnologies. Moreover, composites are the most promising materials for realizing multifunctional structures with active damping, shape control, and health monitoring [80].

Similar to Boeing, Airbus Co. is also in the track for composite fuselages. Most of the A350 fuselage uses carbon fiber composites fabricated in assembled panel sections (Figure 6.10).

The rear section of the plane, however, uses tapers having a smaller diameter, making it best suitable for automated fiber placement (AFP). The Airbus Advanced Composites Center in Illescas, Spain, has begun to produce the 5.5 m rear fuselage, as seen in Figure 6.9, on the A350 XWB. A MAG AFP system was used [103].

6.5
Summary and Outlook

Nowadays, about half of all thermosets are used by the automotive industry, but there is a quickly growing number of metal-replacement applications in aerospace. Airplanes have a large number of small metal components (brackets, clips, window frames, etc.) that are ideal for compression molding of carbon-fiber-reinforced thermosets. Lower-cost carbon fiber (as it becomes "available in mass quantities") will aid these conversion projects [106].

Aerospace industry is also looking for high-temperature thermosets. Reusable launch vehicles, for example, need lightweight materials that can perform well for short duration at temperatures around 500 °C. Current efforts are focused on raising the usage temperatures of PI composites by increasing the T_g of the resin. Realizing this increase in upper usage temperature without sacrificing the processing of polymer composites is a major technical challenge. New thermoset compounds that can handle 260 °C continuous-temperature applications are also under development [106].

The rise of in-line compounding and molding of SMC and BMC over the next few years, and a continuous effort to reduce or eliminate VOC emissions during processing is also predicted.

A big impact of nanocomposites (with nanoclays, carbon nanotubes, silica or titania nanoparticles, etc.) on the thermoset industry is expected, with more high-end opportunities in thermosets than in thermoplastics.

Another important trend is the partial, or in some cases total, substitution of petroleum-based thermosetting materials by renewable-based analogs, especially for epoxy and PU resins. In most cases, the same performances starting from these ''green'' precursors are achieved as compared with petroleum-based thermosetting materials. Renewable-based resources are being viewed not only as a substitute for the existing thermosetting materials, but also as raw materials for producing novel thermosets in more specific applications [2].

References

1. Goodman, S.W. (ed.) (1998) *Handbook of Thermoset Plastics*, 2nd edn, Noyes Publications.

2. Raquez, J.-M., Deléglise, M., Lacrampe, M.-F., and Krawczak, P. (2010) Thermosetting (bio)materials derived from renewable resources: a critical review. *Prog. Polym. Sci.*, **35**(4), 487–509.

3. Forsdyke, K.L. and Starr, T.F. (2002) *Thermoset Resins – Market Report*, Rapra Technology Ltd, Shawbury.

4. Witten, E. and Schuster, A. (2010) Composites Market Report: Market Developments, Challenges, and Chances. AVK – Industrievereinigung Verstärkte Kunststoffe (Federation of Reinforced Plastics).

5. United Soybean Board (2008) Market Opportunity Summary: Soy-Based Thermoset Plastics. USB Report.

6. Fried, J.R. (1995) *Polymer Science and Technology*, Prentice Hall, Englewood Cliffs, NJ.

7. Parkyn, B. (1972) Chemistry of polyester resins. *Composites*, **3**(1), 29–33.

8. Scheirs, J. and Long, T.E. (eds) (2003) *Modern Polyesters: Chemistry and Technology of Polyesters and Copolyesters*, John Wiley & Sons, Ltd, Chichester.

9. Burns, R.B. (1992) *Polyester Moulding Compounds*, Marcel Dekker, New York.

10. Levy Neto, F. and Pardini, L.C. (2006) *Compósitos Estruturais – Ciência e Tecnologia*, Editora Edgard Blücher, São Paulo.

11. Lucintel (2009) The Global Unsaturated Polyester Resin (UPR) Market 2010–2015: Trends, Forecast and Opportunity Analysis.

12. Varma, I.K. and Gupta, V.B. (2000) Thermosetting resin – properties, in *Comprehensive Composite Materials*, Vol. 2 (eds A. Kelly and C. Zweben), Elsevier, Oxford.

13. Ellis, B. (1993) in *Chemistry and Technology of Epoxy Resins* (ed B. Ellis), Blackie Academic, London, pp. 1–35.

14. Bauer, R.S. and Corley, S. (1989) in *Reference Book for Composites Technology* (ed S.M. Lee), Technomic, Lancaster, pp. 17–48.

15. Mika, T.F. and Bauer, R.S. (1988) in *Epoxy Resins – Chemistry and Technology*, 2nd edn (ed C.A. May), Marcel Dekker, New York, pp. 465–537.

16. Stover, D. (1991) Composites use increases on new commercial transports. *Adv. Compos.*, **6**(5), 30–40.

17. Harrington, H.J. (1988) in *Engineered Materials Handbook*, Vol. 2 (eds J.N. Epel, J.M. Margolis, S. Newman, and R.B. Seymour), ASM International, Metals Park, OH, pp. 242–245.

18. Nair, C.P. (2004) Advances in addition-cure phenolic resins. *Prog. Polym. Sci.*, **29**(5), 401–498.

19. Gardziella, A., Pilato, L., and Knop, A. (eds) (2000) *Phenolic Resins, Chemistry, Applications, Standardization, Safety and Ecology*, 2nd edn, Springer, Heidelberg, pp. 1–560.

20. Matsumoto, A. (1999) in *Concise Polymeric Materials Encyclopedia* (ed J.C. Salamone), CRC Press, Boca Raton, FL, pp. 1017–1019.

21. Narayan, R., Chattopadhyay, D.K., Sreedhar, B., Raju, K.V., Mallikarjuna, N.N., and Aminabhavi, T.M. (2006) Synthesis and characterization of cross-linked polyurethane dispersions based on hydroxylated polyesters. *J. Appl. Polym. Sci.*, **99**(1), 368–380.

22. Caraculacu, A.A. and Coseri, S. (2001) Isocyanates in polyaddition processes, structure and reaction mechanisms. *Prog. Polym. Sci.*, **26**(5), 799–851.

23. Mathews, F.L. (1994) in *Handbook of Polymer Composites for Engineers* (ed L. Hollaway), Woodhead Publishing Ltd., Cambridge, pp. 71–102.

24. Hayes, B.S. and Seferis, J.C. (2001) Modification of thermosetting resins and composites through preformed polymer particles: a review. *Polym. Compos.*, **22**(4), 451–467.

25. Amico, S. and Lekakou, C. (2001) An experimental study of the permeability and capillary pressure in resin-transfer moulding. *Compos. Sci. Technol.*, **61**(13), 1945–1959.

26. Fernando, G.F. and Degamber, B. (2006) Process monitoring of fibre reinforced composites using optical fibre sensors. *Int. Mater. Rev.*, **51**(2), 65–106.

27. Raston, C. (2005) Renewables and green chemistry. *Green Chem.*, **7**(2), 57.

28. Frattini, S. (2008) Demand is increasing for renewable resourced resins. *JEC Compos. Mag.*, **45**(38), 32–33.

29. Stewart, R. (2008) Going green: eco-friendly materials and recycling on growth paths. *Plast. Eng.*, **64**(1), 16–23.

30. Kaplan, D.L. (1998) *Biopolymers from Renewable Resources*, Springer, New York, pp. 1–417.

31. Zia, K.M., Bhatti, H.N., and Bhatti, I.A. (2007) Methods for polyurethane and polyurethane composites, recycling and recovery: a review. *React. Funct. Polym.*, **67**(8), 675–692.

32. Bunsell, A.R. and Renard, J. (2005) *Fundamentals of Fibre Reinforced Composite Materials*, Institute of Physics Publishing, London.

33. Schwartz, M.M. (1983) *Composite Materials Handbook*, McGraw-Hill, New York.

34. Hollaway, L. (1994) *Handbook of Polymer Composites for Engineers*, Woodhead Publishing Ltd., Abington, MA.

35. Eckold, G. (1994) *Design and Manufacture of Composites Structures*, Woodhead Publishing Ltd., Abington, MA.

36. Dawson, D.K. (2008) *Aerospace Composites: A Design and Manufacturing Guide*, Hanser Gardner Publications Inc..

37. Matthews, F.L. and Rawlings, R.D. (1996) *Composite Materials: Engineering and Science*, Chapman & Hall, London.

38. Hyer, M.W. (2009) *Stress analysis of Fiber-Reinforced Composite Materials*, DEStech Publications Inc., Lancaster.

39. Kotsilkova, R. (2007) *Thermoset Nanocomposites for Engineering Applications*, Smithers Rapra Technology Limited, Shawbury.

40. Liu, A., Guo, M., Gao, J., and Zhao, M. (2006) Influence of bond coat on shear adhesion strength of erosion and thermal resistant coating for carbon fiber reinforced thermosetting polyimide. *Surf. Coat.Technol.*, **201**(6), 2696–2700.

41. Rudd, C.D., Long, A.C., Kendall, K.N., and Mangin, C. (1997) *Liquid Moulding Technologies*, Woodhead Publishing Ltd., Abington, MA.

42. Canning, M.S. and Stenzenberger, H. (1986) Compimide bismaleimide resins – high performance thermosetting systems. *Mater. Des.*, **7**(4), 207–211.

43. Lubin, G. (1998) *Handbook of Composites*, Chapman & Hall, London.

44. Wang, J.J. and Yi, X.S. (2003) Preparation and the properties of PMR-type polyimide composites with aluminum nitride. *J. Appl. Polym. Sci.*, **14**(89), 3913–3917.

45. Johnson, L.L., Eby, R.K., and Meador, M.A.B. (2003) Investigation of oxidation profile in PMR-15 polyimide using atomic force microscope (AFM). *Polymer*, **44**(1), 187–197.

46. Park, S.J., Lee, H.Y., Han, M., and Hong, S.K. (2004) Thermal and mechanical interfacial properties of the DGEBA/PMR-15 blend system. *J. Colloid Interface Sci.*, **270**(2), 288–294.

47. Chrétien, G. (1986) *Matériaux Composites à Matrice Organique*, Technique et Documentation (Lavoisier), Paris.

48. Ratto, J.J. and Hamermesh, C.L. (1982) Addition curing polystyryl pyridine. US Patent 4362860.

49. Huanzhi, X. and Liangmin, Y. (2005) Method for preparing polystyryl-pyridine resin and its application. China Patent 1624012.

50. Silicone Resin Chemistry, *http://www.srep.com/srep/en/siliconharzfarben/siliconchemie/siliconharzchemie/siliconharzchemie.jsp* (accessed 25 January 2011).

51. Rosato, D.V. (2004) *Reinforced Plastics Handbook*, Elsevier, Kidlington.

52. Fascinating Silicone, The Versatile Silicone Toolbox Resins, *http://www.dowcorning.com/content/discover/discovertoolbox/forms-resins.aspx* (accessed 25 January 2011).

53. Robitaille, S. (2001) Cyanate ester resins, in *ASM Handbook Composites*, Vol. 21 (ed. S.L. Donaldson), ASM International, Materials Park, OH.

54. Hamerton, I. (1994) *Chemistry and Technology of Cyanate Ester Resins*, Chapman & Hall, Lincolnwood, IL.

55. Ibeh, C.C. (1999) Amino and furan resins, in *Handbook of Thermoset Plastics* (ed S.H. Goodman), William Andrew Inc., Westwood, NJ.

56. Huntsman Catalogue (2009) Araldite® Benzoxazine Thermoset Resins.

57. Jubsilp, C., Damrongsakkul, S., Takeichi, T., and Rimdusit, S. (2006) Curing kinetics of arylamine-based polyfunctional benzoxazine resins by dynamic differential scanning calorimetry. *Thermochim. Acta*, **447**(2), 131–140.

58. Tsotra, P., Weidmann, U., and Christou, P. (2009) Benzoxazine chemistry: a new material to meet fire retardant challenges of aerospace interiors applications. Proceedings of the 13th European Conference on Composite Materials – ECCM13, 2008, Stockholm, Sweden.

59. Engineer's Handbook (2004) Urea Formaldehyde (UF) & Melamine Formaldehyde (MF)/Aminos, *http://www.engineershandbook.com/Materials/amino.htm* (accessed 25 January 2011).

60. Hardrict, S.N. (2003) Novel novolac-phthalonitrile and siloxane-phthalonitrile resins cured with low melting novolac oligomers for flame retardant structural thermosets. Thesis. Department of Chemistry, Virginia Polytechnic Institute and State University.

61. Keller, T.M. (1983) Phthalonitrile resin from diphthalonitrile monomer and amine. US Patent 4408035.

62. Ryntz, R.A. (2006) Bring back the steel? The growth of plastics in automotive applications. *J. Coat. Technol. Res.*, **3**(1), 3–14.

63. Schweitzer, P.A. (2006) *Paint and Coatings, Applications and Corrosion Resistance*, CRC Press and Taylor & Francis Group, Boca Raton, FL.

64. Sørensen, P.A., Kiil, S., Dam-Johansen, K., and Weinell, C.E. (2009) Anticorrosive coatings: a review. *J. Coat. Technol. Res.*, **6**(2), 135–176.

65. van Haveren, J., Oostveen, E.A., Micciche, F., Noordover, B.A.J., Koning, C.E., van Benthem, R.A.T.M., Frissen, A.E., and Weijnen, J.G.J.

(2007) Resins and additives for powder coatings and alkyd paints, based on renewable resources. *J. Coat. Technol. Res.*, **4**(2), 177–186.

66. Bierwagen, G. (2008) The physical chemistry of organic coatings revisited—viewing coatings as a materials scientist. *J. Coat. Technol. Res.*, **5**(2), 133–155.

67. Allie, L., Thorn, J., and Aglan, H. (2008) Evaluation of nanosilicate filled poly (vinyl chloride-co-vinyl acetate) and epoxy coatings. *Corros. Sci.*, **50**(8), 2189–2196.

68. Yeh, J.M., Huang, H.Y., Chen, C.L., Su, W.F., and Yu, Y.H. (2006) Siloxane-modified epoxy resin-clay nanocomposite coatings with advanced anticorrosive properties prepared by a solution dispersion approach. *Surf. Coat.Technol.*, **200**(8), 2753–2763.

69. Ashhari, S., Sarabi, A.A., Kasiriha, S.M., and Zaarei, D. (2011) Aliphatic polyurethane-montmorillonite nanocomposite coatings: preparation, characterization, and anticorrosive properties. *J. Appl. Polym. Sci.*, **119**(1), 523–529.

70. Heidarian, M., Shishesaz, M.R., Kassiriha, S.M., and Nematollahi, M. (2010) Characterization of structure and corrosion resistivity of polyurethane/organoclay nanocomposite coatings prepared through an ultrasonication assisted process. *Prog. Org. Coat.*, **68**(3), 180–188.

71. Sabzi, M., Mirabedini, S.M., Zohuriaan-Mehu, J., and Atai, M. (2009) Surface modification of TiO2 nano-particles with silane coupling agent and investigation of its effect on the properties of polyurethane composite coating. *Prog. Org. Coat.*, **65**(2), 222–228.

72. Sauvant-Moynot, V., Gonzalez, S., and Kittel, J. (2008) Self-healing coatings: an alternative route for anticorrosion protection. *Prog. Org. Coat.*, **63**, 307–315.

73. Samadzadeh, M., Hatami Boura, S., Peikari, M., Kasiriha, S.M., and Ashrafi, A. (2010) A review on self-healing coatings based on micro/nanocapsules. *Prog. Org. Coat.*, **68**(3), 159–164.

74. Dunn, D.J. (2004) Engineering and structural adhesives. *Rapra Rev. Rep.*, **15**(1), 3–8.

75. Chasseaud, P.T. (1999) Adhesives, syntactics and laminating resins for aerospace repair and maintenance applications from Ciba Specialty Chemicals. *Int. J. Adhes. Adhes.*, **19**, 217–229.

76. Montimer, J. (2004) Adhesive bonding of car body parts by industrial robot. *Ind. Robot Int. J.*, **31**(5), 423–428.

77. Hutchinson, A.R., Winfield, P.H., and McCurdy, R.H. (2010) Automotive material sustainability through reversible adhesives. *Adv. Eng. Mater.*, **12**(7), 646–652.

78. Schmid, A., Sutton, L.R., Armes, S.P., Bain, P.S., and Manfre, G. (2009) Synthesis and evaluation of polypyrrole-coated thermally-expandable Microspheres: an improved approach to reversible adhesion. *Soft Matter*, **5**(2), 407–412.

79. Loureiro, A.L., da Silva, L.F.M., Sato, C., and Figueiredo, M.A.V. (2010) Comparison of the mechanical behaviour between stiff and flexible adhesive joints for the automotive industry. *J. Adhes.*, **86**(7), 765–787.

80. Brandt, J. (2004) The Research Requirements of the Transport Sectors to Facilitate an Increased Usage of Composite Materials, Part I: The Composite Material Research Requirements of the Aerospace Industry, European Commission under Contract G4RT-CT-2001-05054, EADS Deutschland GmbH, Corporate Research Centre.

81. Mangino, E. (2004) The Research Requirements of the Transport Sectors to Facilitate an Increased Usage of Composite Materials, Part II: The Composite Material Research Requirements of the Automotive Industry, European Commission under Contract G4RT-CT-2001-05054, Centro Ricerche Fiat.

82. Carruthers, J. (2004) The Research Requirements of the Transport Sectors to Facilitate an Increased Usage of Composite Materials, Part III: The

Composite Material Research Requirements of the Rail Industry, European Commission under Contract G4RT-CT-2001-05054, NewRail - Newcastle Centre for Railway Research.

83. Takeda, F., Hayashi, K., Suga, Y., Nishiyama, S., Komori, Y., and Asahara, N. (2005) Research in the application of the VaRTM technique to the fabrication of primary aircraft composite structures. *Mitsubishi Heavy Ind. Tech. Rev.*, **42**(5), 1–6.

84. Weber, B.W. (2007) BMC – taking automotive composites to a New dimension. *Compos. Res. J.*, **1**(4), 53–59.

85. Rao, M.S. (2007) Composite materials and processes for civil aircraft structures. Symposium on Aircraft Design, Bangalore, India.

86. Du Preez, W.B., Damm, O.F.R.A., Trollip, N.G., and John, M.J. (2001) Advanced Materials for Application in the Aerospace and Automotive Industries, CSIR Materials Science and Manufacturing, Pretoria.

87. Yagci, Y., Kiskan, B., and Ghosh, N.N. (2009) Recent advancement on polybenzoxazine—a newly developed high performance thermoset. *J. Polym. Sci., Part A: Polym. Chem.*, **47**(21), 5565–5576.

88. Fuchs, E.R.H., Field, F.R., Roth, R., and Kirchain, R.E. (2008) Strategic materials selection in the automobile body: economic opportunities for polymer composite design. *Compos. Sci. Technol.*, **68**(9), 1989–2002.

89. Robinson, M. and Carruthers, J. (2005) Lightweighting of rail vehicles using composite materials. 10th Technical Forum of the Composites Processing Association, Newcastle Centre for Railway Research-NewRail, Derby, United Kingdom.

90. Mcdermott, J.S. Automotive and Industrial Applications, *http://www.wtec.org/loyola/polymers/c3_s1.htm* (accessed 28 November 2010).

91. National Composites Network *http://www.cogent-ssc.com/research/Publications/NCN_Automotive_Road_Map.pdf* (accessed 5 December 2010).

92. Cirincione, R.J. (2008) A study of optimal automotive materials choice given market and regulatory uncertainty. Master Thesis. Massachusetts Institute of Technology.

93. The Corvette Story *http://www.webcars.com/corvette/beginning.php* (accessed 12 December 2010).

94. Centre for Environmental Assessment of Product and Material Systems (2004) Design for Recycling in the Transport Sector – Future Scenarios and Challenges. CPM Report 2004:7, Chalmers University of Technology, Götenborg University, Götenborg.

95. Roy, R., Colmer, S., and Griggs, T. (2005) Estimating the cost of a New technology intensive automotive product: a case study approach. *Int. J. Prod. Econ.*, **97**(2), 210–226.

96. Mangino, E., Carruthers, J., and Pitarresi, G. (2007) The future use of structural composite materials in the automotive industry. *Int. J. Vehicle Des.*, **44**(3/4), 211–232.

97. Marsh, G. (2004) Can composites get firmly on the rails? *Reinforced Plast.*, **48**(7), 26–28.

98. The French Group of Industrial Composite Processors (GPIC) Composite Materials in the Railway Industry, *http://www.eucia.org/uploads/0758ee9780ed2817ae34377df854994a.pdf* (accessed 15 November 2010).

99. Nangia, S., Mittal, A., Srikanth, G., and Biswas, S. Towards Faster Trains – Role of Composites, *http://www.tifac.org.in/index.php?option=com_content&view=article&id=551:towards-faster-trains-role-of-composites&catid=85:publications&Itemid=952* (accessed 1 December 2010).

100. Deo, R.B., Starnes, J.H., Jr.,, and Holzwarth, R.C. (2001) Low-cost composite materials and structures for aircraft applications. RTO AVT Specialists' Meeting on "Low Cost Composite Structures", Loen, Norway, published in RTO-MP-069(II). (SM1) 1/1-12.

101. Bauer, J. (2001) Increasing productivity in composite manufacturing. RTO AVT Specialists' Meeting on "Low Cost Composite Structures", Loen, Norway,

Published in RTO-MP-069(II). (SM1) 6/1-18.

102. Baker, A. A., Callus, P. J., Leong, K. H., Georgiadis, S., Falzon, P. J., and Dutton, S.E. (2001) An affordable methodology for replacing metallic aircraft panels with advanced composite materials. RTO AVT Specialists' Meeting on ''Low Cost Composite Structures'', Loen, Norway, Published in RTO-MP-069(II). (SM1) 8/1-16.

103. CompositesWorld *http://www.compositesworld.com/news/ airbus-begins-a350-xwb-fuselage- manufacturing* (accessed 15 December 2010).

104. Cinquin, J. (2001) Aeronautical composite structure cost reduction from the material aspect. RTO AVT Specialists' Meeting on ''Low Cost Composite Structures'', Loen, Norway, Published in RTO-MP-069(II). (SM1) 19/1-8.

105. Kaufmann, M. (2008) Cost/Weight optimization of aircraft structures. Licentiate Thesis. KTH School of Engineering Sciences, Stockholm, Sweden.

106. HighBeam Business *http:// goliath.ecnext.com/coms2/gi_0199- 7575996/Thermosets-stay-forever-young- plenty.html* (accessed 25 February 2011).

7
Elastomers

Krzysztof Pielichowski and James Njuguna

7.1
Introduction

Elastomers are polymers that display viscoelastic behavior because of their low cross-linkage density and the presence of flexible segments. They show low Young's modulus and high yield strain, as compared with other materials. Elastomers are generally amorphous polymers that undergo segmental motions above their glass transition temperature. However, they may contain crystalline domains, as well as cross-linked regions. Depending on the response to heat, elastomers can be classified as thermosetting or thermoplastic polymers. Thermosets require vulcanization, using for example, sulfur, peroxides, or urethane compounds. Vulcanization (curing) itself is a slow, irreversible process in which the material is transformed into thermoset structure. Elastomers display large ability for deformations and elongation (up to 1200%), while keeping intact the elastic properties. The elasticity of elastomeric materials is associated with the ability of the macrochains to change configuration and distribute an applied stress. When the stress is removed, elastomers, because of the covalent cross-linkages, return to their original configuration. On the other hand, without the cross-linking sites or with short chains, which uneasily undergo reconfiguration, permanent deformation would occur due to applied stress. Depending on the chosen criteria, different classification schemes of elastomeric materials are utilized.

7.2
Classification of Elastomers

There are different classification schemes of elastomers that depend on the criteria used. For instance, if the presence of unsaturated bonds in the structure is considered, one may distinguish between unsaturated and saturated rubbers. Unsaturated rubbers can be cured in the vulcanization process and they constitute a large class of elastomers, such as

Structural Materials and Processes in Transportation, First Edition.
Edited by Dirk Lehmhus, Matthias Busse, Axel S. Herrmann, and Kambiz Kayvantash.
© 2013 Wiley-VCH Verlag GmbH & Co. KGaA. Published 2013 by Wiley-VCH Verlag GmbH & Co. KGaA.

- *cis*-1,4-polyisoprene (natural rubber),
- *trans*-1,4-polyisoprene (gutta-percha),
- synthetic polyisoprene,
- butadiene rubber,
- chloroprene rubber,
- butyl and halogenated butyl rubbers,
- styrene-butadiene rubber (SBR),
- nitrile and halogenated nitrile rubbers.

Saturated rubbers (cannot be cured) are

- ethylene-propylene rubber (EPR),
- ethylene-propylene-diene monomer rubber (EPDM),
- fluoroelastomers,
- silicone rubber,
- epichlorohydrin rubber,
- chlorosulfonated polyethylene.

Another class of materials are thermoplastic elastomers (TPEs), which combine rubber elastic behavior with all the advantages of thermoplastic materials. There are two homogeneously dispersed phases in these materials – a crystalline hard phase and a soft elastomeric phase, which yield thermoplastic and elastomeric properties.

A different classification scheme is based on the rubber origin – natural or man-made (synthetic rubber).

7.3
Natural Rubber

Natural rubber (mainly *cis*-1,4 polyisoprene) is a polymer composed of 320–35 000 isoprene molecules and it is produced commercially from the latex of the *Hevea brasiliensis* tree, cultivated in plantations in the tropical regions of Malaysia, Indonesia, Brazil, and Africa. This kind of rubber displays a set of excellent performance properties, such as elasticity, high resilience, and abrasion resistance, as well as efficient heat dispersion, that originate from its molecular structure and high molecular mass [1–3]. Natural rubber contains, apart from poly(*cis*-1,4 polyisoprene), about 6% non-rubber components such as proteins, lipids, sugars, and ash. Proteins in NR have long been considered to be essential components that affect the characteristic properties of NR, particularly cured rubber properties. Moreover, fatty acids in NRs have also been regarded to play an important role in governing their physical properties and to act as natural antioxidants for the rubber [4]. Recently, it has been proposed that NR is composed of linear poly-isoprene with two terminal groups [5]. Both terminal groups are active and they can react with natural impurities such as proteins and phospholipids. These reactions can lead to extensions of two linear poly-isoprene segments, connections of three or more linear segments (so-called branches or star), forming a network of different

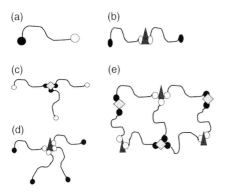

Figure 7.1 Schematic models of naturally occurring network. ○: Initiating terminal (ω), ●: terminating terminal (α), ▲, ◆: natural impurities, (a) linear rubber chain unit, (b) extension of rubber chains, (c) branch of rubber chains, (d) star of rubber chains, and (e) network of rubber chains. Source: Reprinted from Ref. [6], with permission from Elsevier.

chain connections. As a result, raw NR has been considered as a mixture of connected linear poly-isoprene segments ("naturally occurring network") with different connectivity, as shown in Figure 7.1 [6].

To increase the tensile strength of natural rubber vulcanization (curing) process is applied. This process was invented by Hancock and Goodyear in 1839, and made it possible to apply rubber as an engineering material. In this process, sulfur and basic lead carbonate were used. Sulfur as a curing agent still remains in use, but the chemistry of reactions between sulfur and rubber is very complex and still not completely understood. In technological practice, chemical accelerators are applied to shorten the curing time at elevated temperatures since sulfur reacts with rubber very slowly even at elevated temperatures, so that chemicals are usually compounded with rubber along with other materials. Physical properties of the filled rubber that contains additives (fillers, plasticizers, accelerators, and antioxidants) depend mainly on the properties of principal filler – carbon black, such as particle size and shape, surface area of the aggregated or agglomerated structures and their distribution in rubber matrix, as well as rubber–filler interactions. A wide variety of particulate fillers are used in the rubber industry to improve and modify the physical properties of elastomeric materials. The addition of filler usually leads to an increase in modulus and significant abrasion and tear resistance. Although the mechanisms of reinforcement are not fully understood, there is a general agreement about the basic process contributing to the stress strain behavior of the filled vulcanizate [7].

Usually, vulcanized soft rubbers contain about 2–3 wt% of sulfur and are heated in the 140–190 °C range for vulcanizing. The cross-linking density increases with an increase of sulfur content leading to more stiff material; ultimately, a fully cross-linked rubber structure can be formed if sulfur content reaches about 45 wt%. Natural rubber is characterized by high strength and resistance to fatigue and can easily be processed with other materials. As a tire material for bicycles, cars, and airplanes it has excellent adhesion to brass-plated steel cord and has low rolling resistance, as well as shows low hysteresis leading to low heat generation. These features help to maintain tire's service integrity and to extend retreadability [8]. Besides, natural rubber has high resistance to cutting, chipping, and tearing, but a drawback is moderate resistance to environmental damage by heat and light. Proper stabilization measures must therefore be taken into account to design formulations that are stable enough in the service life.

It is estimated that natural rubber production accounts for about 30% of the total world's rubber market.

With the development of the rubber industry, considerable amounts (in the range of millions of tons) of both natural and synthetic waste rubber are produced worldwide from discarded tires, pipes, belts, shoes, scraps, and waste products from rubber processing. Substantial research effort is currently focused on optimization of main recycling approaches which are (i) energy recycling, (ii) thermal decomposition into useful products, (iii) mechanical grinding into a powder and using it as filler in the fresh rubber compounds (iii) modification, and (iv) regeneration (devulcanization). However, the issue of rubber recycling is out of scope of this chapter; interested readers may refer to Refs. [9–11].

7.4
Synthetic Rubbers

Synthetic rubbers, the production of which started on a commercial scale in 1930s, offer several important advantages over natural rubber, for example, resistance to corrosion by fluids and gases, electrical insulation, and large elasticity [12]. At the processing stage a large variety of compounds can be designed and fabricated based on synthetic rubbers, and this calls for rubber materials entering different application areas, such as tires, shock absorbers, seals, and belts. By integrating various additives, including nanoadditives, processing chemicals, or creating alloys with natural rubber or other thermoplastic materials, a wide spectrum of synthetic rubbers can be created. Numerous chemical processing agents include accelerators, activators, vulcanizing agents, stabilizers, flame retardants, and antioxidants. In the area of additives, the widely applied reinforcement by carbon black is effective, because of the presence of active polar groups – carboxyl, phenol, quinone, and lactone – on the carbon black surfaces [13]. These polar groups in the carbon black surfaces interact with rubbers, and the interaction is higher with polar rubbers than hydrocarbon rubber, which is due to the polar–polar interaction. Hydrocarbon rubber such as SBR provides a possible interaction with carbon black and is physical and physicochemical through a double bond of the main chain. The interaction between the hydrocarbon rubber and carbon black can be improved by the introduction of polar groups in the rubber through chemical modification or by some particular additives. A promising approach toward modification of rubber is based on surface layer treatment by chemical or radiation-based methods. The latter group of methods encompasses ultraviolet, corona discharge, low-pressure

RF plasma, and ion beam bombardment, and enables to enhance the role of the surface layer in general rubber performance.

Among synthetic rubbers, SBR is an important synthetic rubber and also the most widely used one. Butadiene unit can have three different components, 1,2-, *cis*-1,4-, and *trans*-1,4-units. Thus, SBR can have various microstructures depending on the component ratios of the styrene, 1,2-, *cis*-1,4-, and *trans*-1,4-units. Polymerization of SBR is performed by solution and emulsion processes [14, 15]. SBR obtained by the solution process has a more complex microstructure compared to the emulsion SBR. The solution SBR is generally characterized by a narrower molecular weight distribution than the emulsion SBR. Generally, SBR has good mechanical properties and good abrasion resistance, higher ozone resistance, and better weatherability than natural rubber. Besides, the SBR rubber fabrication is economically more viable than natural rubber production, and so it is used in many rubber applications, apart from those where oil resistance is required.

Nitrile rubbers are butadiene/acrylonitrile copolymers with the proportions ranging from 55 to 80% butadiene and 20 to 45% acrylonitrile. These rubbers are more costly than ordinary rubbers, so they are limited to special applications such as fuel hoses and gaskets where high resistance to oils and solvents is required. Hydrogenated nitrile, a product for which demand is growing, allows manufacturers to more easily meet environmental emissions requirements and to offer products with superior thermo and mechanical properties. For instance, hydrogenated nitrile can withstand temperatures of more than 150 °C, compared to non-modified nitrile, which remains stable only till about 100 °C. The polychloroprene (neoprene) show lower temperature flexibility, but are characterized by good fuel and oil resistance and increased strength over that of ordinary rubbers. These properties make them suitable candidates for specialty applications such as wire and cable coverings, industrial hoses and belts, and automotive seals and diaphragms [16]. For over 70 years, polychloroprene is still the industry workhorse for underhood, chassis, and adhesives. Even with today's broad selection of specialty elastomers, polychloroprene is the choice for engineering applications that require a mid-performance polymer with a good all-around balance of usage properties. Silicone rubbers have the highest useful temperature range up to 320 °C, but some other properties – strength and resistance to wear and oils – are lower than those in other elastomers. Silicone rubbers are used as electrical insulations, auto-ignition cable sealants, gaskets, or in electronic devices.

Blends of plastics and rubbers offer alternatives to conventional rubber materials [17, 18]. As the commonly applied elastomers, such as EPR, EPDM, or SBR are immiscible with large-scale produced polyesters (poly(ethylene terephthalate) (PET) or poly(butylene terephthalate) (PBT)), there is a need for compatibility to increase interfacial adhesion and to reduce the interfacial tension between the blend's components. Hence, binary blends with PBT/PP applying various *in situ* compatible agents were reported by Sun *et al.* [19] and Holsti-Miettinen *et al.* [20]. PBT/EPDM blends with and without glycidyl methacrylate (GMA) functionality were described in Ref. [21], while Okamoto *et al.* [22] reported on PBT/poly(ethylene-*co*-glycidyl methacrylate) (EGMA) blends.

Significant improvements of the rubber performance can be achieved by dynamic curing of the elastomeric phase during melt mixing. This process generates thermoplastic vulcanizate blends, which have the processing characteristics of a plastic, yet perform with the flexibility and durability of a thermoset rubber. Under dynamic shear, the elastomer is preferentially cross-linked to generate fine particles in a melt processable thermoplastic matrix. Improvements in properties include higher tensile strength and impact resistance, better elastic recovery, improved property retention at high temperatures, greater resistance to chemical attack and swelling by solvents, improved fatigue resistance, greater stability of morphology, and more consistent processability [23]. Papke and Karger-Kocsis [24] investigated TPEs consisting of PET, compatibilizer (EPR-g-GMA), and various rubbers. Authors found that both NBRs containing thermoplastic dynamic vulcanizates (TDVs) showed the highest complex modulus over the entire temperature range (Figure 7.2). The $|E^*|$ of all TDVs begin to drop sharply at temperatures higher than 180 °C. Each TDV is clearly phase separated as suggested by the multiple glass transition temperatures (T_g) in the tan δ versus T traces (Figure 7.3). All TDVs containing EP(D)M or E/αO show two major loss factor maxima indicating the T_g of the respective phases (PET and EPR). On the other hand, three relaxation peaks can be resolved for the NBR-containing blends.

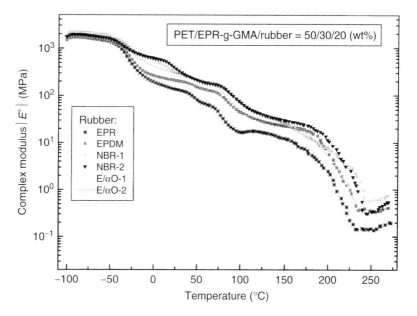

Figure 7.2 $|E^*|$ versus T traces for the PET-TDVs with various elastomers. EPR: Tafmer-P 0480 (Mitsui Chemicals); EPDM: Buna AP 447 (Bayer AG); NBR-1: Perbunan NT 2831 (Bayer AG); NBR-2: Perbunan NT 3946 (Bayer AG); E/αO-1: Engage 8200 (DuPont Dow Elastomers); E/αO-2: Engage 8445 (DuPont Dow Elastomers); EGMA: Lotader AX 8840 (Elf Atochem); and PET: Eastapak 9921W (Eastman Chemical Co). Source: Reprinted from Ref. [24], with permission from Elsevier.

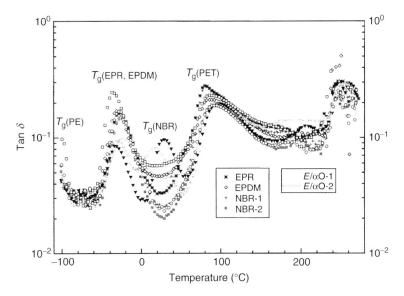

Figure 7.3 tan δ versus *T* traces for the PET-TDVs with various elastomers. Sample codes as in Figure 7.2. Source: Reprinted from Ref. [24], with permission from Elsevier.

7.5
Thermoplastic Elastomers

TPEs constitute a scientifically and commercially interesting class of polymeric materials that combine properties of thermoplastic and elastomeric materials [25–28]. TPEs are thus biphasic materials that possess the properties of glassy or semicrystalline thermoplastic polymers and soft elastomers, and enable rubbery materials to be processed as thermoplastics. In TPE macromolecules discrete thermoplastic segments capable of forming rigid nanoscale domains are bonded through covalent linkages to rubbery segments, which constitute a soft matrix with embedded rigid domains. As the chemically dissimilar segments are joined by covalent bonds, a three-dimensional network of the rigid domains can be formed. The TPE, consequently, exhibits mechanical properties comparable to those of a vulcanized (covalently cross-linked) rubber, with the exception that the network is thermally reversible and can be applied in high-throughput thermoplastic processes, such as melt extrusion and injection molding [29].

Owing to their properties, TPEs have been widely used as the materials in the automobile, engineering, and electrical products. Examples of uses of TPEs in automotive applications include constant velocity joint (CVJ) boots, bumper fascias, body side moldings, grommets, strut cover bellows, interior skins, and clean air ducts.

An important research issue in the TPEs area deals with the origins of the elastomeric stress–strain behavior of these materials. Along this line of interest, Kikuchi *et al.* [30] performed the elastic–plastic analysis using two-dimensional finite element method (FEM), which showed that the matrix between rubber domains in the stretching direction acts as an interconnector between adjacent rubber domains. However, this model qualitatively describes elasticity for one loading condition only, and it does not take composition, cure state, molecular weight, or domain size into account. In another modeling approach, Boyce *et al.* [31] use finite element modeling to describe the micromechanisms of deformation and recovery in thermoplastic vulcanizates. Their simulations indicate that on yielding, a pseudo-continuous rubber phase develops as a result of the drawing of iPP ligaments and shear of rubber particles. Wright *et al.* [32] proposed a microcellular modeling approach to describe the steady-state behavior of dynamically vulcanized blends of EPDM and isotactic polypropylene. Three types of deformation, such as elastic and plastic deformation of iPP, elastic deformation of EPDM, and localized elastic and plastic rotation about iPP junction points were accounted for. The viability of the constitutive model was tested in terms of iPP concentration and EPDM cure state.

Asami and Nitta [33] focused on the morphological changes of an iPP/EPDM system during the uniaxial drawing process to explore the underlying mechanisms of elastic mechanical response of the immiscible TPEs. Figure 7.4 displays the

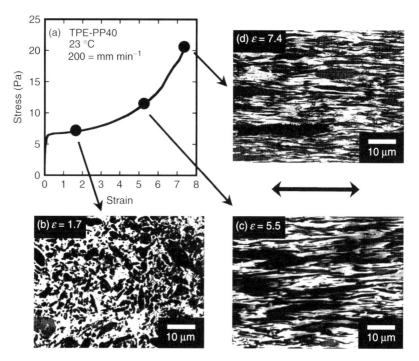

Figure 7.4 Scanning electron microscope pictures of the drawn TPE-PP40 (composition: iPP: 40; EPDM: 43; oil: 17 (wt%). (a) Load–elongation curves of TPE-PP40 measured at 23 °C and at the elongation speed of 200 mm min^{-1} are included. The SEM pictures of TPE-PP40 drawn at (b) strain = 1.7, (c) strain = 5.5, and (d) strain = 7.4. Source: Reprinted from Ref. [33], with permission from Elsevier.

SEM pictures of the central part of the drawn TPE-PP40 specimens, which were uniaxially stretched at strains of 1.7, 5.5, and 7.4. The offset in deformations is a consequence of the permanent set imposed by the yield process of iPP matrix, whereas the dispersed EPDM domains were found to be highly elongated as shown in Figure 7.4b,c. The localized strain in the EPDM domains (according to rough estimation of their extension) was found to be considerably greater than the applied overall strain – it would suggest that the strain concentrated on the rubber domains.

The effects of dynamic cross-linking on rheological, mechanical, and dynamic mechanical properties of high density polyethylene and acrylonitrile butadiene rubber blends have been evaluated by George *et al.* with special reference to the effect of dosage of cross-linking agent. Both, the microscopic and dynamic mechanical analysis indicated that the blends are immiscible and form two-phase structures. But it is evident that dynamic vulcanization can be employed as a technological compatibility technique to get finer and stable morphology, and improved mechanical properties. Besides, the rheological studies reveal that the dynamically vulcanized samples could be processed like thermoplastics [34].

TPEs have proven themselves in meeting the wide range of demanding engineering requirements for transportation purposes. It is expected that the scale of these applications will continue to grow because of the cost savings provided, and the performance delivered by the new tailor-made products. One can assume that TPEs will continue to replace thermoset rubber for applications in which they offer cost advantages and design flexibility, mostly in the automotive sector.

7.6
Fluorine-Containing TPEs

Fluorine-containing TPEs are particularly attractive, because of their unique combination of useful properties, such as high thermostability (the higher the content of fluorine, the more thermostable the polymer), hydrophobicity, lypophobicity, oleophobicity, resistance to aging and to oxidation, hydrolytic stability, chemical inertness, low permeability to gases, and low flammability [35, 36]. These materials are regarded as high value added polymers used in high-tech applications and have found many applications in automotive industry, microelectronics, aeronautics, aerospace, and optics. F-elastomers are routinely used over the past 50 years as sealing devices for gas turbine engines, air cycle machines, auxiliary power units, filters, and actuators.

Two main classes of F-elastomers are distinguished: those that are required to be cross-linked and those that are not. Usually, fluorocarbon elastomers are synthesized from the radical copolymerization of fluoroalkenes – vinylidene fluoride (VDF), hexafluoropropylene (HFP), tetrafluoroethylene (TFE), perfluoromethyl vinyl ether (PMVE), chlorotrifluoroethylene (CTFE), pentafluoropropylene (HPFP), and propylene (P) [37]. Additional fluorinated functional co-monomers containing a cross-linking group (called *cure site monomers*) can be optionally introduced.

Table 7.1 Main commercially available fluoroelastomers.

	HFP	PMVE	CTFE	P	HPFP
VDF	Daiel® 801 (*Daikin*); Fluorel® (3 M/*Dyneon*); Tecnoflon® (*Solvay Sol*); SKF-26 (*Russia*); Viton® A (*DuPont*)	—	Kel F® (*Dyneon*); SKF-32 (*Russia*); Voltalef® (*Elf Atochem*)	—	Tecnoflon® SL (*Solvay Sol*)
TFE	—	Kalrez® (*DuPont*)	—	Aflas® (Asahi Glass); Extreme® (DuPont Dow Elastomers)	—
VDF + TFE	Daiel® 901 (*Daikin*); Fluorel® (*Dyneon*); Tecnoflon® (*Solvay Sol*); Viton® B (*DuPont*) + ethylene: Tecnoflon® (*Solvay Sol*) + X: Viton® GH (*DuPont*)	Viton® GLT (*DuPont*)	—	—	Tecnoflon® T (*Solvay Sol*)

[a]CTFE, chlorotrifluoroethylene ($F_2C=CFCl$); HFP, hexafluoropropene ($F_2C=CFCF_3$); HPFP, 1-hydro-pentafluoropropene ($FHC=CF-CF_3$); P, propene ($H_2C=CHCH_3$); PMVE, perfluoromethyl vinyl ether ($F_2C=CFOCF_3$); TFE, tetrafluoroethylene ($F_2C=CF_2$); VDF, vinylidene fluoride (or 1,1-difluoroethylene) ($F_2C=CH_2$); X, cure site monomer ($XCY=CZ-R-G$); G, function.
Source: Reprinted from Ref. [38], with permission from Elsevier.

In Table 7.1, the crossings between lines and columns represent elastomeric copolymers containing the corresponding co-monomers, as comprehensively reviewed [38].

The first commercial use for the pioneer fluoroelastomer (Viton®) occurred in 1957 when the US Air Force was seeking more durable O-rings that had the ability to seal hot engine lubricants and hydraulic fluids. There is no need to mention that in the air absolute sealing integrity is a must.

7.7
Bio-Based TPEs

Sustainable chemistry promotes the replacing of petroleum derived raw materials with renewable raw materials in the synthesis of polymers. In this way, bio-based materials obtained from renewable resources are receiving considerable attention for an increasing amount of applications, including those in automotive industry [39, 40]. Among bio-based materials, vegetable oils are one of the most abundant, annually renewable natural resources that are available in large quantities from various oilseeds; they are economical and offer the possibility of biodegradation. In the recent years, bio-based materials derived from natural oils, such as castor, palm, canola, and soybean oils were used to synthesize natural polyols, which were then used as raw materials in the preparation of bio-based elastomeric polyurethanes (TPUs) – a sub-class of TPEs [41]. TPUs derive most of their useful properties from the incompatibility of the hard segments made from a diisocyanate and a chain extender, and soft segments that are usually composed of long-chain diols. As the soft segments form a flexible matrix between the hard domains that in turn provide physical cross-linking, a complex, phase-segregated nanostructured morphology, and elastomeric behavior can be observed that varies with the temperature [42–47]. Additional effects may be caused by specific interactions between phases via hydrogen bonding, connected with phase intermixing, and dissociation-reorganization processes. Such a complex microdomain structure strongly influences the properties of TPUs materials for which applications include, for example, automotive exterior body panels and flexible tubings.

The commercialized materials include Spain's Merquinsa Pearlthane® the world's first TPU derived from bio-based materials such as vegetable or fatty acid-based polyols. Recently Ford Motor Company has chosen to apply this bio-based TPU for the Lincoln MKZ tambour console/tambour door which is an overmold of Bio TPU onto recycled ABS. Reduction of the carbon footprint of this part by as much as 40% was reported [48].

GLS introduced a bio-based TPU OnFlex™ BIO, derived from soybean oils. Renewable content is claimed to be at least 20% and hardness ranges from 70 to 80 Shore A [49].

Cerenol™ from DuPont is a family of high-performance polyether diols (molecular weight of 500–3000) derived from renewably sourced 1,3-propanediol; these products can be applied for TPUs synthesis as a replacement of poly(tetra-methylene glycol) (PTMG) [50].

Bio-based TPUs area is very attractive for fundamental research as well. There has been a continuous growth of a number of publications in this area since few years, and new TPUs have been designed [51–55]. In one of the approaches, Corcuera *et al.* [56] synthesized bio-based elastomeric polyurethanes derived from castor oil on the soft segment structure and an aliphatic diisocyanate as 1,6-hexamethylene diisocyanate in the hard segment structure. The hard segment structure was varied by means of different chain extenders, petrochemical-based 1,4-butanediol (BD) and bio-based 1,3-propanediol (PD). Authors have discussed the properties of the obtained TPU from the viewpoint of hard/soft microdomain phase separation and also the hard segment nature and formed structure.

Concerning the thermal properties, TPUs show several thermal transitions: (i) a glass transition of either the hard or soft segments, (ii) an endotherm of the hard segments attributable to annealing, and (iii) endotherm associated with the

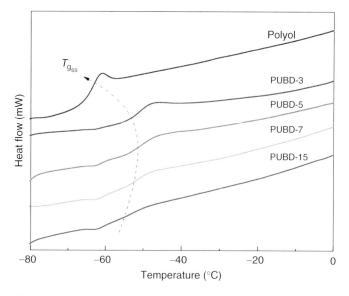

Figure 7.5 DSC thermograms of the low temperature behavior of the synthesized PUBD polyurethanes with different hard segment contents. Source: Reprinted from Ref. [56], with permission from Elsevier.

long-range order of crystalline portions of either the soft or hard segments. In the work under discussion, DSC curves for each polyurethane series with varying hard segment content are shown in Figures 7.5 and 7.6a,b. For each PUBD and PUPD polyurethane series several transitions associated with soft segment in the low temperature range (Figure 7.5) and with the hard segment in the high temperature range (Figure 7.6a,b) can be seen.

7.8
Conclusions

Elastomers constitute an indispensable group of materials that are widely used in automotive, aerospace, and railway applications. They have a well-established position among other materials and the future outlook is promising, as there is a lot of research to develop new products, which are specifically tailored for the challenging transportation market. Elastomeric nanomaterials, bio-based composites, and surface-modified products, along with hybrid materials are either under development or have already entered initial production phase; fabrication and processing will be offering cost advantages and design flexibility to the automotive/aerospace engineers who apply high performance materials. They may have a considerable impact on such applications as engine and transmission seals, engine mounts and suspension bushings, radiators, and heater hoses.

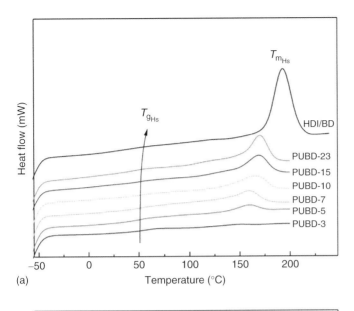

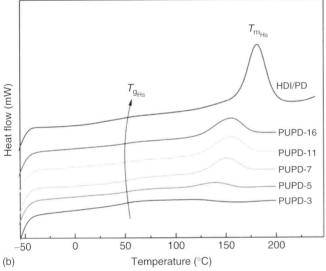

Figure 7.6 DSC thermograms of the high temperature behavior of the synthesized polyurethanes with different hard segment content: (a) PUBD series and (b) PUPD series. Source: Reprinted from Ref. [56], with permission from Elsevier.

Market reports foresee that growth prospects, for example, in TPEs area through 2016 will be very strong in Brazil, Russia, India, and China (BRIC countries). Some highly developed nations such as United States and Canada will see strong growth in their TPE sectors due to a projected recovery in their automotive industries from heavy declines experienced during 2007–2008 crisis. However, the forecasts are done with large error margin due to today's unstable world's economic situation. In the automotive industry, specific areas of growth include soft-touch trims, airbag covers, body seals, two-shot TPE overmolding of larger parts such as door panels, weather seals, and exterior panels that eliminate the need for paint.

An important issue is associated with the comprehensive utilization of discarded rubber (mostly tires) to achieve the goals of protecting environment and recycling resources. Among various methods of disposal of scrap/waste rubber products, nowadays, recycling or reclaiming of rubber is considered as the most positive approach, because it not only saves limited resources of fossil feedstock, but also maintains environmental quality.

References

1. Backhaus, R.A. (1985) Rubber formation in plants – a mini-review. *Isr. J. Bot.*, **34**, 283–293.
2. Kang, H., Kim, Y.S., and Chung, G.C. (2000) Characterization of natural rubber biosynthesis in Ficus benghalensis. *Plant Physiol. Biochem.*, **38**, 979–987.
3. Alk-Hwee, E. and Tanaka, Y. (1993) Structure of natural rubber. *Trends Polym. Sci.*, **3**, 493–513.
4. Sakdapipanich, J.T. (2007) Structural characterization of natural rubber based on recent evidence from selective enzymatic treatments. *J. Biosci. Bioeng.*, **103**, 287–292.
5. Tanaka, Y. (2001) Structural characterization of natural polyisoprene based on structural study. *Rubber Chem. Technol.*, **74**, 355–362.
6. Toki, S., Hsiao, B.S., Amnuaypornsri, S., and Sakdapipanich, J. (2009) New insights into the relationship between network structure and strain-induced crystallization in un-vulcanized and vulcanized natural rubber by synchrotron X-ray diffraction. *Polymer*, **50**, 2142–2148.
7. Vijayaram, T.R. (2009) A technical review on rubber. *Int. J. Des. Manuf. Technol.*, **3**, 1–12.
8. White, J., De, S., and Naskar, K. (2009) *Rubber Technologist's Handbook*, Vol. 2, iSmithers Rapra, Shawbury.
9. Fang, Y., Zhan, M., and Wang, Y. (2001) The status of recycling of waste rubber. *Mater. Des.*, **22** (2), 123–128.
10. Li, Y. (1998) Recycling of rubber. *World Rubber Ind.*, **1** (1), 29–36.
11. Xu, P. (1998) Recycling of discarded tire. *Environ. Prot. Chem. Process Ind.*, **18** (2), 79–81.
12. Bhowmick, A.K. (ed.) (2008) *Current Topics in Elastomers Research*, CRC Press, Boca Raton, FL.
13. Ramesan, M.T. (2004) The effects of filler content on cure and mechanical properties of dichlorocarbene modified styrene butadiene rubber/carbon black composites. *J. Polym. Res.*, **11**, 333–340.
14. Distler, D. (1999) *Wässrige Polymerdispersionen*, Wiley-VCH Verlag GmbH, Weinheim.
15. Kohnle, M.V., Ziener, U., and Landfester, K. (2009) Synthesis of styrene–butadiene rubber latex via miniemulsion copolymerization. *Colloid Polym. Sci.*, **287**, 259–268.
16. Habeeb Rahiman, K. and Unnikrishnan, G. (2006) The behaviour of styrene butadiene rubber/acrylonitrile butadiene rubber blends in the presence of chlorinated hydrocarbons. *J. Polym. Res.*, **13**, 297–314.
17. Utracki, L. (1990) *Polymer Alloys and Blends*, Carl Hanser Verlag, Munich.
18. Tinker, A.J. and Jones, K.P. (1998) *Blends of Natural Rubber*, Chapman & Hall, London.
19. Sun, Y.J., Hu, G.H., and Lambla, M. (1996) Effects of processing parameters on the in situ compatibilization of polypropylene and poly (butylene terephthalate) blends by one-step reactive extrusion. *J. Appl. Polym. Sci.*, **61**, 1039–1047.

20. Holsti-Miettinen, R.M., Heino, M.T., and Seppälä, J.V. (1995) Use of epoxy re-activity for compatibilization of PP/PBT and PP/LCP blends. *J. Appl. Polym. Sci.*, **57**, 573–586.

21. Moffet, A.J. and Dekkers, M.E.J. (1992) Compatibilized and dynamically vulcanized thermoplastic elastomer blends of poly(butylene terephthalate) and ethylene propylene diene rubber. *Polym. Eng. Sci.*, **32**, 1–5.

22. Okamoto, M., Shiomi, K., and Inoue, T. (1994) Structure and mechanical properties of poly (butylene terephthalate)/rubber blends prepared by dynamic vulcanization. *Polymer*, **35**, 4618–4622.

23. Akiba, M. and Hashim, A.S. (1997) Vulcanization and crosslinking in elastomers. *Prog. Polym. Sci.*, **22**, 475–521.

24. Papke, N. and Karger-Kocsis, J. (2001) Thermoplastic elastomers based on compatibilized poly(ethylene terephthalate) blends: effect of rubber type and dynamic curing. *Polymer*, **42**, 1109–1120.

25. Holden, G. (2000) *Understanding Thermoplastic Elastomers*, Hanser Publishers, Munich.

26. Dufton, P.W. (2001) *Thermoplastic Elastomers*, Rapra Technology Ltd., Shawbury.

27. Drobny, J.G. (2007) *Handbook of Thermoplastic Elastomers*, William Andrew Inc., Norwich, NY.

28. Spontak, R.J. and Nikunj, P.P. (2000) Thermoplastic elastomers: fundamentals and applications. *Curr. Opin. Colloid Interface Sci.*, **5**, 334–341.

29. Kear, K.E. (2003) *Developments in Thermoplastic Elastomers*, Rapra Technology Ltd., Shawbury.

30. Kikuchi, Y., Fukui, T., Okada, T., and Inoue, T. (1991) Elastic–plastic analysis of the deformation mechanism of PP-EPDM thermoplastic elastomer: origin of rubber elasticity. *Polym. Eng. Sci.*, **31** (14), 1029–1032.

31. Boyce, M.C., Yeh, O., Socrate, S., Kear, K., and Shaw, K. (2001) Micromechanics of cyclic softening in thermoplastic vulcanizates. *J. Mech. Phys. Solids*, **49**, 1343–1360.

32. Wright, K.J., Indukuri, K., and Lesser, A.J. (2003) Microcellular model evaluation for the deformation of dynamically vulcanized EPDM/iPP blends. *Polym. Eng. Sci.*, **43** (3), 531–542.

33. Asami, T. and Nitta, K. (2004) Morphology and mechanical properties of polyolefinic thermoplastic elastomer. I. Characterization of deformation process. *Polymer*, **45** (15), 5301–5306.

34. George, J., Varughese, K.T., and Thomas, S. (2000) Dynamically vulcanised thermoplastic elastomer blends of polyethylene and nitrile rubber. *Polymer*, **41**, 1507–1517.

35. Scheirs, J. (1997) *Modern Fluoropolymers*, John Wiley & Sons, Inc, New York.

36. Hougham, G., Cassidy, P., Johns, K., and Davidson, T. (1999) *Fluoropolymers*, Kluwer Academic Publishers, New York.

37. Ameduri, B., Boutevin, B., and Kostov, G. (2001) Fluoroelastomers: synthesis, properties and applications. *Prog. Polym. Sci.*, **26**, 105–187.

38. Ameduri, B. and Boutevin, B. (2005) Update on fluoroelastomers: from perfluoroelastomers to fluorosilicones and fluorophosphazenes. *J. Fluorine Chem.*, **126**, 221–229.

39. Shogren, R.L., Petrovic, Z., Liu, Z., and Erhan, S.Z. (2004) Biodegradation behavior of some vegetable oil-based polymers. *J. Polym. Environ.*, **12** (3), 173–178.

40. Park, S.J., Jin, F.L., and Lee, J.R. (2004) Synthesis and thermal properties of epoxidized vegetable oil. *Macromol. Rapid Commun.*, **25** (6), 724–727.

41. Koberstein, J.T. and Russell, T.P. (1986) Simultaneous SAXS-DSC study of multiple endothermic behavior in polyether-based polyurethane block copolymers. *Macromolecules*, **19**, 714–720.

42. Chen, T.K., Sieh, T.S., and Chui, J.Y. (1998) Studies on the first DSC endotherm of polyurethane hard segment based on 4,4′-Diphenylmethane Diisocyanate and 1,4-Butanediol. *Macromolecules*, **31**, 1312–1320.

43. Pielichowski, K. and Njuguna, J. (2005) *Thermal Degradation of Polymeric Materials*, Rapra Technology Ltd., Shawbury.

44. Pielichowski, K. and Słotwińska, D. (2003) Morphological features and flammability of MDI/HMDI-based segmented polyurethanes containing

3-chloro-1,2-propanediol in the main chain. *Polym. Degrad. Stab.*, **80**, 327–331.

45. Pielichowski, K., Słotwińska, D., and Dziwiński, E. (2004) Segmented MDI/HMDI-based polyurethanes with lowered flammability (thermal decomposition study). *J. Appl. Polym. Sci.*, **91** (5), 3214–3224.

46. Pielichowski, K. and Slotwinska, D. (2004) Flame-resistant modified segmented polyurethanes with 3-chloro-1,2-propanediol in the main chain – thermoanalytical studies. *Thermochim. Acta*, **410** (1–2), 79–86.

47. Lewicki, J.P., Pielichowski, K., Tremblot De La Croix, P., Janowski, B., and Liggat, J.J. (2010) High temperature thermal degradation studies of polyurethane/POSS nanohybrid elastomers. *Polym. Degrad. Stab.*, **95**, 1099–1105.

48. Merquinsa Technical Information *http://www.merquinsa.com/news/pdfs/ PRESS_RELEASE_FORD.pdf* (accessed 10 January 2011).

49. GLS Technical Information *http://www.glstpes.com/products_onflex.php* (accessed 10 January 2011).

50. DuPont Technical Information *http://www2.dupont.com/Cerenol_Polyols/ en_US/index.html* (accessed 10 January 2011).

51. Benes, H., Vlcek, T., Cerna, R., Hromadkova, J., Walterova, Z.,

and Svitakova, R. (2012) Polyurethanes with bio-based and recycled components. *Eur. J. Lipid Sci. Technol.*, **114** (1), 71–83. doi: 10.1002/ejlt.201000123.

52. Rio, E., Lligadas, G., Ronda, J.C., Galia, M., Meier, M.A.R., and Cadiz, V. (2011) Polyurethanes from polyols obtained by ADMET polymerization of a castor oil-based diene: characterization and shape memory properties. *J. Polym. Sci.: A: Polym. Chem.*, **49**, 518–525.

53. Lu, Y. and Larock, R.C. (2011) Synthesis and Properties of grafted lattices from a soybean oil-based waterborne polyurethane and acrylics. *J. Appl. Polym. Sci.*, **119**, 3305–3314.

54. Wik, V.M., Aranguren, M.J., and Mosiewicki, M.A. (2011) Castor oil-based polyurethanes containing cellulose nanocrystals. *Polym. Eng. Sci.*, **51** (7), 1389–1396. doi: 10.1002/pen.21939.

55. Bueno-Ferrer, C. *et al.* (2012) Relationship between morphology, properties and degradation parameters of novative biobased thermoplastic polyurethanes obtained from dimer fatty acids. *Polym. Degrad. Stab.* doi: 10.1016/j.polymdegradstab.2012.03.002.

56. Corcuera, M.A., Rueda, L., Fernandez d'Arlas, B., Arbelaiz, A., Marieta, C., Mondragon, I., and Eceiza, A. (2010) Microstructure and properties of polyurethanes derived from castor oil. *Polym. Degrad. Stab.*, **95**, 2175–2184.

Part III
Composites

Composite materials are not a development of the modern age, but they are an evolutionary solution for lightweight designs from nature. Over millions of years of evolution, nature formed the principle of lightweight design by using fibers with high strength (HST) to carry loads.

Most materials have higher stiffness and strength when they are used as fiber. There are numerous examples of composites in nature, such as plant stalk or bamboo, the latter being a filamentary composite (Figures P3.1 and P3.2). Through the years, wood has been a commonly used natural composite (cellulose fiber embedded in lignin) whose properties vary significantly with and against the grain. Such directional anisotropic properties have been mastered by design approaches that take advantage of the superior properties while suppressing the undesirable ones through the use of lamination. Plywood, for example, is made with a number of laminates. Such a stacking arrangement is necessary in order to prevent warping.

The combinations of different materials resulting in superior products started in antiquity and these have been in continuous use down to the present. In early times, mud bricks were reinforced with straw to build houses. More recently, artificial stone was reinforced with steel bars (reinforced concrete, developed by a French gardener, Monier, in 1848 [1]) to build modern buildings and bridges, and so on. At present, composites of matrix (polymeric, metallic, and ceramic) reinforced with fibers are used to build traffic carriers.

The emergence of boron filaments gave birth to a new generation of composites in the early 1960s. The composites that employ high modulus (HM) continuous filaments, such as boron and carbon, are referred to as *advanced composites*. This remarkable class of materials is cited as a most promising development that has profoundly impacted current and future technologies of traffic carriers. The term *composites* or *advanced composites material* is defined as a material consisting of small diameter (around 6–10 μm), HST and HM fibers embedded in an essentially homogeneous matrix. This results in a material that is anisotropic (it has mechanical and physical properties that vary with direction). The lightweight construction principle can be supported by orientation of the fibers in the direction of the load. By choosing the single components – the fibers and the matrix, the fiber volume content, the fiber orientation, and the stacking sequence of

Structural Materials and Processes in Transportation, First Edition.
Edited by Dirk Lehmhus, Matthias Busse, Axel S. Herrmann, and Kambiz Kayvantash.
© 2013 Wiley-VCH Verlag GmbH & Co. KGaA. Published 2013 by Wiley-VCH Verlag GmbH & Co. KGaA.

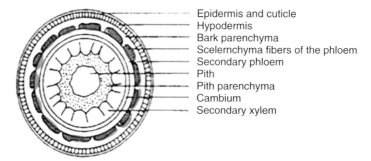

Epidermis and cuticle
Hypodermis
Bark parenchyma
Scelernchyma fibers of the phloem
Secondary phloem
Pith
Pith parenchyma
Cambium
Secondary xylem

Figure P3.1 Crosscut of a flax stalk [2].

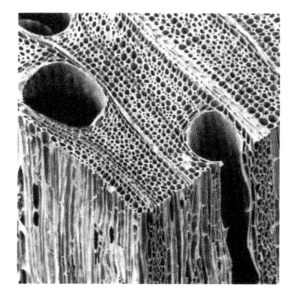

⊢———⊣ 0.1 mm

Figure P3.2 Microscope photo of balsa wood [3].

the laminate, it is possible to design the physical and chemical properties of the composite. To be able to fully exploit this potential, understanding of the anisotropic behavior of the material and knowledge of methods of computation are necessary.

While discussing preforming and production technologies in this chapter, several components have been described. The aim is to provide a clear picture, with specific emphasis on the dependency between process technology and components design.

References

1. Schwaberlapp, M. (2001) Leichtbau und reibungsminderung, werkstofftrends im motorbereich. 2. Deutscher IIR – Werkstoffkongress imat, Stuttgart, January 22–23, 2001.

2. Franck, R. R. *Bast and other plant fibres*, Woodhead Publishing Limited, ISBN: 1-85573-684-5.

3. Vural, M. (2003) Dynamic response and energy dissipation characteristics of Ravichandran, G. balsa experiment and analysis, ELSEVIER. *International J. Solids and Structures*, **40**(9), 2147–2170.

8
Polymer Matrix Composites

Axel S. Herrmann, André Stieglitz, Christian Brauner, Christian Peters, and Patrick Schiebel

8.1
Introduction

For economic and ecological reasons, lightweight construction is important in transportation technology. The highest potential for weight reduction of traffic carriers lies in the applications using polymer fiber composites, especially those of carbon-fiber-reinforced plastics (CFRPs). A weight reduction of up to 30% is possible with CFRP when compared with aluminum designs.

The aircraft industry has always been a pioneer in technical innovations that combine the aspects of lightweight design, safety, and system integration. The lifetime of an aircraft is a number of decades. Hence current design has to continue to be competitive in 20 years. The aim is to develop a CFRP fuselage for the next generation, long-range aircraft, such as the Airbus A350XWB or the Boeing 787 ("Dreamliner") with an entry into service in 2013–2014. The share of CFRP will be dominant in the structure of these new aircraft (Figure 8.1). In the 20-year-long operation of a commercial aircraft, 1 kg saved weight can save approximately 3 tons of fuel consumption.

The synergy aspects grow considerably with regard to lightweight construction requirements of the different traffic carriers of the future. The EU guidelines already plan to reduce the CO_2 emission of automobiles to less than $140\,g\,km^{-1}$, which corresponds to a consumption of approximately 5.7 l of fuel per 100 km. However, the automotive industry has met the challenge of reduced CO_2 emission by, for example, developing electric cars. This will require lightweight structures as the batteries currently envisaged are very heavy. In the near future, the necessary reduction in the consumption will not be achieved by improved drive concepts and aerodynamics alone [1]. A solution can only be found by lightweight construction. Similar scenarios are found in truck and rail vehicle engineering [2] and in shipbuilding [3].

Structural Materials and Processes in Transportation, First Edition.
Edited by Dirk Lehmhus, Matthias Busse, Axel S. Herrmann, and Kambiz Kayvantash.
© 2013 Wiley-VCH Verlag GmbH & Co. KGaA. Published 2013 by Wiley-VCH Verlag GmbH & Co. KGaA.

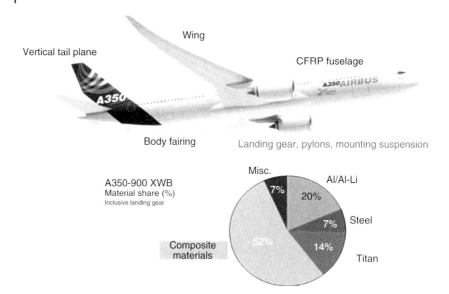

Figure 8.1 Overview of materials distribution in the A350XWB. (Copyright Airbus S.A.S. Reproduced with permission.)

8.1.1
Materials for Fiber Composites

8.1.1.1 Fibers

While composite materials owe their unique balance of properties to the combination of matrix and reinforcement, it is the reinforcement system that is primarily responsible for such structural properties as strength and stiffness. The reinforcement dominates the field in terms of volume, properties, and design versatility. Almost all fibers in use today are solid and have a near circular cross section. Hollow fibers have been developed, are commercially available, and show promise for improved mechanical properties of composites, especially in compressive strength.

Reducing weight by increasing strength and stability is the requirement in lightweight design. Therefore lightweight engineers do not refer to strength and modulus to characterize a material but refer to the specific strength and specific modulus. This means strength and stiffness divided by density and gravity.

Using these dimensions, the rupture length will increase by decreasing density and it is a known fact that reinforced polymers offer a high potential in lightweight structure design.

Figure 8.2 compares different types of reinforcement.

8.1.1.2 Glass Fibers

Glass fibers were the first reinforcements that had already been developed by 1950. Compared to metallic materials, they improved the strength but not the stiffness. Glass fiber is undoubtedly the most widely used fiber; it has gained acceptance

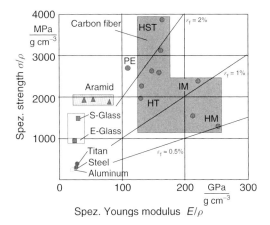

Figure 8.2 Specific strength and specific stiffness for fiber materials in comparison to metallic materials [4].

because of its low costs, lightweight, and HST. Glass fibers are extensively used in primary structures of, for example, sport and utility aircraft as well as helicopter or wind turbine rotor blades. The two most common grades of glass fibers are "E" (for electrical board) and "S" (high strength (HST) for structural use). E-Glass provides a high strength-to-weight ratio, good fatigue resistance, and excellent chemical, corrosion, and environmental resistance. While E-glass has proved highly successful in aircraft secondary structures such as wing fairings and wing-fixed, trailing-edge panels, some applications require higher properties. To fill these demands, S-glass was developed, which offers up to 25% higher compressive strength, 40% higher tensile strength, 20% higher modulus, and 4% lower density. The use of other glass types such as A-glass, C-glass, and even D-glass has been limited, because they are of lower strength and not suitable for structural purposes.

8.1.1.3 Carbon Fibers

Carbon fibers were used by Edison in the nineteenth century as light bulb filaments. The research that resulted in the use of carbon fibers in modern composites is attributed to the work of Shindo in Japan, Watt in England (Royal Aircraft Establishment), and Bacon and Singer in the United States (Union Carbide) in the early 1960s.

As soon as stability problems are involved, glass fiber reinforcements are insufficient because of their low Young's modulus. Carbon fibers were developed during the mid-1970s. They offer both high specific strength and high specific stiffness. Carbon fibers are manufactured by pyrolysis of an organic precursor such as rayon, PAN (polyacrylonitrile), or petroleum pitch. Generally, as the fiber modulus increases, the tensile strength decreases. The area covered by those fibers in the diagram (Figure 8.2) is quite large, because carbon fibers are available with

Table 8.1 Properties of various fibers.

Glass fiber type designation	E	C	D	S/R	M	O
Chemical composition						
SiO_2 (%)	51–57	60–65	72–75	62–65	—	99.5
Al_2O_3 (%)	12–15	<6	—	20–25	—	—
B_2O_3 (%)	5–9	<7	<23	0–1.2	—	—
CaO (%)	17–22	~14	—	—	—	—
MgO (%)	<5	<3	0.5–0.6	10–15	—	—
K_2O, Na_2O (%)	0–0.7	8–10	<4	0–1.1	—	—
F_2 (%)	0–0.6	—	—	—	—	—
Fe_2O_3 (%)	0.5	—	—	—	—	—
Other oxide	—	<1.5	<1	—	—	0.5
Density (g cm^{-3})	2.55–2.60	2.4–2.45	2.14–2.16	2.49–2.55	2.89	2.1–2.2
Filament diameter (µm)	5–13	—	—	—	10	—
Young's modulus (GPa)	73	71	55	87	125	62–70
Specific tensile strength (GPa × cm^3 g^{-1})	1.38–1.4	1.3	—	—	2.8	—
Specific modulus (GPa × cm^3 g^{-1})	28.8	29	—	—	50.3	—
Tensile strength (GPa)	2.5–3	—	—	4.05	—	—
Breaking strain (%)	3.5–4.5	3.5	4.5	5.4	~5.5	—
Heat conductivity (W m^{-1} K^{-1})	0.9–1	—	0.8	—	—	0.9
Coefficient of thermal expansion (10^{-6} K^{-1})	5.1–6	7.2	2–3.5	—	—	1.4

extremely HST or high modulus (HM). Also so-called intermediate-modulus (IM) fibers were developed, which offer HST and high stiffness as well.

8.1.1.4 Aramid Fibers

Aramid fibers have been used for structural applications since the early 1970s. Combining extremely high toughness and energy-absorbing capacity (very good projectile and ballistic protection characteristics has led to their use in bulletproof vests), tensile strength, and stiffness with low density (the lowest in recently developed, advanced composite materials), aramid fibers offer very high specific tensile properties. Low compressive strength is one of the weaknesses of the aramid fiber. But where the highest compressive strength is needed, hybrids of aramid and carbon fibers are generally used (Tables 8.1–8.3).

Details of the properties of various fibers are given in Table 8.1.

8.1.1.5 Resin

The role of the resin in composites is vital. Resin selection controls process ability, upper use temperature, flammability characteristics, and corrosion resistance of

Table 8.2 Properties of carbon fibers.

Carbon fiber type designation	HT	HST	IM	HM	UHM/ HMS	Isotropic	HM type 1	HM type 2
Precursor	PAN						Pitch	
Density (g cm^{-3})	1.74–1.8	1.78–1.81	1.7–1.8	1.78–2.06	1.8–2.18	1.5	2.15	2
Filament diameter (µm)	~7	—	~5	~6.5	~5	—	~10	~11
Tensile strength (GPa)	3.4–3.6	3.85–5	3.1–5.6	1.9–2.75	2.1–2.2	1.0	3.3–3.5	3.8–5.0
Young's modulus (GPa)	228–240	235–250	290–310	330–520	440–830	50	700–800	840
Compression strength (GPa)	2.5	—	4.2	1.5–1.6	1.1–1.8	0.7	0.7	—
Breaking strain (%)	1.0–1.6	1.65–2.1	1.07–1.93	0.4–0.85	0.38	0.57	2.3	0.4–0.5
Rupture length (km)	206	—	311	125	194	—	—	—
Heat conductivity (W m^{-1} K^{-1})	—	—	17	—	—	120	—	640
Heat capacity (J kg^{-1} K^{-1})	0.71	—	—	0.71	—	—	—	—
Electrical resistivity (10^{-6} Ω mm^2 m^{-1})	20	—	—	8	—	—	10	—
Coefficient of thermal expansion (10^{-6} K^{-1})	−0.5 to −0.1	—	—	−0.5	−1.5	—	—	—

the composite. Although the loads are carried by the fiber, composite mechanical performance depends to a large extent on the resin modulus and failure strain and ability of the resin to bind to fiber. Resins that are used to make structural parts can be divided into two classes: thermosets and thermoplastics.

Thermoset matrices react directly in the reinforcing textile to a solid, which becomes idealized into one single molecule network. The educts can be injected or the textile can be impregnated if the viscosity of the educts is sufficient high at low temperatures. The permeability of the matrices inside the textile is generally good because of the low viscosity of the relatively short monomer molecules at reaction temperature. A disadvantage of using thermosets is the control of a chemical process from the FRP-part manufacturer. Depending on the system applied, the manufacturer has to handle mixing chemicals, temperature and pressure controls, as well as treating solvents.

Typical systems are unsaturated polyester and epoxy resins. The polymerization of unsaturated polyester is started by adding radicals, that is, peroxide, and generally needs only room temperature. This simplifies the requirements for the equipment, especially the mold. The disadvantages are chemical shrinkage and

Table 8.3 Properties of aramid fibers.

Aramid fiber type designation	Kevlar			Twaron		
Quality	29	49	149	LM	SM	HM
Density (g cm^{-3})	1.44	1.44	1.47	1.45	1.39	1.45
Filament diameter (μm)	12	12	12	12	12	12
Tensile strength (GPa)	3.62	3.62	3.44	2.8	2.8–34	2.85–3.15
Young's modulus (GPa)	67–83	124–140	186	89	65–71	100–130
Breaking strain (%)	3.3–4	2.3–2.9	2	4.3	3.4	1.3–3
Heat conductivity (W m^{-1} K^{-1})	—	—	—	—	—	0.04–0.05
Coefficient of thermal expansion (10^{-6} K^{-1})	−2 to −2.6	−5.2	−1.49	−2	−3.5	−4

the release of styrene. Epoxy resins react exothermally as a polyadditivation of two components, which requires activating temperature. Preformulated systems are often used; these systems avoid mixing failures, but need a cool storage facility. This is also applied for resin preimpregnated textiles ("prepregs"). The reaction is started upon passing a critical temperature. Using high temperatures leads to disadvantages as good temperature control is needed. Problems with expansion at high temperatures can occur, resulting in increased expenditure on equipment. The advantages of epoxy resins are their excellent mechanical properties and reduced volumetric shrinkage during cure. Another thermoset matrix system is phenolic resin, which is used in flame-critical areas such as generators or aircraft interior applications. The processing is ambitious because of the release of water during cure.

There is a growing interest in thermoplastic matrices because of their short processing time, which offers the chance to realize high volume production as needed in, for example, the automotive industry. Thermosets solidify instantaneously and theoretically allow cycle times of within minutes. In addition, they have a better impact tolerance than thermosets, can be welded, and are easier to recycle. The relatively high viscosity and processing temperatures, however, lead to only a limited number of special production processes, that is, thermoforming or pultrusion. This reduces the possible part size and geometry. Several thermoplastic polymers are available for fiber-reinforcing applications (Figure 8.3), for example, commodities such as polypropylene, technical polymers such as polyamide, and high performance polymers such as polyetheretherketone (PEEK), polyetherimide (PEI), or polyphenylensulfide (PPS), which are used for aircraft parts.

8.1.1.6 Intermediate Forms of Reinforcement

The textiles that are usually used as composites are woven fabrics. These fabrics can be used in a wide range of applications. Woven fabrics are composed of the warp yarns in the production direction (0°) and the weft yarn in the cross direction (90°).

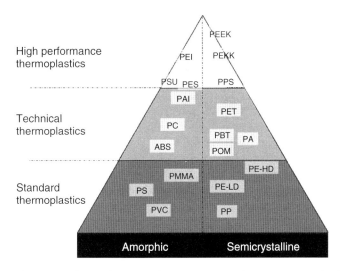

Figure 8.3 Thermoplastics used as matrix material in FRP. (Copyright Faserinstitut Bremen e.V.)

The construction in both directions is mostly balanced. With suitable arrangement of fiber in the fabric, the required shear strength, drapability, and stiffness can be modified to suit the application. Plain weaves have a better shear strength, especially during the manufacturing process, and harness satins have a higher stiffness for tensile and compression loads, as well as a better drapability. To create a unidirectional (UD) fabric, a very thin weft yarn is used compared to the warp yarn.

The waviness of the yarns inside the fabrics reduces the stiffness and strength of the reinforcement. The so-called spread tow fabric offers the advantage of relatively lower crimp. The tows are spread in thin and flat UD tapes and woven into a spread tow fabric. This technique is used to reduce the weight of composites. From the mechanical point of view, thinner yarns in a fabric lead to a better material performance. The textile production runs at higher length per time but at reduced mass per unit area. This reduced mass per unit area causes a higher number of layers in the final composite part with additional time for the layup process.

Other more specific weaves are, for instance, spiral ribbons, which are used in curved structures and 3D woven fabrics to reduce the layup time during composite manufacturing and enhance the out-of-plane properties, but reducing the in-plane properties.

For tubes and other three-dimensional profiles, braidings are often used. The mechanical performance of braidings is of the same quality as that of woven fabrics but with a lower productivity than in the textile process. The advantage of braidings is that they are fitted close to the final contours by automated braiding technology. Even on complex curves, thread displacement can be avoided with this technique. The fiber angles can be adjusted between $\pm 10°$ and $\pm 80°$ for symmetrical and unsymmetrical setups. A combination with fibers at $0°$ is also possible.

Wide applications can be found for composites made from noncrimp fabrics (NCF). They are available as UD, bidirectional, and multiaxial-stitched, bonded constructions. Fiber orientations can be adapted for each layer individually. Compared to woven fabrics, the yarns are more stretched inside the individual layers, leading to laminates with a higher strength and stiffness. The manufacturing of NCF needs major machinery but the textile process runs faster than the weaving process. Bidirectional and also often multiaxial NCF have an unsymmetrical layup. For a balanced and symmetrical laminate, different types of NCF have to be combined. The use of NCF leads to reduction of the manufacturing cost because of fewer layers caused by the higher layer surface area weight.

8.1.1.7 Fiber Volume Fraction

An important parameter in all composites is the regulation of the volume or mass fraction of subcomponents. Fiber-reinforced polymers offer a wide range of material settings for tailored material characteristics, for example, the variation of fiber volume content leads to a high stiffness or excellent damping performance.

Depending on the chosen manufacturing process, the fiber volume fraction varies from approximately 25 to 70%. Hand layup processes are often used for shipbuilding industries, for wind blade manufacturing and railway transportation systems. The fiber volume content is about 30%. For components used in aircraft, the aim is to reach 60% fiber volume for structural strength and high stiffness. The preforming process leads to compaction of the fiber-wrought material. Depending on the manufacturing process, the fiber volume content can be set. Using prepreg in an autoclave process, for example, material is produced with a defined matrix content and is cured under adjusted pressure. In a closed mold process, the fiber volume content can be set by the amount of fiber material put in the cavity.

8.1.1.8 Fiber Orientation Angle

The fiber orientation angle is of vital importance when it comes to the design and manufacture of fiber-reinforced composites. This is the main reason for the anisotropic characteristics of the material and disregarding it will either lead to components that do not fully utilize the potential of the material or, in the worst case, to components that fail under load. The anisotropic characteristics induced by the fiber orientation can best be visualized in a polar diagram (Figure 8.4).

The fibers should be aligned along the principal stress direction. For components that are mostly loaded with tensile or compressive stresses, this is in the direction coaxial to the fibers in the case of UD loading. Component stiffness and strength will be highest if the load flux can travel along the fibers without interruption. For components that mostly experience shear loads, the fibers must be aligned at an angle of $45°$ to the shear load. If the designed and actual loads differ by a large degree, the component efficiency is greatly reduced.

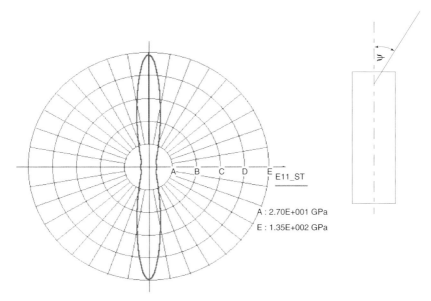

A : 2.70E+001 GPa
E : 1.35E+002 GPa

Figure 8.4 Polar diagram of the UD composite T400/FIBERDUX6376. (Reproduced from A.S. Herrmann, 2010, Mechanik der Faserverbundwerkstoffe, lecture at University of Bremen.)

Distortions of the fiber orientation angle can also arise during manufacture. This is mostly an issue during the preforming of the component. The drapability of the material is greatly influenced if deviations occur and it depends on the magnitude of the deviations.

The principal effect is that of shearing of the fabric when a $0°/90°$ fabric is formed into a double curvature shape. The resulting deviations occur on a large scale, with the degree of the deviation dependent on the fabric and the geometry. Smaller scale fiber angle deviations are in-plane waviness and out-of-plane wrinkles. Not all defects can be avoided during preforming; the shearing of fabric, especially, is an unavoidable phenomenon. Therefore, fiber angle deviations of up to $\pm 4°$ are usually tolerated in aircraft.

For example, the swept wings of a commercial airliner are loaded differently inboards and outboards, which is reflected in the laminate stacking. Closer to the wingtip, the loading is shear dominated and closer to the center it is bending dominated. Hence, close to the wingtip, the laminate consists predominantly of $\pm45°$-plies, whereas closer to the center, the laminate is enriched with $0°$-plies to enhance bending stiffness and strength. Furthermore, the spars, which run along the wingspan, also incorporate different fiber orientations. The caps, as in any I-beam, are largely responsible for the tensile and compressive stresses, whereas the shear forces are born by the eponymous shear webs. As expected, the laminate in the caps hence consists primarily of $0°$-plies and in the web, it consists of $\pm45°$-plies (Figure 8.5).

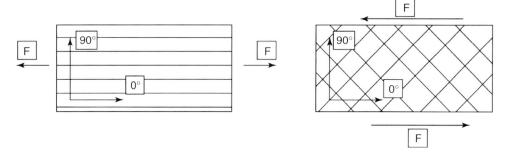

Figure 8.5 Example for different fiber orientations. (Reproduced from A.S. Herrmann, 2010, Mechanik der Faserverbundwerkstoffe, lecture at University of Bremen.)

8.1.2
Processes

The production process for CFRP structures can be classified into various stages from raw material to the finished component. The first step of processing leads from carbon fiber roving to a fabric, noncrimp fabric, braid, or stitched fiber orientation. The aim is to produce a defined fiber preform that respects the load-specific fiber orientation and components geometry. The fixed form is placed in a molding tool for resin penetration and curing. In the following, the preforming and curing processes described have specific advantages and are usable for different component types. These processes differ in manufacturing quality, flexibility, surface behavior, and productivity, as well as in production costs. This chapter gives a brief overview of production technologies for different product quality and manufacturing rates. Some examples of products are given for these technologies and specific restrictions as well as benefits are described.

8.1.2.1 **Preforming Processes**
Preforming a textile component is a process where the fiber material is fixed in a shape that can be handled and placed in a curing tool. This process – where the fiber material becomes draped – is significant for the value creation of composite manufacturing. The process reliability during following impregnation depends on localized behavior of the material. Disorientation of fiber material or stacking variation results in a change in permeability and this influences the impregnation. For complex structures, preforming is an important step in value creation. Nevertheless, preforming processes are mainly carried out manually, because automation of textile positioning and draping is quite challenging. The following technologies give a brief view of innovative automation applications in fiber preforming.

Tailored Fiber Placement (TFP) The mechanical characteristics of FRP are dominated by the type of reinforced fiber and the internal fiber orientations of the material. A minimal difference between the fiber direction and the maximum stress direction significantly reduces potential material performance [5, 6].

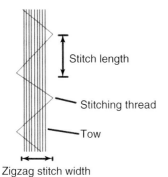

Stitch length

Stitching thread

Tow

Zigzag stitch width

Tailored Fiber Placement (TFP) technology makes it possible to place rovings along any desired direction, for example, according to principal stresses or strains. In this manner, fiber reinforcement can be aligned as closely as possible to the optimal mechanical direction to tailor the local stiffness and strength of the composite material [7, 8]. This highly automated and reproducible manufacturing process enables fabrication of specialized preforms for advanced composites used, for example, in aircraft parts [9].

Exact positioning of dry rovings on a base material is done by the use of modified embroidery machines. Different base materials can be used. Fabrics or nonwoven fabrics made of reinforcement fibers as well as non-load-bearing materials such as thin films or removable base materials are in application. The amplitude of the zigzag stitches, which are composed of an upper and a lower thread, fixes single rovings (Figure 8.6).

The flexibility of this textile-manufacturing process allows the production of dry performs for two- and three-dimensional parts. Without unclamping the base material, the rovings can be laid up one by one. The result is a near-net-shape preform with several layers one on top of another, which is easy to handle. When several TFP layers are laid one on top of another, the lower layers are affected by more stitches then the upper ones. In-plane properties are reduced by fiber undulations and filament fractures. Hence the maximum thickness of the preform is limited. Depending on the application, an optimum layer count has to be defined to reach a balance between economical manufacture and mechanical performance. Currently, different new layup technologies are under development that avoid the stitching process during the fiber placement [10].

TFP technology can be used for local reinforcements of big parts as well as for preforms for complete parts. TFP technology has clear advantages especially for parts with complex shapes or complex stress fields. TFP machines are numerically controlled and they work with high reproducibility. In order to optimize productivity, a group of knitting heads can work parallel to each other on the same base material. Small batch series, as well as mass production, can be operated economically with TFP (Figure 8.7).

The main improvement of TFP is that the material properties can be designed by adapting the fiber orientation in the direction of the stress flow inside the part.

Figure 8.7 Preform manufacturing with TFP technology. (Copyright Faserinstitut Bremen e.V., CTC Stade.)

To orientate the roving on defined curves, the production properties have to be adapted locally. Roving paths with small radii require more fixing points through stitching yarn than a linear path. The manufacturing speed is inversely proportional to the density of the stitches. Average layup speed of a TFP head is about 0.5 kg h^{-1}. Consequently, a typical eight-head TFP machine produces 4 kg of perform per hour.

The consolidation of carbon fiber TFP preforms to composites can be done using different standard methods for thermoset matrices such as resin transfer molding (RTM), resin film infusion (RFI), or vacuum infusion (VI). The main use of TFP is for thermoset plastics but thermoplastic parts can also be produced by stamp forming of hybrid preforms. For this purpose, thermoplastic fibers or foils have to be integrated into the TFP process.

Continuous Preforming Based on Fabrics and Braids   Dry textile materials offer significant better drapability to complex shapes than prepreg materials. Dry fibers are processed to so-called preforms for highly curved profiles with more sophisticated fiber architectures, especially with variations in fiber direction and cross section. Such preforms can be applied by using automation principles because of their original relations to the textile processing of clothing products. A specific automation principle with the intention of fast processing at process-stable conditions is that of continuous fully automated preforming. After preforming the textiles to the desired fiber shape and architecture, the preforms can be impregnated by accelerated RTM or infusion processes to get the desired fully cured CFRP component.

UD Braiding   The braiding process is a continuous preforming for complex curved profiles. In a rotational braiding machine, a curved tooling core is moved through a ring where fibers are pulled from braiding bobbins that rotate on another larger ring surrounding the inner ring (Figure 8.8). Fibers are laid down on the core by moving it continuously through the inner ring. Because the bobbins in the outer ring rotate in the opposite direction, it is possible to lay up two fiber orientations (e.g., $\pm45°$) at the same time. If one band of the bobbins is equipped with multifilament carbon or glass rovings and the other band is equipped with monofilament yarn (e.g., PES)

Figure 8.8 The core is moved through the inner ring of the braiding machine [20].

Figure 8.9 Two braiding machines in one production line for different fiber orientations [13].

with much lower linear density, the process is then called the UD braiding, in which the reinforcing fibers have significant lower undulations comparable to so-called UD fabrics [11, 12]. In order to lay up different fiber orientations it is necessary to move the core after one layer layup in the opposite direction. Another option for a highly automated process scenario is the stringing of two or more braiding machines working with different rotations in one production line (Figure 8.9).

The process limits the achievable products to hollow preforms. Desired profile shapes can be realized by using additional draping and folding steps off-line the braiding machines.

The winding–draping process was developed for closed 360° frames with significant but varied curvatures such as those of aircraft window frames. Flat multiaxial textiles such as NCFs are wound around a mandrel. In order to fix shape and position of this preform on the mandrel, a reactive binder activated by heat is

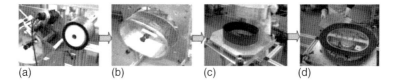

(a)　　　　　　　(b)　　　　　　　(c)　　　　　　　(d)

Figure 8.10 Process steps at winding drape process for preforms of aircraft window frames [13]: (a) pulling of textiles from creels; (b) winding around mandrel; (c) pulling the mandrel partially out of preform; (d) 90° folding of textile to L-shape [20].

required on the textile. This mandrel is then partially pulled out of this preshaped preform. The resulting protruded part of the preform is then folded by 90°, so that an L-shaped preform frame is obtained (Figure 8.10) [13].

To obtain the complete preform, it is often required to decompose it to generic preforms with lower complexity, which can be separately produced and afterward assembled to the desired preform. A similar principle was applied for an integral aircraft fuselage frame with so-called LCF cross section. The LCF was split in three preforms: a large "Z" (subpreform I), a small "Z" (subpreform II), and a small "C" (subpreform III) cross section (Figure 8.11).

The continuous preforming process for curved fuselage frames with layers of 0° and 45° was developed on the basis of the winding–draping process. Flat textiles stocked on creels are continuously pulled through a two-step draping device. The cross section is formed in the first step and the curvature, in the second. In order to fix the preform's shape, to stiffen it for following handling processes,

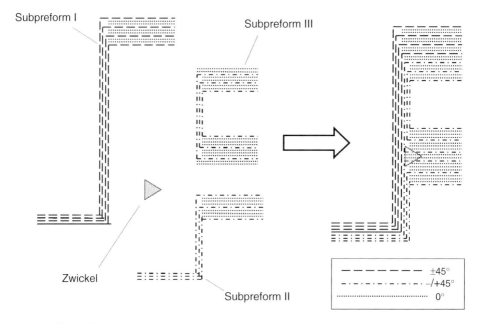

Figure 8.11 Layup [20].

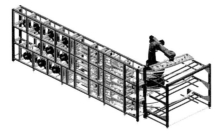

Figure 8.12 Pultrusion plant [21].

to avoid fiber disorientations, and to stabilize the cutting edge of the textile, it is equipped on one or two sides with a reactive or thermoplastic binder. The binder is activated by temperature using infrared (IR) or induction heating zones inside the machines between the draping steps. The continuous preforming machine (Figure 8.12) consists of three lines that can operate independently of each other with different preform shapes and curvatures. In the current configuration, the three lines produce the three main subpreforms of the integral fuselage frame.

8.1.2.2 Processes for Composite Manufacturing with Thermoset Matrix Systems

Thermoset matrix systems offer great conditions for lightweight design and various processes for component manufacturing tailored to multiple component design needs. Epoxy resins are used for most aircraft and automotive composite products because of their good weight-specific mechanical performance, temperature stability, low costs, and processability. The thermoset matrix systems can be designed for specific needs where a low viscosity makes them usable for most injection or infusion processes.

Epoxy resins are used to make structural components because of their good mechanical properties. For blade making of industrial wind energy components, unsaturated polyester resins are used because of the low curing temperature required for large components. With regard to safety in an aircraft cabin, for example, fire resistance is very important. Phenol resins offer a self-quenching behavior.

Liquid Resin Infusion Processes The modified vacuum infusion (MVI) process is part of the larger group of liquid resin infusion (LRI) processes. This group contains many different infusion processes, including the Seeman Composites Resin Infusion Molding Process (SCRIMP) and the vacuum-assisted processes (VAPs). Although all these processes have different names, many differ just in minor details. The MVI process itself is an established one within the aerospace industry and is used for the manufacturing of flying parts.

The process uses a tool on just one side of the part, while the other side is covered by a vacuum bag. As a result, the capability of the process is limited to two-dimensional parts and just one side of the produced part has a geometrically determined surface. The infusion and the curing normally

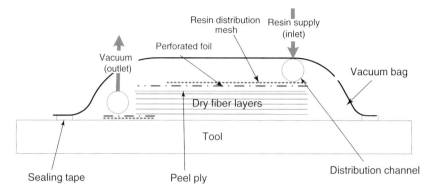

Figure 8.13 Schematic layout of an MVI process arrangement. (Copyright Faserinstitut Bremen e.V.)

take place in an oven. The schematic layout of an MVI process is shown in Figure 8.13.

Although the auxiliary materials and their arrangement are very similar to other infusion processes, the MVI process requires attention in some details. As in all infusion processes, the resin inlet and outlet are positioned to minimize the infusion length. With the MVI process, the resin inlet has to be positioned on top of the dry fiber material, while the position of the outlet is next to the part. A key material of the arrangement is the so-called resin distribution mesh, which supports the flow of the resin and leads to a uniform resin front.

At the beginning of the infusion process, the resin is pulled by the applied vacuum from the resin reservoir into the vacuum bag. During this first phase of the infusion, the preheated resin fills the inlet channel before starting to create a linear resin front within the resin distribution mesh. While the front is moving along the surface, the resin also infuses the fiber material from top to bottom. As a result, a diagonal flow front is established through the thickness. The distribution mesh does not cover the whole surface of the part, but continues on the bottom of the part. This arrangement allows the resin front to straighten before reaching the outlet, minimizing the risk of dry spots at the edges of the part. As soon as the part is fully infused, the resin outlet is connected to the inlet; this creates equilibrium within the vacuum bag and stops any resin flow. The process continues with increase in the temperature in the oven to start the curing of the resin.

Automated Layup Processes of Prepregs With the increasing use of prepreg – preimpregnated fiber materials – it became obvious, that a hand-laminating process is insufficient in terms of quality and process time. Therefore different automation methods have been developed for prepreg materials. The most common automated deposition processes are automated tape laying (ATL) and the more sophisticated automated fiber placement (AFP) processes. In combination with the curing in an autoclave, these processes produce high quality parts with high fiber volume fractions and low porosity.

An ATL machine normally consists of a gantry and a movable head. The material for the deposition is stored as a roll within the head. The head incorporates, among others, a cutting device and an application roll. During the deposition, the head can depose the material, which is normally about 300 mm wide, in any direction on a tool, which is positioned below the gantry. The part is now built up by applying the prepreg material layer by layer onto the tool. The machine uses a certain amount of pressure during the application process, which consolidates the laminate already in this step; this is one reason for the high fiber volume fraction achievable with this process [14]. The material is automatically cut as soon as the ATL head reaches the programmed edge of the part. As most parts are not simple rectangles and consist of ply with a $45°$ fiber direction, a typical zigzag pattern appears on the edges of the part; therefore, the part needs to be trimmed in a later process step. Although the ATL process is a fast and efficient process, it has its advantages in the manufacturing of large two-dimensional parts. Complex parts, which for example, have double-curvature geometries, require a more flexible process like the AFP process.

The AFP process shares much commonality with the ATL process, but it can be used for the production of fairly complex and 3D-shaped parts. Instead of deposing a 300 mm wide strip of prepreg, an AFP machine deposes up to 32 parallel strips of material, each just 6 mm wide, the so-called tows. Each of the tows can be individually controlled and cut. This allows the automatic layup of small local reinforcements within the laminate and also reduces the unused material at the part edges. The large number of tows makes it difficult to store the material in the head, which is why most machines have a separate storage compartment. In addition, the design of an AFP machine can be different from that of an ATL machine. In addition to the gantry system, there are columns systems, where the deposition head is positioned sideways, and systems where the head is mounted on an industrial robot. Some of the systems also include a rotatable mounting device for the tool, which further increases the capability of the machine to produce complex parts. Each of the different configurations has its advantages and disadvantages. For example, a column configuration limits the ability of the AFP process for the deposition on female tools (Figures 8.14 and 8.15).

Different types of technologies are available and under investigation for an increased layup rate. Available systems are based on heavy gantries to reach the required accuracy, while new systems focusing on mobile lightweight robots (Figure 8.16).

Under investigation is a new system using several communicating robots for a parallelized manufacturing process of large structures at the DLR, Stade, with the aim to reach a significantly increased layup rate.

Although the automated deposition process has been developed mainly for use with thermoset prepegs, they can be adapted for the use with thermoplastic resin systems. The main obstacle is the temperature required to process thermoplastic prepreg, which for aerospace-grade materials is in excess of $300°$C. This makes the process and therefore the equipment more complicated, but provides the opportunity for an *in situ* manufacturing process.

Figure 8.14 Torres automated tape layer: 20 kg h^{-1} (Torres). (Copyright Airbus S.A.S. Reproduced with permission.)

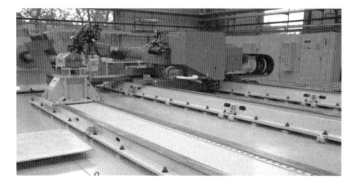

Figure 8.15 MAG Cincinnati tow placement: 8 kg h^{-1} (CTC, Stade).

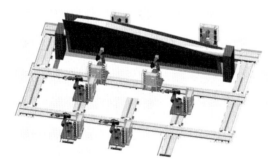

Figure 8.16 Robots MAG fiber placement: 100 kg h^{-1} (DLR, Stade).

Thermoset prepreg requires a curing process after the deposition. In most cases, the curing is done in a pressured and heated vessel, the autoclave. The uncured component is prepared for curing with the application of a vacuum bag. During the cure, vacuum is applied to the bag, evacuating trapped air and volatiles from the part in order to minimize the likelihood of porosity. In order to start the chemical reaction of the resin, the temperature is increased. For standard high performance resin systems, the cure temperature is in the region of 180°C. Simultaneously with the elevation of the temperature, the pressure is increased. Normal pressure level during the cure is around 7 bar [15].

The disadvantages of this process are the high costs because of energy consumption and the process time involved. Many autoclaves are filled with nitrogen in order to reduce the fire risk caused by the exothermic reaction of the resin. In addition, a large of amount of energy is required to heat the volume of the autoclave and the tool with the part. This leads to slow heat-up rates normally not exceeding 3°C min^{-1}. Depending on the part thickness and the resin system, the cure cycle might require a dwell at a lower temperature in addition to the cure time itself. Together with the slow heat-up and cool-down rates, the curing process takes several hours even for small parts.

The combination of an ATL or an AFP deposition process with an autoclave cure leads to a state-of-the-art manufacturing process for aerospace parts with unmatched quality and reproducibility. These technologies are used to manufacture major components of the Airbus A350XWB and the Boeing 787, including the wing covers, fuselage panels, and barrels.

Pultrusion   Pultrusion[1] is a well-known process for the continuous production of FRP profiles. Pultrusion was invented in 1956 in the United States [16]. The variety of different cross sections led to high market shares of pultruded FRP profiles in the construction (window frames, tubes, and slats) as well as in the sports and leisure sectors (ski equipment and tent poles). Low production costs for the profiles are achieved by the intrinsic fully automatic pultrusion process and material costs are lowered when glass fiber rovings and unsaturated polyester are used.

The fibers are continuously pulled out of a creel store and through the whole process chain by so-called pullers at the end of the machine. These pullers are alternately operating gripping devices (Figure 8.17). In a first step, the fiber materials are preoriented to a shape close to the intended cross section. Following this, the fiber rovings are pulled through a bath with the matrix in a liquid state of a low viscosity so that a good impregnation of the fibers is possible. An alternative impregnation process is the direct injection of the resin in the forming tool, called a *die*. The die gives the impregnated fibers the final shape in cross section and the resin gets cured because of different temperature zones in the die. The velocity of the whole process is optimized to obtain a good impregnation in the resin bath as well as a good curing in the pultrusion die, which is, of course, also influenced by the resin type and its reactivity and by the temperatures in the heating zones of the die. In order to finalize the curing of the profile, that is, to temper it, especially if epoxy resin is used, the machine is sometimes equipped with a passing oven, through which the profiles moved after they leave the curing die. At the end of the machine, just behind the gripping pullers, an automatic cutting device is installed, which cuts the profiles to their desired length.

---

1) *Pultrusion: Artificial expression, combined of "pull" and "extrusion".*

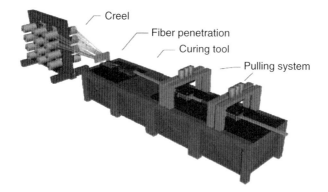

Figure 8.17 Machine setup of a pultrusion line [17].

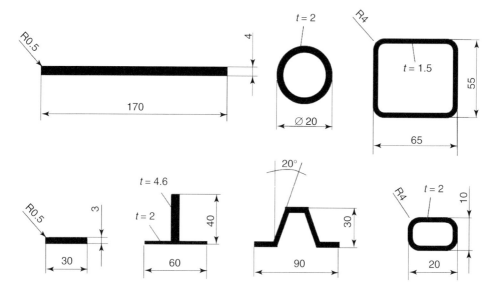

Figure 8.18 Examples of possible pultruded cross sections [20].

The pultrusion process is applicable for different profiles with open and hollow constant cross sections (Figure 8.18) consisting of different fiber materials (mainly rovings, but also fabrics).

PRTM Process The PRTM process combines the pultrusion with the RTM process for the continuous and fast production of CFRP profiles with high fiber volume content, high glass transition temperature, and excellent properties of the laminates, which can be applied in aeronautic structures. In the PRTM process, dry flat textiles consisting of different fiber orientations (e.g., ±45° or ±60°) are pulled by conventional pultrusion pullers through a preforming system in the first step (Figure 8.19). After getting the intended cross section, the preform is continuously

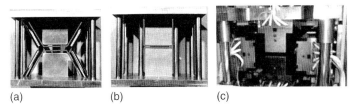

(a)   (b)   (c)

Figure 8.19 PRTM machine – preforming system (a,b) and CFRP profile press (c). (CTC, Stade.)

pulled into the heated injection die (which is equipped with a cooled tool entrance), where an adapted epoxy system is injected at a pressure between 2 and 3 bar. Upon leaving the injection die, the impregnated profile has a prepreg morphology. The main curing of the resin is achieved in the RTM press, which is moved with the same velocity as are the profiles by the pullers. Friction between tooling surface and FRP profile is very low because of the low viscosity of the resin in the injection die and the synchronized movement of the pressing die. After the press process, the profile is pulled through an additional tempering zone in an oven [18].

High process and product quality was validated on straight profiles with rectangular, T- and Ω-cross sections, so that the process can be used for the production of aircraft stringers, for panels in box and fuselage assemblies. A main characteristic of the process is that components at a different grade of completion can be produced on demand such as dry preforms, prepregs, and final cured parts.

Continuous Panel Manufacturing The ACROSOMA process is a continuous pultrusion technology for large composite panels. A light foam core is continuously equipped with cover sheets made of dry textiles on the upper and lower side of the foam, which are stitched together by an additional yarn-tufting process. The tufting sticks the components together and integrates impregnation flow channels for the resin through the foam core. At the end of the process chain is a convection heating zone that initiates a resin curing. At this stage, a consolidated and continuously produced sandwich panel is available (Figures 8.20 and 8.21). The main advantage of this technology is the integral connection between the upper and lower cover sheet, which reduces the tendency of delamination under heavy concentrated loads. Therefore, ACROSOMA panels can be used in applications with the possibility of concentrated loads under several service conditions such as in the trailer of trucks.

Fast-RTM The philosophy of FAST-RTM is based on the principle of fast-paced temperature change close to the cavity in order to achieve a quick and homogeneous heat up. Therefore, in the first heating phase, the maximum fluid temperature of $250\,^\circ$C and a high volume flow of up to $30\,\mathrm{m}^3\,\mathrm{h}^{-1}$ of the thermal fluid will be used to heat the metal RTM tool with a heating rate of approximately $30\,^\circ$C min^{-1} up to the injection temperature of $100\,^\circ$C. Different areas of the tool have distinct thermal requirements, hence three independently controlled heating systems are available each of which can cycle through a temperature

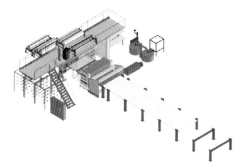

Figure 8.20 Machine setup for continuous panel manufacturing. (Source: ACR2009.)

Figure 8.21 ACROSOMA machine in service. (Source: ACR2009.)

profile specific to the heated region inside the tool. A volume flow of 10 m³ h⁻¹ each enables an efficient control of the different temperatures of the tool. Once the cavity temperature has crossed the lower injection temperature threshold in almost all adjacent areas, a short homogenization phase is initiated by a quick transition to a lower fluid temperature. For this, the entire volume flow can also be facilitated. Subsequent to the injection, the fluid temperature inside the tool is matched to the thermal requirement of the reactive resin, which results in a quite homogeneous heating of the cavity. The cooling phase can be significantly reduced compared to conventional RTM processes through the use of the thermal fluid at room temperature and the full 30 m³ h⁻¹ volume flow (Figure 8.22).

A facility of this type can, depending on size and mass of the tool, heat or cool four to eight tools in parallel. This makes the technology feasible for application in series production. As the heated fluid can be kept at a heightened temperature in the process to heat several tools, the overall thermal energy required can be reduced. This ensures that – in contrast to heating the tool in an oven or a heated press – merely the thermal energy to heat the tool is required while the remainder of the energy is saved in the fluid.

As an example for the FAST-RTM process chain, the production of an integral airplane frame structure is described in the following.

In airframe design, it is customary to tailor each frame to the individual loads it is subjected to and therefore design each frame with a different geometry and

Figure 8.22 FAST RTM machine. (Copyright Faserinstitut Bremen e.V.)

laminate layup to achieve maximum weight saving. The tool for the production of an integrated frame is equipped with the FAST-RTM technology in the outer areas, which enables a targeted temperature control between the injection (about 100 °C) and the cure temperature (about 180 °C). During serial production, the tool will not be cooled down below the injection temperature. The heating process is hence separate from the geometry-defining elements of the tool, so that the needed variability of some frames can be achieved. The tool inlay defines the geometry and allows for individual laminate designs. It can also act as a preforming tool for the layup process. Furthermore, the inlay is mostly thin walled to facilitate a fast heating of the cavity through the metallic surface of the outer tublike tool. Only the thin-walled inlay must be heated during each production cycle (Figure 8.23).

The inlay was designed so that it can be changed at high temperatures (180 °C) in a quick and risk-free manner.

Furthermore, jigs were designed, which allow the usage of water-soluble mandrels instead of metallic ones for the inner part of the frame. In addition, the tool can be used in combination with the heated COREON mandrels, developed by Faserinstitut Bremen [18]. The tool lids were designed with inlet ports, which allow the sensors and the electrical supply of the heated mandrels to enter the tool.

The approach of this method is to use the water-soluble mandrel material itself as an electrical resistance heater. The outer area of the mandrel is reinforced with an additive, which enables setting a defined or specifically varying electrical conductivity in the mandrel material. The amount of locally adjusted electrical power which is converted to heat emission results from the locally adjustable thickness and locally adjustable electric conductivity of the electrically conductive layer, which can be adapted to suit the local heat requirement. The variation of the supplied electrical power results in a regulated heating profile on the mandrel surface. Owing to a multilayered construction, the heat in the edge layer is also

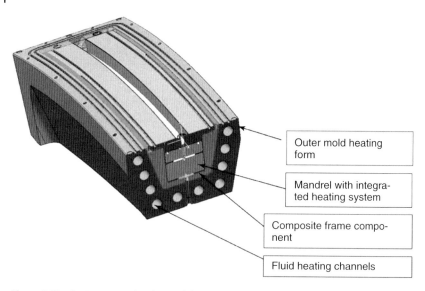

Figure 8.23 Design principle of a modular FAST RTM tooling. (Copyright Faserinstitut Bremen e.V.)

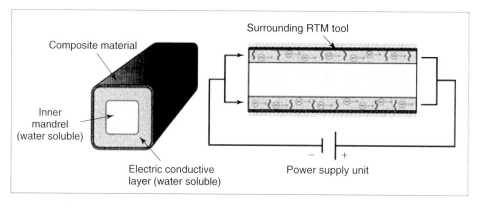

Figure 8.24 Heating principle heating of water-soluble mandrel material. (Copyright Faserinstitut Bremen e.V.)

directly generated at the preform. The inner mandrel is heat insulated for an efficient and fast process (Figure 8.24).

The use of lost mold mandrels enables a production process where these mandrels can be released without residue by using plain water. The material is made of a ceramic-based granulate combined with hollow glass balls and a chemical binder to form the geometry. An additional advantage is that complex geometries can be realized without a division of the mandrel. Mandrels from a porous molding material are used to obtain a lower thermal conductivity than that of metallic tools. The heating process can therefore be positively influenced. When warming the cavity, a high temperature gradient in the mandrel material results in a heated outer layer directly on the fiber preform. To additionally counteract inhomogeneous heating through the outer cavity, the Faserinstitut Bremen has developed an integrated heating system called *COREON* that generates a regulated heat emission at the surface of the water-soluble mandrel.

High Pressure RTM   The high pressure resin transfer molding (HP-RTM) process is used for high production rates and is often combined with the use of very reactive bicomponent matrix materials. This combination makes the technology very attractive for industries with very high production rates.

High injection pressure is used for accelerated bicomponent mixing in an injection blast pipe – which results in a variable resin injection pressure of about 80–100 bar. The fiber preform needs to be locked into position when the tool becomes closed. This clamping process is required because of increased flow rate and allows the usage of more viscous resins or lower injection temperature. In addition to the higher pressure and fiber clamping device, the HP-RTM technology is often combined with a new injection philosophy called *compression-RTM*. During the injection process, the tool is not completely closed – a small gap is used to spread the resin over the preform. On closing this small gap, the preform gets penetrated in the Z-direction for a very fast form-filling process. Usually, the tool needs to have some small interfuse channels at the resin inlet to mix the two reactive resin components. The tools are used almost isotherm on a temperature level that induces the curing reaction very quickly. Owing to the combination of changes compared to conventional RTM, the process is accelerated to a few minutes.

For fast reuse of tools, the cured component is released by a mechanical dropout system integrated in the tool. As a result of high pressure injection, the surface quality seems to be improved and in case of occurring porosity, the voids are compacted to very small structural defects.

Fully Automated RTM-Production Line   The ongoing use of fiber-reinforced plastics and the permanently increasing rates in the civil aircraft manufacturing leads to more automated production lines for composite production. Therefore a public-funded project was started at the CTC GmbH, Stade, in 2003 to point out ways for a fully automated CFRP production, a simulation of the possible processes, and a conversion into hardware.

An investment in automation technologies has to be justified by high production rates – one good product for such automation is that of the aircraft attach fitting for vertical tail planes of the A320 family. This component is produced for each single aisle vertical tail plane in 12 different characteristics. The connection between fuselage and vertical tail plane is later made by these attach fittings (Figure 8.25). For a production rate of 40 a/c per month, about 5800 components are produced per year.

Figure 8.25 AIRBUS Single Aisle VTP attach fitting (forward outside right). (Copyright Airbus S.A.S. Reproduced with permission.)

As an example of 1 of the 12 attach fittings, a fully automated process chain had to be developed for the RTM process. The selected demonstrator component consists of nearly 50 single layers. The whole process is divided into a preform and a curing process. Thus holding time of the RTM tools is shortened and it is possible to parallelize processes. See Figure 8.26 for an overview of the RTM process chain and tooling cycle.

The preform process starts with the automated cutting of the layer. One such machine with material feeder was installed for blank production. For the automated pickup of the blanks, the cutter was provided with a robot that is equipped with an adjustable vacuum effector. In order to use the full capacity of material supply, the single blanks are waste-optimized distributed and placed close together. With the handling device thus realized, it is possible to decollate every single blank without impairment to the flanking blanks and without changing the effector. At the same time, a draping of the single layer takes place at the grab of a blank by the effectors' 3D-contoured effective area. The picked-up blanks are engaged in the correct order into a microwave-transparent preform tool. The consolidation of

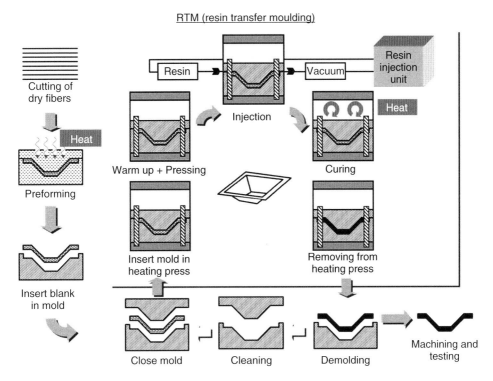

Figure 8.26 Principle of the resin transfer molding process (RTM) [19].

preform takes place under vacuum in a 12 kW strong microwave. Thus the process time is shortened significantly.

The finished preform is taken out of the preform mold with a vacuum-assisted effector by a second robot and put into the opened curing mold. The curing mold is fully automatically transferred on a roller conveyor to the lid station and supplied with the lid of the RTM tool. The closed mold is moved into a heating press and supplied with hoses for resin and vacuum connection by a docking station especially developed for this purpose. The advantage of this automatic docking station is that no manual interference becomes necessary and that the volume of throwaway parts is remarkably reduced. With the developed system, only two copper tubes of about 15 cm length are wasted per injection, while during the serial process two resin-filled hoses of minimum 2 m length including screwed fittings have to be thrown away. After the press is closed and the resin connection is completed, the injection is started over a fully automatic injection system with a resin trap. Further characteristics of this injection system are seen in the preliminary microwave heating of the resin and in the controlling of the duration of the injection by the determination of the resin mass flowing through the preform. Thereby, the adherence to injection pressures and temperatures are constantly monitored and logged. After curing, the automatic separation of injection lines takes place. The utilization of the thermal characteristics of the single-component resin used makes possible an absolutely clean and drip-free separation.

The tool is transferred to the lid-opening station and opened up again by electrically driven mold-opening cotter pins. Applying the required force for separating the lid from the mold is the major challenge in this step. The lid is taken off over elaborated kinematics and a third robot releases the finished part from mold with a suction gripper. After releasing the part from the mold the robot changes its effector for cleaning the mold. This cleaning effector works with a combination of an air jet and a vacuum cleaner. After this, the curing mold is available for a further process.

In summary, the following processes were carried out:

- vacuum effector for sorting and stacking the single blanks;
- microwave-assisted preform process;
- automated lid station for opening and closing the curing mold;
- automatic docking unit for resin and vacuum supply;
- fully automatic injection system with resin trap;
- automated cleaning method for RTM molds.

The total procedure could be optimized to the extent that single process steps are implemented into serial production at the AIRBUS plant in Stade. Thereby, it is possible to shorten the production time for one attach fitting at Stade plant by about 50%. At the same time, capacities in RTM production are increased. In addition to the future expandability of the manufacturing activity, it is substantially improved toward automation.

8.1.2.3 Processes for Composite Manufacturing with Thermoplastic Matrix

Future aircrafts and vehicles will contribute significantly to energy and CO_2 savings by using intelligent lightweight constructions. The use of thermoplastic composites with carbon woven fabrics will play an increasing role to achieve these ambitious weight specifications [14, 20–22].

The potential of thermoplastic CFRP parts inside the thermoforming process is the ability of *in situ* consolidation in just one process step. There is increased interest in reinforced composites on the basis of thermoplastic polymers because of their superior producibility and formability. Thermoplastics offer several benefits such as uncritical and unlimited storage time, a better impact tolerance compared to thermoset matrices, a good ultimate strain performance, reduced crack propagation, excellent chemical resistance, and quick forming process availability. They are recyclable, and a wide range of customized laminates and matrices are available. At present, a barrier to comprehensive dissemination of thermoplastic polymer structures is the lack of economic, quick, and reliable component manufacture processes. To overcome this deficit, fully automated process chains for the manufacturing and assembly of thermoplastic composites have to be developed in order to achieve the production rates and cycle times required by the automotive and aircraft industries as well as the civil engineering sector [20, 22].

Qualified thermoplastics for aircraft application are only those high-efficiency polymers that have high application temperatures, high melting points, and a good chemical consistency. In this case, thermoplastic polymers with matrices from PPS, PEI, and PEEK have been well accepted and certified for aircraft components at Airbus and Boeing.

Automated Stamping of TP Blanks The sheet-forming process with low cycle times (<2 min) produces parts for different industrial applications using amorphous or semicrystalline thermoplastic composites. There exist several manufacturing methods with different process types, for example, the compression molding, stamp-, and diaphragm-forming processes. All the processes use reheated preconsolidated sheets ("organic blanks," Figure 8.27) with woven fabrics or UD tapes for the press process in common, tempered molds.

The following steps describe the process of producing parts by thermoforming (Figure 8.28):

- material clamping with spring holders inside an open frame;
- homogeneous IR-heating of defined melting temperature of, for example, 300–420 °C, with a tolerance of ±5 °C for high efficiency thermoplastics (PEI, PPS, and PEEK);
- stamping process under pressure (e.g., 4.5 MPa) with passive and active cooling down process inside a low end and automated tempered tooling device;
- deforming process;
- quality and material inspection.

Figure 8.27 Aircraft fuselage thermoplastic carbon composite clip examples [22].

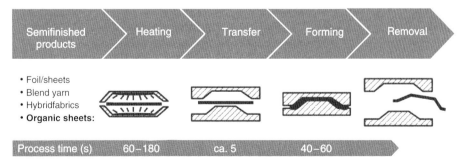

Figure 8.28 Thermoforming process and forming cycle [23].

The process to date is a part-automatized manufacturing process with high material wastage. This is necessary at present, in order to guarantee secure textile handling during the heating and the press steps. At the end of the process, the fiber orientation, the geometry of the part, the degree of crystallization, and, finally, the inner quality determines the quality of the parts. Nondestructive testing (NDT) methods such as ultrasound and the final contour cutting are not integrated inside a thermoforming process chain now. Every part is individually 100% tested for inner quality using nondestructive ultrasound processes.

The matrix is transformed during the forming process from a melted, highly viscous liquid to a solid at solidification temperature inside common tempered tools. Between the two types of polymers that exist inside the mold, there are different solidification temperatures: for semicrystalline polymers, the crystallization temperature is important, and for amorphous polymers, it is the glass temperature. The drapability of the heated blanks inside the tool are regulated *inter alia* by polymer viscosity and the friction connected with the blank temperature and the layer architecture. In this context, it is necessary to understand the interactions and the thermal effects of the thermoforming process conditions and the mechanical performance connected to the component quality of the pressed parts.

The characteristics of fiber-reinforced, thermoplastic, semifinished products are directly combined with the thermoforming manufacturing processes and the crystallization process inside the tooling of semicrystalline polymers (as an example). The crystalline phase has to be considered in this case as a kind of self-reinforcement. The essential performance characteristics of the semicrystalline thermoplastics accordingly affect the part in the crystalline phase and the degree of crystallization [22].

The drapability properties of the reinforced thermoplastics with woven fabrics or UD tapes are regulated by different deformation modes and flow mechanisms. The flow mechanism is a mixture of matrix resin percolation, transverse flow, interply shear, and interplay rotation and shear effects, and it depends on the fiber orientation [24]. Constrained to the matrix viscosity, the woven fabrics enable the shear, slide, and stretch effects. These shear effects compress the matrix orthogonal to the fiber direction (Figure 8.29), with a consequent increase in shearing deformation at the matrix melting temperature [25]. The woven fabric drapability changes with the thermal gradient of the matrix are related to its viscosity and succeeding local residual stresses and shape deviations inside the thermoforming parts. It starts at the first contact of the fabric surfaces with the mold, the matrix freeze at the surface with a high thermal gradient of about $\Delta T \sim 130\,°C$, and the part cooled rapidly down (about $130–150\,°C\ min^{-1}$) in the mold. In addition, these effects become more and more complex in 3D components (Figure 8.30).

The angle of the final parts are smaller than the (e.g., $90°$) angle of the mold. This phenomenon is commonly known as the *spring-in* or *spring-forward effect*. This is an effect of the material anisotropy and the thermal properties of the materials (fiber and matrix). The effect depends primarily on the differences between the in-plane

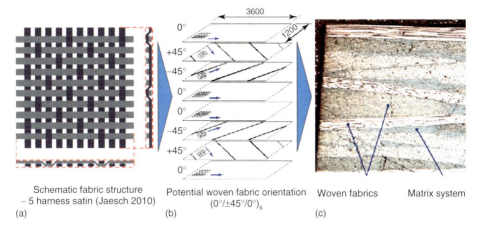

Figure 8.29 (a) Fabric structure; (b) potential textile orientation; (c) cross section of a Cetex® organic sheet with a fiber volume constant of 50% [26, 27].

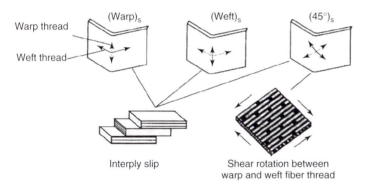

Figure 8.30 Deformation mechanisms of fabric-reinforced laminate [24].

and through-thickness shrinkage properties of the material Figures 8.31 and 8.32 show two arc sections made from isotropic and anisotropic materials [24, 28].

The goal of the automated thermoforming process is the implementation and link of all process elements to guarantee a cycle time of 1 min. An important key element is the automated NDT and NDI (nondestructive-inspection) element using ultrasonic and digital image analysis methods. The objects for quality inspection are pores and delaminations as well as fiber orientation, degree of crystallization, and geometry. The automated thermoforming process cell gives the aircraft, automotive, and civil engineering industry as well as the textile engineering a potential instrument to produce parts of thermoplastic CFRP or GFRP with a high level of automation of consistently high quality without neglecting the demands of low production prices. The thermoplastic process cell can be adjusted according to individual requirements. Thus we are able to anticipate the total cost of ownership

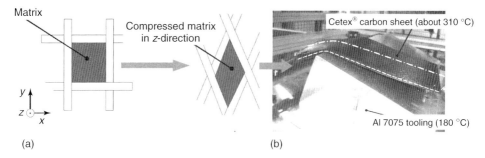

Figure 8.31 Matrix compressed about fabric shape distortions inside the thermoforming process by Berthold [25] (left) and Peters [22] (right).

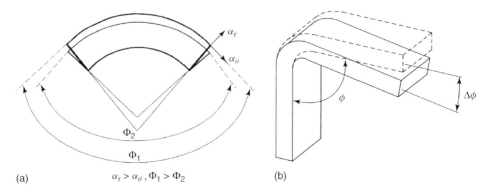

Figure 8.32 Spring-in effect of anisotropic materials ((a) [24] and (b) [28]).

as well as cycle times and average unit costs of the entire system with low energy consumption. As all individual process elements are integrated, the cycle time can be reduced with the automated thermoforming process cell. The number of parts integrated into the process device can be automatically adapted in flexible process devices [22, 27].

Thermoplastic Laminates Made of Hybrid Textiles   For complex shapes, the use of hybrid textiles is often more cost efficient than thermoplastic laminates made of thermoplastic foils, thermoplastic nonwoven fabrics, or thermoplastic spray substrates. High drapability is obtained through matrix integration in fiber form during the textile production.

   Feedstocks can either be a plied side-by-side or as commingled yarn as a combination of carbon fiber with thermoplastic multifilaments. The thermoplastic matrix filaments are produced on a melt spinning plant that specializes in technical multifilaments with a low degree of shrinkage and without or little sizing, which is required for better fiber–matrix adhesion. The hybrid fabrics can be produced with common textile technologies such as weaving, knitting, and braiding; the geometrical flexibility of the TFP process can also be used. The subsequent drapery and consolidation

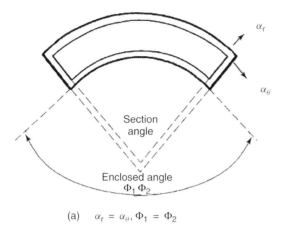

(a) $\alpha_r = \alpha_\theta, \Phi_1 = \Phi_2$

Figure 8.33 Non-spring-in effect of isotropic materials [24].

Figure 8.34 Automated clip demolding and handling after the thermoforming process [22].

takes place in a thermoforming press. Here the thermoplastic multifilament results in the matrix fraction (Figure 8.32). Heating and cooling rates can be set much higher than for thermoset matrix systems. The basic parameters for the consolidation process depend on material and textile types. CF–PPS hybrid preforms made from commingled yarn need a time of 5 min (cooling speed 90 K min^{-1}) for consolidation at 340 °C and 40 bar (Figures 8.33 and 8.34). For molding and demolding, the laminate temperature should be below the melting point of the thermoplastic but can be above T_g. An overview of the process chain is given in Figure 8.35.

8.1.3
Manufacturing Process Simulation

A distinctive feature of composite materials is that the resulting material properties can be designed by selection of different fiber/fabric and matrix materials and are

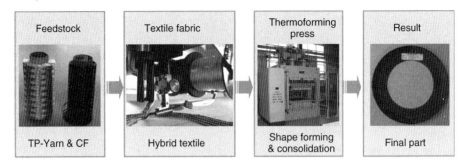

Figure 8.35 Process chain for CFRTP parts made from hybrid textiles [2].

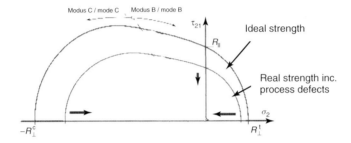

Figure 8.36 Influence of process defects on the material strength [29].

influenced by the manufacturing process itself. As matrix material, epoxy resin as a thermoset polymer is mostly used and because of the high requirements of the aerospace industries, toughened, thermally stable resin systems are used. This leads to the point that resin systems are typically chosen with a high curing temperature around 180 °C to reach glass transition temperatures above 160 °C. During the curing of CFRP materials, inhomogeneous material properties play a major role in the development of process-induced deformations and stresses because of the effect of chemical matrix shrinkage and thermal expansion. The resulting material properties will be influence by process-dependent deformations and stresses. In fact, internal stresses may become large enough to lead to fiber–matrix debonding, matrix failure, or delamination. In general, defects on the microscale such as matrix–fiber debonding or small matrix cracks are hardly measurable and are therefore only noticeable in reduced stiffness or strength properties (Figure 8.36).

The superposition of process-induced and mechanical stresses also leads to critical situations. In composite materials, residual stresses can change the failure behavior [29]. Using criteria such as the Puck criterion, a complex stress state can be condensed and displayed by a vector of effort. In general, this vector starts at the origin of the material coordinate system and indicates failure if it touches the engulfing failure curve. In case of present residual stresses, the vector of effort does not start at the origin (Figure 8.37). By neglecting process-dependent stresses, the effort value can change significantly, for example, the

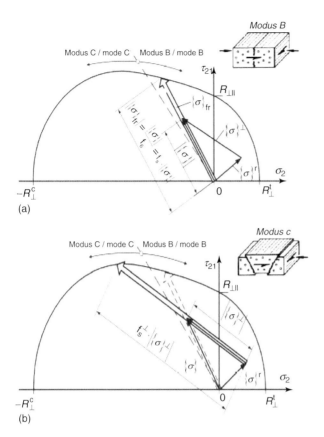

Figure 8.37 Influence of residual stress on the failure behavior [29] (a) without residual stress and (b) change of the material effort due to residual stress.

mode of transverse failure can change from mode B to the more critical failure mode C.

This highlights the need for reliable methods to analyze the level of process-dependent residual stresses and deformations.

In general, a manufacturing process simulation has to account for all relevant mechanisms that occur between the moment the yet uncured structure is draped inside the mold and placed inside an oven/autoclave and the moment the fully cured structure is released from the mold. These imply that a complete manufacturing process simulation has to start by evaluating temperatures at the surfaces of the structure, possibly using CFD tools. Next is the computation of resulting temperatures inside the part by means of thermodynamic analysis, which must include the exothermic heat reaction of the resin for accuracy reasons. The thermodynamic analysis has to be coupled with a curing simulation to determine the degree of cure and resulting heat generation; while a mechanical simulation finally analyzes process-induced deformations and residual stresses. The aim is to determine residual stresses and process-induced deformations and the possibility to perform a sensitivity analysis on material, design, and process parameters. Information about deformations is important in order to reach the stringent tolerance specifications of aerospace structures. Next in importance is the ability to predict the internal loading of the structure which, combined with external loads, determines the long-term behavior and structural integrity of the investigated part.

A manufacturing process simulation of the *whole manufacturing process* is defined as a multiphysics and multiscale problem. Therefore, intelligent idealizations and coupling methods have to be found between different disciplines such as draping, heating, curing, and distortion on the part, at macro, meso, and micro levels. A high expertise about the process itself is needed to take into account all part-quality-relevant parameters. Therefore, the virtual process chain must always be a kind of problem-oriented modeling with capabilities to switch between different disciplines and scales.

In Figure 8.38 a manufacturing process simulation for a load-carrying structure of an aircraft is presented. A sandwich with reinforced CFRP pins in the thickness direction was developed to be a part of the next generation vertical tail plane with improved impact characteristic behavior. During the research project, process-induced stresses occurred; these were visible in the form of low mechanical performance and cracks. Therefore the manufacturing process was been analyzed using numerical methods, starting at the level of thermodynamic analysis and concluding with a sequential, coupled mechanical analysis. The whole process has been simulated using a transient thermodynamic analysis in order to determine the temperature inside the part. During this step, the degree of cure is computed, which

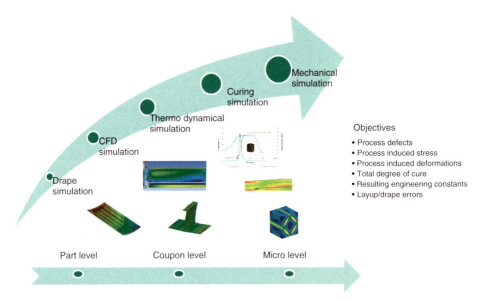

Figure 8.38 Virtual process description. (Copyright Faserinstitut Bremen e.V.)

also applies an exothermic heat flux related to the chemical curing reaction. After this, the temperature results are transferred to a transient mechanical analysis. This weak sequential coupling of the thermodynamic and the mechanical analysis was chosen because temperature has an impact on the mechanical behavior, whereas mechanical deformation has no influence on the temperature. For the mechanical analysis, the material behavior is idealized as linear viscoelastic, dependent on the glass transition temperature. Finally, cure shrinkage of the resin is implemented using an incremental strain formulation also dependent on the degree of cure. The resin engineering properties are dependent on the degree of cure and the glass transition temperature and coupled via analytical micromechanical approaches to compute the homogenized engineering properties on the ply level.

It is useful to divide the development of the resin properties into three separate stages: first, a purely viscous behavior where the resin does not support shear loads and does not develop residual stresses; second a viscoelastic behavior after gelation, where stresses will be generated, but decay to some degree relative to the process time; and third, an elastic behavior after vitrification, where the material behaves nearly linearly. A material model must take into account all three stages. Different models have been developed in the last decades, which couple the resin modulus directly to the degree of cure [30]. Further approaches can be found in the literature on the aspect of how to model the resin modulus development by either linear, incremental linear [31], or nonlinear approaches [32, 33]. In the given example, a nonlinear approach was developed on the basis of shear rheological experiments.

The process simulation can provide the following results: residual degree of cure, glass transition temperature, resulting homogenized engineering constants, process-induced deformations, and process-induced stresses. The validation of these results will be available in case of degree of cure, glass transition temperature, engineering constants, and process-induced deformations. Simulation of stresses is not trivial; in fact, the behavior of the polymer matrix material is thermomechanical and nonlinear even while cooling down from the curing temperature to room temperature.

In the last decades, many researchers investigated the development of process-induced stresses on a macromechanical level and also on the microscopic level. The validation of process-induced stresses on the macro level is very difficult; in fact, there are no nondestructive or destructive validated tests available to measure stress in composites. The development of process-induced deformations is mostly understood and has been modeled, but stresses are influenced by nonlinear effects such as viscoelasticity, microscopic yielding and degradation, temperature, and cure-dependent engineering constants. The question of choosing micro or macro approaches can be addressed on the basis of the physical effects that occur. Hobbiebrunken *et al.* [34] have shown that the behavior of thermoset resins is thermomechanically complex, and effects such as micro yielding and degradation, which are dependent on the temperature, will affect the maximum microscopic stress level significantly. Therefore, an analysis of the development of residual stresses on the micromechanical level can be an important tool to interpret effects and verify homogenized residual stresses on the macro level and to identify the interaction between residual stresses and process-dependent defects.

In summary, the virtual process chain for the manufacturing process of composite materials can provide information to get the full potential of the materials themselves and to reach the stringent quality requirements of the aerospace industry. This is one of the key points to develop robust process cycles and to decrease manufacturing costs.

8.1.4
Merging Technologies

The use of advanced composite materials has steadily increased over the last decades, allowing for both ecologically and economically desirable lightweight solutions in many technical applications. To obtain the best possible solution, optimizing the material alone is, however, only a part of the bigger picture. As real products often have to sustain vastly different loads and environmental stresses during their life cycle, the components as a whole have to be optimized to achieve the best possible performance for all possible usage conditions. This process includes the integration of technologies that enhance a structure's potential, such as allowing the surveillance of the structure's health state or enabling the structure to adapt itself to changes in loads or environment.

8.1.4.1 **Structural Health Monitoring**
Structural health monitoring (SHM) can be described as the use of a permanently integrated sensing (and possibly actuating) capability in a structure to record and analyze loading conditions and/or damage state in a nondestructive manner and use them to gain information about its present state. It is especially desirable for lightweight structures, as it helps to monitor the damage accumulation that occurs when changing the design philosophy from safe-life (no damage at all) to fail-safe (damage can occur, but the growth is controlled) (Figure 8.39).

The most common method to ensure the reliability of a structure is to inspect it at regular time intervals (maintenance cycles), thus ensuring that a particular damage just below a certain threshold cannot grow enough to trigger fatal failure between two cycles. This results in additional costs for both the maintenance itself and the over dimensioning that is needed to ensure survival with a growing damage. To reduce the said costs, an SHM system, which is capable of monitoring the structure, enables both a higher time interval between maintenance cycles, as in-service monitoring is possible, and a more lightweight structure, as damage is detected immediately after the system's resolution is exceeded (Figure 8.40).

An optimal SHM system would be able to fulfill all the tasks outlined by Rytter [35] (and enhanced by Inman *et al.* [36]):

1) damage detection;
2) damage localization;
3) damage type and extent;

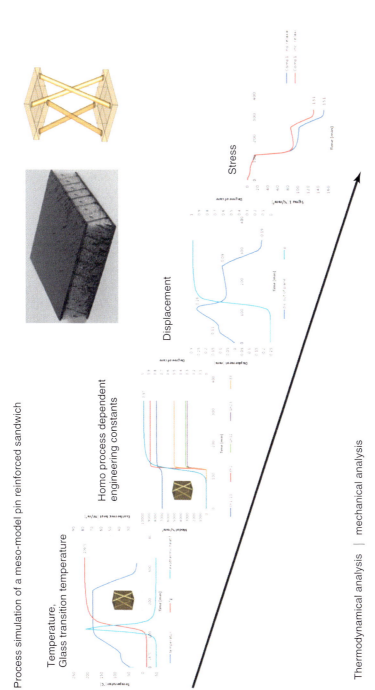

Figure 8.39 Manufacturing process simulation. (Copyright C. Brauner, Faserinstitut Bremen e.V.)

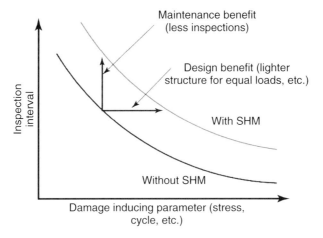

Figure 8.40 Advantages of SHM for a fail-safe structure (following [35]).

4) prognosis of remaining lifetime;
5) self-evaluation, self-control, self-healing, and so on.

In order to gain information about the state of the structure under surveillance, one can either monitor the damage in the structure directly (damage monitoring) or monitor the loads and deduct the health status (loads monitoring) either by predicting the degradation of the structure's properties or by comparing the change in load distribution in a known state because of the results of the changes of the properties.

For the monitoring itself, a lot of different physical phenomena have potential to be used to examine the health status. Many technologies known from NDT applications can be used in SHM, examples being ultrasonic body waves or lamb waves, acoustic emission, impedance measurement, strain measurement using fiber Bragg gratings (FBGs), and many more (Figure 8.41 and [37–39]). Another advantage of combining composites and SHM is the possible use of integrated sensors for online process monitoring during the manufacturing process. As the sensors are integrated into the material, one can gain additional information about process parameters within a structure and, in consequence, structural properties influenced by them.

Furthermore, the potential of SHM as described in the preceding are quite obvious, next to no real application outside of loads monitoring and the surveillance of large, immobile structures is found in regular use. While a number of technologies have shown both damage detection, localization, and quantification ability in laboratory use, the step toward practical use has not been taken in most technical applications. Some of the main hindrances are the broad range of environmental conditions affecting the measurements, the need for a longer lifetime for the monitoring system compared to the monitored structure, and, naturally, the cost. Other problems, which are more related to composite materials and the aviation industry as a driving force in the

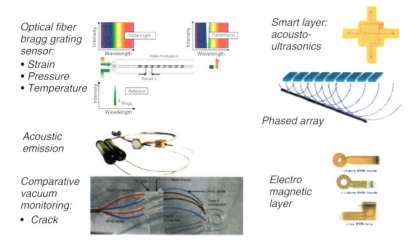

Figure 8.41 Potential technologies for future SHM applications [40].

development of technologies related to them, are concerns about repair and replacement of the used sensors when a structural damage has occurred, an aversion to the integration of local disturbances, complex systems, and the related infrastructure into airplanes and the complex and long approval process. However, all these points are conquerable, and the next decades will see a rising application of SHM technologies to monitor advanced composite materials.

8.1.4.2 Adaptive Structures

Active or adaptive structures, often referred to as *smart structures*, are structures that react actively to changes in their environment. The main idea is to directly and actively influence form, configuration, or properties of the structure hosting an adaptive system by integrating suitable materials with actuating potential. To obtain this ability, both a way to sense these changes (sensors) and a way to react (actuators) are needed. The possible applications are endless: from active noise control and vibration dampening (already in service) to adaptive blades for wind turbines and morphing wings for airplanes, anything involving vibrations or deformation is a potential target for adaptronics. Structures made from advanced composite materials, when compared to traditional metal structures, have the advantage of the ability to integrate active elements into the material. In addition, it is possible to use their anisotropy to support an adaptive deformation, thus opening even more areas of application.

For a functioning adaptive system, several components are necessary (Figure 8.42). At first, the stimulation of the structure due to sound, load, temperature, or other factors has to be sensed. Possible sensing mechanisms include, but are not limited to, those mentioned in Section 8.1.4.1. The main function of the sensor is to convert the stimulation into a signal that is then used to create a reaction. This can either be done by direct coupling or via an integrated

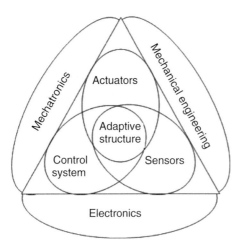

Figure 8.42 Components of an adaptive structure. (Copyright Faserinstitut Bremen e.V.)

electronic control system, a processor. To react actively to the influence, an active material, the actuator, has to be excited. Actuator positioning and the actual actuation process require in-depth understanding of the influenced structure to obtain maximal results with a minimum of energy requirements [41, 42]. It is not uncommon to use the same material as both sensor and actuator, as for instance piezoelectric materials can be used to both measure and induce strain.

When used in combination with advanced composite materials, the advantages of merging both technologies are obvious. For once, it is possible to place actuators inside a composite structure, allowing to optimize the position for the desired task. In addition, integrating parts of the adaptive system shields them from undesirable influences of the surroundings. Secondly, the use of composite material allows the utilization of their anisotropic material behavior. This allows to adapt the structures for certain loads by actively modifying them, and in addition, there is a potential to switch a composite material from one stable state to another, changing the structure's configuration without the need of constant energy feeding (called *multistable laminates*, see [43]).

In summary, adaptive structures offer many opportunities to optimize a structure that is exposed to changing influences or has to fulfill multiple tasks. Combined with the superior mechanical properties of composite materials, this allows for lighter, safer structures even better adapted to fulfill their function and the economical and ecological needs of our time.

8.2
Further Reading

Further information to the subject of polymer composites in general and the application of aerospace structures can be found in:

- Lorenz, T. (2001) Kosteneffektive CFK Fertigungsverfahren der nächsten generation. Proceedings of the 7th Nationales Symposium SAMPE, Werkstoff- und Fertigungstrends bei Verbundwerkstoffen, Erlangen, Deutschland e.V., February 22–23, 2001.
- Hinz, B. (2000) Der VARI-Prozess (vacuum assisted resin infusion) für großflächige Luftfahrtbauteile. Proceedings 6 Nationalee Symposium SAMPE, Stuttgart, Deutschland e.V., März 16, 17, 2000.
- Herrmann, A.S. (2000) Strukturelles Nähen – Eine Maßnahme zur realisierung C. sickinger von hochleistungsfaserverbundstrukturen. Tagungsband H. Wilmes der DGLR Jahrestagung 2000, Leipzig, September 18–21, 2000.
- Herrmann Kostengünstige, A.S. (2000) Faserverbundstrukturen – Eine frage neuer produktionsansätze. Tagungshandbuch zur 31. Internationale AVK-Tagung, Baden-Baden, Oktober 1–2, 2000.
- Herrmann, A.S. (1999) Optimierte textiltechniken für hochleistungs-C. Sigle, A. Pabsch verbundstrukturen. Proceedings, Techtextil, 1999.
- Schappe Techniques (2009) N.N. ENSEMBLE, Fiches BDEF GB Issue 03–2009.
- Schiebel, P. (2011) Friedrich new fiber placement technology for a resource efficient production of thermoplastic high-performance fiber reinforced plastics. ICMAC 2011, Belfast Waterfront, Belfast, UK, March 22–24, 2011.
- Berthold, U. (2001) Beitrag zur Thermoformung gewebeverstärkter Thermoplaste mittels elastischer Stempel. Dissertation. Technische Universität Chemnitz.
- Van Dreumel, W. (2004) Thermoplastic composites for aircraft applications. Coronet Seminar, Universität Delft, 2004.
- Friedrich, K. *et al.* (1997) in *Thermoforming of Continuous Fiber Thermoplastic Composite Sheets, Composite Sheet Forming*, Vol. 9 (ed. D. Bhattacharyya), Elsevier Science B.V., Amsterdam.
- Icardi, G (2000) European SMC/BMC market review. Proceedings Automotive Seminar, SMC/BMC for the Automotive Industry, Coventry, England, p. 1–1 ff.
- (a) Peters, C. *et al.* (2008) Integrated thermoforming process for CFRP-Parts mass production. JISTES 2008, Japan International SAMPE Technical Seminar at Doshisha University Kyoto, Kyoto, July 15, 16, 2008; (b) Peters, C. (2008) Untersuchung fertigungsbedingter Schadstellen an thermoplastisch umgeformten Faserverbundstrukturen auf Basis von C/PPS, Master thesis. Private Fachhochschule Göttingen, Oktober 2008.
- Oelgarth, A. (1997) Langfaserverstärkte thermoplaste, potentiale und zukunftsperspektiven. Dissertation. RWTH Aachen.
- Kunststoffe (2009) Gewicht und Kosten reduzieren. Kunststoffe (Ausgabe 9/2009).
- Wijskamp, S. (2005) Shape distortions in composites forming. PhD thesis. University of Twente, Enschede (NL), May 2005.
- Zobeiry, N. (2006) Viscoelastic constitutive models for evaluation of residual stress in thermoset composite during cure. PhD thesis. University of British Columbia.

- Partridge, I.K. and Skordos, A. (2001) Modelling the cure of a commercial epoxy resin for applications in resin transfer moulding. *Polym. Eng. Sci.*, **41**.
- Hahn, H.T. (1976) Residual stresses in polymer matrix composite laminates. *J. Compos. Mater.*, **10**, 266–278.
- Schürmann, H. (2008) *Konstruieren mit Faser-Kunststoff Verbunden*, 2nd edn, Springer-Verlag, Berlin.
- Wijskamp, S. (2005) Shape distortions in composite forming. PhD thesis. University Twente, The Netherlands.

Conference Series

- SAMPE – Conference of Society for the Advancement of Material and Process Engineering.
- JEC Show – Annual Exhibition and Conference in Paris, Singapore.
- Composites Europe – Conference and Exhibition.
- ITHEC – Conference for Thermoplastic Composites.
- CFK-Valley Convention Stade.

Literature

- Bernhard, W. (2010) *Stahlfaserbeton. Grundlagen und Praxisanwendung.* 2. Auflage Vieweg & Teubner, Wiesbaden, ISBN: 978-3-8348-0872-1.
- Ehrenstein, W. (Hrsg.) (2006) *Faserverbund-Kunststoffe – Werkstoffe, Verarbeitung, Eigenschaften*, Hanser, ISBN: 3-446-22716-4.
- Flemming, M. and Roth, S. (2003) *Faserverbundbauweisen*, Springer, ISBN: 3-540-00636-2.
- Fowler, P.A., Hughes, J.M., and Elias, R.M. (2006) Review biocomposites: technology, environmental credentials and market forces. *J. Sci. Food Agric. Ausgabe*, **86**, 1781–1789.
- Griffith, A.A. (1920) The phenomenon of rupture and flow in solids. *Philos. Trans. R. Soc. London*, **221A**, 163–198.
- Puck, A. (1996) *Festigkeitsanalyse von Faser-Matrix-Laminaten*, Hanser, ISBN: 3-446-18194-6.

References

1. Schwaberlapp, M. (2001) Leichtbau und reibungsminderung, werkstofftrends im motorbereich. 2. Deutscher IIR – Werkstoffkongress imat, Stuttgart, January 22–23, 2001.
2. Altmann, O. Modulare Hybridbauweise.
3. Herbeck, L. (2000) Technologietransfer aus dem flugzeugbau in den schiffbau. Symposium, Neue Materialien Niedersachsen, Braunschweig, April 2000.
4. Michaeli, W., Huybrechts, D. und Wegener, M. (1994) *Dimensionieren mit Faserverbundkunststoffen*, Hanser, ISBN: 3-446-17659-4.
5. Li, R., Kelly, D., and Crosky, A. (2002) Strength improvement by fiber steering around a pin loaded hole. *Compos. Struct.*, **57**(1–4), 377–383.

6. Tosh, M.W. and Kelly, D.W. (2001) Fiber steering for a composite C-Beam. *Compos. Struct.*, **53**, 133–141.

7. Mattheck, C., Baumgartner, A., Kriechbaum, R., and Walther, F. (1993) Computational methods for the understanding of biological optimization. *Comput. Mater. Sci.*, **1**(3), 302–312.

8. Ogale, A. and Mitschang, P. (2004) Tailoring of textile preforms for fiber-reinforced polymer composites. *J. Ind. Text.*, **34**, 77.

9. Schiebel, P., Block, T., and Herrmann, A.S. (2008) "Prospects and risks of tailored preforms for advanced composite applications" TEXCOMP9 (annual report). *Recent Adv. Text. Compos.*, 215–220.

10. Schiebel, P. and Friedrich, L. (2011) New fiber placement technology for a resource efficient production of thermoplastic high-performance fiber reinforced plastics. ICMAC 2011, Belfast.

11. Willden, K. *et al.* (1997) Advanced Technology Composite Fuselage – Manufacturing. NASA Contractor Report 4735, NASA Langley Research Center, Hampton, VA.

12. Staub, M. *et al.* Verfahren zur herstellung eines Bauteils in Faserverbundbauweise. DE 102 10 517 B3, German Patent 29.01.2004.

13. Jörn, P. (2004) Entwicklung eines produktionskonzeptes für rahmenförmige CFK-Strukturen im Flugzeugbau, science report aus dem Faserinstitut Bremen e.V. Band 2. Dissertation der Universität Bremen, Logos Verlag, Berlin, 2004, ISBN: 3-8325-0477-X.

14. Oelgarth, A. (1997) Langfaserverstärkte thermoplaste, potenziale und zukunftsperspektiven. Dissertation RWTH Aachen.

15. Ermanni P. (2007) Composites Technologien, Script Version 4.0, ETH Zürich, August 2007.

16. Meyer, R.W. *et al.* (1985) *Handbook of Pultrusion Technology*, Chapman & Hall, New York.

17. Krieger, M. (2007) Weiterentwicklung eines Prozesses zur Herstellung gekrümmter Profile mit dem Pultrusionsverfahren, Diploma Thesis at University of Applied Science Osnabrück in cooperation with Faserinstitut Bremen e.V., Bremen, 12. Dezember 2007.

18. Hillmeyer, R. *et al.* (2004) Pultrusion resin transfer molding of advanced aerospace structures. SAMPE Long Beach (USA) Material Characterisation and Process Simulation, May 16, 2004.

19. Icardi, G. (2000) European SMC/BMC market review. Proceedings Automotive Seminar, SMC/BMC for the Automotive Industry, Coventry, England, p. 1–1 ff.

20. Purol H. *et al.* (2007) Fertigungsmethoden und Prozesse in der Kontinuierlichen CFK Produktion. Abschlussbericht Projekt EMIR (AP 6.5.3), Forschungsberichte aus dem Faserinstitut, Bremen, January 18, 2007.

21. Reinhold, R. (2010) *Lösungen für anspruchsvolle Faserverbundbauteile*, Artikel in *Kunststoffe 5/2010*, S. 38–40, Carl-Hanser-Verlag, München.

22. Peters, C. *et al.* Integrated thermoforming process for CFRP-parts mass production. JISTES 2008, Japan International SAMPE Technical Seminar at Doshisha University Kyoto, Kyoto, July 15–16, 2008.

23. Ermanni, P. (2007) Composites Technologien, Script Version 4.0. ETH, Zürich, August 2007.

24. Friedrich, K. *et al.* (1997) in *Thermoforming of Continuous Fibre Thermoplastic Composite Sheets, Composite Sheet Forming*, Vol. 9 (ed. D. Bhattacharyya), Elsevier Science B.V.

25. Berthold, U. (2001) Beitrag zur Thermoformung gewebeverstärkter Thermoplaste mittels elastischer Stempel. Dissertation, Technische Universität Chemnitz.

26. Kunststoffe (2009) Gewicht und Kosten reduzieren. Kunststoffe (Ausgabe 9/2009).

27. Peters, C. (2008) Untersuchung fertigungsbedingter Schadstellen an thermoplastisch umgeformten Faserverbundstrukturen auf Basis von C/PPS. Master thesis. Private Fachhochschule Göttingen, October 2008.

28. Wijskamp, S. (2005) Shape distortions in composites forming. PhD thesis. University of Twente, Enschede (NL), May 2005.

29. VDI (2006) *VDI-2014, Part 3*, Beut Verlag.

30. Svanberg, J. *et al.* (2004) Prediction of shape distortions Part I. FE-implementation of a path dependent constitutive model. *Composites Part A*, **35**, 711–721.

31. Johnston, A.A. (1997) An integrate model of the development of process-induced deformation in autoclave processing of composites structures. PhD thesis. The University of British Columbia.

32. White S.R. and Hahn H.T. (1992) Process modeling of composite materials: residual stress development during cure. Part I. model formulation, *J. Compos. Mater.*, **26**, 2423–2453.

33. Abou Msallem, Y. *et al.* (2010) Material characterization and residual stresses simulation during the manufacturing process of epoxy matrix composites. *Composites Part A*, **41**, 108–115.

34. Hobbiebrunken, T. *et al.* (2005) Microscopic yielding of CF/epoxy composite and the effect on the formation of thermal residual stresses. *Compos. Sci. Technol.*, **65**, 1626–1635.

35. Rytter, A. (1993) Vibrational based inspection of civil engineering structures. PhD thesis. University of Aalborg.

36. Inman, D.J. *et al.* (eds) (2005) *Damage Prognosis for Aerospace, Civil and Mechanical Systems*, John Wiley & Sons, Ltd, Chichester.

37. Balageas, D. (2006) *Structural Health Monitoring*, ISTE Ltd, London.

38. Staszewski, W. (2004) *Health Monitoring of Aerospace Structures – Smart Sensor Technologies and Signal Processing*, John Wiley & Sons Ltd, Chichester.

39. Giurgiutiu, V. (2008) *Structural Health Monitoring with Piezoelectric Wafer Active Sensors*, Associated Press, Burlington, VT.

40. Boller, C. (2010) SHM in Action. Advanced Course on SHM, Teaching Materials, Hamburg.

41. Gawronski, W. (2004) *Advanced Structural Dynamics and Active Control of Structures*, Springer, New York.

42. Clark, R.L. *et al.* (1998) *Adaptive Structures – Dynamics and Control*, John Wiley & Sons, Inc., New York.

43. Hufenbach, W. *et al.* (2002) Design of multistable composites for application in adaptive structures. *Compos. Sci. Technol.*, **62**, 2201–2207.

9
Metal Matrix Composites

Maider García de Cortázar, Pedro Egizabal, Jorge Barcena, and Yann Le Petitcorps

9.1
Introduction

Metal matrix composites (MMCs) can be considered to be new as well as old materials. P. Paufler, in his recent work on Damascus sabres from the sixteenth century, found that iron was reinforced by carbon nanotubes [1]. From that point of view, MMCs, especially nanocomposite materials, are very old. The reinforcement effect of iron by carbon addition has also been known for a long time. By looking at the micrographs of a eutectoid composition of steel, we could consider that because of the hard cementite wires, which precipitate inside the soft iron matrix, the composition is effectively an MMC. The ferritic matrix is extremely soft and ductile as is the case in most metals, and the cementite wires can be considered to be the ceramic reinforcements. At the eutectoid composition, the volume fraction of cementite is in the range 10 vol%, the aspect ratio (length/diameter) is >10, and there is no chemical reaction between the matrix and the carbide reinforcement.

On the basis of the following definition of composite materials, steel can be considered as a composite material except for the fact that there is no synergistic characteristic between iron and cementite (e.g., the Young's modulus of pure iron and cementite are similar):

> A combination of two or more materials (reinforcing elements, fillers, and composite matrix binder), differing in form or composition on a macroscale. The constituents retain their identities, that is, they do not dissolve or merge completely into one another although they act in concert. Normally, the components can be physically identified and exhibit an interface between one another. Examples are cermets and metal−matrix composites [2].

The history of modern MMCs begins during the 1960s with the reinforcement of a copper matrix by tungsten wires. In contrast to the two previous examples, the reinforcement is not obtained by an *in situ* precipitation. The copper matrix is very soft and ductile, and has a high thermal conductivity, whereas the tungsten wires are stiff and strong, with an acceptable thermal conductivity. A synergistic effect can

Structural Materials and Processes in Transportation, First Edition.
Edited by Dirk Lehmhus, Matthias Busse, Axel S. Herrmann, and Kambiz Kayvantash.
© 2013 Wiley-VCH Verlag GmbH & Co. KGaA. Published 2013 by Wiley-VCH Verlag GmbH & Co. KGaA.

be found between the two constituents. A unique aspect of this association between copper and tungsten is that there is a good wettability of copper on tungsten and no chemical reaction between the two constituents. This is an example of an MMC made up of a metallic matrix and a metallic reinforcement. Because of these features, both the liquid and solid manufacturing routes can be employed for preparing the composite. Tungsten, however, does not have the typical properties of a metal. It is strong, stiff, and has a low coefficient of thermal expansion. Its physical properties (except for density) are very close to the properties of a ceramic reinforcement such as SiC. Thus, in modern MMCs, the reinforcement is a ceramic compound. Concerning the example of copper, however, the use of a continuous silicon carbide fiber will not be possible as copper will react with SiC to form copper silicides. As a consequence of this chemical reaction both the fiber and the matrix will be damaged. However, it is possible to reinforce copper with SiC, if the surface of the fiber is protected by a carbon coating. These examples of steel or copper composites show the difficulties encountered when designing an MMC.

Many requirements, such as the thermodynamic equilibrium between the matrix and the reinforcement, have to be taken into account while designing an MMC. Most of the metals form boride, carbide, nitride, silicide, and oxide compounds; however, most of the ceramic reinforcements have to be light and contain carbon, oxygen, nitrogen, and boron atoms, which will react with the matrix. Owing to the fact that a chemical reaction is sometimes necessary in order to strengthen the load transfer between the fiber and the matrix, the atomic diffusion coefficient and the kinetic of reaction have to be taken into account. To do this, sacrificial diffusion barriers have been designed to protect the reinforcement and to promote the load transfer. The melting point of metals is low enough to prepare the composite by a liquid route; however, most ceramics are not wetted by liquid metals. To overcome this, the fiber surface can be modified and some alloying elements added to the matrix to permit the natural infiltration of a porous preform. Alternatively, pressure has to be applied to force the metal to infiltrate the preform. The matrix usually has a lower Young's modulus and a higher coefficient of thermal expansion than the reinforcement. As a consequence, after cooling, the matrix is in tension, the fiber is in compression, and the interface is sheared. Complex interphases have to be designed, particularly when the matrix has a low ductility. The variability in strength of the ceramic fibers (Weibull modulus) has to be taken into account; a large variability makes the rupture of the fiber unpredictable. However, a predamaging state can be observed before the catastrophic rupture of the composites. Recently, nonscientific or technical questions, such as the availability of the constituents parts (mainly the reinforcement), the continued activity of the composite manufacturer as well as the manufacturing cost, have had to be solved. The question of the recyclability of these materials has only emerged more recently.

Numerous interesting properties can be expected from this combination of a metal matrix with a ceramic reinforcement such as the following:

- high temperature properties (Young's modulus, yield strength, fatigue life, lower creep);
- low coefficient of linear thermal expansion (CTE);
- high wear resistance, high thermal conductivity with stiffness or lightness.

Among the available matrix materials, several combinations can be proposed:

- light and ductile metals or alloys (Al-, Mg-, and Ti-based);
- stiff intermetallic compounds (Ti_3Al, TiAl, NiAl, FeAl);
- metal with specific properties (Cu, Ag).
- strong and stiff monofilaments (m), particulates (p), continuous fibers (f), short fibers (s), or whiskers (w) ($Al_2O_{3m,s,f,p}$ $C_{s,w}$, $SiC_{m,p}$, TiC_p, TiB_{2p}).

In order to produce these composites, different fabrication routes can be considered:

- solid-state routes (powder metallurgy, foil/fiber/foil (FFF) diffusion bonding, hot isostatic pressuring (HIPing));
- semi-solid-state routes (plasma spraying, matrix sputtering (electron beam, EB));
- liquid-state routes (preform infiltration, squeeze casting, melt stirring (stir casting), *in situ* precipitation).

In contrast to polymer matrix composites and the long historical knowledge of metals, the rise of MMCs is relatively new, mainly due to the recent development of the ceramic reinforcements besides the considerable number of problems that needed to be addressed. The development of MMCs during the last five decades can be summarized as follows:

- 1960s – reinforcement of copper by tungsten wires;
- 1980s – reinforcement of aluminum and titanium by nonmetallic reinforcements (C, B, SiC);
- 1980s – discovery of the problems at the reinforcement/matrix (R/M) interface (reactivity, notch effect, residual stresses, corrosion mechanisms, etc.);
- 1980–1990 – first industrial applications (transport, mainly automotive, leisure, and sport);
- 1990–2000 – investigations to improve manufacturing and reliability, and to reduce costs.
- 2010 – assessments of other industrial applications, besides the transport industry (e.g., energy, nuclear plant).

Historically, MMCs were designed for use in aeronautic and space applications because of their lightness and high mechanical properties. However, the first commercial applications were at room temperature in the automotive industry, for improving the wear properties of car engines. Aluminum/SiCp brake or drum disks were proposed by different car makers VW (Lupo Tdi), Toyota (RAV-4EV), Chrysler (Plymouth Prowler), Lotus (Elise), and Ford (Prodigy). In all cases, the composite material was prepared by stir casting. The matrix is a foundry aluminum alloy (A359), the diameter of the SiC_p particulates is around 10 μm and the volume fraction in the range 20%. In order to overcome the lack of wettability of SiC_p within the aluminum alloys, the SiC particulates are oxidized at 500 °C in air in order to form a silica layer at the surface. Within the alloy, the silica layer reacts with the magnesium present in the aluminum alloy. MMC pistons have also been proposed by Toyota (Celica), Honda (Prelude), and Porsche (911 Boxter). For those parts, the composite is prepared by pressure infiltration of an alumina–silica (Saffil™) preform using foundry alloy (AK12 (AlSi12CuNiMg)). Owing to the presence of silicon, the castability is good and owing to the presence of magnesium, the infiltration and the adhesion to the fiber is improved. Thermal stability and stiffness were key requirements for this particular application.

These composite materials have a better stiffness (70 GPa – unreinforced matrix; 100 GPa – composite); however, either because of the large size of the SiC_p or the diameter of the alumina fibers, the ductility is very weak (<1% elongation in tension). A better ductility can be obtained with smaller particles (ø ~ 1 μm) and a two-phase aluminum wrought alloy ($2009/SiC/15_p/T4$). The incorporation of the particles is only possible in the solid state for the matrix and the composite is prepared by the powder metallurgy route. Such composites are presently used instead of titanium by Eurocopter for the manufacture of rotor blade sleeves of helicopters.

9.1.1
General Aspects of MMCs

Metals or alloys of interest for the fabrication of MMCs (Table 9.1) can be ranked in three categories:

- **Light metals**: These are Mg-, Al-, and Ti-based matrices. Magnesium and aluminum alloys are not only very light but also ductile materials. Therefore an improvement in stiffness and strength to rupture is expected regardless of the reinforcement. Owing to their low-medium melting point, these composites can be prepared either by the liquid or solid routes. Though magnesium is chemically stable with carbon, it reacts with silicon. Titanium alloys are strong and stiff; they can only be reinforced with very strong and stiff reinforcements. However, the titanium will chemically react with most of the reinforcements. The solid route has always been preferred for the fabrication of the composite in order to limit the reaction zone between the fiber and the matrix. TiC and TiB could be good candidates for the reinforcement of titanium because they do not react with the matrix and that they also have a high stiffness. However, if they are not present as fiber, the resulting composite will ultimately have a poor strength at room temperature, which improved at medium temperatures (400–500 °C).
- **High thermal conductivity metals**: These are Al, Cu, and Ag matrices. Pure aluminum has high thermal conductivity but low stiffness; it can be reinforced by pitch carbon fibers or C-diamond in order to maintain a high thermal conductivity. However, during the fabrication or for long exposures at high temperature, some deleterious aluminum carbides can be expected at the interface. Copper and silver are more interesting for that purpose as they do not react with carbon. The incorporation of diamond in these matrices will improve the thermal conductivity

Table 9.1 Metals or alloys of interest for the fabrication of MMCs.

Matrix	ρ (kg m^{-3})	T° (°C)	E (GPa)	UTS (MPa)	$\sigma^{Y0.2}$ (MPa)	CTE α' (10^{-6} K^{-1})	k (W m^{-1} K^{-1})
Al (1100)	2710	643–655	69	90	35	23.6	222
Al–Cu (2024)	2780	503–638	73	O: 185 T6: 495	O: 95 T6: 415	23.2	O: 193 T6: 151
Al–Si (356)	2750	570–650	70	225	200	19.3	121
Cu (electronic grade)	8941	1084	115	221–455	69–365	17.7	392
Ag	10500	957	70	180	150	19	420
Mg–Al (AZ91A F)	1810	470–595	45	165–230	97–150	27	72
Mo	10220	n.a.	325	550–700	400–550	5.1	137
MoSi$_2$	6260	1870	407	165	n.a.	8.12	58.9
Inconel 600	8420	n.a.	207	655	310	11.5–13.3	14.8
Ti-64	4420	1649	106–114	897–1205	808–1075	9.2	7.2
Ti-6242	4540	1649	115	930–1100	830–920	9.9	6
γ-TiAl	3900	1460	176	600	—	10	—
Ti$_3$Alp Superα2	4740	1600	150	880	—	10	—

ρ: density, T°: melting or liquidus range, E: Young's modulus, UTS: ultimate tensile strength, $\sigma^{Y0.2}$: yield strength 0.2% proof., CTE: coefficient of linear thermal expansion, k: thermal conductivity.

and the stiffness and will also decrease the density. Owing to the lack of wettability between the carbon and the metal, if the liquid route is preferred, either pressure will be necessary to infiltrate the preform or the matrix must be alloyed.

- **Refractory or intermetallics**: These alloys (Mo) or compounds (TiAl, MoSi$_2$) are very strong and stiff, they have nearly no ductility at room temperature, and they can only be reinforced by stronger continuous fibers. Their mechanical behavior is similar to that of ceramic at room temperature and metal at high temperatures. To limit chemical reaction with the reinforcement, the composite is prepared by the solid route. In a few cases, a chemical equilibrium can be found between the two constituents (Y$_2$O$_3$-coated C-fibers are in equilibrium with TiAl; SiC is in equilibrium with MoSi$_2$). In order to improve toughness, small-diameter fibers or particulates are preferred (Table 9.2).

Reinforcements (Table 9.2) are stiffer, lighter, and have a lower CTE than the metals. They can be ranked in three categories according to their shape or aspect ratio:

Table 9.2 Reinforcements available.

Reinforcements	ρ (kg m^{-3})	$\varnothing$ (μm)	E (GPa)	UTS (MPa)	CTE α (10^{-6} K^{-1})	k (W m^{-1} K^{-1})
Particulates and whiskers						
SiC(α)	3160	1–50	390–410	—	4.4–4.6	42.5
SiC(β)	3160	1–50	300–450	—	4.5	135
TiC	4938	1–50	450–500	—	8.8	17–21
B$_4$C	2512	1–50	480	—	5.5	27
TiB$_2$	4500	1–50	350–550	—	8.1	100
TiB	4500	1–50	480–550	—	n.a.	n.a.
C-diamond	3515	1–50	900–1200	—	1.2–2.3	1000–2000
Fibers						
W metallic wire	19300	25–400	405	1700–4000	4.5	170
Alumina FP (Dupont)	3900	20	380	1300–2000	8.3	35
C (PAN) high strength	1900	7	230	4500–4800	−1.2 (axial.) 10 (transversal.)	—
C (PAN) high modulus	—	7	415	2400–2600	−1.2 (ax.) 10 (trans.)	—
SiC high Nicalon S	3050	12	400	2500	—	18
Monofilament						
B (W) CVD	2600	50–100	400	3000	5	38
SiC (C) CVD SCS-6™	3290	145	390	4200	4	16
SiC (W) CVD SM1140$^+$	3300	105	415	4000	4	16

- **Particulates (p)**: The aspect ratio is in the order of 1. The expected properties for the composite are a better hardness, a higher stiffness at elevated temperature, intermediate performances in tension, and a poor ductility – the choice is very versatile.
- Generally, the composite is prepared by powder metallurgy and sometimes by stir casting or squeeze casting. There is good availability of these particles at a low cost.
- **Fiber tows (f)**: the diameter is in the order of 7 μm for the carbon fibers and 15–20 μm for the SiC and alumina fibers. The tow is made of 1000–12 000 individual fibers. Owing to the infinite aspect ratio, the load transfer between the fiber and the matrix is excellent. Taking into account the properties required for the reinforcement of metals, the most promising fibers are carbon fibers, owing to their high strength and stiffness and the nearly pure silicon carbide fibers. However, these fibers are very expensive (i.e., US$13 000 kg^{-1} for the Hi Nicalon™ S fiber). In order to individually coat each fiber within the tow, the composite is prepared by liquid infiltration. This is only possible for low melting point matrices (based on aluminum, magnesium, copper, or silver).

- **Monofilaments (m)**: Monofilaments are the most accomplished reinforcements and also the most expensive ($\sim$US\$10 000 kg^{-1}). They combine a very high strength ($\sim$4000 MPa) and a high Young's modulus ($\sim$400 GPa). Owing to its large diameter (100–140 μm), the monofilament can be individually coated by the matrix either by a liquid or solid route. These monofilaments are expected to be used for high value components (aeroengine parts, nuclear components). Most of these monofilaments have an external pyrocarbon coating, which plays the role of a sacrificial diffusion barrier and a crack deflection interphase at the filament/matrix interface. Only two or three manufacturers in the world are capable of producing these monofilaments.

Taking into account these considerations, the following give some examples of recent applications of MMCs:

1) MMCs for thermal and structural properties in electronic applications. The property required for the matrix and the reinforcement is a high thermal conductivity. For electronic applications (heat sink component, packaging materials), the composite material has to be light and so the volume fraction of the ceramic phase is important (more than 50 vol%). Cu/diamond or Ag/diamond and Al/SiC are good candidates for that purpose. There is no chemical reaction between Cu or Ag and carbon (no carbides); the liquid route works for making the composite but there is a lack of wettability, so the infiltration of the preform has to be done under pressure.

2) MMCs for thermal and structural properties in nuclear environments. For fission applications, the requirements are a thermal conductivity higher than 200 W m^{-1} K^{-1} and a good resistance to creep at 550 °C. Pure copper meets only the thermal requirements and strong fibers are necessary to avoid the creep. Cu/W or Cu/SiC composites are good candidates. Concerning the Cu/SiC system, the silicon carbide monofilament is protected by a carbon layer (4 μm in thickness) that avoids the formation of copper silicides at the interface. However, in order to get a composite effect (load transfer), the reinforcement–matrix adhesion has to be improved. This can be done by a thin layer (<0.1 μm) of titanium, which is a strong carbide former. This titanium layer is deposited by physical vapor deposition process (PVD). Over this titanium layer, another thin copper layer is deposited by PVD in order to allow the electrical conductivity of the monofilament. The copper matrix is then deposited by electroplating to reach a volume fraction in the range 30% and the coated filaments are gathered and then HIPed. During the compaction at 600 °C, titanium diffuses toward the carbon and copper layers, improving the bonding. Owing to the production of waste products from nuclear power, its storage is now a considerable concern. Boron is one of the most prolific neutron absorbers. An aluminum/boron system could be of interest for such an application. A monofilament of boron could be coated by, liquid aluminum, and the coated filament then wound and HIPed. As the boron monofilament has a high ultimate tensile strength, the composite is very resistant and is able to absorb free neutrons as well.

3) MMCs for thermal and structural properties in spatial environment. Owing to the negative coefficient of thermal expansion of carbon fibers in the axial direction, the lightness of magnesium, and the lack of chemical reaction between carbon and magnesium, Mg/C composites have been studied for 30 years for manufacturing antenna tubes for satellites. The difficulties in preparing the composite by liquid infiltration have not yet been solved. It would be very interesting to reinforce a metal with carbon fibers.

4) MMCs for structural properties in aeronautics. Titanium is widely used in aeronautic and the reinforcement of titanium by a large variety of ceramic has been studied for 40 years. Titanium alloy matrices (α, β, $\alpha + \beta$), titanium-based intermetallic matrices (γ-TiAl, α_2_Ti$_3$Al) or semimetallic matrices (MoSi$_2$) have been reinforced by nearly all the ceramics available. At present, the systems of interest are 1D-Ti($\alpha + \beta$)/SiC$_m$ composites prepared by the FFF diffusion-bonding process or more recently by rapid coating and HIPing. Isotropic Ti($\alpha + \beta$)/TiB$_w$ or TiC$_p$ composites are prepared by powder metallurgy or by *in situ* precipitation after cooling of the titanium matrix from a liquid state. TiAl/TiB$_2$ or MoSi$_2$/SiC composites prepared by powder metallurgy could be of interest for the high stiffness of the ceramic and for the thermodynamic equilibrium between the constituents.

5) MMCs as self-lubricating materials. The inclusion of solid lubricants to obtain tailor-made engineering composites, with the main requirements being a low friction coefficient and wear rate at a certain mechanical strength [3]. This self-lubricating material can provide cost-efficient solutions in the transport sector, in terms of bearings, bushes, clutches, gears, and other sliding elements [4]. They are produced by conventional powder metallurgy. The most common matrices are based on copper and iron alloys, while the typical reinforcements are graphite, MoS$_2$, WS$_2$, and BN. Their use in the automotive sector may be limited to the cost of the filler. There is a strong potential for the development of new metal/filler combinations and the synthesis and modification of novel self-lubricating particulates.

The objectives of MMC research programs in *transport applications* generally fall into two main areas – structural and thermal management applications.

9.1.2
Structural Applications

Usually, structural components were intended to address one of two critical issues: weight/mass reduction and increased durability/lifetime.

• Weight/mass reduction
  In general, any mass reduction program requires increased strength or stiffness coupled with lowered density (i.e., an increase in specific stiffness over other alternative materials).
  The driver for weight savings in transportation applications is also twofold:

– In response to widespread increasing fuel prices, improvements in vehicle fuel economy are increasingly required.

Many programs were attempted, fostered to a large extent by the vehicle builders, in the belief that some of the value of any weight savings would be passed down to the component manufacturers. Any of the available composite technologies produces materials costs higher than that of conventional material, however the value of the saved weight is a critical part of the ultimate viability of these composite parts.

In general, the value of unsprung weight should be significantly higher than weight saved in sprung weight. However in almost all composite programs independent of the component features (brake, engine, transmission, or bodywork) the value of around US\$10 kg^{-1} is still commonly used as a benchmark. This value is low when it comes to unsprung weights such as brake components. An additional issue with brake components programs (disk, caliper body, etc.) was their "safety critical" status; much more testing and failure mode analysis was required for these components to be placed in service than for those applied to parts used in the body, transmission, and engine applications.

– In the commercial road/rail/aircraft fields, a requirement for increased range and load-carrying capability.

Any reduction in the weight of the vehicle, could add to load capacity or extend the range of the vehicle between stops.

Actually, the limiting factor for most commercial vehicles is the volume capacity of the vehicle and not the weight capacity and there are limits to increase the external dimensions of a commercial road vehicle or railcar.

• Increase in durability/lifetime

In most cases, increased durability beyond some acceptable minimum does not have a known positive value to the manufacturer, as the manufacturer is normally in the business of providing spares and replacement parts and this is usually as profitable as or more so than the sale of the original parts. Spares or replacement parts are priced much higher than so-called original equipment manufacturer (OEM) parts and are an important part of the manufacturer's business.

However, there are some exceptions that have emerged in this area of durability. This could be the case of brake components, specifically, disks. The conventional materials have not been able to effectively address the demand for a longer lasting product and consequently there has been renewed interest in the use of MMC materials for disks (especially in the case of lightweight vehicles).

Another example driven by a need for increased durability is the turbo rotor. The cold side rotor is normally an aluminum casting but increasingly tight emission specifications have required some bleeding of exhaust gases back into the cold side of the turbo. This causes the cold side rotor lifetime to be reduced; typically, they are replaced at every service, which is considered unacceptable by the end users, who would prefer a solution using a material with a higher operating temperature. Aluminum matrix composites are considered to extend the upper use temperatures as compared with aluminum casting alloys and this formed the rationale for the use of this kind of materials in these applications.

9.1.3
Thermal Management Applications

A large area of current interest in MMC materials industry is in thermal management of electronic systems. Thermal management has become an important aspect to be considered, especially for the current and future electrical vehicle (EV) and hybrid electrical vehicle (HEV) concepts. Heat management of electronics (electric inverter, battery) must also be studied.

The inner temperature of electronic devices such as transistors increases cyclically as a consequence of their own operation. Traditionally, heat dissipation from electronic devices (chips, boards, etc.) is carried out by passive strategies, through the use of high thermal conductivity metals, such as copper or silver, as heat sinks. Nevertheless, a large mismatches in the coefficient of thermal expansion between the heat sink and the electronic substrate lead to thermomechanical fatigue in the junction, and consequently a possible premature failure of the device. The CTE and thermal conductivity of heat sinks have to be engineered for every specific application; otherwise, the lifetime of electronic devices can be dramatically limited.

Among the large number of different MMCs, the most efficient and therefore the most commonly developed for heat dissipation purposes, are those based on copper or aluminum matrices. The main advantage of aluminum over copper is its lighter weight, which makes it interesting for transport applications.

In general, MMCs show several improvements in comparison with currently used materials:

- Lower and tailorable CTE (the higher the volume fraction of ceramic added to the metal, the lower the CTE). This is of special interest for electronic applications, where electronic packaging materials must have a CTE as similar as possible to those of the neighboring materials. This reduction of the CTE mismatch also suggests a reduction of s tresses between the materials during thermal cycling.
- High heat dissipation capability. This is very important in heat dissipation applications. It is of interest in electronic applications and some components of heat exchangers and thermal protection systems (TPSs).
- Lightweight, suitable for transport applications.
- High stiffness at high temperatures, which is a useful characteristic that assures dimensional stability to components working at high temperatures.
- Better mechanical properties (mainly at high temperatures) for a wide variety of applications.

9.2
Relevant MMC Systems

Starting from the assumption that the reader knows basic aspects and main characteristics of MMCs, the objective is not to provide an introduction or review but rather to inform about the most relevant MMC systems, those of commercial importance. They will be classified in discontinuously and continuously reinforced MMCs, taking into consideration the commercial importance of each.

9.2.1
Discontinuously Reinforced Metal Matrix Composites and Associated Processes

Discontinuously reinforced aluminum alloys (DRA) represent the most commonly used MMC in commercial applications. In 2000, they represented nearly 70% of the MMC market in mass [5]. The balance of properties and cost provided by the combination of aluminum alloys and different particulates such as SiC, TiC, Al_2O_3, C, or B_4C allows them to compete with conventional metals in applications where weight saving, thermal management, or wear resistance are crucial requirements. Other metals such as Cu-, Ti-, and Ni-based superalloys and ferrous metals have also been reinforced with particulates but their commercial application has been much more limited so far. SiC- Al_2O_3-, and TiC-reinforced commercial DRA are presently commercially available and B_4C-reinforced alloys are used for nuclear containment applications and bicycle frames [6]. Al/SiCp has been used in some of the most known MMC applications such as the brake disks of the German high speed train IC-2, the rear brake drums of the VW Lupo TDI 3L model, the rotor blade sleeves of the Eurocopter EC120 and N4 helicopters and electronic packages [7–11].

The largest development of discontinuously reinforced MMCs took place in the 1980s when it was seen that even though MMCs based on continuous reinforcements and whiskers could provide better performances, they were only affordable for some limited applications. The processing and secondary operations with particulate-reinforced aluminum presented fewer difficulties and actual potential applications were identified. Research efforts were focused on solving aspects related to compatibility of the reinforcements and matrices, improvement of the interphase, development of affordable primary and secondary operations, and improvement of the homogeneity in the distribution of the reinforcement within the matrix. A large array of particulate-reinforced, Al-based composites have been developed in the last 30 years even though the number of producers of DRA around the world is still very limited.

Table 9.3 presents an overview of the main producers of Al-based, particulate-reinforced alloys and some of the available materials.

Some of the materials represented in the table have already been used in commercial applications. The 359/SiC/10$_p$/T6 alloy from Alcan was used in the production of the brake rotors of the German high speed train. Alcan also supplied an Al–Si alloy reinforced with 20–30% of SiC particulates for the rear brake drums of the VW Lupo-3L TDI model. Helicopter blades were made of the 2009/SiC/15$_p$/T4 material and the 6092/SiC/17.5$_p$ material from DWA was used for the production of ventral fins of the military F-16 airplane. Products from TTC materials produced by a variant of the Lanxide™ pressureless infiltration of SiC$_p$ preforms concept have been used in thermal management applications in space, and ground transport applications such as the Toyota Prius and Japanese

Table 9.3 Main suppliers and examples of commercially available particulate-reinforced aluminum matrix composites.

DRA producers	Alloy/reinforcement system/ thermal treatment	$\sigma^{Y0.2}$ (MPa)/UTS (MPa)/E (GPa)/ elongation (%)
Alcan (Canada) DURALCAN™ [12, 13]	6061/Al_2O_3/10_p/T6	296/352/81.4/10
	359/SiC/10_p/T6	303/338/86.2/1.2
	339/SiC/10_p/T6	358/372/87.6/0.3
AMC Aerospace Metal Composites (UK) [14]	2124/SiC/25_p/extruded/T4	480/680/115/5
	6061/SiC/40_p/extruded/T1	440/560/140/3
	2124/B_4C/20_p/extruded/T4	468/630/111/2.9
DWA Technologies Inc. (USA) [15]	2009/SiC/15_p/extruded/T6	361/500/90.3/7.6
	6092/SiC/17.5_p/extruded/T6	448/510/105/6
	7050/SiC/15_p/extruded/T6	510/547/94/1.5
Goodfellow Cambridge Ltd. (UK) [15]	2124/SiC/15_p	400/610/100/6
	6092/SiC/15_p	500/550/100/2.3
ULTALITE® Cyco Systems (Australia) [16]	ULTALITE/fly ash/20_p/T6	296/331/81.4/2
	ULTALITE/fly ash/20_p/T6	296/338/81.4/2.5

References [12–17] material property data obtained from commercial brochures and reports.

Shinkansen 700 train baseplates or the Motorola's printed wiring board cores for the telecommunication Iridium system.

Particulate-reinforced aluminum alloys are presently an established material and there exists a large literature body on both their properties and processing conditions [18, 19]. Nevertheless, most of this open literature has been produced before 2000 and they usually present low values of fracture properties of particulate-reinforced MMCs. Traditionally, MMCs have been associated with low values in properties such as toughness, ductility, and fatigue resistance and, added to this, their high costs have hindered their use in many potential structural applications. Since then, the MMC sector has focused its research activities on trying to minimize these technical and cost-related problems [6]. The continuous research activities of the last 10 years have made it possible to minimize these drawbacks by the application of new alloy/reinforcement combinations, improved selection of reinforcement sizes, distribution, and morphologies, and production process variants.

In the last 10 years, the research in new materials has been devoted to the study of the nature and morphology of particulates as well as the development of alloys other than those of aluminum such as those with magnesium or titanium and Ni-based superalloys. New reinforced aluminum alloys, reinforced with TiC, B_4C, and fly ash [6, 20], have extended the range of commercially available MMCs that had been mainly based on SiC and Al_2O_3 particulates. Furthermore, much research activities are at present focused on the application of nanocarbon particulates and fibers. So far, the problems of segregation and tendency of clustering of submicron-sized particulates have not yet been solved so as to make it possible to produce such MMCs for commercial applications but important technical advances are continuously being obtained at lab scale [21].

Research activities are also focused on the improvement of primary and secondary processes. Liquid metal routes such as stir casting represent the most applied processing methods for the production of DRA, reaching up to 67% of the total. Alcan is the major producer with a capacity of 30 000 MT [6]. Materials with reinforcement content of 10–40% are produced for applications where high production volumes and low cost are required. In the case of liquid metal infiltration of preforms, reinforcement volumes are usually higher (40–70%). The application of the pressureless infiltration concept of preforms developed by Lanxide is an established technology for the production of high content MMCs and the main applications belong to the thermal management sector [18]. Powder metallurgy is also much used for the production of advanced materials that are subsequently extruded or forged for the aerospace industry.

In the field of MMCs, there are newer material combinations that are worth researching and even though the research effort is currently lower than in the 1980s and 1990s, many leading companies and research institutions continue investing efforts and resources.

Aerospace companies are looking for alternatives to currently available Al/SiCp that may further reduce the density or increase the strength of the components. B_4C- and nano-SiC-reinforced aluminum alloys are currently being researched. The MMC producer aerospace metal composite (AMC) in the United Kingdom has recently issued an aluminum alloy 2124 reinforced with B_4C with the reference AMC220BC that presents outstanding specific mechanical properties. TiB_w- and TiC_p-reinforced titanium alloys, to substitute steel or titanium alloys in select turbine and aircraft applications are other examples. Interest in these composite materials, especially titanium alloys reinforced with *in situ* formed TiB_w is growing significantly. Different production routes have been developed over the past few years: liquid-based routes such as casting and ingot metallurgy to obtain cost-effective composites or powder-based routes to control the microstructure for example (for more details see Section 9.3.3).

The automotive industry continues looking for particulate-reinforced aluminum alloys that present better cost/properties relations. The advent of the EV for high volume applications in the next years may be a really important factor in the development and application of MMCs as lightness and high performance materials are a must for these applications because of the current limitations of batteries. Pressureless infiltration processes based on licenses from Lanxide technology are being further developed by American and European companies such as TTC and Tecnalia in Spain. The potential of MMCs for the substitution of ferrous alloys in wear-resistant applications such as pistons and brake components has already been established and now the research efforts are mainly centered in the development of lower cost materials. Fly ash reinforced aluminum alloys developed by the Australian company Cyco Systems, in close collaboration with large automotive companies represent an example of this trend (for more details, see Section 9.3.2). The application of functionally graded materials (FGM) for valve trains, brake disks, and pads is also being currently researched in Europe but this seems to be a long-term research activity. Preform-based drums for heavy-duty vehicles such as trucks are also being developed.

9.2.2
Continuously Reinforced Metal Matrix Composites and Associated Processes

Continuous reinforcements exist as tows or as monofilaments. The former are less expensive and easier to handle during processing, but show lower strength than the more expensive and large diameter monofilaments. These reinforcements are very sensitive to chemical attack by the matrix since even a small amount of interaction can significantly decrease the mechanical properties of composites; this effect is more significant than the effect of the diameter of the fiber. During the last two decades, several continuous reinforced matrix composites have been investigated such as Al/B, Al/C, Mg/C, Al/SiC, Cu/C, and TiAl/C.

All these composites have presented many drawbacks such as corrosion at the fiber–matrix (F/M) interface, poor mechanical properties, difficulties in preparing the composite (lack of infiltration, excessive F/M reaction, etc.) and thus all of these have been given up today.

The composite system made up of titanium alloy as matrix material and silicon carbide monofilaments as reinforcements has been by far the most studied system for continuously reinforced composites and it is the only one commercially available now. The mechanical properties of continuously reinforced composites depend highly on the reinforcements used and are of anisotropic nature, that is, they offer the highest value of strength and stiffness along the axis of the reinforcements, though the transverse properties may be even lower than those of unreinforced alloys. Thus, potential applications are limited to components that are subjected to largely uniaxial stress. To avoid the high cost of monofilament-reinforced titanium matrix composites (TMCs), research efforts have been made to produce titanium composite-reinforced with lower cost tow-based fibers available in the market, such as carbon fibers or alumina fiber tows (Table 9.2). Both were quickly given up to reinforce titanium alloys because of the severe chemical reaction problems. In addition, any type of coating attempt became very difficult and expensive because of the disposition of fibers in the form of tows.

All the efforts today are focused on the reinforcement of titanium by protected silicon carbide monofilaments. SiC monofilaments have been adapted to reinforce titanium alloys by the incorporation of a pyrocarbon coating to reduce the reaction zone between the silicon carbide monofilament and the titanium matrix. Among the SiC monofilaments, the SCS-6™ has a diameter of ~140 μm and the silicon carbide is deposited on a 33 μm carbon core, whereas in the Sigma series, a smaller-diameter (15 μm) tungsten wire is used as the substrate and the fiber diameter varies from ~100 to 140 μm. The SCS-6 has essentially been established as the industry standard, in particular for titanium alloys, while the SM1140$^+$ has been developed as an alternative for those applications that require much higher strengths. A real diffusion barrier to avoid this chemical reaction has not been found yet and a brittle interface results during processing or subsequent exposure to elevated temperature. The F/M interface is a critical constituent of composites because load transfer from the matrix to the high modulus fiber and vice versa occurs through the interface. In addition, as the CTE of titanium is nearly twice that of the fiber, residual stresses are originated during cooling from the processing temperature ($\sim$950 °C) and consequently microcracks are initiated at the reaction interface.

The development of this composite system has matured over 30 years and many investigations have been performed to improve the processing and understand the behavior of this composite material. An intense effort has been directed toward the understanding, characterization, and life prediction of these composites, to reduce cost, and to improve production routes. In the United States, through a combination of corporate funding and a variety of US-government-sponsored development efforts, the three most significant programs supporting these investigations have been the following:

- National Aerospace Plane (NASP) program (1988–1992).
- Integrated High Performance Turbine Engine Technology (IHPTET) program (1988–2005).
- Titanium Matrix Composite Turbine Engine Component Consortium (TMCTECC) program (1994–2000).

The first program was conducted to evaluate advanced materials in support of the NASP airframe design. TMC evaluation was a significant part of the program and was a combined effort by the five prime NASP contractors to develop high temperature, SiC-reinforced TMC for use as a primary structure on the NASP airframe. The NASP airframe design was based on the assumption of a metallic outer skin and substructure for most surfaces with the TMC acting as the TPS. The IHPTET program, a three-phase program, started in 1988 with the objective of doubling the gas turbine propulsion capability by the end of 2005. The program has been successful in achieving the stated goals, and technology developed and demonstrated has been transferred to production engine programs. The third program (TMCTECC), with a budget of US$26 million, was built on the lessons learned in the previous two. It started in 1994 and finished in 2000. The program was focused on reasonable TMC components for gas turbine engine applications through the development of improved processing methods. Through a combination of lean manufacturing principles and Six Sigma improvement methods, processes were developed that reduced cost and improved reliability.

The result of the improvements enabled the transition of the technology from development to production status:

- **Ti6242/SiC nozzle actuator piston rod**
  The nozzle actuator piston rod of the two Pratt & Whitney F119 engines used on the F-22 aircraft. The material used in this application was a solid rod of

13-8Cr–Ni precipitation-hardened stainless steel. Now it has been replaced by a Ti6Al–2Sn–4Zr–2Mo alloy reinforced with SiC monofilaments. This Ti/SiC$_m$ piston represents the first aeronautic application of this composite system

- **Ti6242/SiC nozzle actuator link**
  On the basis of the experience gained from the actuator piston rod, the Ti/SiC was certified for nozzle links on the General Electric F110 engine, used in the F-16 aircraft. The material used in this application was a square tube of Inconel 718. Now it has been replaced by a Ti6Al–2Sn–4Zr–2Mo alloy reinforced with SiC monofilaments. This link has been certified for the F110 engine in the F-16 aircraft.

Outside the United States, in Europe, continuously reinforced TMCs also became the subject of significant research efforts, in private projects, collaborations between renowned R&D centers and universities, as well as within the framework of large European projects financially supported by the European Commission. Some of the activities worth mentioning are those started in the 1980s at DLR in Germany in collaboration with MTU (Daimler Chrysler) and those initiated in the United Kingdom at the Defence Evaluation and Research Agency (DERA) who designed, developed, and engineered this outstanding material for construction of bladed rings (BLINGs) in the low pressure compressor of the turbine. Other applications such as landing gear components have also been investigated and tested. In June 2003, SP Aerospace and the Dutch government performed the world's first test flight of a primary structural landing gear component in TMC. A lower drag brace for the F16 main landing gear was developed by SP Aerospace. The TMC-based material was supplied by FMW Composite Systems (USA) and the Royal Netherlands Air Force (RNLAF) provided full support and flight clearance for the test flight on their F16 "Orange Jumper" test aircraft. The flight demonstrated that the use of TMC material is feasible for primary structural landing gear applications. However, the TMC technology is expensive, costing about three times more than the reference component.

For space applications, programs financially supported by the European Space Agency (ESA) through the Technology Research Program (TRP) commenced in 1998 (ESTEC/Contract No. 12590/97/NL/MV and ESTEC/Contract No. 15509/01/NL/CK). During these investigations, titanium intermetallic matrix composites (IMCs) were recognized as promising materials for future "metallic" TPS applications in reusable launch vehicles (RLVs), representing a good alternative to the dense nickel-based alloys.

The major drawback of these materials has been the high cost. This problem has been approached mainly by American companies, in particular within the TMCTECC program.

The three primary areas of development were the following

1) SiC monofilament cost reduction
2) Development of preprocessing concepts
3) Simplified manufacturing methods.

9.2.2.1 SiC Monofilament Cost Reduction

The cost of the SCS monofilaments has been considerably reduced, mainly owing to processing and manufacturing improvements. In a time frame of 10 years, it has come down by a factor of 4, the SCS monofilament cost based on a 3000 lb year rate is in the range US$5000 kg^{-1}).

9.2.2.2 Development of Preprocessing Concepts

The second area of improvement has been in the preparation or manipulation of the Ti/SiC$_m$ before its consolidation by HIPing or hot pressing. These improvements have been achieved with the development of preprocessing concepts that eliminate the cost inherent to the primary method for producing this type of composites: the FFF method (for more details of new manufacturing processes, see Section 9.3.3).

9.2.2.3 Simplified Manufacturing Methods

The third area in which there has been improvement is the complete manufacturing cycle. The application of lean manufacturing methods, that is, combined operations or tooling to decrease manufacturing complexity, reduce defects, and increase quality has led to a general reduction of the cost.

9.3
Case Studies

9.3.1
Case Study 1: Aluminum Matrix Composites for IGBTs

This case study provides an analysis of the steps for the implementation of MMCs in the power control module of a HEV that CPS Technology Corporation has used in view of the substitution of metal baseplates in insulated-gate bipolar transistor (IGBT) power modules [22].

9.3.1.1 Introduction

MMC-based heat sinks have gained especial interest during the last eight years because of their potential performance in the thermal management of the power train and recovery systems of HEVs, where the power is generated by conventional internal combustion motors and recovered from braking. In this application, IGBT power modules are used to control the switching in the electronic power DC/DC conversion [23]. Although the IGBT technology has been employed for the past 20 years in industrial applications, it was only recently that higher performance and levels of reliability (10-year lifetime) have allowed their use in automotive-based applications [24].

One of the main drivers has been the employment of a new, high class materials such as MMCs. In order to meet the increasing thermal management challenges, the size and weight of the power train are being reduced [25]. The Toyota Prius hybrid vehicle has incorporated Al/SiC as a baseplate in 1997 [26]. Materials manufacturers such as CPS Technologies have continued meeting the increasing demand by the mass production of Al/SiC baseplates.

This case study provides some guidelines for the selection of materials for the baseplate of IGBT power modules for HEV/EV applications.

9.3.1.2 Requirements and Materials Selection

IGBT power switching generates a large amount of heat that needs to be dissipated from the module housing. The IGBT assembly incorporates a metal heat sink baseplate to release heat by thermal conduction and often includes a pin fin design in order to promote heat dissipation to the environment by convection and/or radiation. This is usually accomplished with extra active cooling (air or liquid cooling) for enhanced performance. Figure 9.1 shows a schematic of IGBT power assembly. The electronics (the actual IGBT) is attached to a metalized ceramic substrate that provides the required electrical insulation and electrical connections. The IGBT and ceramic substrate assembly is attached to the baseplate by soldering or brazing technologies. The baseplate provides the functional attachment of the IGBT assembly and electronics to a cold plate that will ultimately provide the heat sink for thermal dissipation.

Dielectric substrate materials are used for electrical insulation of the electronics to the rest of the IGBT housing. Their strength and CTE values are relatively low. So choosing compatible CTE values is necessary when considering the heat sink attachment. Usually, these dielectric materials are metalized with a very thin ($\sim$300 μm) copper layer that provides the electrical connection between devices and also a solder or braze Cu attachment surface on the side opposite to the electronics to integrate these substrates with a heat sink. These copper layers are attached by an activated attachment process called direct bond copper (DBC). DBC substrates are commonly used in IGBTs.

The metallic baseplate must provide long-term reliability by minimizing the induced stresses during the cyclic loading (power and temperature). Two main requirements must be fulfilled to achieve an efficient thermal management: high

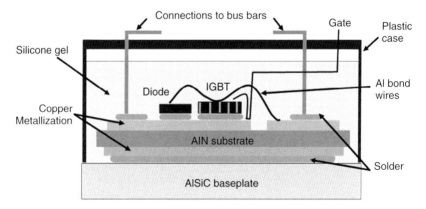

Figure 9.1 Schematic of IGBT module assembly. (Source: Courtesy of CPS.)

thermal conductivity and matching of the coefficient of thermal expansion. The baseplate and the insulating ceramic substrate have different thermal expansion rates and a temperature increase/decrease causes thermal stresses that could induce cracking at the soldering layer. Furthermore, for an accurate design, other secondary assets must also be considered, such as weight (density), strength and stiffness, surface finishing, and machinability, as they also have a strong influence on the system reliability.

As the materials for this application need to have high thermal conductivity, metals such as copper (400 W m^{-1} K^{-1}) and aluminum (200 W m^{-1} K^{-1}) are the obvious choice, but suffer from CTE mismatch with the ceramic substrate (Table 9.4). Other metallic alloys, such as those based on W (W/Cu and W/Ni/Cu) and Mo (Mo/Cu) exhibit a very attractive combination of thermal conductivity (150–200 W m^{-1} K^{-1}) and low CTE (6–9 ppm). Unfortunately, they are very expensive because of the need for machining from large parts; in addition, they

Table 9.4 Physical properties of materials used in electronic systems.

Matrix	Reinforcement type	Vol%	ρ (kg m^{-3})	CTE α (10^{-6} K^{-1})	k (W m^{-1} K^{-1})
Semiconductors					
Silicon	—	—	2 300	4.1	150
SiC	—	—	3 100	3.5–5	300
Gallium arsenide	—	—	5 300	5.8	54
GaN	—	—	6 100	3.2	150
Ceramic substrates					
Alumina	—	—	3 900	6.7	20
Aluminum nitride	—	—	3 300	4.5	150–180
Metals					
Cu (electronic grade)	—	—	8 900	17	400
Al	—	—	2 700	23	200
Tungsten	—	—	19 300	4.5	180
Beryllium	—	—	1 860	13	150
Kovar	—	—	8 300	5.9	17
Reinforcements					
C-diamond	—	—	3 500	1.2–2.3	1000–2000
C$_f$	—	—	1 900	−1.2 (ax.)	100–500
SiC$_p$	—	—	3 100	3.5–5	100–300
Composites					
Cu	C-diamond	50	5 350	5.5	420
Cu	C$_f$	28	7 200	6.5	290 (XY)
Al	C$_f$	75	2 020	6.5	290 (XY)
Al	SiC$_p$	60	2 900	9–10	170–220

(Source: Adapted from Ref. [27].)

are also dense, which is clearly a negative aspect for their use in automotive components where vibration is taken into consideration. Other materials with relatively high success for electronic components are those combining Cu/Invar/Cu and Cu/Mo/Cu in a sandwich-like structure. As with the previous family of metals, they are very heavy, with different heat dissipation capabilities in-plane or through thickness directions.

Overall, tailored CTE materials such as MMCs combine higher mechanical reliability with high thermal conductivity and lightness. Diamond-based composites provide an excellent solution, but their very high cost prohibits their use in automotive applications. Among MMCs, AlSiC materials have a good thermal dissipation and CTE that is compatible with the electronics and dielectric substrates (Table 9.4). In addition, these materials are lightweight and the fabrication technology allows for functional shapes to be considered. AlSiC is ideal for these applications because of its low density, high strength, and stiffness, so that it meets the weight budget and is tolerant to shock and vibration.

9.3.1.3 Manufacturing Process, Design, and Performance Analysis

Al/SiC parts can be manufactured by two main kinds of techniques. The first method involves a powder metallurgy method, where aluminum powders are mixed with SiC particles. The composites mixture has to be sintered in order to obtain a dense billet. The volume content of SiC_p is limited because of the need to follow a pressureless sintering in order to obtain a net-shape specimen with an acceptable cost. In the second method, the amount of reinforcement can be increased when a liquid infiltration technique is taken into account [28]. The SiC particles can be manufactured into a preform using an injection-molding process. Then, the SiC preform is placed in a form that defines the final part shape and is infiltrated with molten aluminum metal under pressure.

AlSiC parts are manufactured at CPS using a near-net-shape approach, by creating a SiC preform by injection molding of a SiC slurry with a binder. This is followed by sublimation to remove the binder and finally the resultant preform, which is usually called "green part" is infiltrated under pressure with molten aluminum. Parts can be made with sufficiently tight tolerances in order to avoid further machining. The material is fully dense (without voids) and is hermetic. High stiffness and low density enable the manufacture of larger parts with thin walls such as large fins for heat dissipation. AlSiC can be plated with nickel and nickel–gold or by other metals by thermal spraying. Ceramic and metal inserts can be incorporated into the preform before aluminum infiltration, resulting in a hermetic seal [29].

The detailed geometry and assembly in the IGBT module is illustrated in Figure 9.2. The AlSiC baseplate can be tailored to the specific assembly application to meet the final assembled bow requirements (Figure 9.2a). Bow optimization, using a mechanical attachment to a cold plate for improved heat dissipation, can improve the thermal interface between the module and the baseplate. In addition, AlSiC baseplates can also be fabricated with a convex bow to enhance the module base/cold plate interface while maintaining a flat surface for dielectric substrate attachment [30]. This bow can be engineered and cast into the final product.

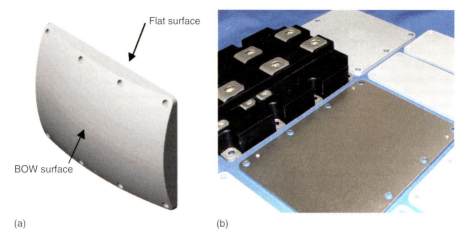

Flat surface

BOW surface

(a) (b)

Figure 9.2 Baseplate configurations: (a) baseplate showing cooler contact bow surface and flat substrate surface and (b) IGBT baseplates for traction and power conversion. (Source: Both courtesy of CPS technologies.)

A schematic illustrating the bow is shown in Figure 9.2. Bow deflection values of 0.2–0.3 mm subject to customer's specifications have been fabricated on baseplates approximately 140 mm². Near-net-shape has also been produced and incorporated in a single step for the holes for the nut-and-bolt attachment of the power module housing to the power control system of the vehicle. This plate has been nickel plated in order to create a suitable finishing and interface for the placement of the solder or brazing attachment of DBC Al_2O_3 or AlN substrates (Figure 9.2b).

The MMC baseplate performance has been compared with that of a plain copper plate [31]. The solder interface between the DBC and the baseplate has been inspected by nondestructive testing (ultrasonic technique) after a relevant number of thermal cycles. In the case of the copper IGBT baseplate, the solder layer fails by delamination after about 400 thermal cycles. The thermal dissipation path continues to degrade as the delamination continues. As a consequence, the electronic system is subjected to higher and higher thermal load that will ultimately cause the device to fail. On the other hand, for an AlSiC IGBT baseplate, the delamination effect is minimized even after 30 000 thermal cycles. In power electronic systems, this equates to long-term reliability. AlSiC (200 W m⁻¹ K⁻¹) has a lower thermal conductivity than Cu (400 W m⁻¹ K⁻¹), but as already illustrated in the Cu system, the designer must consider stress compensation for these systems to manage the CTE difference between the Cu and the attached substrates. Stress compensation layers increase the length of the thermal dissipation path, so the benefit of the high thermal conductivity of Cu will hardly be realized. The AlSiC system can often be optimized in design to have very thin solder layers for a shorter thermal dissipation path. As a result, the AlSiC baseplate solution can have equal or better thermal dissipation when compared to Cu baseplate solutions.

9.3.1.4 Conclusions

AlSiC has enormous potential as a material solution for IGBT modules baseplates for power control thermal management in HEV. The compatible CTE value results in improved reliability of IGBT assemblies by reducing the magnitude of thermally induced stresses. In particular, the near-net-shape fabrication infiltration process provides baseplates that are cost competitive. Functional geometrical features for improved thermal interfaces between module base and heat sink can be cast during fabrication as net shape (without machining). A main conclusion is that AlSiC is a proven solution for a reliable thermal management of IGBTs in HEV/EV applications, as the material has higher thermal conductivity than traditional thermal management materials (pure aluminum) and lower processing cost than other MMC competitors [28, 44].

9.3.2
Case Study 2: Aluminum Matrix Composites for Automotive Brake Disks

This case study deals with the development of brake drum prototypes produced with flash-reinforced materials manufactured by Cyco Systems, Australia [20].

9.3.2.1 Introduction

The potential application of particulate-reinforced aluminum alloys in braking systems was readily identified in the early stages of the development of MMCs. The main requirements of components such as rotors, disks, and calipers seemed to match perfectly with the main advantages of particulate-reinforced aluminum alloys. The high thermal conductivity required to dissipate the heat generated during braking, low density compared to ferrous metals, large wear resistance provided by the hard ceramic particulates, as well as good damping properties, seemed to represent the ideal application for MMCs. Even though the higher cost compared to cast iron as well as the concerns related to hot spots that might appear in the MMC disks because of their heterogeneity have hindered its progress, it continues to be one of the most clear potential applications of particulate-reinforced MMCs. The train industry was the first in applying MMCs for brake applications. Brake rotors for the German high speed train ICE-2 were developed by Knörr Bremse AG with a reinforced aluminum alloy AlSi7Mg + 20% SiC particulates supplied by Alcan. Brake pads were also developed that could be used with the new disks as the nature of the abrasion produced during the braking was different with the new materials. Compared to conventional parts made out of cast iron weighing 120 kg per piece, the 76 kg rotor offered an attractive weight-saving potential. It was calculated that a saving of up to 500 kg per bogie could be achieved. Furthermore, heat dissipation was greatly improved and the aluminum-based component presented an improved wear resistance. Even though the cost of the material was higher than the previously used cast iron concept, the long-term performance was better. Furthermore, the weight saving and the reduction in the number of disks needed to comply with the braking specifications was considered to be enough to compensate the higher cost of the material. The successful development of the train disks boosted the idea of the first automotive MMC brake disk.

The first commercial applications were the brake disks of both the Lotus Elise and Plymouth Prowler cars. The rear brake drums of the Volkswagen Lupo 3L and Ford Prodigy followed thereafter. The Toyota RAV4-EV also used these applications and the feasibility of using particulate-reinforced aluminum for low weight cars was confirmed. The challenge now is to apply these materials for commercial vehicles.

9.3.2.2 Requirements and Materials Selection

Cyco Systems is an Australian company devoted to the production and delivery of advanced materials. In 2005, they presented a new MMC based on the incorporation of fly ash particulates into aluminum alloys. The commercial name of the new material is ULTALITE® Al-MMC and it was developed as a low cost MMC for automotive applications.

ULTALITE is a combination of Al–Si alloys and ceramic particles derived from fly ash, a by-product from the burning of black coal in power stations. The cost of this material is significantly lower than that of other traditional reinforcements such as SiC or Al_2O_3. In addition, a proprietary production process based on semisolid stir casting has been optimized that makes it possible to distribute the particulates in secondary alloys at low costs. Furthermore, die life is enlarged and machining is possible with conventional tools. According to tests carried out by industrial companies such as Teksid in Europe and Aisin Takaoka (Division of Toyota) in Japan, the material does not lose any significant property after going through to different recycling and reuse processes [20].

CYCO commenced the development of ULTALITE Al-MMC in the early 1990s. Toward the end of 2001, PSA Peugeot Citroen contracted Cyco to produce 50 UlLTALITE prototype brake drums similar in design to their existing cast iron brake drums. Delivery of the prototype brake drums for Peugeot was completed in early 2003. Peugeot undertook in-house testing of these prototypes, and although the results were positive in qualifying the potential of ULTALITE, it was clear that further work would be required to improve fly ash distribution and consistency of the formulation. This would be necessary in order to achieve the optimum level of wear and heat resistance required for the brake drum application. This work culminated in the production of new batches of ULTALITE Al-MMC brake drum prototypes in early 2004 in preparation for brake dynamometer testing. A series of brake dynamometer tests were conducted at Brake Testing International (BTI) in the United Kingdom to evaluate performance of ULTALITE drum brake in comparison with standard OE cast iron brake drums. The ULTALITE brake drums achieved a very positive result.

The reinforcement of the ULTALITE composites is Ceramatec ceramic particulates derived from fly ash, a by-product of the power generation industry. They consist mainly of a mixture of SiO_2 and Al_2O_3 particulates. The raw fly ash material solidifies while suspended in the exhaust gases and forms an amorphous material with spherical particles with a size range between 0.5 and 150 μm, and Ceramatec particulates of the appropriate size are extracted from the raw fly ash. More than

100 million tons of fly ash is produced each year, in the United States alone and most of it is directly disposed of in ash ponds or landfills.

Potential competitors of the ULTALITE composites for brake disks are other aluminum-based MMCs and the extensively used cast iron material. The former have already been used in previous similar applications but they are more expensive and therefore have only been used in low volume special applications. According to data provided by Cyco, the cost of ULTALITE is <60% of that of conventional composites.

9.3.2.3 Manufacturing Process, Design, and Performance Analysis

ULTALITE composites are produced by stir casting. Cyco systems acquired the rights of a patent produced by Professor Bob Pond of Johns Hopkins University and has further improved the production method by applying new concepts in the stirring stage. The ceramic particles are introduced under the surface of the semisolid aluminum by feeding through the internal passageway of a hollow internal tube. The lower end contains teeth that produce high shear forces that help avoid the formation of clusters and provide a uniform mixing. The short stirring period together with the elimination of the fine fly ash particulates minimizes the exothermic chemical reaction between the molten metal and the reinforcing phase (Figure 9.3).

Casting Brake Drums The density of fly ash-derived particulates is similar to that of aluminum alloys and therefore no stirring is required in the holding furnace to avoid the segregation of particulates before the casting step. In fact, casting equipment and processes may be kept unchanged and there is no need for any extra investment for primary and secondary operations.

Brake drums were produced by squeeze casting – a process in which the introduction of the melt into the mold is made under slow injection speeds and pressure is applied on the surface of the metal during the entire solidification step. The objective is to produce near-net-shape components and to eliminate the porosity produced by the turbulent flow and solidification shrinkage common to the high

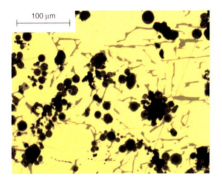

Figure 9.3 Optical micrograph of the ULTALITE® material showing the presence of spherical-shaped reinforcement particulates embedded in the aluminum matrix.

Figure 9.4 Ventilated brake drum redesigned by Cyco Systems and produced with ULTALITE®.

pressure die casting (HPDC) process. Figure 9.4 presents an image of a brake drum prototype produced by this method.

The ULTALITE brake drum composed of an Al–Si7Mg2 alloy reinforced with 20% of particulates weighs 2.3 kg, less than half the weight of a conventional cast iron drum. Furthermore, the developed prototype has been redesigned by Cyco Systems and presents venting features incorporated to further dissipate the heat generated during the successive brakes.

Testing of Brake Drums Table 9.5 compiles the data of mechanical properties obtained from specimens drawn from the squeeze cast drums. Two different reinforcement contents, 5 and 20 wt%, are shown.

Figure 9.5 presents data obtained from dynamometer tests. Tests were carried out on squeeze cast drums with a diameter of 20.32 cm. The results were compared to those obtained on equivalent cast iron drums following the procedure established

Table 9.5 Tensile properties of ULTALITE® materials.

Material	Reinforcement (vol%)	$\sigma^{Y0.2}$ (MPa)	UTS (MPa)	Elongation (%)
ULTALITE-2OII (T6)	20	—	331	2.0
ULTALITE-S5 (T6)	5	—	338	2.5

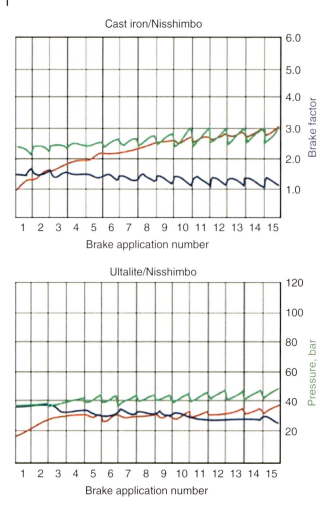

Figure 9.5 The plots show brake factor and variations in pressure and temperature measured across 15 stops. Blue curve = brake factor, red = temperature changes, and green = variations in the applied pressure.

in SAE specification J25222. Standard off-the-shelf Nisshimbo brake linings were used in the tests.

The results show that the DRA material presents higher brake factors with an average improvement of around 30%. Furthermore, the maximum temperature measured in the DRAs did not exceed 185 °C compared to 300 °C in the conventional cast iron brakes, as foreseen to be caused by the higher heat dissipation capacity of aluminum. This fact resulted in an operating temperature down to 100 °C lower than for the cast iron brakes. Eventually, the wear of the off-the-shelf linings after the tests was also checked and the results showed that the DRA drums produced even lower wear than the cast iron ones.

9.3.3
Case Study 3: Titanium Matrix Composites (TMCs) for Aerospace Applications

9.3.3.1 **Introduction**
In aerospace applications, where new materials for higher thrust levels, lighter weight and increased efficiency are major requirements for designers, TMCs stand out from other composites systems. Titanium alloys are widely used in the aeronautic industry for their relatively high specific properties (stiffness, yield strength, strength to rupture, etc.). The reinforcement of titanium is very interesting for improving the creep at elevated temperature (above 500 °C) as well as the fatigue life. However, it is difficult to produce a composite for at least two reasons: (i) the properties of the matrix are already high and (ii) this metal or the alloy reacts with nearly all materials. The fabrication of the composite is also difficult because of the high melting point as well as the fact that this metal is also highly reactive with the oxygen. The atmosphere has to be controlled taking into account that the vapor pressure of titanium begins to increase above 1000 °C. Argon or vacuum are the most suitable atmospheres in which to prepare the composite. The common alloys are the Ti-64 or Ti-6242 grades, which combine the properties of the two crystallographic structures bcc(β) and hcp(α) of titanium. Usually, the unreinforced alloys are forged at a temperature close to the transus temperature, heated to the two-phase field, quenched, and followed by a subsequent aging treatment at a moderate temperature. However, this dual-phase structure $\alpha + \beta$ of titanium alloys may be changed during the fabrication of the composite. If the liquid route is chosen to make the composite, the microstructure will be completely changed; if the solid route is used, grain growth could be expected. Moreover, the reinforcements are usually made of light alloys such as those of carbon, boron, nitrogen, or oxygen, which stabilize the α-phase, so the F/M interface will mainly be surrounded by this less ductile phase. The presence of the alloying elements within the alloy also has the advantage to decrease the kinetics of the reaction between the matrix and the reinforcement. Before making the composite, the first question that needs to be addressed is what is the reinforcement that will thermodynamically fit with the titanium alloy?

9.3.3.2 **Requirements and Materials Selection**
For continuous TMCs, the only reinforcements available are the SiC chemical vapor deposition (CVD) monofilaments. These monofilaments have been specially designed for the purpose. They present a very high tensile strength, a high modulus, and they have an outer layer coating of carbon. The carbon is not stable with the titanium matrix and during the fabrication of the composite, a continuous TiC reaction zone is formed at the interface. Even though TiC has a large field of nonstoichiometry, this carbide plays the role of a partial diffusion barrier. These composites are supposed to be used in the temperature range 500–600 °C, for which the reaction kinetics are very low and the material can be considered to be stable. The other possibility is to find a ceramic compound that has the adequate mechanical performances and does not react with the titanium matrix.

Table 9.6 Room temperature (RT) properties of reinforced and unreinforced high strength metals.

Material	UTS (MPa)	$\sigma^{Y0.2}$ (MPa)	Elongation (%)	E (GPa)	ρ (kg m^{-3})
Ti-64 grade 5 wrought bar	900–1200	800–1050	10–18	105–115	4420
Ti-64 grade 5 composite 33 vol%	1650	1450	1.2	196	4150
17-7PH-S17700 stainless steel	1170–1650	965–1590	1–7	204	7800

The answer can found by looking at the binary or ternary phase diagrams. As an example, understoichiometric TiC or nearly stoichiometric TiB fit this requirement. Moreover, TiC and TiB have a high Young's modulus and have a CTE close to the CTE of Ti, which is better for reducing the thermal residual stresses. TiC is commercially available in powder form (1–50 µm in size), whereas TiB is not commercially available and will need to be prepared by an adequate process. A minimum ductility is necessary for the components used in aeronautics. In order to combine ductility and improved mechanical properties of the TMC, an adequate volume fraction of reinforcement has to be defined. It has been shown that for a continuous Ti/SiC composite with a fiber volume fraction of fibers in the range 30–35%, the properties of the resulting composites are close to those of a high strength steel but with only half the density of the steel (Table 9.6).

The situation is different as regards discontinuous TMCs. The properties of the composites are influenced by the volume fraction and by the aspect ratio of the reinforcement. The aspect ration of TiB can be higher than 20, whereas the aspect ratio of TiC is closer to 1 or 2. The properties of the composite can be computed by the shear-lag model (Figure 9.6).

Ductility is also function of particle size and it is accepted that better ductility can be obtained from particulates with a mean diameter <1 µm and a volume fraction <10%. As a consequence, a high volume fraction of TiC is necessary to get a composite effect but the ductility will be poor. For Ti/TiB composite, a compromise between ductility and improved performances can be found for <9 vol% of TiB.

9.3.3.3 Manufacturing Process, Design, and Performance Analysis

Owing mainly to the high reactivity of titanium alloys with the SiC monofilaments, production routes are limited to different preprocessing steps to set up fibers and matrix that are finally consolidated by diffusion bonding. Foremost among them are the FFF, tape cast, and matrix-coated fiber techniques, as also a technique known as the *ultrarapid liquid fiber-coating technique* (Figure 9.7). In the FFF technique, fiber mats are placed between matrix foils, stacked to a multilayer arrangement, and finally consolidated by hot pressing. The hot-pressing temperature is just below the α–β transition temperature, the pressure is on the order of 100 MPa and the duration of pressing is around 1 h. During degassing of the binder, the pressure is maintained so as to avoid the displacement of the fibers and to maintain a

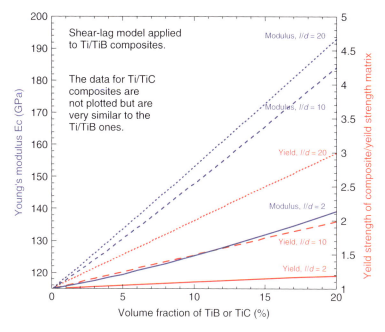

Figure 9.6 Computed Young's modulus and yield strength for different aspect ratios and different volume fraction of the reinforcement.

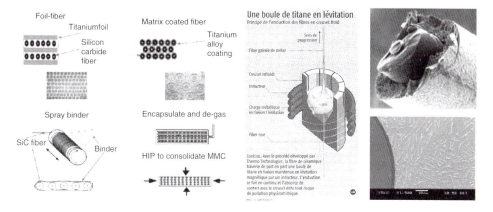

Figure 9.7 Manufacturing routes for TMCs.

nearly hexagonal distribution. The volume fraction is controlled by the thickness of the foils and by the distance between tow fibers. The tape cast approach is different in that the matrix is added by plasma spray in a powder form, setting up monotapes. Both are mature techniques and have been used to manufacture a range of component forms; however, they present some drawbacks. In the former method, the process is limited to those titanium alloys that can be fabricated as thin foils. This limitation as well as the high cost can be reduced with the latter approach, which uses the matrix in powder form. However, in both cases, the limitations of the final shapes that can be obtained and, more importantly, the difficulty in obtaining a homogeneous fiber distribution are major challenges. In the matrix-coated fiber approach, metal coating is applied in a vapor phase directly to individual silicon carbide monofilaments using deposition processes, such as magnetron sputtering or EB-PVD. The former is a versatile process to obtain coatings of different alloys unlike the latter, which presents the advantage of greater productivity because the deposition rate is higher. The coated fibers are then arranged for final consolidation by an HIP technique. As fibers are individually coated by the matrix, the consolidation occurs between metallic matrix and little (or the lack of) contact between the molten matrix and fibers, which is an advantage when reactive metals such as titanium alloys are used. Composites with excellent fiber distribution can be obtained although the cost of the process itself is still a major challenge. In the ultrarapid liquid fiber-coating technique, the coating is applied from the melt. By induction, titanium is melted and an amount of titanium is maintained in magnetic levitation. Once titanium is in this liquid stage, the fibers rapidly cross the melt, thereby producing a constant thickness of titanium over the individual fibers. The coating step is faster than in previous techniques (150 times faster than EB-PVD). The speed of the withdrawal has to be tuned to account for the reactivity and the wettability of the fibers. The faster the withdrawal, the thicker is the coating produced. The withdrawal speed must be chosen so that it is above the visco-inertial regime (Figure 9.8). Similar to the previously described techniques, the individual coated fibers are arranged to be finally consolidated by HIPing. This technique is under evaluation for preparing the BLING discs.

All these achievements of production status and cost reduction for the Ti/SiC systems made it possible for the emergence of the first contract to supply the material in a commercial application. In 2007, GKN Aerospace was awarded a

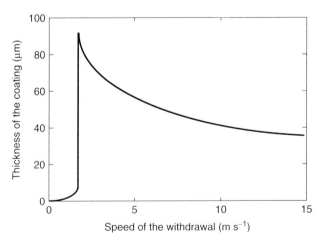

Figure 9.8 Different regimes versus the speed of withdrawal for a SiC monofilament in a liquid titanium [32].

Figure 9.9 Ti/SiC thrust link for the Boeing 787 [33].

contract by Boeing to develop and supply Ti/SiC composite thrust links for the Boeing 787 [33] (Figure 9.9). The purpose of a thrust link assembly is to transmit the tremendous thrust generated by the engine to the airframe. It is designed with a special joint at each end to eliminate the transmission of twisting and bending loads. It consists of a Ti/SiC center tube, which is plasma welded to two titanium end lugs, machined, and then assembled. The Ti/SiC center tube is produced by a tape cast technique, silicon carbide monofilaments and titanium powder mats are rolled onto a mandrel to finally obtain the composite by diffusion bonding.

It can be concluded that Ti/SiC composites for uniaxial applications are available at present. Flight-qualified components are currently in production and further expansion of the material market is proceeding with similar unidirectionally reinforced parts. However, it is not applicable for multiaxis-loaded components because of the low transverse properties; this is a limiting design factor for new components with more complex loadings.

Discontinuously reinforced TMCs can be prepared either by the liquid route or by the powder metallurgy route. For Ti/TiC$_p$ composites, the powder metallurgy technique or *in situ* precipitation from carbon particulates is recommended for small diameter of TiC particulates. The synthesis of Ti/TiC$_p$ composites has been conducted by mechanical alloying and followed by spark-plasma-sintering (SPS) from elemental Ti powder with graphite. Sintering by SPS can proceed at substantially lower temperatures compared with the conventional powder metallurgy process. For Ti/TIB$_w$ composites, the situation is different as TiB$_w$ is not commercially available. The composite material can be prepared either from the solid route by reaction of fine TiB$_2$ particulates ($<1\,\mu$m in diameter) with titanium or by the liquid route from boron or TiB$_2$ particulates. The TiB needles precipitate upon solidification and the boron content has to be <8 at% in order to obtain fine proeutectic TiB needles. From this content of boron, the volume fraction of TiB in the composite is around 9 vol%. Significant enhancements have been achieved with this low TiB$_w$ concentration; these include extraordinary grain refinement that results in increases in specific strength and stiffness while maintaining ductility and improving fatigue resistance. This grain refinement effect is also very interesting for titanium alloys in general as they present grain sizes of the order of a few millimeters in the as-cast condition and thus expensive thermomechanical processing is used to refine the microstructure (Figure 9.10).

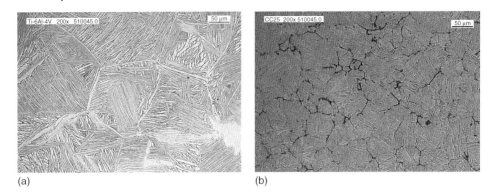

(a) (b)

Figure 9.10 The microstructures of cast Ti6Al4V: (a) without TiB_w and (b) with $TiB_w < 9\,vol\%$ [34].

Despite Ti/TiBw composites having been already used in some commercial applications to replace steel (i.e., intake and exhaust valves for the engine of Toyota Altezza car [35], recently aerospace programs are assessing their excellent properties including the so-called "second-tier" mechanical properties, fatigue, fatigue crack growth, or fracture toughness. Ti/TiB_w composites are a very promising material that could open the door for a wide variety of fatigue driven aerospace applications (Figure 9.11).

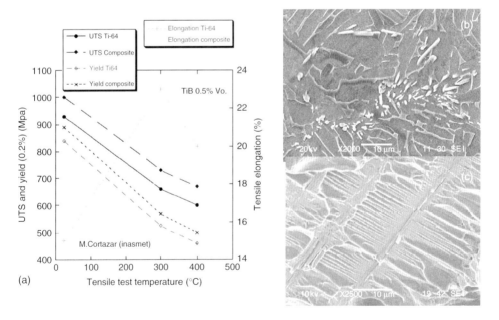

(a)

Figure 9.11 Tensile properties of casted TiB(0.5 vol%)/Ti64 composites tested in temperature (a). The TiB whiskers precipitate at the β-grain boundary (b); the rupture of the composite, however, is controlled by the stacking faults in the TiB crystals (c).

9.4
Summary and Outlook

After more than 50 years of development, structural MMCs may now be considered as a mature technology. Design engineers have access to a large body of material properties data, information on secondary operations, and long-term performance of a large array of different matrix/reinforcement systems. The cost and the need for increasing the performance of MMCs are now the major obstacles for the use of these materials in new applications. As a consequence, it is foreseen that the future of the research activities related to MMCs will be linked to these aspects. Research on low cost preforms and on technologies for the production of locally reinforced components will continue in the following years as well. The development of metal matrix nanocomposites (MMNCs) is also seen as a promising field [36]. Challenges related to the cost and the tendency for agglomeration of the nano reinforcements must yet be solved before the great potential of MMNCs for structural applications in industry may be realized.

Regarding thermal management, future trends in transport will not only be focused on already ongoing applications such as low power (air conditioning and heating) and high power (power traction) in HEV and full EVs, such as the Al/SiC heat sinks for IGBT dissipation in the Toyota Prius [44] or the case study of CPS Technology Corporation already described. They will also be focused on thermal management of LED lighting arrays [37, 38] and battery modules [39]. A very promising scenario has opened for full electric vehicles as the power requirements and sales are expected to increase considerably [40]. Other important markets will come from railway transport (i.e., commercial success on the Shinkansen 700 bullet train in Japan [41] where a market size of $40M at a $40 per part, with a10% growing rate has been foreseen [42]. Finally, construction and agricultural EVs, where the power consumption is higher than the passenger cars [43] are also considered as future markets.

9.5
Further Reading

A list of relevant handbooks on MMCs as a starting point to or to find in-depth information about an specific topic:

- Chawla, N. and Chawla, K.K. (2006) in Metal Matrix Composites (eds N. Chawla and K.K. Chawla), Springer, New York.
- Clyne, T.W. and Withers, P.J. (1993) in An Introduction to Metal Matrix Composites (eds E.A. Davis and I.M. Ward), Cambridge University Press, New York.
- Evans, A., San Marchi, C., and Mortensen, A. (2004) Metal Matrix Composites in Industry: An Introduction and a Survey, Kluwer Academic Publishers.
- Everett, R.K. and Arsenault, R.J. (eds) (1991a) Metal Matrix Composites: Processing and Interfaces, Academic Press, San Diego, CA.

- Everett, R.K. and Arsenault, R.J. (eds) (1991b) Metal Matrix Composites: Mechanism and Properties, Academic Press, San Diego, CA.
- Kainer, K.U. (2006) Metal Matrix Composites: Custom-Made Materials for Automotive and Aerospace Engineering, Wiley-VCH Verlag GmbH, Weinheim.
- Suresh, S., Mortensen, A., and Needleman, A. (1993) in Fundamentals of Metal Matrix Composites (eds S. Suresh, A. Mortensen, and A. Needleman), Butterworth Heinemann, Boston, MA.
- Taya, M. and Arsenault, R.J. (1989) Metal Matrix Composites: Thermomechanical Behaviour, Pergamon Press, Oxford.

Acknowledgments

The authors wish to thank Mr Occhionero, Vice President of Marketing and Technical Sales at CPS Technologies and Mr Graham Withers, president of Cyco Systems, for their disposition to provide permission in using data and figures for the case of studies on IGBTs and brake drums, respectively, as well as the following people for providing their vision and information on the present state and insights into the future of commercial MMCs materials and applications: T.W. Clyne, A. Mortensen, D. L. Saums, J. Cornie, M. J. Hollins (Lanxide Technology Company L.P.), S. Midson (The Midson Group Inc.), S. Gourdet (EADS France), G. Requena (TUWien-Austria), M. Basista (IPPT-Poland), B. Coleman (Engineered fiber solutions Inc.-USA) J. Coleto, I. Obieta, and R. Seddon (Tecnalia).

References

1. Reibold, M., Paufler, P., Levin, A.A., Kochmann, W., Pätzke, N., and Meyer, D.C. (2006) Materials: carbon nanotubes in an ancient damascus sabre. *Nature*, **444**, 286.

2. About.com *http://metals.about.com/library/bldef-Composite-Material.htm* (accessed 21 December 2012).

3. Schatt, W. and Wieters, K.P. (1997) *Powder Metallurgy: Processing and Materials*, EMPA.

4. German, W.R.M. (1997) *Powder Metallurgy Science*, Metal Powder Industries Federation, Princeton, NJ.

5. Rittner, M. (2000) *Metal Matrix Composites in the 21st Century: Markets and Opportunities*, BCC, Inc., Norwalk, CT.

6. Miracle, D.B. (2005) Metal matrix composites-from science to technological significance. *Compos. Sci. Technol.*, **65**, 2526–2540.

7. Egizabal, P. Metal ceramic composites in technology: needs and applications. *State of the Art Reports*, Vol. 2, Part VI, KMM, e-library, *http://www.kmm-noe.org* (accessed 21 December 2012).

8. Requena, G. (2004) Applications of Metal Matrix Composites, *http://mmc-assess.tuwien.ac.at/mmc/*.

9. Zeuner, T., Stojonov, P., Sham, P.R., Ruppert, H., and Engels, A. (1998) Developing trends in disc brake technology for rail applications. *Mater. Sci. Technol.*, **14**, 857–863.

10. Sonuparlak, B., Lehigh, M.D., and Robbins, M.A. Silicon Carbide Reinforced Aluminum for Performance Electronic Packages, PCC Advanced Forming Technology. Metal Matrix Composites Business Unit.

11. N.C. and Chawla, K.K. (2000) Metal-matrix composites in ground transportation. *JOM*, 67–70.

12. Duralcan® Composites for Wrought Products. Property Data. Alcan. Technical Brochure.

13. Duralcan®. Alcan. Technical Brochure.

14. Materion Corporation AMC Materials, *http://www.amc-mmc.co.uk/materials.htm* (accessed 21 December 2012).

15. *http://www.matweb.com/search/datasheet.*

16. ULTALITE® (2005) *http://www.ULTALITE.com* (accessed 21 December 2012).

17. DWA Aluminium Composites, *http://www.dwa-dra.com/.*

18. Rosso, M. (2006) Ceramic and metal matrix composites: routes and properties. *J. Mater. Process. Technol.*, **175**, 364–375.

19. NCN (2006) Technology Roadmap for the Metal Matrix Composites Industry. National Composites Network, March 2006.

20. Withers, G. and Zheng, R. (2008) ULTALITE®-a Low Cost, Light Weight Aluminum Metal Matrix Composite for Braking Applications. Autoengineer, March 2008.

21. Rohatgi, P.K. and Schultz, B. (2007) Advanced metals and alloys. Lightweight metal matrix nanocomposites – stretching the boundaries of metals. *Mater. Matters*, **2**(4), 16.

22. CPS Technologies – AlSiC Solutions Reliability with Smart Composite Products *www.alsic.com* (accessed 21 December 2012).

23. Larminie, J. and Lowry, J. (2003) *Electric Vehicle Technology Explained*, Wiley VCH-Verlag GmbH, p. 157.

24. Nishiura, A., Soyano, S., and Morozumi, A. (2007) IGBT modules for hybrid vehicles. *Fuji Electric Rev.*, **53**(3), 65–68.

25. Sleasman, T. and Sonuparlak, B. (2010) Cooling of IGBT Based Power Modules for Hybrid Electric and Electric Vehicles. Power System Design Europe (Jul./Aug.), pp. 50–52.

26. Evans, A., San Marchi, C., and Mortensen, A. (2003) *Metal Matrix Composites in Industry: An Introduction and a Survey*, Springer, p. 329.

27. Barcena, J., Merveille, C., Maudes, J., Vellvehi, M., Jorda, X., Obieta, I., Guraya, C., Bilbao, L., Jiménez, C., and Coleto, J. (2008) Innovative packaging solution for power and thermal management of wide-bandgap semiconductor devices in space applications. *Acta Astronaut.*, **62**, 422–430.

28. Mallik, S., Ekere, N., Best, C., and Bhatti, R. (2011) Investigation of thermal management materials for automotive electronic control units. *Appl. Therm. Eng.*, **31**, 355–362.

29. Occhionero, M., Adams, R., Fennessy, K., and Hay, R. A. (1998) Aluminum silicon carbide (AlSiC) for advanced microelectronic packages. IMAPS May 1998, Boston Meeting.

30. Occhionero, M. A., Adams, R.W., Fennessy, K. P., and Sundberg, G. (2001) AlSiC Baseplates for Power IGBT Modules: Design, Performance and Reliability.

31. Schuetze, T., Berg, H., and Schilling, O. (2001) 6.5kV IGBT Module Delivers Reliable Medium-Voltage Performance. Power Electronics Technology (Sept. 1).

32. Feigenblum, J. (2002) Thesis INP Grenoble, France Procédé inductif d'enduction métallique de fibres par voie liquide.

33. NetComposites *http://www. netcomposites.com/news.asp?4657* (accessed 21 December 2012).

34. García de Cortazar, M., Agote, I., Silveira, E., Egizabal, P., Coleto, J., and Le Petitcorps, Y. (2008) Titanium Composite Materials for Transportation Applications. *JOM*, **60**(11), 40–46.

35. Saito, T. (2004) The automotive application of discontinuously reinforced TiB-Ti composites. *JOM*, **56**(5), 3–36.

36. Rohatgi, P.K. and Schultz, B. (2007) Lightweight metal matrix nanocomposites – stretching the boundaries of metals. *Mater. Matters*, **2**, 16–19.

37. Scotch, A.M. (2010) Thermal evaluation of materials for LEDs. IMAPS New England Chapter Meeting, 2010.

38. Christensen, A. and Graham, S. (2009) Thermal effects in packaging high power light emitting diode arrays. *Appl. Therm. Eng.*, **29**, 364–371.

39. Zweben, C. (2001) in *ASM Handbook, Composites*, Vol. 21 (eds D.B. Miracle

and S.L. Donaldson), ASM International, Materials Park, pp. 1078–1084.

40. Yole Devéloppement (2009) Power EV-HEV 2010. Power Electronics in Electric and Hybrid Cars a 10-Year Market Forecast, Executive Summary, 2009.

41. Evans, A., Marchi, C.S., and Mortensen, A. (2003) *Metal Matrix Composites in Industry: An Introduction and a Survey*, Kluwer Academic Publishers, Dordrecht.

42. White, D. (2009) AlSiC part II, a once exotic material makes a comeback.

IMAPS San Diego Chapter, San Diego, CA, June 2009.

43. Luniewski, P. (2009) System and power module requirements for commercial, construction and agriculture vehicles CAV. ECPE – Workshop, Munich, Germany, March 2009.

44. Sleasman, T. and Sonuparlak, B. (2010) IGBT Gate Driver for Automotive Power Inverters. Power System Design Europe (Jul./Aug.), pp.26–28.

10
Polymer Nanocomposites

James Njuguna and Krzysztof Pielichowski

10.1
Introduction

High performance structural design imposes a number of restrictions on the properties of materials required to be used. It also follows that nowadays, lighter, thinner, stronger, and cheaper structures are very important goals of materials science and engineering. The restricted parameters could be seen as a subset of the design parameters. Formulation and establishment of the design parameters build up a process of certification with emerging new materials intended for high-performance applications. For instance, launching a heavy lift system into low Earth and geosynchronous orbits generally accounts for €5000–15 000 and 28 000 kg^{-1}, respectively. Hence, considering the growing oil/gas prices, the demand for lightweight materials in the aerospace industry is tremendous. Even in general aviation, fuel costs account for ∼50% of the operation costs. As a result, over the last three decades, the usage of fiber-reinforced polymer (FRP) composites in these applications has increased from <5% by structural weight (Boeing 737) to 50% (Boeing 787), with a gain of over 20% fuel efficiency. However, in these conventional structural materials, the fiber-orientation is usually in-plane (*x*- and *y*-direction) resulting in fiber-dominated material properties in these directions, whereas matrix dominates in the *z*-direction. Therefore, FRPs are very sensitive to intrinsic damage, such as delamination (in particular), matrix cracking, and fatigue damage [1–3]. So far, several approaches have been adopted to tackle these composites' deficiencies including improving the fracture toughness of the ply interfaces via epoxy/elastomer blends and reducing the mismatch of elastic properties (and stress concentrations) at the interfaces between the laminated plies.

However, conventional composite materials lack other required functional properties such as high electrical/thermal conductivities to account for electrostatic dissipation and lightening strike protection. At present, it is believed that the best route to achieve multifunctional properties in a polymer is to blend with nanoscale fillers. The fundamental bases behind this idea are the three main characteristics

Structural Materials and Processes in Transportation, First Edition.
Edited by Dirk Lehmhus, Matthias Busse, Axel S. Herrmann, and Kambiz Kayvantash.
© 2013 Wiley-VCH Verlag GmbH & Co. KGaA. Published 2013 by Wiley-VCH Verlag GmbH & Co. KGaA.

that define the performance of polymer nanocomposites: nanoscopic confinement of matrix polymer chains; nanoscale inorganic constituents and variation in properties, that is, many studies have reported that mechanical, conductivity, optical, magnetic, biological, and electronic properties of several inorganic nanoparticles significantly change as their size is reduced from macroscale to micro and nanolevels; and nanoparticle arrangement and creation of large polymer/particle interfacial area.

Large quantity of nanomaterials such as carbon nanotubes (CNTs), nanofibers, SiO_2, and montmorillonite (MMT) are presently available because of the establishment of well-developed manufacturing technologies such as chemical vapor deposition method, ball milling, and electrospinning. With ease of manufacturing in bulk, fiber-reinforced polymer nanocomposites are finding increasingly more practical applications (e.g., in the manufacture of composite aero parts, structures, and microelectronics). The improvements that are being identified for the high-performance structures and payloads are related with the type of property that is expected to be modified: primarily mechanical, thermal, and electrical properties. In reality, the high-performance structures and payloads are multifunctional elements with various design drivers that cannot be separated because of the limited vehicular size and weight. For instance, to demonstrate potential airframes' applications, O'Donnell *et al.* [4] conducted a mass analysis study on a CNT-reinforced polymer structured aircraft. The analysis presented considered notional Boeing 747-400 and 757-200, Airbus A320, and Embraer E145 with CNT-reinforced polymer as the primary structural material, replacing the entire volume of structural aluminum with CNT-reinforced polymer, without including any modifications to the geometry or design of the airframe. Each airframe modeled saw an average of 17.32% weight reduction in the low initial takeoff mass category, with a minimum mass reduction of over 10% in the high initial takeoff mass category. The average fuel savings for all CNT-reinforced polymer-structured airframes was 9.8%. Although the probability of CNT-reinforced polymer-structured contemporary airframes is unlikely, this type of analysis provides insight into a small group of benefits seen by a nanostructured material applied on the macroscale.

At present, CNTs, carbon nanofibers, layered silicates, and polyhedral oligomeric silsesquioxanes (POSS) are presently being used commercially in nanocomposites [5, 6]. Nanoclays, however, are the most dominant commercial nanomaterials, accounting for nearly 70% of the total volume of nanomaterials commercially used. In particular, the design concept of nanostructured nanoclay materials has gained widespread importance in automotive industry mainly because of low cost and availability, as compared to other nanomaterials such as POSS and CNTs. Notably, automotive and packaging market segments are presently to account for nearly 80% of total nanocomposite consumption [7]. The use of nanomaterials in current and future transport structures will allow environmentally friendly materials with higher performance and low manufacture costs, as discussed in the following sections.

10.2
Fiber-Reinforced Nanocomposites

10.2.1
Natural Fiber-Reinforced Nanocomposites

The incorporation of nanoclays and natural fibers in resin systems thus provides reinforcements to resin systems at two scales. The nanoclay enhances the bio-based polymer system in stiffness and hygro-thermal properties, while the natural fibers provide the main stiffness and strength. In addition, the enhanced barrier properties of the nano-reinforced resin retard moisture from reaching the natural fibers and thereby providing a synergetic effect between scales for an efficient bio-based composite. Hybrid bio-based composites that exploit the synergy between natural fibers (industrial hemp) in a nano-reinforced bio-based polymer can lead to improved properties while maintaining environmental appeal. Bio-based resins obtained by partial substitution of unsaturated polyester (UPE) with epoxidized soybean oil (ESO) increase toughness but compromise stiffness and hygro-thermal properties. Reinforcement of the bio-based resin with nanoclays permits to retain stiffness without sacrificing toughness, while also improving barrier and thermal properties.

Longkullabutra *et al.* [8] recently reported an improvement in the tensile strength of epoxy resin and hemp/epoxy resin composites using CNTs. The CNTs adding nanopowder were vibrated via the vibration milling technique for 6–48 h. Different volume percentages of CNTs were dispersed for hemp/epoxy resin composites. The results indicate that adding the milled CNTs can improve tensile properties of composites. Elsewhere, Liu and Erhan [9] fabricated epoxidized soybean oil (ESO)-based composites reinforced with flax fibers and nanocomposites reinforced with organoclay. The flexural modulus and tensile modulus of the developed composites increased proportionally with the amount of epoxy resin, 1,1,1-tris(p-hydroxyphenyl)ethane triglycidyl ether (THPE-GE). The flexural modulus also increased with fiber contents lower than 10 wt%, but showed a decrease beyond 10 wt%. The tensile modulus increases with fiber content until a maximum, at 13.5 wt%, and then it decreases. X-ray diffraction (XRD) and by transmission electron microscopy (TEM) data indicated that the organophilic clay was well dispersed in the matrix. An intercalated structure of the composite is developed. Dynamic mechanical study shows the ESO/clay nanocomposites with 5–10 wt% clay content possess storage modulus ranging from 2.0 to 2.70 MPa at 30 °C. As T_g, about 20 °C was measured from a dynamic mechanical study. The mechanical study predicates these materials to be promising as an alternative to petrochemical polymers.

In another example, Faruk and Matuana's [10] investigations aimed at identifying the best approach to incorporate nanoclay into wood–plastic composites (WPCs) to enhance their mechanical properties. Two different methods of introducing nanoclays into HDPE-based WPCs were examined. The first method involved the reinforcement of HDPE matrix with nanoclay, which was then used as a matrix in the manufacture of the (melt blending process). The second method consisted of a direct addition of nanoclay into HDPE/wood–flour composites during conventional dry compounding (direct dry blending process). The melt blending process, in which nanoclay/HDPE nanocomposite was used as matrix, appeared to be the best approach to incorporate nanoclay in WPCs. The experimental results indicated that the mechanical properties of HDPE/wood–flour composites could be significantly improved with an appropriate combination of the coupling agent content and nanoclay type in the composites.

CaCO$_3$/wood cellulose nanocomposite materials have been prepared by controlled generation of carbonate in aqueous solution from an organic precursor (DMC, dimethyl carbonate) in the presence of wood cellulose fibers and CaCl$_2$ [11]. The work demonstrated that the quantity and morphology of CaCO$_3$ particles deposited at the surface of cellulose fibers were strongly influenced by the hydrolysis conditions. The amount and size of CaCO$_3$ deposited on the cellulose fibers increased with increasing reaction time. Besides, from the reactions performed at room temperature originated nano-sized CaCO$_3$ particles with spheroid morphology, while at 70 °C micrometric aggregates of Ca(OH)$_2$ were obtained. Additionally, lower mass fractions of fibers in the reacting suspensions favored the formation of spheroid particles of CaCO$_3$. Finally, the presence of carboxyl groups at the substrate surface increased the selectivity of precipitation of CaCO$_3$ particles on the surface of the fibers. It is acknowledged that the mechanism by which CaCO$_3$ particles are retained at the surface of cellulosic fibers is not understood yet. Nevertheless, CaCO$_3$/cellulose nanocomposites might be considered as potential reinforcing fillers in PE matrix-based composites as preliminary dynamic mechanical analysis (DMA) studies demonstrated that PE composites with CaCO$_3$/cellulose fibers showed a much higher mechanical performance as shown on Figure 10.1.

It follows that silane coupling chemicals present three main advantages: (i) they are commercially available in large quantities; (ii) at one end, they bear alkoxysilane groups capable of reacting with OH-rich surface, and (iii) at the

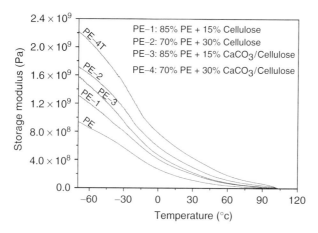

Figure 10.1 The dynamic storage modulus curves of PE and PE-based composites [11].

other end, they have a large number of functional groups which can be tailored as a function of the matrix to be used [12]. The last feature ensures, at least, a good compatibility between the reinforcing element and the polymer matrix or even good covalent bonds between the functional groups. The reaction of silane coupling agents with lignocellulose fibers (mainly cellulose and lignin) was found to be quite different compared with that observed between the coupling agents and glass surface, in the sense that with cellulose macromolecules only prehydrolyzed silanes underwent the reaction with cellulose surface. Research work by Abdelmouleh *et al.* [13] clearly shows that cellulose fibers can be effectively used as reinforcing elements in thermoplastic low-density polyethylene and natural rubber matrices. Cellulose fibers were incorporated into the matrices, as such or after chemical surface modification involving three silane coupling agents, namely γ-methacryloxypropyltrimethoxy-silane (MPS), γ-mercaptoproyltrimethoxy-silane (MRPS), and hexadecyltrimethoxy-silane (HDS). As expected, the mechanical properties of the composites increased with increasing the average fiber length and the composite materials prepared using both matrices and cellulose fibers treated with MPS and MRPS displayed good mechanical performances. On the other hand with HDS bearing merely an aliphatic chain, only a modest enhancement of composite properties was observed which was imputed to the incapacity of HDS to bring about covalent bonding with matrix.

In a different study, characterization of different hybrid composites have verified this synergistic behavior in which systems with 10% EMS and 1.5 wt% nanoclay retained the original stiffness, strain to failure (Figure 10.2), and hygro-thermal properties of the original resin while improving toughness [14]. Optimum designs that maximize the synergy of the constituents are thus possible and the presented

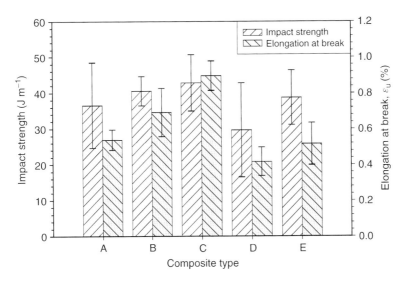

Figure 10.2 Impact strengths from notched Izod tests and tensile strains at failure [14].

results provide an initial benchmark to identify such balance, and thus increase the potential applications of bio-based composites.

In conclusion, the environmental concerns linked to the use of halogen-based fire retardants open new routes to develop fire-resistant composites. The elaboration of polymer–silicate clay nanocomposites has already addressed these fireproof issues. Indeed, the nanoscale dispersion of clay layers forms a passive barrier, a sort of inorganic labyrinth, which hinders the out-diffusion of volatile decomposition products from thermal cracking. For instance, studies by Guigo *et al.* [15] on nanocomposites composed of isolated lignin, natural fibers, and natural additives are elaborated through conventional thermoplastic processes such as injection molding. The obtained composites have wood-like mechanical behavior and some properties lie in the range of those of polyamides (PAs). For certain practical applications such as automotives, the thermal stability and fire resistance of the nanocomposites need further improvement.

10.2.2
Polyamide/Layered Silicates Nanocomposites

The electrospun nanocomposite fibers have great potential for application where both high surface-to-volume ratio and strong mechanical properties are required, such as the high-performance filters and fiber reinforcement materials [16]. As the mechanical properties of fibers, in general, improve substantially with decreasing fiber diameter, there is considerable interest in the development of continuous electrospun polymer nanofibers. In this respect, Lincoln *et al.* [17] reported that the degree of crystallinity of PA-6 annealed at 205 °C increased substantially with the addition of MMT. This implied that the silicate layers could act as nucleating agents and/or growth accelerators. In contrast, the study of Fong *et al.* [18] showed a very similar overall degree of crystallinity for electrospun PA-6 and PA-6/Cloisite-30B nanocomposite fibers containing 7.5 wt% of organically-modified montmorillonite (OMMT) layers.

Fornes and Paul [19] have found that OMMT layers could serve as nucleating agents at 3% concentration in PA-6/OMMT nanocomposite, but retarded the crystallization of PA-6 at a higher concentration of around 7%. In addition, the differences in the molecular weight of PA-6 and the solvent used for electrospinning were also expected to have different impacts on the mobility of PA-6 molecular chains and the interactions between the PA-6 chains and OMMT layers, which may also affect the crystallization behavior of PA-6 molecules during the electrospinning. Li *et al.* [20] manufactured PA-6 fibers and nanocomposite fibers with average diameters of around 100 nm by electrospinning using 88% aqueous formic acid as the solvent. The addition of OMMT layers in the PA-6 solution increased the solution viscosity significantly and changed the resulting fiber morphology and sizes. TEM images of the nanocomposite fibers and ultra-thin fiber sections and the wide-angle X-ray diffraction (WAXD) results showed that OMMT layers were well exfoliated inside the nanocomposite fibers and oriented along the fiber axial direction. The degree of crystallinity and crystallite size were both increased for the nanocomposite fibers and more significantly for the fibers electrospun from 15% nanocomposite solution, which exhibited the finest average fiber size. As a result, the tensile properties of electrospun nanocomposites were greatly improved. The Young's modulus and ultimate strength of electrospun nanocomposite fibrous mats were improved up to 70 and 30%, respectively, when compared with PA-6 electrospun mats. However, the ultimate strength of the nanocomposite fibrous mats electrospun from 20% nanocomposites solution was decreased by about 20% because of their larger fiber sizes. The Young's modulus of PA-6 electrospun single fibers with a diameter of around 80 nm was almost double the highest value that had been reported for the conventional PA-6 fibers and could be improved by about 100% for the electrospun nanocomposite single fibers of similar diameters.

In another interesting study [21], a range of polymer matrices were examined including polyvinyl alcohol, poly(9-vinyl carbazole), and PA. To compare production methods, polymer composite films and fibers were produced. It was found that by adding various mass fractions of nanofillers, both the Young's modulus and hardness increased significantly for both films and fibers. In addition, the thermal behavior was seen to be strongly dependent on the nanofillers that were added to the polymer matrices. Wu *et al.* [22] prepared carbon fiber and glass fiber reinforced PA-6 and PA-6/clay nanocomposites. The fabrication method involved first mechanically mixing PA-6 and PA-6/clay with E-glass short fiber (6 mm long) and carbon fiber (6 mm long), separately. A twin-screw extruder at a rotational speed of 20 rpm extruded the fibers. The temperature profiles of the barrel were 190–210–230–220 °C from the hopper to the die. The extrudate was pelletized, dried, and injection molded into standard test samples for mechanical properties test. The injection-molding temperature and pressure were 230 °C and 13.5 MPa, respectively. The study found that the tensile strength of PA-6/clay containing 30 wt% glass fibers was 11% higher than that of PA-6 containing 30 wt% glass fiber, while the tensile modulus of the nanocomposite increased by 42%. Flexural strength and flexural modulus of neat PA-6/clay was found to be similar to PA-6 reinforced with 20 wt% glass fibers. It was concluded that the effect of nanoscale clay on toughness was more significant than that of the fiber. Heat distortion temperatures of PA-6/clay and PA-6 were 112 and 62 °C, respectively. Consequently, the heat distortion temperature of fiber-reinforced PA-6/clay system was almost 20 °C higher than that of fiber-reinforced PA-6 system. Notched Izod impact strength of the composites decreased with the addition of the fiber. The scanning electron microscopy (SEM) microphotographs showed that the wet-out of glass fiber was better than carbon fiber. The study concluded that the mechanical and thermal properties of the PA-6/clay nanocomposites were superior to those of PA-6 composites in terms of the heat distortion temperature, tensile and flexural strength, and modulus without sacrificing their impact strength. This was attributed to the nanoscale effects, and the strong interaction force existed between the PA-6 matrix and the clay interface.

In case of short fibers, Akkapeddi [23] prepared PA 6-nanocomposites using chopped glass fibers. In a typical experiment, a commercial grade PA-6 of MW = 30 kg mol^{-1} and specially designed functional organo-quaternary ammonium–clay complexes (organoclays) based on MMT or hectorite type clays were used. Freshly dried PA-6 (moisture < 0.05%) was blended with 3–5 wt% of a selected organoclay powder and extruded at 260 °C in a single step, under high shear mixing conditions. Alternatively, the organoclay was master-batched first into PA-6 (at 25 wt% loading) and then re-extruded in a second step with more PA-6 to dilute the clay content to ≤ 5 wt%. Conventional chopped glass fiber with 10 μm diameter and about 3 mm length was then added, as an optional reinforcement through a downstream feed port at zone 6 of the twin screw extruder. The glass fiber was compounded with the molten, premixed PA-6 nanocomposite either as a one-step extrusion process or in a second extrusion step. The extrudate was quenched in a water bath and pelletized. The pellets were dried under vacuum at 85 °C, and injection molded into standard ASTM test specimens. As shown in Figure 10.3, significant improvements in modulus were achievable in both the dry and the moisture conditioned state for PA-6 nanocomposites compared to standard PA-6, at any given level of glass fiber reinforcement.

In particular, a small amount (3–4 wt%) of nanometer-scale dispersed layered silicate was capable of replacing up to 40 wt% of a standard mineral filler or 10–15 wt% of glass fiber to give equivalent stiffness at a lower density. In addition, improved moisture resistance, permeation barrier, and fast crystallization/mold cycle time contribute to the usefulness of such composites.

Vlasveld *et al.* [24] developed a three-phase thermoplastic composite, consisting of a main reinforcing phase of woven glass or carbon fibers and a PA-6 nanocomposite matrix. The nanocomposite used in this research had moduli that were much higher than unfilled PA-6, also above T_g and in moisture conditioned samples. Flexural

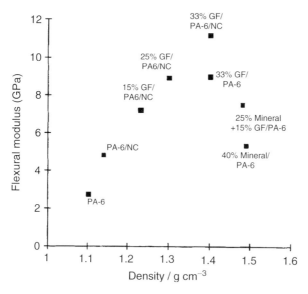

Figure 10.3 Modulus versus density of glass fiber (GF), PA-6/nanoclay (PA-6/NC) versus PA-6 molding resins [23].

tests on commercial PA-6 fiber composites showed decrease of the flexural strength upon increasing temperature. The researchers claimed that the strength of glass fiber composite can be increased by more than 40% at elevated temperatures and the temperature range at which a certain minimum strength is present can be increased by 40–50 °C. Carbon fiber composites also showed significant improvements at elevated temperatures, although not at room temperature. On the basis of flexural tests on PA-6-based glass and carbon fiber composites over a large temperature range up to near the melting point, it became clear that for these fiber composites it is important to have a reasonably high matrix modulus: Both glass and carbon composites were very sensitive to a decrease of the matrix modulus below values around 1 GPa. At higher moduli, carbon fiber composites are more sensitive to the matrix modulus than glass fiber composites. The modulus of unfilled PA-6 decreased below the (arbitrary) 1 GPa level just above T_g, it is noteworthy that the nanocomposites used in this research had moduli that were much higher and stayed above the 1 GPa level up to 160 °C, which was more than 80 °C higher than for unfilled PA-6. The nanocomposites also showed much higher moduli in moisture conditioned samples, and even in moisture conditioned samples tested at 80 °C the modulus was much higher than that of the dry unfilled PA-6, again well above 1 GPa. DMA measurements showed that the nanocomposites did not show a change of T_g, and that the reduction of the modulus upon absorption of moisture was due to the T_g decrease.

The mechanical properties of APA-6 and HPA-6 (Akulon® K222D, low MW injection-molding grade hydrolytically polymerized PA-6) nanocomposites were compared with injection molded neat HPA-6. As expected, the HPA-6 nanocomposite had the highest modulus over the entire range of temperatures (20–160 °C) and moisture contents (0–10 wt%) tested. However, APA-6 came close and had the highest maximum strength because of its characteristic crystal morphology that was directly linked to the reactive type of processing used. This same morphology, it was claimed, also made APA-6 slightly less ductile compared to melt processed HPA-6. Compared to the melt processed HPA-6, APA-6 polymerized at 150 °C and the HPA-6 nanocomposite offered a higher modulus at similar temperature, or similar modulus at a higher temperature (40–80 °C increase). It is noteworthy that such an increase in maximum use temperature, related to the heat distortion temperature, can seriously expand the application field of PA-6 and PA-6 composites. For all PAs, temperature and moisture absorption reduced the modulus and the strength and increased the maximum strain that was directly related to the glass transition temperature. While with increasing testing temperature at a certain moment, the T_g of the dry polymer was exceeded and moisture absorption reduced the T_g at a certain point below the testing temperature. However, the effect of both was in essence the same. Retention of mechanical properties of APA-6 after conditioning at 70 °C for 500 h and subsequent drying was demonstrated. Conditioning submersed in water at the same temperature, however, resulted in a brittle material with surface cracks, as is common to most PAs. Continued crystallization and removal of unreacted monomer caused this behavior. Given the fact that submersion at elevated temperatures is usually not an environment in which PA-6 and its composites are applied, the encountered property reduction was therefore not detrimental for application of these materials. The overall conclusion of the comparative study for application of the PAs as matrix material in fiber composites was that both APA-6 and the HPA-6 nanocomposites outperformed the melt processed HPA-6 in terms of modulus and maximum strength. Therefore, the researchers concluded that both "improved" PAs may be expected to enhance the matrix dominated composite properties such as compressive and flexural strength, provided that a strong fiber-to-matrix interphase is obtained.

Another comparative study was conducted by Sandler *et al.* [25] on melt spun PA-12 fibers reinforced with CNTs and nanofibers. A range of MWNT and carbon nanofibers were mixed with a PA-12 matrix using a twin-screw microextruder, and the resulting blends spun to produce a series of reinforced polymer fibers. The work aimed to compare the dispersion and resulting mechanical properties achieved for nanotubes produced by the electric arc and a variety of chemical vapor deposition techniques. A high quality of dispersion was achieved for all the catalytically grown materials and the greatest improvements in stiffness were observed using aligned, substrate-grown, CNTs. The use of entangled MWNT led to the most pronounced increase in yield stress, most likely as a result of increased constraint of the polymer matrix due to their relatively high surface area. The degrees of polymer and nanofiller alignment and the morphology of the polymer matrix were assessed using XRD and differential scanning calorimetry (DSC). The CNTs were found to act as nucleation sites under slow cooling conditions, the effect of scaling with effective surface area. Nevertheless, no significant variations in polymer morphology as a function of nanoscale filler type and loading fraction were observed under the melt spinning conditions applied. A simple rule-of-mixture evaluation of the nanocomposite stiffness revealed a higher effective modulus for the MWNT compared to the carbon nanofibers, as a result of improved graphitic crystallinity. In addition, this approach allowed a general comparison of the effective nanotube modulus with those of nanoclays, as well as common short glass and carbon fiber fillers in melt-blended PA composites. The experimental results further highlighted the fact that the intrinsic crystalline qualities, as well as the straightness of the embedded nanotubes, were significant factors influencing the reinforcement capability.

10.2.3
Epoxy/Layered Silicates Nanocomposites

In the early 1990s, Toyota Research Group synthesized PA-6-based clay nanocomposites that demonstrated the first use of nanoclays as reinforcement of polymer systems [26]. They concluded that nanoclays not only influenced the crystallization process, but that they were also responsible for morphological changes. Recognizing these benefits, many researchers, using a variety of clays and polymeric matrices, have produced nanocomposites with improved properties [27]. Haque *et al.* [28], using a similar manufacturing process as Chowdhury *et al.* [30] (i.e., vacuum-assisted resin infusion method (VARIM)), showed large improvement of the mechanical properties of their S2-glass fiber laminates and at very low layered silicate content. They showed that by dispersing 1 wt% nanosilicates, S2-glass/epoxy-clay nanocomposites exhibited an improvement of 44, 24, and 23% in interlaminar shear strength, flexural strength, and fracture toughness, respectively. Similarly, the nanocomposites exhibit approximately 26 °C higher decomposition temperatures than that of conventional composites. The increased properties at low loading were associated to several factors: enhanced matrix properties due to lamellar structures, synergistic interaction between the matrix, clay, and fibers, and enhanced matrix–fiber adhesion promoted by the clay.

The clays were also presumed to decrease the coefficient of thermal expansion mismatch, significantly reducing residual stresses, and leading to higher quality laminates [29]. An increased interfacial bonding, matrix agglomeration, and coarse morphology were observed from the fractured surface of low-loading nanocomposites. The degradation of properties at higher clay loadings was believed to be caused by phase-separated structures and also by defects in the cross-linked structures. However, the authors acknowledged that further work is necessary in order to achieve fully exfoliated structure in clay–epoxy nanocomposites.

Similarly, Chowdhury et al. [30] employed VARIM process to manufacture woven carbon FRP matrix composites. They investigated the effects of nanoclay particles on flexural and thermal properties. Different weight percentages of a surface modified MMT mineral were dispersed in SC-15 epoxy using sonication route. The nanophased epoxy was then used to manufacture 6000 fiber tow-plain weave carbon/epoxy nanocomposites using VARIM technique. Flexural test results of thermally postcured samples indicated a maximum improvement in strength and modulus of about 14 and 9%, respectively. DMAs results of thermally postcured samples showed a maximum improvement in storage modulus by about 52% and an increase in glass transition temperature of about 13 °C. A 2 wt% nanoclay seems to be an optimum loading for carbon/SC-15 epoxy composites in terms of mechanical and thermal properties. Microstructural studies revealed that nanoclay promotes good adhesion of fiber and matrix, thereby increasing the mechanical properties.

Lin et al. [31] successfully prepared layered silicate/glass fiber/epoxy hybrid composites using a vacuum-assisted resin transfer molding (VARTM) process. Figure 10.4 shows a schematic of the experimental set-up for the closed-mold VARTM process.

They focused on such issues by selecting clay and short length glass fibers to reinforce an epoxy resin. To study the effects of the fiber direction on the clay distribution in the hybrid composites, unidirectional glass fibers were placed in two directions: parallel and perpendicular to the resin flow direction. The intercalation behavior of the clay and the morphology of the composites were investigated using XRD and TEM. The complementary use of XRD and TEM techniques revealed an intercalated clay structure in the composites. Dispersion of clay in the composites was also observed using SEM; the observed clays were dispersed between both the bundles of glass fibers and within the interstices of the fiber filaments. The mechanical properties of the ternary composites were also

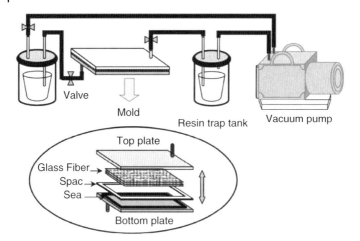

Figure 10.4 Schematic of the preparation of nanoclay/glass fiber/epoxy composites using VARTM [31].

evaluated. The results indicated that introducing a small amount of organoclay to the glass fiber/epoxy composites enhanced their mechanical and thermal properties, confirming the synergistic effects of glass fibers and clays in the composites. Elsewhere, Aktas *et al.* [32] developed a novel approach for characterization of nanoclay dispersion in polymeric composites using electron microprobe analysis (EMPA). Dispersion analysis was performed on three sets of center-gated discs fabricated by resin-transfer-molding (RTM). The first set was neat epoxy polymer without reinforcement, whereas the second set comprised 17 vol% randomly-oriented chopped glass fiber preforms. The last set, in addition to the glass fiber reinforcement, contained 1.7 wt% Cloisite 25A nanoclay. On completion of curing, a sample along the radius of a nanoclay reinforced disc was analyzed on an electron microprobe analyzer. The results from scanning electron micrographs indicated that nanoclay exists in clusters of various sizes ranging from over 10 µm down to submicrometer scale. Nanoclay clusters larger than 1.5 µm were analyzed by digital image processing on the scanning electron micrographs taken along the part's radius. The dispersion of nanoclay smaller than 1.5 µm was quantified by compositional analysis via wavelength dispersive spectrometry (WDS). Distribution of nanoclay clusters larger than 1.5 µm was found to be approximately constant along the radius with an average value of 1.4% by volume. Similarly, nanoclay clusters smaller than 1.5 µm were found to be distributed evenly with an average value of 0.41 wt%. In addition, the glass transition temperature improved by 11% with the addition of nanoclay.

Gilbert *et al.* [33, 34] and Timmerman *et al.* [35] demonstrated that fracture toughness and mechanical properties are increased by incorporation of metal and inorganic particles. In these studies, they have been developing the concept of La PolynanoGrESS (layered polynanomeric graphite epoxy scaled system) that utilizes the nanoparticle effect in an epoxy matrix and scales to a continuous carbon-fiber-reinforced composites systems. Typically, Timmerman *et al.* [35] modified the matrices of carbon fiber/epoxy composites with layered inorganic clays and a traditional filler to determine the effects of particle reinforcement, both on micro and nanoscale, on the response of these materials to cryogenic cycling. The mechanical properties of the laminates studied were not significantly altered through nanoclay modification of the matrix. The incorporation of nanoclay reinforcement in the proper concentration resulted in laminates with microcrack densities lower than those seen in the unmodified or macro-reinforced materials as a response to cryogenic cycling. Lower nanoclay concentrations resulted in a relatively insignificant reduction in microcracking and higher concentrations displayed a traditional filler effect. In another development, Brunner *et al.* [36] exploited the work of Timmerman *et al.* [35] on the use of epoxy with a relatively small amount of nano-size filler as matrix in fiber-reinforced laminates. They [36] focused on investigating whether a nano-modified epoxy matrix yields improved delamination resistance in a fiber-reinforced laminate compared to a laminate with neat epoxy as matrix material. To start with, neat and nano-modified epoxy specimens without fiber reinforcement were prepared for a comparison of the fracture toughness of the matrix material itself. Additional properties of the neat and nano-modified epoxy were also determined (partly taken from Timmerman *et al.* [35]) and compared. The study reported fracture toughness improvement up to about 50% and an increase in energy release rates by about 20% for an addition of 10 wt% of organosilicate clay.

10.2.4
Epoxy/CNT Nanocomposites

Nanoscaled fillers such as CNTs and carbon nanofibers (CNFs) offer new possibilities toward low-weight composites of extraordinary mechanical, electrical, and thermal properties. Taking into consideration their high axial Young's modulus, high aspect ratio, large surface area, and excellent thermal and electrical properties, these fillers can be used as modifiers for the polymer matrices of the FRP composites leading to advanced mechanical behavior. The size scale, high aspect ratio, low density, and other exceptional properties of nanotubes are generally advantageous when they are applied in a variety of applications. However, in the case of nanotube-reinforced polymer composites, there has only been a moderate strength enhancement that is significantly below the theoretically predicted potential. To achieve the full reinforcing potential of nanotubes, there remain two critical issues that have to be firstly solved: the dispersion of nanotubes in a polymer matrix and the interfacial bonding between the nanotubes and the polymer matrix.

Nevertheless, very recently, based on the scaling argument by correlating the radius (r), fiber strength (σ), and interface strength (τ) with the energy absorbed per unit cross-sectional area by fiber pull-out (i.e., $G_{\text{pull-out}} \sim r\sigma^2/\tau$), it was shown that the improvements in toughness in polymer/CNT nanocomposites cannot be attributed to nanotube pull-out mechanism as the pull-out energy significantly decreases when the fiber radius is scaled down to nanoscale. In line with this argument, there are many studies that reported reductions in toughness with the incorporation of CNTs, even at low loadings. Besides, with other nanoscale fillers, it is realized that conventional toughening mechanisms cannot be transferred to polymer nanocomposites directly.

In general, weakly interacted nanotube bundles and aggregation of nanotubes would result in a poor dispersion state that significantly reduces the aspect ratio of the reinforcement. The reason for the weak interfacial bonding behavior lies in the atomically smooth, nonreactive surface of the nanotubes that cannot ensure efficient load transfer ability from the polymer matrix to the nanotube lattice. To solve this problem, a number of methods have been developed to maximize the benefits of nanotubes in polymer composites, that is, surfactant assisted dispersion [37], sonication with high power [38], *in situ* polymerization [39], electric field or magnetic-induced alignment of nanotubes [40, 41], plasma polymerization [42], and surface modification such as inorganic coating [43], polymer wrapping [44], as well as protein functionalization [45].

One key area where nanocomposites can make a significant impact is in addressing interlaminar toughness in the fiber-reinforced composites. Interlaminar toughness improvement of fiber-reinforced composites has been in the research focus for a considerable time, as it is directly related to the dynamic as well as the damage tolerance performance of the composite. The problem has been addressed in various ways: stitching, Z-pinning, or interleaving with a notable increase in toughness while also providing improvement in mechanical properties, such as fatigue life. Other approaches focus on tailoring the matrix or interface properties in order to provide the necessary interlaminar fracture toughness. Notably, matrix toughening may be performed through chemical modification or, more recently with the incorporation of fillers in the matrix material. Interface modification can also be performed by grafting in order to tailor the chemical compatibility between the fibers and the matrix.

Gojny *et al.* [46] investigated on interlaminar shear strength of nano-reinforced FRPs and also reported about an efficient technique (mini-calendering) to disperse carbon-based nanoparticles in epoxy resins. The application of a mini-calender to disperse CNTs (and carbon black) proved to be an efficient approach to reach a good state of dispersion and enabled to manufacture high volumes of nanocomposites. (This method is an established and common technique to disperse microparticles in different matrices, for example, color pigments for cosmetics or lacquers.) A major advantage of the calendering method is, besides the improved dispersion results, the efficiency in manufacturing of larger amounts of nanocomposites. The produced nanotube/epoxy composites exhibit a significant increase in fracture toughness, as well as an enhancement of stiffness already at low nanotube contents. Later on, Gojny *et al.* [47] investigated the influence of CNTs on the interlaminar shear strength of a glass-fiber-reinforced polymer composite (GFRP). They reported an increase of +19% in interlaminar shear strength with a weight fraction as low as 0.3 wt% of amino-functionalized double wall carbon nanotube (DWCNT-NH$_2$) in the epoxy matrix, Figure 10.5.

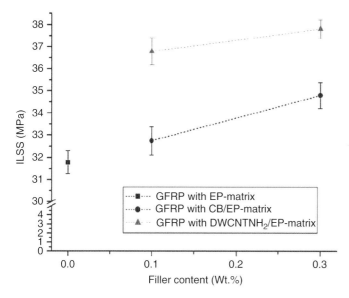

Figure 10.5 Interlaminar shear strength (ILSS) of the (nano-reinforced) GFRPs (EP – epoxy, CB – carbon black, and DWCNT – double wall carbon nanotubes) [47].

It was claimed that the nanometer size of the particles enabled their application as modifiers in fiber-reinforced polymers. The composites were produced via the RTM process and the particles were not filtered by the glass-fiber bundles. A follow-up work by the same research team reported that the interlaminar shear strengths of the nanoparticle-modified GFRP were significantly improved (+16%) by adding only 0.3 wt% of CNTs [48]. The interlaminar toughness (G_{Ic} and G_{IIc}) was not affected in a comparable manner. The laminates containing CNTs exhibited a relatively high electrical conductivity at very low filler contents.

Zhao *et al.* [49] fabricated CNTs and continuous carbon fiber (T300) reinforced unidirectional epoxy resin matrix composites. They prepared CNTs by catalytic decompose of benzene using floating transition method at 1100–1200 °C. Benzene was used as carbon source and ferrocene as catalyst with thiophene. The CNTs used were straight with diameter 20–50 nm, internal diameter 10–30 nm, and length 50–1000 μm. The volume fraction of continuous carbon fiber (first filler) in the composites without second filler (CNT) was 60%. The flexural strength of the composites reached the maximum value of 1780 MPa when the weight percent of CNT in epoxy resin matrix was only 3%. The study concluded that flexural strength and modulus of the composites increased firstly and then decreased with the increasing of CNT contents in epoxy resin matrix.

Hsiao *et al.* [50] and Meguid and Sun [51] investigated the tensile and shear strength of nanotube-reinforced composite interfaces by single shear-lap testing. They observed a significant increase in the interfacial shear strength for epoxies with contents between 1 and 5 wt% of MWNT when compared to the neat epoxy matrix. In particular, instead of processing and characterizing CNT/polymer

(a)

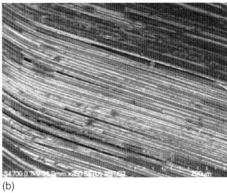

(b)

Figure 10.6 SEM picture (a) of the fracture surface of the bonding area of the 5 wt% MCNT + epoxy case; failure at the graphite fiber of the adherends was observed. SEM picture (b) of the fracture surface of the bonding area of the epoxy-only case; failure occurred at the epoxy surface of the adherends and no significant graphite fiber fracture was observed [50].

composites, Hsiao *et al.* [50] explored the potential of CNT to reinforce the adhesives in joining two composite structures. In the study, different weight fractions of MWNT were dispersed in epoxy to produce toughened adhesives. The reinforced adhesives were used to bond the graphite fiber/epoxy composite adherends. This experimental study showed that adding 5 wt% MWNT in the epoxy adhesive effectively transferred the shear load from the adhesive to the graphite fiber system in the composite laminates and improved the average shear strength of the adhesion by 46% (±6%). Significant enhancement of the bonding performance was observed as the weight fraction of CNTs was increased. As shown in Figure 10.6a, the 5 wt% MWNT effectively transferred the load to the graphite fibers in the adherends and the failure was in the graphite fiber system. On the other hand, for epoxy adhesives containing no MWNT (Figure 10.6b), the failure occurred at the epoxy along the bonding interface and no significant graphite fiber fracture was observed.

Despite the promising results, the researchers concurred that further experiments involving increasing MWNT weight fractions and more detailed SEM observations are required in order to understand and model the role of the MWNT in enhancing adhesion.

Various studies can be found in the literature regarding the incorporation of CNFs in polymeric matrices and the final mechanical and/or electrical properties of these materials. As in all cases where nano-sized fillers are involved, the development of high performance CNF/polymer composites requires homogeneous dispersion of CNFs in the polymeric matrix which is crucial to the composite performance. The quality of the stress transfer between the nanofibers and the matrix material play an important role in the composite properties interface quality in order to achieve efficient load transfer from the matrix to the CNFs. Early studies by Hussain *et al.* [52] reported that matrix reinforcement with nanowhiskers can damage the fibers in composite materials. As such, he incorporated micro and nanoscale Al_2O_3 particles in filament-wound carbon fiber/epoxy composites. He observed an increase in modulus, flexural strength, interlaminar shear strength, and fracture toughness when the matrix was filled at 10 vol% with alumina particles (25 nm diameter). This effect stemmed largely from the large surface area of the filler and the ability of the particles to mechanically interlock with the fibers. Hybrid reinforced composites consisting of two or more different types of reinforcing fibers have also been studied in the polymer matrix composite systems. It was also reported that hybridization by incorporating whiskers into the matrix causes fiber damages resulting in a decrease in ultimate strength. However, the work claimed that the incorporation of rigid spherical filler, especially fine or nano-sized filler, did not cause serious damages to the fiber surfaces.

10.3
Sandwich Structures

Sandwich composites are used in a wide range of applications from aircraft, ships, ballistic vests, and helmets through to racing car and high-end sports cars to provide a range of functions including structural stiffness, crash energy management, heat shielding, and many others. These structures, composed of a core of cellular material and outer composite skins, are lightweight and yet offer high resistive stiffness against traction, compressive, and bending loads. These properties are utilized to produce functional structures that must sustain high stresses under normal conditions. During severe impact loads in automotive applications, for example, these structures must dissipate impact energy to protect either the rest of the structure or the vehicle occupants.

Research has shown that damage initiation thresholds and damage size in sandwich composites depend primarily on the properties of the core materials and facings and the relationship between them. Much of the early research on sandwich composites under impact focused on honeycomb core (Nomex, glass thermoplastic, or glass-phenolic) sandwich constructions. A key problem in honeycomb sandwich construction is the low core surface area for bonding. Consequently, expanded foams (often thermoset) are now preferred to achieve reasonably high thermal tolerance, although thermoplastic foams are also used. In turn, the response of foam core sandwich constructions to impact loading has been studied by many researchers. Accordingly, it is now well understood that the response of foam core sandwich composites strongly depends on the density and the modulus of the foam.

A possible way of improving the foam density and modulus properties of foam materials is through the inclusion of small amounts of nanoparticles (CNTs and nanofibers, TiO_2, nanoclay, etc.). Up to now, MMT nanoclays have been the best candidate for foam reinforcement because of ease in processing, enhanced thermal–mechanical properties, wide availability, and cost. Likewise, polyurethanes (PUs) are core materials of choice because of their tailorable and versatile physical properties, ease of manufacture, and their low costs. The use of polyurethanes filled with nanoparticles to construct either laminates or foams is relatively new. Moreover, the use of nanoparticles in such laminates, or foams in sandwich composite construction, is in its infancy but has been found to be both realistic and beneficial. For instance, by using <5% by weight of nanoclay loadings, significant improvement in foam failure strength and energy absorption has been realized with over a 50% increase in the impact load carrying capacity when compared to a neat foam sandwich. However, as most current research concentrates on the processing and characterization of nanophased foams and evaluation of static properties only, dynamic materials data on impact failure mechanisms and impact property relations is missing. For the application of nanophased foams in sandwich constructions for ballistic resistance, a proper understanding of their impact behavior at both high and low-velocity impact is required.

Therefore, by taking advantage of the emerging new materials, nanophased sandwich structures have been fabricated and tested for low-velocity impact resistance in the literature. In a recent development, Njuguna *et al.* [53] fabricated and characterized a series of nanophased hybrid sandwich composites based on PU/MMT. Polyaddition reaction of the polyol premix with 4,4′-diphenylmethane diisocyanate was applied to obtain nanophased polyurethane foams that were then used for fabrication of sandwich panels. It has been found that the incorporation of MMT resulted in higher number of PU cells with smaller dimensions and higher anisotropy index. The obtained materials exhibited improved parameters in terms of thermal insulation properties. Importantly, these foams can also be selectively stiffened to meet specific requirements. The results also showed that nanophased sandwich structures are capable of withstanding higher peak loads than those made of neat polyurethane foam cores when subject to low-velocity impact despite their lower density than that of neat PU foams. This is especially significant for multi-impact recurrences within the threshold loads and energies studied as shown on Figure 10.7. A feasible application for these lightweight structures is in energy absorbing structures or as inserts in hollow structures.

Subramaniyan and Sun [54] reported that core shell rubber (CSR) nanoparticles having a soft rubber core and a glassy shell significantly improved the fracture toughness of an epoxy vinyl ester resin more than MMT nanoclay particles, having the same weight fraction. However, hybrid blends of CSR and nanoclay were found to yield the best balance of toughness, modulus, and strength. The same investigators highlighted that when the nanoclay particles were used to enhance the polymer matrix in a conventional GFRP composite, the interlaminar fracture toughness of the composite was less than that of the unenforced composite. As a possible reason for this result the arrangement of the nanoclay particles along the fiber axis was suggested.

A study by Kireitseu *et al.* [55] on rotating fan blades of turbine engine represents another feasible aerospace or defense heavy applications. They considered a large rotating civil engine blade, illustrated on Figure 10.8, which are typically hollow and usually have stiff rib-like metallic structures in order to increase the rigidity and maintain cross-sectional profile of the blade.

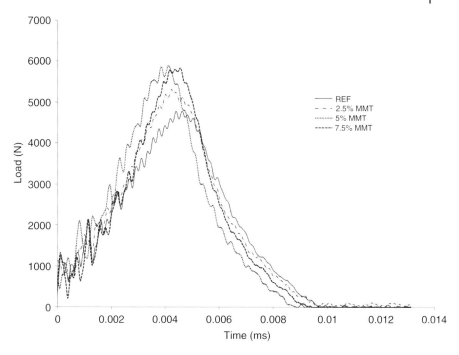

Figure 10.7 Load versus time graph obtained from second impact test.

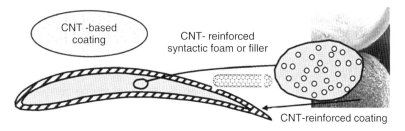

Figure 10.8 Design concepts for damping material [55].

They suggested a foam filled fan concept to replace the metal structure or traditional fillers with CNT-reinforced syntactic foam and also, a CNT-reinforced composite layer to be on the top. Results on damping behavior and impact toughness of the composite sandwiches showed that CNT-reinforced samples have advanced impact strength and vibration damping properties over a wide temperature range. Experiments conducted using a vibrating clamped beam with the composite layers indicated up to a 200% increase in the inherent damping level and a 30% increase in the stiffness with some decrease (20–30%) in density of the composite. The cross-links between nanotubes and composite layers also served to improve load transfer within the network resulting in improved stiffness properties. The critical issues to be considered include choice between nanotubes and related matrix materials for vibration damping, tailoring nanotube/matrix interface with respect to a matrix, and orientation, dispersion, and bonding of the nanotubes in matrix. It is anticipated that significant weight, thickness, and manufacturing cost reductions could be achieved in this way.

10.4
High-Temperature Fiber-Reinforced Nanocomposites

Selected few examples of high temperature fiber-reinforced nanocomposites discussed herein include poly(p-phenylene benzbisoxazole) (PBO), AS-4/poly(ether ether ketone) (PEEK), polyimide, and polyarylacetylene (PAA). PBO, a rigid-rod polymer, is characterized by high tensile strength, high stiffness, and high thermal stability. Kumar *et al.* [56] found out that PBO/CNT-reinforced fibers exhibited twice the energy absorbing capability than the plain PBO fibers. The nanocomposites were prepared as follows: into a 250 ml glass flask, equipped with a mechanical stirrer and a nitrogen inlet/outlet, were placed ~4.3 g (0.02 mol) of 1,4-diaminoresorcinol dihydrochloride, ~4 g (0.02 mol) of terephthaloyl chloride, and ~12 g of phosphoric acid (85%). The resulting mixture was dehydrochlorinated under a nitrogen atmosphere at 65 °C for 16 h and subsequently at 80 °C for 4 h. At this stage, 0.234 g of purified and vacuum-dried HiPco nanotubes was added to the reaction flask. The mixture was heated at 100 °C for 16 h while stirring and then cooled to room temperature. P_2O_5 (8.04 g) was added to the mixture to generate poly(phosphoric acid) (PPA; 77% P_2O_5). The mixture was stirred for 2 h at 80 °C and then cooled to room temperature. Further P_2O_5 (7.15 g) was then added to the mixture to bring the P_2O_5 concentration to 83% and the polymer concentration to 14 wt%. The mixture was heated at 160 °C for 16 h with constant stirring. Stir opalescence was observed during this step. The mixture was finally heated at 190 °C for an additional 4 h while stirring. An aliquot of the polymer solution was precipitated, washed in water, and dried under vacuum at 100 °C for 24 h. An intrinsic viscosity of 14 dl g^{-1} was determined in methanesulfonic acid at 30 °C. A control polymerization of pure PBO was also carried out under the same conditions without adding SWNT. For PBO/SWNT (90/10) composition, 0.47 g of purified HiPco tubes (SWNT) was added to the mixture. The sequence of steps and polymerization conditions remained the same as those for PBO/SWNT (95/5) composition. Intrinsic viscosity values of PBO and PBO/SWNT (90/10) were 12 and 14 dl g^{-1}, respectively. Single-walled nanotubes were well dispersed during PBO synthesis in PPA. PBO/SWNT composite fibers were successfully spun from the liquid crystalline solutions using dry-jet wet spinning. The addition of 10 wt% SWNT increased PBO fiber tensile strength by about 50% and reduced shrinkage and high-temperature creep. The existence of SWNT in the spun PBO/SWNT fibers was evidenced by the 1590 cm^{-1} Raman peak.

Jen *et al.* [57] manufactured carbon fibre/PEEK (AS-4/Advanced PEEK Composite APC-2) (APC-2) nanocomposite laminates and also studied their mechanical responses. The experimental procedure was as follows: firstly, the nanoparticles were diluted in alcohol (50 ml alcohol:2 g SiO_2) and stirred uniformly, then 16 plies of $[0/90]_{4s}$ cross-ply and $[0/\pm45/90]_{2s}$ quasi-isotropic prepregs were cut, SiO_2

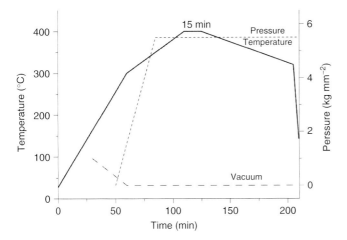

Figure 10.9 Pressure–temperature profile of the curing process of AS-4/PEEK APC-2 nanocomposites [57].

solution was then spread on the prepreg in a temperature-controlled box, and later the nanoparticles were weighed after evaporation of alcohol in the range of 111–148 mg ply. A repeat on spreading for 5, 8, 10, and 15 plies was the next step followed by curing (the curing process is shown in Figure 10.9) the stacked plies in a hot press to form a laminate of 2 mm thick.

Next, the laminates were cut into specimens and tested according to ASTM D3039M. The tensile tests were repeated at 50, 75, 100, 125, and 150 °C to receive respective stress–strain curve, strength, and stiffness, and the obtained data compared with the original APC-2 laminate (no SiO_2 nanoparticles) to find the optimal SiO_2% by weight. From tensile tests it was found out that the optimal content of nanoparticles (SiO_2) was 1% by total weight. The ultimate strength increased by about 12.48% and elastic modulus 19.93% in quasi-isotropic nano-laminates, while the improvement of cross-ply nanocomposite laminates was less than that of quasi-isotropic laminates. At elevated temperatures, the ultimate strength decreased slightly below 75 °C and the elastic modulus reduced slightly below 125 °C; however, both properties degraded highly at 150 °C ($\approx T_g$) for the two lay-ups. Finally, after the constant stress amplitude tension–tension (T–T) cyclic testing, it was found that both the stress cycles (S–N) curves were very close below 10^4 cycles for cross-ply laminates with or without nanoparticles, and the S–N curve of nano-laminates slightly bent down after 10^5 cycles.

Sandler *et al.* [58] produced poly(ether ether ketone) nanocomposites containing vapor-grown CNF using standard polymer processing techniques. Macroscopic PEEK nanocomposite master batches containing up to 15 wt% vapor-grown CNF were prepared using a Berstorff co-rotating twin-screw extruder with a length-to-diameter ratio of 33. The processing temperatures were set to about 380 °C. The strand leaving the extruder was quenched in a water bath, air dried, and then regranulated followed by drying at 150 °C for 4 h. Tensile bars according to the ISO

179A standard were manufactured on an Arburg Allrounder 420 injection molding machine at a processing temperature of 390 °C, with the mold temperature set to 150 °C. Before mechanical testing, all samples were heat treated at 200 °C for 30 min followed by 4 h at 220 °C in an attempt to ensure a similar degree of crystallinity of the polymer–matrix. Macroscopic tensile tests were performed at room temperature with a Zwick universal testing machine. The cross-head speed was set to 0.5 mm/min^{-1} in the 0–0.25% strain range and was then increased to 10 mm min^{-1} until specimen fracture occurred. Evaluation of the mechanical composite properties revealed a linear increase in tensile stiffness and strength with nanofiber loading fractions up to 15 wt% while matrix ductility was maintained up to 10 wt%. Electron microscopy confirmed the homogeneous dispersion and alignment of nanofibers. An interpretation of the composite performance by short-fiber theory resulted in rather low intrinsic stiffness properties of the vapor-grown CNF. DSC showed that an interaction between matrix and the nanoscale filler could occur during processing. However, such changes in polymer morphology due to the presence of nanoscale filler need to be considered when evaluating the mechanical properties of such nanocomposites.

Ogasawara *et al.* [59] directed their investigations toward the improvement of heat resistance of a relatively new phenylethynyl terminated imide oligomer (Tri-A PI) by loading of MWNT. They fabricated the MWNT/Tri-A PI composites containing 0, 3.3, 7.7, and 14.3 wt% MWNT using a mechanical blender without any solution (dry condition) for several minutes. The volume fraction of MWNT was calculated to be 2.3, 5.4, and 10.3 vol% from the density of the MWNT (1.9 g cm^{-3}) and the cured polyimide (1.3 g cm^{-3}). Scanning electron micrographs showed the particle size of the imide oligomers to be in the range of 0.1–10 μm, and MWNT were not dispersed uniformly in the mixture. The loss of aspect ratio during the mechanical blending was not significant; therefore the MWNT were flexible for mechanical blend process with the imide oligomers. The preparation of the nanocomposite involved the melt mixing of MWNT/imide oligomer at 320 °C for 10 min on a steel plate in a hot press, and then curing at 370 °C for 1 h under 0.2 MPa of pressure with polytetrafluoroethylene (PTFE) spacer (thickness 1 mm). The resulting composites containing 3.3, 7.7, and 14.3 wt% MWNT exhibited relatively good dispersion in macroscopic scale. Tensile tests on the composites showed an increase in the elastic modulus and the yield strength, and decrease in the failure strain. Figure 10.10 shows the effect of the MWNT concentration on Young's modulus of the composites.

DMA showed an increase in the glass transition temperature with incorporation of the CNTs. The experimental results suggested that the CNTs were acting as macroscopic cross-links, and were further immobilizing the polyimide chains at elevated temperature. As to the reason why dispersed MWNT increased the heat distortion temperature, the researchers explained that the dispersed MWNT impedes the molecular motion in polyimide network at elevated temperature. The other property improvements in this material are that MWNT showed some potential for controlling electric conductivity and electromagnetic wave absorbability. Although static properties were obtained, discussions were not given, and it is evident that

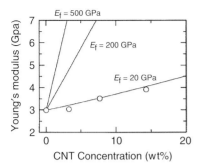

Figure 10.10 Effect of the MWCNT concentration on Young's modulus of the composites [59].

more research work would be required to prove that the suggested phenomenon is a true cause of higher glass transition temperatures.

PAA is going through increasing development in the field of advanced heat-resistant composites owing to its outstanding heat resistance and excellent ablative properties. Fu *et al.* [60] have reviewed the advantages of PAA resin over the state-of-the-art heat-resistant resin. The main potential applications of PAA resin are used in conventional resin matrix composites with ultra-low moisture out-gassing characteristics and improved dimensional stability suitable for spacecraft structures, as an ablative insulator for solid rocket motors, and as a precursor for carbon–carbon composites. Carbon fiber-reinforced PAA composites (carbon fiber/PAA) undoubtedly play a very important role in all these fields. Unfortunately, the mechanical properties of the carbon fiber/PAA material are not yet sufficiently satisfactory to replace the widely used heat-resistant composites such as carbon or graphite-reinforced phenolic resin. The mechanical properties of carbon fiber-reinforced resin matrix composites depend on the properties of carbon fiber and matrix, especially on the effectiveness of the interfacial adhesion between carbon fiber and matrix.

PAA has high content of benzene ring and hence a highly cross-linked network structure that renders the material brittle. Moreover, the chemical inert character-istics of the carbon fiber surface lead to weak interfacial adhesion between fibers and nonpolar PAA resin. To ensure that the material could be used safely in complicated environmental conditions and to exploit the excellent heat resistant and ablative properties more effectively, it is necessary to improve the mechanical properties of the carbon fiber/PAA composites. To achieve this purpose, two kinds of methods can be used. One method is to improve the properties of PAA resin by structural modification or by intermixing other resins, such as phenolic resin. The other is treatment of carbon fiber surface. The treatment of carbon fiber surface has been studied for a long time and several methods such as heat treatment, wet chemical or electrochemical oxidation, plasma treatment, gas-phase oxidation, and high-energy radiation technique have been demonstrated to be effective in the modification of the mechanical interfacial properties of composites based on polar resins such as epoxy. In investigations by Zhang *et al.* [61], for instance, carbon fibers were treated with oxidation–reduction followed by vinyltrimethoxysilanes-silsesquioxane (VMS–SSO) coating method to improve the interfacial mechanical properties of the carbon fiber/PAA composites.

Polar functional groups, including carboxyl and hydroxyl, on carbon fiber surface were imported after the oxygen plasma oxidation treatment. The quantity of carboxyl on carbon fiber surface was decreased and that of hydroxyl on carbon fiber surface was increased after the $LiAlH_4$ reduction treatment. The $LiAlH_4$ reduction time was decided according to the experimental parameter of Lin *et al.* [62]. The VMS–SSO coating was grafted onto the carbon fiber surface by the reaction of the hydroxyl in VMS–SSO and that on carbon fiber surface. The VMS–SSO coating concentrations and treatment time were decided according to Zhang *et al.* [63] who had optimized VMS–SSO coating treatment parameters. The investigation found out that interlaminar shear strength of the carbon fiber/PAA composites was increased by 59.3% at the end of treatment [61]. The conclusion that carbon fiber surface oxidation–reduction followed by silsesquioxane coating treatment is an effective method to improve the interfacial mechanical properties of carbon fiber/PAA composites were drawn. This kind of method could be widely used in different resin matrix composites by changing the functional groups on silsesquioxanes according to that on the resin.

10.5
Age and Durability Performance

The study of degradation and stabilization of polymers is therefore an extremely important area from the scientific and industrial point of view and a better understanding of polymer degradation will ensure the long life of the product [64]. Polymer degradation in broader terms include biodegradation, pyrolysis, oxidation, mechanical, photo, and catalytic degradation. According to their chemical structure, polymers are vulnerable to harmful effects (e.g., temperature, chemicals, light, water and moisture) of the environment. In the following paragraphs, epoxy and PAs are studied. It is important to note that presently there is little attention that has been given to the study of durability of polymer nanocomposites as compared to their preparation techniques and evaluation of mechanical properties.

The presence of MWNTs improves the thermal stability of PA-6 under air obviously, but has little effect on the thermal degradation behavior of PA-6 under nitrogen atmosphere. The thermal degradation mechanism of PA-6 has been proposed by Levchik *et al.* [65]. Vander Hart *et al.* [66] observed that in the presence of clay, the α-phase of PA-6 transforms into the γ-phase. The effect of a modifier on the degradation of nanocomposite was studied by ^{13}C NMR. In the presence of modifier (dihydrogenated-tallow ammonium ion) the nylon nanocomposite begins to degrade at 240 °C, whereas the virgin polymer does not. They concluded that the organic modifier is less stable. The combination of shear stress and temperature may lead to extensive degradation of the modifier and the extent of clay dispersion may not depend on the modifier.

In an intumescent ethylene vinyl acetate (EVA)-based formulation, using PA-6 clay nanocomposite instead of pure PA-6 (carbonization agent) has been shown to improve the fire properties of the intumescent blend. Using clay as "classical" filler enabled the same level of flame retardant (FR) performance to be obtained in the first step of the combustion as when directly using exfoliated clay in PA-6. But in the second half of the combustion, the clay destabilizes the system and increases the flammability. Moreover, a kinetic modeling of the degradation of the EVA-based formulations shows that adding clay to the blend enables same mode of degradation and the same invariant parameters as for the PA-6 clay nanocomposite containing intumescent blend. The increase in the flammability by the K-10 in the second half of the combustion shows the advantages of using nanoclay rather than microclay in an intumescent system [65].

The efficiency of the self-protective coatings that form during the pyrolysis and the thermo-oxidative degradation (in presence of oxygen) of PA-6 clay nanocomposite are also investigated [67]. The nanocomposites itself can be protected from fire/flame/oxygen by coating the organosilicon thin films. A PA-6 and PA-6 clay nanocomposite (PA-6 nano) substrates were coated by polymerizing the 1,1,3,3-tetramethyldisiloxane (TMDS) monomer doped with oxygen using the cold remote nitrogen plasma (CRNP) process. The thermal degradation behavior of deposits under pyrolytic and thermo-oxidative conditions shows that the residual weight evolution with temperature depends on the chosen atmosphere.

Organically modified clay-reinforced PA-6 was subjected to accelerated heat aging to estimate its long-term thermo-oxidative stability and useful lifetime compared to the virgin material [68]. Changes in molecular weight and thermal and mechanical properties were monitored and connected to the polymer modification encountered during aging. Generally, the strong interaction between the matrix and the clay filler renders the polymer chains, mainly that adjacent to silicates, to be highly mechanically restrained, enabling a significant portion of an applied force to be transferred to the higher modulus silicates. This mechanism explains the enhancement of tensile modulus that the non-aged clay-reinforced PA-6 exhibited (1320 MPa) with regard to the neat polymer (1190 MPa), as shown in Figure 10.11.

The effects of hydrothermal aging on the thermomechanical properties of high-performance epoxy and its nanocomposites are also reported in the literature [69]. It was found that the storage modulus and relaxation behavior were strongly affected by water uptake, while the fracture toughness and Young's modulus were less influenced. Dependence of tensile strength and strain at break on water uptake was found to be different in neat epoxy and epoxy–clay systems. Further improvement of the flame retardancy using combinations of the nanofiller and traditional FR-additives (e.g., aluminum trihydrate) was observed. The nanocomposites based on nanofillers and aluminum trihydrate passed the UL 1666 riser test for fire-resistant electrical cables [70]. Becker *et al.* [71] have found that the water uptake (in aquatic environment) was considerably reduced in epoxy nanocomposites with a particular clay loading percentage.

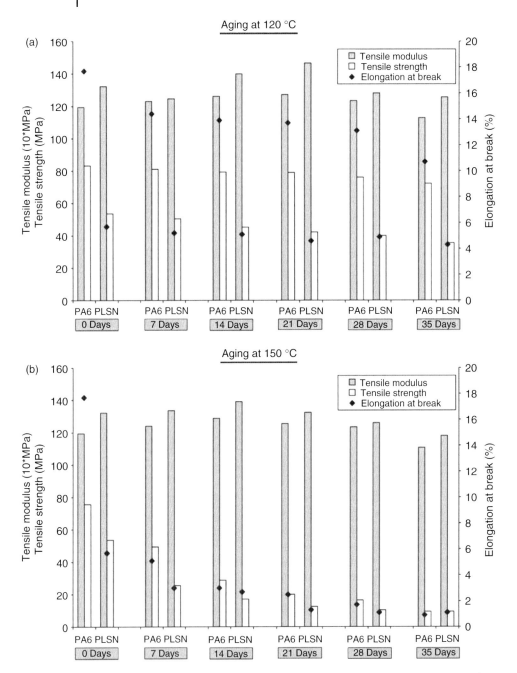

Figure 10.11 Tensile properties of PA-6 and PA-6 nanocomposite, oven-aged at (a) 120 °C and (b) 150 °C [68].

There are two factors which have opposite influences on the thermal stability of epoxy–clay nanocomposites. First factor is that the addition of clay to epoxy decreases the curing reactivity of epoxy resin. Lower reactivity of the resin generally results in lower cross-linking density of the cured resin and the longer polymer chains among the cross-linking points. It is known that a longer polymer chain is thermally less stable than a shorter chain, so both the nanocomposites are easier to degrade than the pristine epoxy resin. Secondly, silicate layers have a good barrier to gases such as oxygen and nitrogen, they can insulate the underlying materials and slow the mass loss rate of decomposition products. Moreover, exfoliated nanocomposites have better barrier properties and thermal stability than intercalated ones. In the case of intercalated nanocomposites (10 wt% clay), the first factor is dominant, whereas for exfoliated nanocomposites (2 wt% clay), the second factor is dominant.

10.6
Concluding Remarks

In the last two decades, some studies have shown the potential improvement in properties and performances of FRP matrix materials in which nano and microscale particles were incorporated. This technology of nano and microscale particle reinforcement can be categorized into inorganic layered clay technology, single walled, and multi-walled CNT, carbon nanofiber technology, and metal particle technology. To date, nanofiller reinforcement of fiber-reinforced composites has been shown to be a possibility, but much work remains to be performed in order to understand how nanoreinforcement results in major changes in material properties. The understanding of these phenomena will facilitate their extension to the reinforcement of more complicated anisotropic structures and advanced polymeric composite systems.

One of the technological drawbacks is that the mechanical reinforcement by the application of nanofillers as a structural element in polymers is more difficult to realize and still remains a challenging task. An efficient exploitation of the nanofiller properties in order to improve the materials performance is generally related to the degree of dispersion, impregnation with matrix, and to the interfacial adhesion. The advantage of nanoscaled compared to micro-scaled fillers is their enormous surface area, which can act as interface for stress transfer. The methods published so far on the improvement of mechanical properties of polymer composites have mainly focused on the optimization of the manufacturing process of the composites, that is, with the use of prepared nanofillers. The property and performance enhancements made possible by nanoparticle reinforcement may be of great utility for carbon or glass fiber-epoxy composites that are used for the high-performance and high-temperatures applications such as aerospace engine components and nacelle, storage of cryogenic liquids, and motorsports. Then again, precautionary measures should be observed in high temperatures as the structure and properties of these materials can change radically when they are exposed to extreme temperatures, especially in a cyclical manner.

Further, through nanofiller reinforcement, an electrically conductive matrix could provide enhanced feasibilities including stress–strain monitoring or damage detection. The application of an electrical field is known to orient the nanoparticles in the in-field-direction that may result in an increased efficiency of the z-reinforcement of the laminates. As a further benefit, the electrical conductivity in z-direction should be increased with this approach. Nanoparticles and especially CNTs provide a high potential for the modification of polymers. They are very effective fillers in regard to mechanical properties, especially toughness. Besides, they allow the implication of functional properties that are connected to their electrical conductivity, into polymeric matrices. The electromicromechanical technique had been studied as an economical nondestructive evaluation method for damage sensing, characterization of interfacial properties, and nondestructive behavior because conductive fiber can act as a sensor in itself as well as a reinforcing fiber.

However, nanotechnology might also have detrimental effects to the environment and it is paramount to understand both the hazards associated with nanomaterials and the levels of exposure that are likely to occur, while still taking advantage of the technology.

References

1. Njuguna, J. and Pielichowski, K. (2003) Polymer nanocomposites for aerospace applications: properties. *Adv. Eng. Mater.*, **5**, 769–778.
2. Njuguna, J. and Pielichowski, K. (2004) Polymer nanocomposites for aerospace applications: characterisation. *Adv. Eng. Mater.*, **6**, 204–210.
3. Njuguna, J. and Pielichowski, K. (2004) Polymer nanocomposites for aerospace applications: fabrication. *Adv. Eng. Mater.*, **6**, 193–203.
4. O'Donnell, S.E., Sprong, K.R., and Haltli, B.M. (2004) Potential impact of carbon nanotube reinforced polymer composite on commercial aircraft performance and economics. AIAA 4th Aviation Technology, Integration and Operations (ATIO) Forum, Chicago, Illinois, September 20–22, 2004.
5. Markarian, J. (2005) Automotive and packaging offer growth opportunities for nanocomposites. *Plast. Addit. Compd.*, **7**, 18–21.
6. Edser, C. (2002) Auto applications of drive commercialization of nanocomposites. *Plast. Addit. Compd.*, **4**, 30–33.

7. Njuguna, J., Pena, I., Zhu, H. *et al.* (2009) Opportunities and environmental health challenges facing integration of polymer nanocomposites: technologies for automotive applications. *Int. J. Polym. Technol.*, **1**, 113–122.
8. Longkullabutra, H., Thamjaree, W., and Nhuapeng, W. (2010) Improvement in the tensile strength of epoxy resin and hemp/epoxy resin composites using carbon nanotubes. *Adv. Mater. Res.*, **93–94**, 497–500.
9. Liu, Z. and Erhan, S.Z. (2008) "Green" composites and nanocomposites from soybean oil. *Mater. Sci. Eng., A*, **483–484**, 708–711.
10. Faruk, O. and Matuana, L.M. (2008) Nanoclay reinforced HDPE as a matrix for wood-plastic composites. *Compos. Sci. Technol.*, **68**, 2073–2077.
11. Vilela, C., Freire, C.S.R., Marques, P.A.A.P. *et al.* (2010) Synthesis and characterization of new $CaCO_3$/cellulose nanocomposites prepared by controlled hydrolysis of dimethylcarbonate. *Carbohydr. Polym.*, **79**, 1150–1156.

12. Xie, Y., Hill, C.A.S., Xiao, Z. *et al.* (2010) Silane coupling agents used for natural fiber/polymer composites: a review. *Composites Part A: Appl. Sci. Manuf.*, **41**, 806–819.

13. Abdelmouleh, M., Boufi, S., Belgacem, M.N. *et al.* (2007) Short natural-fibre reinforced polyethylene and natural rubber composites: effect of silane coupling agents and fibres loading. *Compos. Sci. Technol.*, **67**, 1627–1639.

14. Haq, M., Burgueño, R., Mohanty, A.K. *et al.* (2008) Hybrid bio-based composites from blends of unsaturated polyester and soybean oil reinforced with nanoclay and natural fibers. *Compos. Sci. Technol.*, **68**, 3344–3351.

15. Guigo, N., Vincent, L., Mija, A. *et al.* (2009) Innovative green nanocomposites based on silicate clays/lignin/natural fibres. *Compos. Sci. Technol.*, **69**, 1979–1984.

16. Njuguna, J., Pielichowski, K., and Desai, D. (2008) Nanofiller-reinforced polymer nanocomposites. *Polym. Adv. Technol.*, **19**, 947–959.

17. Lincoln, D.M., Vaia, R.A., Wang, Z.-G. *et al.* (2001) Secondary structure and elevated temperature crystallite morphology of nylon-6/layered silicate nanocomposites. *Polymer*, **42**, 1621–1631.

18. Fong, H., Liu, W., Wang, C.S. *et al.* (2002) Generation of electrospun fibers of nylon 6 and nylon 6-montmorillonite nanocomposite. *Polymer*, **43**, 775–780.

19. Fornes, T.D. and Paul, D.R. (2003) Crystallization behavior of nylon 6 nanocomposites. *Polymer*, **44**, 3945–3961.

20. Li, L., Bellan, L.M., Craighead, H.G. *et al.* (2006) Formation and properties of nylon-6 and nylon-6/montmorillonite composite nanofibers. *Polymer*, **47**, 6208–6217.

21. Cadek, M., Le Foulgoc, B., Coleman, J.N. *et al.* (2002) Mechanical and thermal properties of CNT and CNF reinforced polymer composites. *AIP Conf. Proc.*, **633**, 562–565.

22. Wu, S., Wang, F., Ma, C.M. *et al.* (2001) Mechanical, thermal and morphological properties of glass fiber and carbon fiber reinforced polyamide-6 and polyamide-6/clay nanocomposites. *Mater. Lett.*, **49**, 327–333.

23. Akkapeddi, M.K. (2000) Glass fiber reinforced polyamide-6 nanocomposites. *Polym. Compos.*, **21**, 576–585.

24. Vlasveld, D.P.N., Bersee, H.E.N., and Picken, S.J. (2005) Nanocomposite matrix for increased fibre composite strength. *Polymer*, **46**, 10269–10278.

25. Sandler, J.K.W., Pegel, S., Cadek, M. *et al.* (2004) A comparative study of melt spun polyamide-12 fibres reinforced with carbon nanotubes and nanofibres. *Polymer*, **45**, 2001–2015.

26. Usuki, A., Kawasumi, M., Kojima, Y. *et al.* (1993) Swelling behavior of montmorillonite cation exchanged for ω-amino acids by ε-caprolactam. *J. Mater. Res.*, **8**, 1174–1178.

27. Gao, F. (2004) Clay/polymer composites: the story. *Mater. Today*, **7**, 50–55.

28. Haque, A., Shamsuzzoha, M., Hussain, F. *et al.* (2003) S2-Glass/epoxy polymer nanocomposites: manufacturing, structures, thermal and mechanical properties. *J. Compos. Mater.*, **37**, 1821–1837.

29. Njuguna, J., Pielichowski, K., and Alcock, J.R. (2007) Epoxy-based fibre reinforced nanocomposites. *Adv. Eng. Mater.*, **9**, 835–847.

30. Chowdhury, F.H., Hosur, M.V., and Jeelani, S. (2006) Studies on the flexural and thermomechanical properties of woven carbon/nanoclay-epoxy laminates. *Mater. Sci. Eng., A*, **421**, 298–306.

31. Lin, L., Lee, J., Hong, C. *et al.* (2006) Preparation and characterization of layered silicate/glass fiber/epoxy hybrid nanocomposites via vacuum-assisted resin transfer molding (VARTM). *Compos. Sci. Technol.*, **66**, 2116–2125.

32. Aktas, L., Hamidi, Y.K., and Altan, M.C. (2004) Characterisation of nanoclay dispersion in resin transfer moulded glass/nanoclay/epoxy composites. *Plast. Rubber Compos.*, **33**, 267–272.

33. Gilbert, E.N., Hayes, B.S., and Seferis, J.C. (2002) Metal particle modification of composite matrices for customized density applications. *Polym. Compos.*, **23**, 132–140.

34. Gilbert, E.N., Hayes, B.S., and Seferis, J.C. (2002) Variable density composite systems constructed by metal particle

modified prepregs. *J. Compos. Mater.*, **36**, 2045–2060.

35. Timmerman, J.F., Hayes, B.S., and Seferis, J.C. (2002) Nanoclay reinforcement effects on the cryogenic microcracking of carbon fiber/epoxy composites. *Compos. Sci. Technol.*, **62**, 1249–1258.

36. Brunner, A.J., Necola, A., Rees, M. *et al.* (2006) The influence of silicate-based nano-filler on the fracture toughness of epoxy resin. *Eng. Fract. Mech.*, **73**, 2336–2345.

37. Gong, X., Liu, J., Baskaran, S. *et al.* (2000) Surfactant-assisted processing of carbon nanotube/polymer composites. *Chem. Mater.*, **12**, 1049–1052.

38. Shaffer, M.S.P. and Windle, A.H. (1999) Fabrication and characterization of carbon nanotube/poly(vinyl alcohol) composites. *Adv. Mater.*, **11**, 937–941.

39. Park, C., Ounaies, Z., Watson, K.A. *et al.* (2002) Dispersion of single wall carbon nanotubes by in situ polymerization under sonication. *Chem. Phys. Lett.*, **364**, 303–308.

40. Martin, C.A., Sandler, J.K.W., Windle, A.H. *et al.* (2005) Electric field-induced aligned multi-wall carbon nanotube networks in epoxy composites. *Polymer*, **46**, 877–886.

41. Shi, D., He, P., Lian, J. *et al.* (2005) Magnetic alignment of carbon nanofibers in polymer composites and anisotropy of mechanical properties. *J. Appl. Phys.*, **97**, 064312.

42. Shi, D., He, P., Lian, J. *et al.* (2004) Plasma coating and magnetic alignment of carbon nano fibers in polymer composites. *JOM*, **56**, 129–130.

43. Olek, M., Kempa, K., Jurga, S. *et al.* (2005) Nanomechanical properties of silica-coated multiwall carbon nanotubes-poly(methyl methacrylate) composites. *Langmuir*, **21**, 3146–3152.

44. Star, A., Stoddart, J.F., Steuerman, D. *et al.* (2001) Preparation and properties of polymer- wrapped single-walled carbon nanotubes. *Angew. Chem. Int. Ed.*, **40**, 1721–1725.

45. Bhattacharyya, S., Sinturel, C., Salvetat, J.P. *et al.* (2005) *Appl. Phys. Lett.*, **86**, 113104.

46. Gojny, F.H., Wichmann, M.H.G., Köpke, U. *et al.* (2004) Carbon nanotube-reinforced epoxy-composites: enhanced stiffness and fracture toughness at low nanotube content. *Compos. Sci. Technol.*, **64**, 2363–2371.

47. Gojny, F.H., Wichmann, M.H.G., Fiedler, B. *et al.* (2005) Influence of nano-modification on the mechanical and electrical properties of conventional fibre-reinforced composites. *Composites Part A: Appl. Sci. Manuf.)*, **36**, 1525–1535.

48. Wichmann, M.H.G., Sumfleth, J., Gojny, F.H. *et al.* (2006) Glass-fibre-reinforced composites with enhanced mechanical and electrical properties – benefits and limitations of a nanoparticle modified matrix. *Eng. Fract. Mech.*, **73**, 2346–2359.

49. Zhao, D.-L., Qiao, R.-H., Wang, C.-Z., and Shen, Z.-M. (2006) in *Anonymous AICAM 2005. Proceedings of the Asian International Conference on Advanced Materials (AICAM 2005), Beijing, China, November 3–5, 2005*, Trans Tech Publications, pp. 517–520.

50. Hsiao, K., Alms, J., and Advani, S.G. (2003) Use of epoxy/multiwalled carbon nanotubes as adhesives to join graphite fibre reinforced polymer composites. *Nanotechnology*, **14**, 791–793.

51. Meguid, S.A. and Sun, Y. (2004) On the tensile and shear strength of nano-reinforced composite interfaces. *Mater. Des.*, **25**, 289–296.

52. Hussain, M., Nakahira, A., and Niihara, K. (1996) Mechanical property improvement of carbon fiber reinforced epoxy composites by Al_2O_3 filler dispersion. *Mater. Lett.*, **26**, 185–191.

53. Njuguna, J., Michalowski, S., Pielichowski, K. *et al.* (2011) Fabrication, characterisation and low-velocity impact on hybrid sandwich composites with polyurethane/layered silicate foam cores. *Polym. Compos.*, **32**, 6–13.

54. Subramaniyan, A.K. and Sun, C.T. (2007) Toughening polymeric composites using nanoclay: crack tip scale effects on fracture toughness. *Composites Part A: Appl. Sci. Manuf.*, **38**, 34–43.

55. Kireitseu, M., Hui, D., and Tomlinson, G. (2008) Advanced shock-resistant and vibration damping of nanoparticle-reinforced composite material. *Composites Part B: Eng.*, **39**, 128–138.

56. Kumar, S., Dang, T.D., Arnold, F.E. *et al.* (2002) Synthesis, structure, and properties of PBO/SWNT composites. *Macromolecules*, **35**, 9039–9043.

57. Jen, M.R., Tseng, Y., and Wu, C. (2005) Manufacturing and mechanical response of nanocomposite laminates. *Compos. Sci. Technol.*, **65**, 775–779.

58. Sandler, J., Werner, P., Shaffer, M.S.P. *et al.* (2002) Carbon-nanofibre-reinforced poly(ether ether ketone) composites. *Composites Part A: Appl. Sci. Manuf.*, **33**, 1033–1039.

59. Ogasawara, T., Ishida, Y., Ishikawa, T. *et al.* (2004) Characterization of multi-walled carbon nanotube/phenylethynyl terminated polyimide composites. *Composites Part A: Appl. Sci. Manuf.*, **35**, 67–74.

60. Fu, H.J., Huang, Y.D., and Liu, L. (2004) Influence of fibre surface oxidation treatment on mechanical interfacial properties of carbon fibre/polyarylacetylene composites. *Mater. Sci. Technol.*, **20**, 1655–1660.

61. Zhang, X., Huang, Y., Wang, T. *et al.* (2007) Influence of fibre surface oxidation–reduction followed by silsesquioxane coating treatment on interfacial mechanical properties of carbon fibre/polyarylacetylene composites. *Composites Part A: Appl. Sci. Manuf.*, **38**(3), 936–944.

62. Lin, Z., Ye, W., Du, K. *et al.* (2001) Homogenization of functional groups on surface of carbon fiber and its surface energy. *J. Huaqiao Univ.*, **22**, 261–263.

63. Zhang, X., Huang, Y., Wang, T. *et al.* (2006) Influence of oligomeric silsesquioxane coatings treatment on the interfacial property of CF/PAA composites. *Acta Mater. Compos. Sin.*, **23**, 105–111.

64. Pielichowski, K. and Njuguna, J. (2005) Thermal Degradation of Polymeric Materials. Rapra Technologies Limited, Shawbury, Surrey.

65. Levchik, S.V., Weil, E.D., and Lewin, M. (1999) Thermal decomposition of aliphatic nylons. *Polym. Int.*, **48**, 532–557.

66. Vander Hart, D.L., Asano, A., and Gilman, J.W. (2001) Solid-state NMR investigation of paramagnetic nylon-6 clay nanocomposites. 2. measurement of clay dispersion, crystal stratification, and stability of organic modifiers. *Chem. Mater.*, **13**, 3796–3809.

67. Gilman, J.W., Kashiwagi, T., and Lichtenhan, J.D. (1997) Nanocomposites: a revolutionary new flame retardant approach. *SAMPE J.*, **33**, 40–46.

68. Kiliaris, P., Papaspyrides, C.D., and Pfaendner, R. (2009) Influence of accelerated aging on clay-reinforced polyamide 6. *Polym. Degrad. Stabil.*, **94**, 389–396.

69. Njuguna, J. and Pielichowski, K. (2010) Ageing and Performance Predictions of Polymer Nanocomposites for Exterior Defence and Aerospace Applications. Polymer Nanocomposites in Aerospace Applications, Hamburg, Germany, March 11–12, 2010, Smithers Rapra Technology Ltd.

70. Beyer, G. (2005) High Performance Fillers 2005, Rapra Technology Ltd., Shrewsbury, pp. P5–P7.

71. Becker, O., Varley, R., and Simon, G. (2004) Thermal stability and water uptake of high performance epoxy layered silicate nanocomposites. *Eur. Polym. J.*, **40**, 187–195.

Part IV
Cellular Materials

Eusebio Solórzano and Miguel A. Rodriguez-Perez

A cellular solid material is necessarily composed by two phases, one continuous solid phase and the other either continuous or discontinuous gaseous phase. The solid phase is generally named matrix [1]. The two phases present in cellular materials can be appreciated in the example materials shown in Figure P4.1.

Wood, cork, and human bones are representative examples of natural cellular materials. The idea of developing and manufacturing this type of material arose when trying to imitate these natural structures that exhibited excellent properties. Nowadays, different cellular structures ranging from simple 2D honeycombs (used as core in sandwich panels for structural applications) to foams, produced from a previous stage in which the solid phase was in liquid state [2], are industrially produced.

In the literature, it is possible to find several classifications for cellular materials [1]. The most widely used one divides them into *closed cell* and *open cell* structures. In closed cell structures the gas is kept inside the pores so the material presents a continuous solid phase and a discontinuous gaseous phase. On the contrary, an open cell structure is characterized by a continuously dispersed gas phase into the solid one. In practice, most of the produced materials present intermediate structures so part of the gaseous phase is occluded inside the pores whereas another fraction of this phase is interconnected. Therefore, most of the cellular materials are characterized by the *open cell content* or *cell interconnection fraction* which expresses the amount of continuous gas phase over its overall content.

In addition, the term *cellular materials* implies a certain classification, this one being the most general term that covers a wide range of structures produced by many different techniques. As an example, the concept "foam," included in the term *cellular material*, is much more restrictive and refers to materials produced from a liquid/melt state, thus, implying a cellular structure of a random nature. In this sense, production routes strongly condition the cellular structure (cell morphology and topology) which can be and associated to different cellular materials categories as done in Figure P4.2. It is important to point out that the classification exhibited in Figure P4.2 has exceptions although it probably fits to most of produced cellular solid materials.

Structural Materials and Processes in Transportation, First Edition.
Edited by Dirk Lehmhus, Matthias Busse, Axel S. Herrmann, and Kambiz Kayvantash.
© 2013 Wiley-VCH Verlag GmbH & Co. KGaA. Published 2013 by Wiley-VCH Verlag GmbH & Co. KGaA.

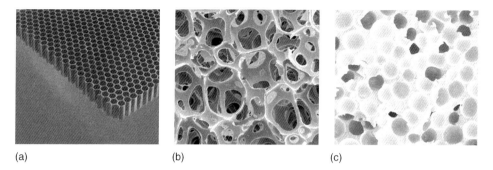

(a)   (b)   (c)

Figure P4.1 Different cellular structures: thermoplastic honeycomb (a), open cell foam (b), and closed cell foam (c).

The presence of a gaseous phase makes cellular materials lighter than their former solids. This simple fact involves weight reduction together with a diminution of the amount of raw materials employed which, in many cases, implies costs savings. The concept of relative density, ρ_r, is often used to express the amount of material used with respect the 100% dense solid (Eq. (P4.1)). Common values range from 0.9 down to 0.02. Similarly, porosity, p, is defined as the solid fraction substituted by the gaseous phase (Eq. (P4.2)):

$$\rho_r = \frac{\rho_{\text{foam}}}{\rho_{\text{solid}}} \tag{P4.1}$$

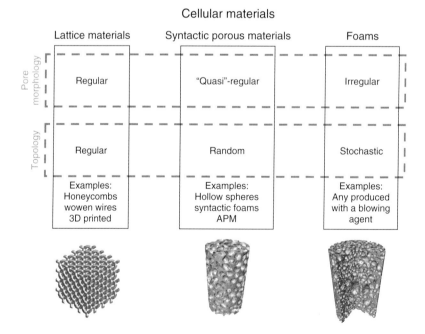

Cellular materials

	Lattice materials	Syntactic porous materials	Foams
Pore morphology	Regular	"Quasi"-regular	Irregular
Topology	Regular	Random	Stochastic
	Examples: Honeycombs wowen wires 3D printed	Examples: Hollow spheres syntactic foams APM	Examples: Any produced with a blowing agent

Figure P4.2 Classification of cellular solids as a function of the morphology and topology.

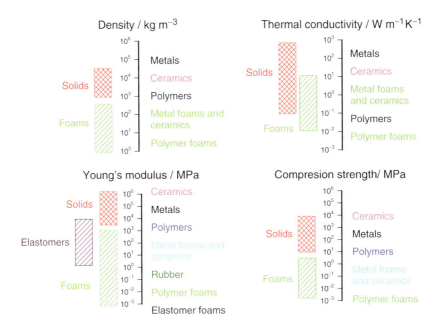

Figure P4.3 Comparative properties of dense and foamed materials.

$$p = 1 - \rho_r \tag{P4.2}$$

Moreover, cellular materials extend the range of various properties of their respective dense solids as can be observed in Figure P4.3. In general, cellular solids are characterized by a nice assortment of properties that solids do not have. Among others, they can exhibit extremely low thermal transport properties, high energy absorption capability, and in the case of presence of a significant amount of open cell fraction, high and easy-tunable acoustic absorption properties. These materials can also be considered as "environmentally friendly products" assuming that less amount of raw materials are necessary for their production.

As it can be expected, relative density, ρ, and properties, P, are intimately related. Thus, a rough approach for the relationship between any property and the relative density of the solid can be obtained by a simple scaling law where A and n are the coefficient and the exponent that are generally obtained by empirical methods:

$$P_{\text{foam}} = A \times P_{\text{solid}} \times \left(\frac{\rho_{\text{foam}}}{\rho_{\text{solid}}} \right)^n \tag{P4.3}$$

Finer approaches for the properties prediction demonstrate, apart from the influence of the relative density, an additional effect of cellular morphology and topology on them [3–5]. Therefore, the material formulation together with a strict control of the processing parameters will condition the final properties through the relative density and structure. Under this point of view, cellular solids can be considered as tailored materials as they can be designed, both in density and structure, to fulfill certain requirements. As an example, considering a specific application with a density constrain it could be possible to find a cellular material to satisfy the conditions by choosing the solid matrix and the processing method to define a particular cellular structure.

Assuming that there exist large differences in the technological approaches to produce cellular materials with different matrix nature (polymeric-elastomeric, metallic, and ceramic) as well as the possible applications for the transport industry, we will separately analyze the cellular materials depending on their matrix constituent. Nevertheless, we have excluded from our treatment foams based on a ceramic matrix, mostly because these materials have a small number of applications in the transport industry.

In the following pages the reader will find information about the production, technologies, new trends, and applications both for polymeric and metallic foams in the transport industry.

References

1. Gibson, L.J. and Ashby, M.F. (1997) *Cellular Solids – Structure and Properties*, 2nd edn, Chapter 1, Cambridge University Press, Cambridge.
2. Banhart, J. (2001) *Prog. Mater. Sci.*, **46**(6), 559–632.
3. Cunningham, A. and Hilyard, N.C. (1994) NC physical behavior of foams – an overview, in *Low Density Cellular Plastics: Physical Basis of Behaviour* (eds N.C.
Hilyard and A. Cunningham) Chapter 1, Chapman & Hall, London.
4. Khemani, K.C. (1997) Polymeric foams: an overview, in *Polymeric Foams: Science and Technology*, ACS Symposium Series (ed K.C. Khemani) Chapter 1, American Chemical Society, Washington, DC.
5. Rodríguez-Pérez, M.A. (2005) *Adv. Polym. Sci.*, **184**, 87–126.

11
Polymeric Foams

Eusebio Solórzano and Miguel A. Rodriguez-Perez

11.1
Introduction

Polymeric foams are extensively used materials that fulfill the necessities of thousands of applications in market sectors such as aeronautics, automotive, construction, packaging/cushioning, biomedical, tissue engineering and renewal energies.

A representative value of their importance is the total consumption of 4.15 billion kg of cellular plastics accomplished in the United States market during 2009. This value represents approximately 10% of the total resin consumption and more than the 50% of the total market in volume [1] (see Figure 11.1).

One of the key points that have permitted plastic foams to succeed in different sectors is the versatility to produce enormous varieties made of different polymers and additives as well as the possibility to be conformed under a wide range of foam structures, densities, etc. This fact is only possible thanks to the high number of existing foam-processing techniques. In this sense, thousands of patents on polymeric foam-processing techniques have been developed since the early 1940s [2].

In this section, we will carry out an overview of the most significant processing techniques, paying also attention to the new trends in processes and materials. In this chapter we will treat separately processing techniques for thermoplastic, thermoset and nanocomposite foams. In a second part, the current and potential applications of these materials in the transport industry will be exposed, by dividing them into three main classes: structural, comfort-security and environmental issues.

11.2
Blowing Agents for Polymer Foams

Plastics need a gas source to expand and be shaped till they reach their "foamed" structure. The adopted name for the component that allows plastics to expand is a *blowing agent*.

Any substance that produces a porous structure in a polymer mass can be defined as a blowing agent [3]. The gaseous phase present in foamed materials

Structural Materials and Processes in Transportation, First Edition.
Edited by Dirk Lehmhus, Matthias Busse, Axel S. Herrmann, and Kambiz Kayvantash.
© 2013 Wiley-VCH Verlag GmbH & Co. KGaA. Published 2013 by Wiley-VCH Verlag GmbH & Co. KGaA.

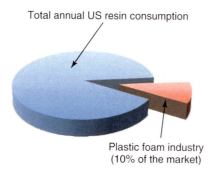

Total annual US resin consumption

Plastic foam industry
(10% of the market)

Figure 11.1 Plastic foam industry represents an important sector of the total resin annual consumption.

is derived from this agent. Blowing agents include gases that expand/segregate when pressure is released, liquids that develop cells when they change to gases and chemical agents that decompose while releasing a gas.

The blowing agent plays a critical role in manufacturing the foam. The choice of blowing agent and the final processing conditions are linked. In the case of foaming of thermoplastics, the blowing agent may modify the melt viscosity and/or the temperature of the polymer and thus the melt rheological properties during foam formation. In cases where polymerization and foaming are developed at the same time, typical of thermosets, the blowing agent may also affect the compatibility, reactivity and mixing of the components.

The blowing agent also plays an important role for properties and structure. The blowing agent controls the final density of the foam and also affects the cellular microstructure and morphology of the foam [2, 4]. If the structure is closed cell-like, the blowing agent is retained at least until it diffuses out, which in some particular cases can take a long time (decades). If an open-cell structure is formed, the blowing agent escapes almost immediately after the foam solidifies. The type of gas, in combination with the structure may also affect the properties. In closed cell materials, the type of gas has a strong effect for heat insulation applications whereas its influence is intermediate for buoyancy, impact resistance and load bearing applications [5].

The blowing agents are generally classified as physical or chemical. Chemical blowing agents (CBAs) are generally solid at standard temperature and pressure (STP) and undergo a chemical transformation when producing gas, while physical blowing agents (PBAs), generally a liquid or gas at STP, either experience a reversible change of state. One important exception to this classification would be water, a liquid used extensively to make polyurethane (PU) foam that reacts with isocyanate to liberate CO_2 gas as will be later explained in detail.

1) **Physical blowing agents**

A PBA provides gas for the expansion of polymers by promoting a change in physical state. The change may involve evaporation (boiling) of a liquid or phase segregation within a polymer–gas system at elevated temperature/pressure after depressurization to atmospheric pressure. Common gaseous blowing agents include: CO_2, N_2 and low boiling liquids such as short-chain (C_2 to C_4) aliphatic hydrocarbons and halogenated C_1 to C_4 aliphatic hydrocarbons. PBAs are used in the production of all types of foamed plastics, both thermoplastics and thermosets over the full range of density. They are mostly used when the foam density is low (<50 kg m^{-3}). PBAs are relatively low in cost but require special equipment for use.

2) **Chemical blowing agents**

CBAs liberate a gas or a certain mixture of gases during the foam-processing conditions, either because of thermal decomposition or chemical reaction. The decomposition of the original molecule gives a result of one or more gases for polymer expansion and one or more solid residues that remain in the foamed polymer. Table 11.1 lists the most common CBAs together with their decomposition temperature range, gases evolved and gas yield. As can be seen from Table 11.1, the gases produced are generally N_2 and CO_2 and once produced, they behave very similar to a PBA but with some effects due to presence of the decomposition solid residues.

CBAs are finely divided solids in the micron range. They can be blended with the plastics before processing or be fed directly into a hopper. The use of CBAs include certain benefits in comparison to PBAs, such as broader operating window, self-nucleation and finer cell size. CBAs are quite extended as they require little modification to any existing thermoplastics processing line. CBAs may be incorporated into any thermoplastic process to produce foams [6, 7], such as extrusion, injection molding, calendering and rotational molding.

Table 11.1 Different chemical blowing agents.

Name	Decomposition temperature (°C)	Heat release	Gas yield at STP (10^{-3} m^3 kg^{-1})	Released gases
Azodicarbonamide (ADC)	200–230	Exo	220–245	N_2, CO, NH_3, O_2
4,4-Oxybis (benzenesulfonylhydrazine) (OBSH)	150–160	Exo	120–125	N_2, H_2O
p-Toluenesulfonylhydrazide (TSH)	110–120	Exo	110–115	N_2, H_2O
p-Toluenesulfonylsemicarbazide (TSS)	215–235	Exo	120–140	N_2, CO_2
Sodium carbonate	120–150	Endo	130–170	CO_2, H_2O
Citric acid derivatives	200–220	Endo	110–150	CO_2, H_2O
Dinitrosopentamethlenetetramine	195	Exo	190–200	N_2, NH_3, HCHO
Polyphenylene sulfoxide (PPSO)	300–340	Exo	80–100	SO_2, CO, CO_2

11.3
Thermoplastic Foams: Conventional Processing Technologies

11.3.1
Injection Molding

Injection molding has enjoyed much success owing to its unique nature of making detailed three-dimensional shaped plastic parts [8]. This technique is, by far, the most employed technique for solid polymer processing in automotive industry.

However, foam injection molding is a more complicated process and has faced important challenges during its development which have restricted its potential applications. These are constraints on part geometry, size and surface quality of produced parts. During the past decades, many special injection molding processes have been developed either to address these problems or to produce parts with special features.

On the other hand, the benefit of obtaining this type of materials is high. Injection process applied to foamed parts has several advantages such as the absence of sink marks, negligible warping and residual stresses, higher rigidity-to-weight ratio, lighter weight and material cost reduction. Because of these benefits injected foams are specially selected in many applications, such as in the manufacturing of electrical appliances, machine enclosures and furniture, as well as for industrial applications and building components.

Injected foamed materials can be divided into two main classes: structural foams, characterized by a smooth transition from the outer solid skin to the inner foamed core, and co-injected foamed materials, in which the transition from solid to the foamed core is abrupt. In principle, coinjected foam processing needs higher machinery investment as it is a more complicated technology. Figure 11.2 shows the typical structure of these two types of materials.

Several injection foaming processes that have evolved from the traditional molding technique are currently being used for foam production. Among others, we can cite gas-assisted injection molding [9], reaction injection foam molding

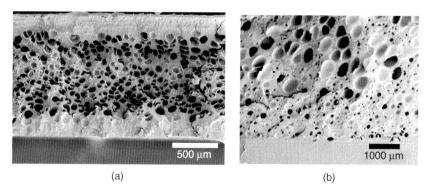

(a)   (b)

Figure 11.2 Foamed materials produced by injection molding: (a) structural foam and (b) coinjection molded material.

[10], low-pressure foam molding [11], high-pressure foam molding [12], coinjection foam molding [13] and microcellular injection foam molding [14–16]. Some of these techniques are briefly described in the subsequent paragraphs:

1) **Structural foam processing**
   a. **Low-pressure foam molding:** A CBA or a PBA can be employed to produce the cellular structure during processing. In this process a mold cavity is partially filled with the foamable mixture using an injection machine. The self-expanding foaming material fills the mold and produces sufficient internal gas pressure to pack out the mold uniformly, permitting shrinkage compensation and preventing sink marks. A schematic of the process is depicted in Figure 11.3. Thanks to the relatively low processing and machinery costs, the low-pressure foam molding process accounts for about 90% of thermoplastic structural foam production worldwide [11]. Some drawbacks include low density reductions (typically not higher than 25%), nonuniform cell size distribution [11] and poor surface quality of the produced items which restrict broader applications.
   b. **High-pressure foam molding:** This alternative technology was developed to improve the surface quality of structural foams [12]. Figure 11.4 shows a schematic of this process. The melt containing a dissolved gas (usually created by the decomposition of a CBA) is first injected into the mold under high pressure with complete mold filling. The mold is packed under high pressure, as in conventional injection molding. The high pressure prevents blowing agent expansion and thus less swirl is created in the part surface. Upon the establishment of a solid skin, the mold expands to allow blowing agent expansion inside the core. This technology is not as popular as the low-pressure one due to practical limitations in the part geometry (only

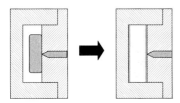

Figure 11.3 Schematic draw of a typical low-pressure foam molding.

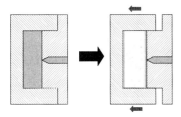

Figure 11.4 Schematic of high-pressure foam molding process.

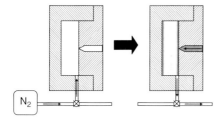

Figure 11.5 Simplified schematic of gas counter process technique.

parts with simple geometries can be produced) and the amplified mold investments [12].

c. **Gas-counter-pressure foam molding:** Gas-counter-pressure foam molding also allows significant improvements in the surface quality of the parts [17, 18]. The concept involves using a gas-pressurized mold, which, through controlled venting, allows the foam expansion to occur after a smooth surface has been formed permitting a better control of the foaming stage. The parts produced typically show 5–10% less density reduction than those produced by the low-pressure technology, but display a more uniform cell structure across the part, which results in improvements in physical properties. A simplification of the gas counter-pressure foam molding process is showed in Figure 11.5. A pressure-tight mold is charged with a pressurized gas before the injection of plastic. After the proper pressure has been reached in the mold, a predetermined shot of polymer containing a dispersed and compressed blowing agent is injected. During the injection sequence, controlled venting occurs to maintain consistent counter-pressure, as the unfilled mold volume is being reduced. Upon the formation of a solid smooth skin of required thickness, the gas pressure in the mold is released to allow foaming expansion in the interior of the molded body. Apart from the mentioned benefits, the main drawbacks of this technique are the increase in the cost of the mold and the longer cycle times.

2) **Coinjection foam molding**

This process is also called sandwich foam molding. It involves the sequential or simultaneous injection of two materials, a solid one that will be in contact with the mold surface forming the "skin," and a foamed material compatible with the "skin" which will be the "core." Thanks to flow effects, the skin layers travel together with the core material and are pushed to the extremities of the mold

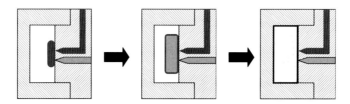

Figure 11.6 Scheme of coinjection foam molding process.

cavity. This finely synchronized process forms a foam core encapsulated in a solid skin. An exemplary schematic of the process is provided in Figure 11.6. This hybrid technology is based on two separate injection units. One injection unit produces the skin component that is the material without the foaming agent and the other provides the foamable mixture. This process offers intrinsic flexibility of using the optimal properties of each material to reduce polymer cost, injection pressure, clamping tonnage and residual stresses.

11.3.2
Extrusion Foaming

Foam sheets can be produced by direct extrusion. Materials with medium and high relative densities (around 0.5 or higher) are produced in conventional extrusion equipments by adding a CBA as raw material and by using a temperature profile that allows the decomposition of the blowing agent in the barrel with pressures high enough to keep the released gas dissolved in the polymeric matrix. By controlling the pressure, temperature at the die and the formulation of the polymer, it is possible to produce foamed sheets with a homogeneous cellular structure from almost any polymeric matrix. Polypropylene and PVC foamed sheets are currently in the market [19]. It is also possible to produce coextruded profiles with two solid skins and a foamed core, improving the surface quality and mechanical properties of the products.

Low-density foams (relative densities up to 0.03) can also be produced, but then the extrusion technology becomes more complicated. The high expansion ratios need a high amount of gas to be dissolved in the matrix, which is usually achieved by using PBAs [20]. For polystyrene, carbon dioxide can be used as the main blowing agent because of the high solubility of this gas in PS. Figure 11.7 shows a typical cellular structure for this type of material. However, for polyolefins, hydrocarbons such as isobutene or pentane are needed. The foams produced using this technology have found applications in thermal insulation and packaging sectors.

The extrusion foaming process for low-density foams can use a single extruder or two extruders operating in tandem. The tandem extruder setup normally allows an excellent control of process variables and is used in industry to produce very low-density foams of PS, PE and PP. Steps in both processes are polymer melting, injection and dissolution of the blowing agent into the polymer melt, cooling of the melt containing the blowing agent to a temperature close to the melting temperature of the base polymer, the nucleation in the die, bubble growth out of the die and stabilization of the resultant cellular structure by cooling and solidification. The schematic of single barrel foam extrusion process is given in Figure 11.7a. In the single extruder, high length-to-diameter ratio (L/D) equipment is normally required to provide sufficient elements for polymer melting, gas dissolution, cooling, pressurization and foaming through the die.

One important advantage of this processing technology is that it allows producing non-cross-linked low-density foams. Therefore, the foams are recyclable and, in

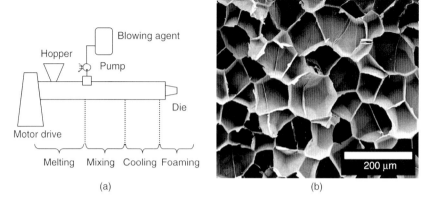

(a)   (b)

Figure 11.7 (a) Single barrel foam extrusion process and (b) extruded polystyrene foam (XPS, 40 kg m^{-3}) produced using carbon dioxide as blowing agent.

fact, most of the foamed industrial products incorporate an important amount of recycled material coming from trimmings of previous productions.

11.3.3
Compression Molding

Thick foams (up to 120 mm) can be made by compression molding methods whereby cross-linking and blowing agent decomposition are carried out in a closed mold. Expansion takes place in a subsequent step [21, 22]. This type of process is used to produce low-density flexible cross-linked polyolefin foams, useful for thermal insulation and energy absorption, but also for the production of low-density rigid cross-linked PVC foams with applications in structural panels and in the core of sandwich panels. Using this process thick foam blocks with relative densities as low as 0.02 are currently produced. The cross-linking of the matrix polymer allows producing foams with fine and homogeneous cellular structures. An example of the typical cellular structure of these materials is showed in Figure 11.8a.

For the production of low-density foams a two-stage process is used [23]. The process (Figure 11.8b) involves firstly compounding the polymer with cross-linker, blowing agent and required additives to produce a solid sheet. Banbury batch mixers and twin-screw extruders are used to this end. A defined mass of compound is then placed in a mold and press-cured. The temperature and time in this first stage are selected to fully cross-link the polymer and to partially decompose the blowing agent. When the mold is opened the product expands directly, literally jumping out of the mold. Molds are designed to evacuate the foam and facilitate the foam's sudden expansion. The second foaming stage heats the pre-expanded cross-linked foam to higher temperatures to complete decomposition of the CBA and expand it up to fill a second mold of the final dimensions.

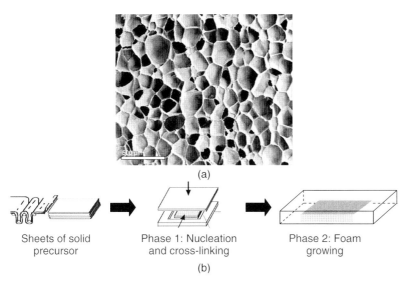

(a)

Sheets of solid precursor   Phase 1: Nucleation and cross-linking   Phase 2: Foam growing

(b)

Figure 11.8 (a) Cross-linked polyethylene foam of 30 kg m^{-3} produced by compression molding and (b) schematics of compression molding technology.

After production blocks are sliced and eventually thermoformed to produce final parts. One important aspect of this technology is that blocks present a characteristic profiles in density and properties caused during the last foaming step in a closed mold [22, 24].

11.3.4
Gas Dissolution Foaming

This technology is based on the dissolution of a gas (usually nitrogen or carbon dioxide) in a polymeric matrix using high pressures in an autoclave, the nucleation of the cellular structure by a rapid pressure drop and the heating of the polymer containing gas to temperatures high enough to expand the material [25, 26] (Figure 11.9a). The process is, for instance, used to industrially produce cross-linked polyolefin foam blocks with very low relative densities (as low as 0.02), fine and homogeneous cellular structures and free of residues of a blowing agent. Figure 11.9b shows an example of the cellular structure of this type of foam. Owing to the high thermal stability (thanks to the cross-linking), possibility of thermoforming, the absence of volatile emissions and the good thermal insulation, these materials are suitable for use in car interiors. In addition, this technology can be used to produce thin microcellular sheets of amorphous polymers with significant weight reduction (up to 50%), excellent surface quality and mechanical properties. Owing to the technical difficulties associated to this technology it is not very popular in the market, only the companies Zotefoams (UK) and Microgreen (USA) use the technology to produce industrial materials.

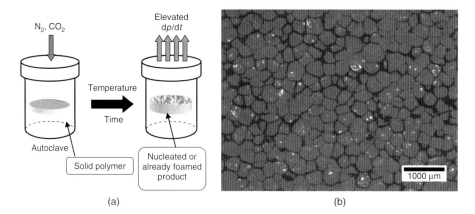

Figure 11.9 (a) Diagram showing the two main steps of the gas dissolution process. (b) Optical micrograph of a 58 kg m^{-3} cross-linked polyethylene foam produced by gas dissolution foaming (cell borders are ink-dyed).

11.4
Thermoplastic Foams: New Trends, Materials and Technologies

11.4.1
Microcellular Foams

In contrast to the cell density of a conventional foam ranging from 10^2 to 10^6 cells cm^{-3}, the cell density of a microcellular foam is 10^6 cells cm^{-3} or higher and it is ideally defined with a uniform cell size of about 10 μm and with a cell density as high as 10^9 cells cm^{-3} [25]. This ultrafine bubble dispersion, characteristic of microcellular plastics, is the result of creating a sudden thermodynamic instability into a polymer–gas solution generating larger number of nuclei within the polymer matrix. An example of the cellular structure of these materials is showed in Figure 11.10.

Historically, microcellular plastics are not new. They existed somehow in the thin transition layer of structural foams. However, the microcellular processing technology was invented at the Massachusetts Institute of Technology in 1980 and the first US patent on microcellular technology was issued in 1984 [27].

In general, any processing technique for microcellular plastics involves the presence of a gas in supercritical conditions (a gas above both its critical pressure and temperature), generally CO_2. At this condition, the solubility and diffusivity of this fluid in most polymers is rather high. Therefore, once a sufficient amount of gas is dissolved and in equilibrium with the polymer, the cell nucleation is promoted by a pressure quenching. Extrusion, injection and solid state gas dissolution are the technologies which allow the production of microcellular foams. Sudden pressure release is produced at the end of the extrusion die or injection nozzle in the first two processes, which involve a molten polymer going out through these cavities and producing shear in the bubbles and material movement until it is solidified.

Figure 11.10   Cellular structure of a foamed TPE produced by extrusion using the Mucell Technology.

In the solid state gas dissolution process, the gas pressure is released from a vessel and the nucleated material foamed in a second step at temperatures above the melting point (Section 11.3.4).

Microcellular processing technology may include two different, almost opposite, concepts:

1) **Solid substitution**: products lighter than a solid (10–30 wt% weight reduction) with similar or upgraded properties, specially looking for improved material toughness [28, 29]. Under this concept, microcellular foams produced with proper cell size have four to five times higher impact strength than the unfoamed counterparts [20]. The fatigue life of microcellular foams is also found to be 14 times that of solid parts [20, 30]. Nevertheless, cell size and cell size distribution determine the differences in properties between conventional foam and microcellular foam. Thus, the improvement of properties can be only achieved when the cell size is smaller than a critical size. Shimbo *et al.* [31] clarified that the critical size depends on the kind of plastic material.
2) **Improved foam properties**: alternative microcellular foams in the medium or low density range with improved compared properties to their respective conventional foams.

Industrial applications of microcellular foams are mostly associated with the concept of solid substitution, that is, limited weight reduction, and make use of microcellular injection molding processing to obtain such materials. This technology has completed the laboratory stage and transitioned to industry application. It is a pretty novel technology considering that the first injection molding machine was developed in 1997 by Trexel Inc. The most popular trade name for this technology is MuCell® and is licensed by Trexel Inc. since 2000. Optifoam® licensed by Sulzer Chemtech [32], Ergocell® licensed by Demag [33] and ProFoam® process [34] developed at IKV, are also microcellular injection molding technologies developed recently.

Apart from obvious advantages associated with injection molding techniques for industrial applications, collateral advantages are found when using microcellular foam injection processing technology, thanks to the fact that the gas fills the interstitial sites between polymer molecules; the viscosity and the glass transition temperature of the polymer melt can be effectively reduced [20, 25]. This enables the material to be processed at much lower pressure and temperature, which would proportionately downsize the long-run production cost with a view to saving energy. This also helps to improve the molding thermodynamics that results in a quicker cycle time. Additionally, the process is a low-pressure molding process and produces stress-free and less warped injection molding products such as conventional injected foams.

11.4.2
Nanofoams

Nanofoams, nano-cellular foams, or sub-microcellular foams – that is, foams with cell sizes in a range of tens to hundreds of nanometers, sufficiently below 1 μm – are the extension of the microcellular foams concept. The idea of creating cells with sizes in the nanoscale in a polymeric material is exciting and only recently explored.

Owing to their unique structure, nano-cellular foams are expected to have properties that are superior to those of currently existing cellular and microcellular materials. It is anticipated that nanofoams would provide novel materials with excellent mechanical properties, ultralow thermal conductivity, ultralow dielectric constant, and interesting damping properties. One of the most interesting characteristics is the possibility of producing materials with thermal conductivities as low as $10–14$ mW mK^{-1}. Currently, one of the best materials for thermal insulation is rigid PU foam with conductivities close to 20 mW mK^{-1} just after production and with values increasing over time owing to the diffusion of the gases enclosed in the cellular structure. Nano-cellular foams with relatively low densities (below 0.1) as a result of the absence of conduction through the solid phase (Knudsen effect) could significantly reduce these values that, in addition, the conductivity will be constant with time. [ref, Pinto, J., Solórzano, E., Rodriguez-Perez, M. A., and de Saja, J. A. (2012) Thermal conductivity transition between microcellular and nanocellular polymeric foams: experimental validation of the Knudsen effect. Proceedings International Conference on Foams and Foams Technology, Foams 2012, Barcelona, September, 12,13, 2012)].

These materials have the potential to be used for any application where foamed polymers are currently used, as well as for new applications in construction, packaging, automotive, microelectronics, and household products industries, offering important technical advantages.

The processing techniques that allow obtaining polymeric nano-cellular materials are currently very limited, and are based on the use of nano-structured block copolymers or on the use of extremely stringent processing conditions as used for obtaining homogenous polymers.

Two different approaches have been used to produce the cellular material starting from a nano-structured block copolymer. The first one consists of using blocks with different thermal stability, a thermally stable block together with thermally labile blocks [36]. While heating, the thermally labile block undergoes thermolysis, leaving intact the pores with the size and shape similar to the labile copolymer morphology. As a result, Highly thermally stable nanofoams have been produced; however, the nanofoams produced have only 15–25% volume void fraction. Materials produced by this method are normally used for microelectronics applications [33].

The alternative method needs one of the block polymers to present a substantially higher-gas solubility and diffusivity in comparison with the polymer constituting the other block, typically in higher proportions [37–39]. In addition, the glass transition temperature of both blocks needs to be different, with the block of the material in higher proportion and lesser affinity with the blowing agent being higher. Supercritical CO_2 gas dissolution allows for dissolving significant amounts of gas in both phases, but particularly higher in the minority high-solubility block. Therefore, the nucleation and growth of cells is preferentially produced within the nanodomains, allowing the production of sub-microcellular materials of significant higher porosities than those obtained by the previous method. Figure 11.11 shows the typical structure of both the nano-structured polymer precursor and the final foams produced using this method [38]. One of the main drawbacks of such materials and methodologies is the need for block copolymers that are rather expensive and those that often involve the use of hazardous chemicals during its production.

Recently, it has been proved that it is possible to produce nano-cellular foams with relatively low densities, starting from polymeric matrices with a homogeneous structure at the nanoscale, using CO_2 as a blowing agent. Stringent processing conditions during the foaming step (high pressures and high pressure drop

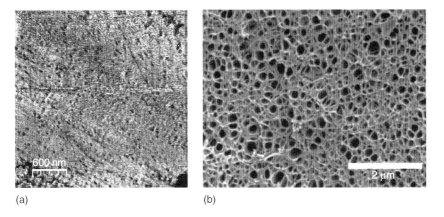

(a)   (b)

Figure 11.11 (a) AFM image of the solid Poly(methyl methacrylate) PMMA precursor with 5% (methyl methacrylate) − (butyl acrylate) − (methyl methacrylate) MAM block copolymer. (b) SEM image a nanocellular foam produced from such precursor. Relative density: 0.5, average cell size: 0.2 μm, cell density: 3.5×10^{14} cells cm^{-3}.

rates) and the postfoaming step are needed to achieve the extremely high cell densities required (10^{15}–10^{16} cell cm^{-3}). [Ref: Costeux, S., Jeon, H., and Khan, I. (2012) Nanocellular foams from acrylic polymers: experiments and modeling. Proceedings International Conference on Foams and Foams Technology, Foams 2012, Barcelona, September, 12th, 13, 2012)].

11.4.3
Bioplastic and Biodegradable Foams

Bioplastic and biodegradable polymers are a new generation of polymers obtained from various natural resources. Interest in environmentally safe and friendly biodegradable foams is expanding considering the increasing prices of plastic resins made from complex hydrocarbons (petroleum, gas, etc.) and the environmental concern of waste reduction. Furthermore, by introducing significant renewable biomass feedstock for the production of such materials the rate of CO_2 fixation would increase and move towards equilibrium.

It is important to make a clear distinction between the term *bioplastic*, which includes any kind of plastic obtained from a natural-renewable resource, and the term *biodegradable polymer*, which refers to the susceptibility of a polymer to be decomposed by living things or by environmental factors. The ASTM defines biodegradable as capable of undergoing decomposition into carbon dioxide, water, methane or biomass resulting from the enzymatic action of microorganisms, which can be measured by standardized tests in a specified period of time reflecting available disposal condition [40]. Natural environmental factors that cause decomposition include thermal or ultraviolet light action as well as bacteria, fungi and yeast.

An exemplary case of a bioplastic but not a biodegradable material is the bio-derived polyethylene, or renewable polyethylene, obtained from the polymerization of ethylene which in turn is produced by the dehydration of ethanol. Ethanol is typically produced by fermentation of agricultural feedstock such as sugar cane or corn. Bio-derived polyethylene is chemically and physically identical to traditional polyethylene. Foaming process of renewable polyethylene is the same as that for the conventional one. This section is more orientated on biodegradable foam processing.

A wide range of biodegradable plastics are available on the market. Table 11.2 lists some of the most important ones.

Biodegradable foaming process is different from conventional thermoplastic foam processing. Biodegradable foams are water soluble, easily thermally degradable and moisture sensitive whereas most thermoplastic foams are not. The foaming process is similar, but the process for biodegradable foams must have low shear and lower temperatures to avoid deterioration of the much more delicate chemical structure. This fact makes the processing window for biodegradable foams quite narrow.

Foams with cell size ranging above 50–100 μm can be produced using biodegradable plastics as raw materials although the microcellular range is further restricted.

Table 11.2 Most used biodegradable polymers.

Polymer	Nomenclature
Starch	—
Polyhydroxybutyrate	PHB
Polylactic acid	PLA
Polycaprolactone	PCL
Polyvinyl alcohol	PVOH
Ethylene vinyl alcohol	EVOH

Water is polar in nature and mixes well with biodegradable resins such as starch and due to this is also used as blowing agent. Nevertheless water may create problems when mixing with conventional hydrocarbons, such as polyolefins, although there is no compatibility problem between polymers. Therefore, the blowing agent selection depends on the nature of the polymer, rheology, solubility and diffusivity characteristics.

Typically biodegradable cellular materials have been produced by the conventional extrusion technique [41–43]. Other techniques are the baking process for starch [44, 45] or the leaching method [46]. Current biodegradable foam parts include food service boxes, cups and food trays. Figure 11.12 shows two SEM micrographs of bioplastic foams produced at laboratory scale.

Biodegradable foams offer several potential advantages versus conventional thermoplastic foams, as an example, their accelerated degradation, the water solubility (PVOH, starch) and the feedstock independency from nonrenewable resources. In addition, some disadvantages and environmental risks associated are claimed, such as possible incorporation of the foam additives and modifiers to landfills and water resources. Nevertheless, further studies and life cycle analyses are needed to study those environmental problems associated with the use of biodegradable foams. Finally separation methods are needed to be developed to

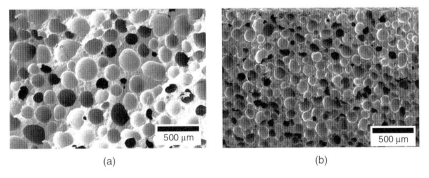

(a)   (b)

Figure 11.12 Cellular structure of typical foams based on bioplastics. (a) PHB foam and (b) starch-based foam.

detect and differentiate biodegradable polymers from others. This will ensure the success of biodegradable foams for new market applications.

11.5
Thermosets Foams: Conventional Processing Technologies

Polyurethanes (PUs) are by far, the most commercialized thermoset foamed materials.

Thermoset PUs are formulated using a wide variety of raw materials and applied to thousands of end uses including both nonfoam and foam applications. The flexibility in the formulation is probably one of the main characteristics of this material. For instance, it is possible to formulate both rigid and flexible PU and, obviously, producing both types of derived foams. Additionally PU-foam formulation also enables a great flexibility in the final cellular structure which allows varying density, cell size and cell characteristics. One of the particular characteristics is the possibility of producing open or closed cell structures. Generally, flexible PU foams are produced under open cell "appearance" whereas rigid PU is formulated to be a closed cell foam. The later applications for each of these materials are intimately related to the foam structure and flexible/rigid nature.

PU foam expansion takes place during the reactive process of a bicomponent mixture (polyol + isocyanate) after turbulent mixing. The reactive process simultaneously induces blowing and polymerization of such a system. The blowing agent is generally present in the polyol component. This blowing gas is commonly CO_2, derived from the chemical reaction with water (Figure 11.13). A liquefied gas of low boiling point is sometimes added to the polyol, depending on the final application, although still keeping some amount of water.

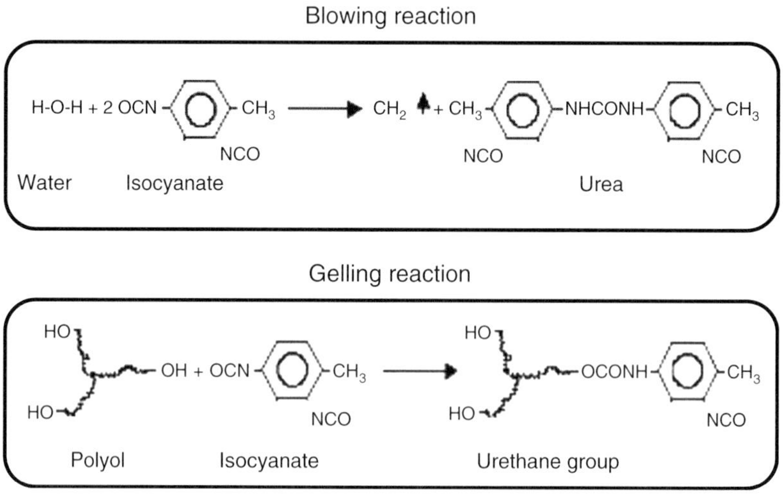

Figure 11.13 Chemical reactions occurring throughout the PU foam formation.

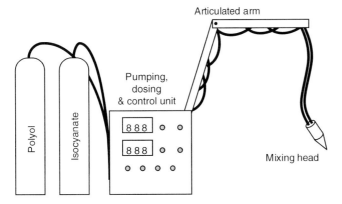

Figure 11.14 Schematic of a typical high-pressure injection machine used in PU industry.

The two simultaneous reactions produced are the blowing reaction-that generates the CO_2 and secondarily forms polyurea-, and the gelling reaction that promotes the urethane formation that produces a covalent polymer network [47, 48]. Usually, the polyol and isocyanate used are polyfunctional.

Industrial production of both main types of PU is generally associated with high-pressure or low-pressure injection machines, which inject/spray the two mixed liquids. Figure 11.14 shows a scheme of such a system.

Although the chemical reactions and the basic chemistry for open cell flexible PU foams and closed cell rigid PU are quite similar, they also present significant differences in chemistry, structure and, obviously, final applications as it is described in subsequent sections.

11.5.1
Flexible PU Foams

In a water-blown flexible PU typical relative densities are below 10%. Reaction process is exothermic and significantly increases the mixture temperature, sometimes over 100 °C. This enables significant additional expansion owing to gas volume increase with temperature.

The typical flexible foam formulation contains several additives to provide the desired properties of the final product. The basic ingredients include polyol, water, isocyanate, silicone surfactants and catalysts. A typical formulation used in flexible PU is showed in Table 11.3. A detailed discussion of these ingredients can be found in the literature [49, 50]. The silicone surfactant is used to achieve stable cells. It retards drainage flow in cell windows and prevent cell coalescence until the material gains sufficient strength to be self-supporting which means that, in practice, the surfactants also help to control the precise timing and the degree of cell-opening. The degree of cell-opening rules the viscoelasticity of the final foam. Without this additive the foam system would suffer catastrophic coalescence. With increasing surfactant concentration, a foam system will show improved stability and cell-size

Table 11.3 Basic components of conventional flexible PU formulations.

Component	Parts by weight
Polyol	100
Water	2–6
Isocyanate	40–60
Silicone surfactant	0.2–5
Amine catalyst	0.1–1.0
Organometallic catalyst	0.0–0.5

control. Today, the majority of flexible foams are made from a class of surfactants identified as polysiloxane–polyoxyalkylene copolymers. For both the blowing and the gelling reactions, either organic or organometallic catalysts are used, primarily to accelerate the reaction of isocyanate with hydroxyl groups (gelling reaction). By adjusting the catalyst's level and ratio, the desired balance between the gelling and blowing reactions can be achieved.

Flexible PU foam can be either molded or produced continuously on a conveyor belt. Among all the applications for PUs, flexible slabstock foam accounts for the largest volume produced. Depending on the formulation, a wide range of densities and mechanical properties are available for flexible PU foam. Such versatility is an advantage shared by no other type of cushioning material. Figure 11.15a shows the structure of an open cell flexible PU foam used for cushioning and Figure 11.15b shows a high resilience partially open cell foam. The cellular structure is basically composed by struts, that is, there are no cell walls or they are practically opened. The low airflow resistance of these materials is one important characteristic used

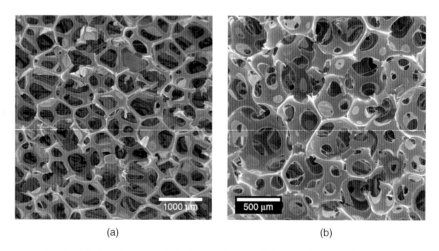

(a)   (b)

Figure 11.15 (a) Typical open cell flexible PU foam and (b) partial open cell flexible PU foam.

in cushioning applications [49]. Foams with high resilience have been designed to have a very low recovery speed after a compression; this is achieved by a combination of a different chemistry of the polymeric matrix and by increasing the airflow resistance of the cellular structure (partially open cell structure). In addition, the open cell cellular structure allows these materials to have excellent acoustic absorption.

Among many others, major applications of flexible PU comprise comfort and cushioning (seats, mattress, etc.) and they are also used as acoustic absorbers in different industries.

11.5.2
Rigid PU Foams

Rigid PU foams are produced with densities in the range 20–200 kg m^{-3}. The materials with densities in the range 20–50 kg m^{-3} have found an important market in thermal insulation while, the materials with higher densities are used in structural applications, such as in the core of sandwich panels and/or fillers in polymeric or metallic hollow structures.

The chemistry used to produce these materials is based on the same concepts used to produce the flexible materials (see previous section). The three key differences [51] are the selection of polyols and isocyanates that on polymerization produce a rigid polymer (i.e., a polymer with a glass transition temperature higher than room temperature), the use of a blowing agent with low thermal conductivity and low diffusivity trough the solid matrix (oriented to thermal applications of low-density closed cell foams) and finally the different surfactants used to provide a higher stability of the cell walls suppressing the opening of the cell windows.

Figure 11.16a shows the typical structure of a medium–high-density foam of 120 kg m^{-3} used for structural applications. A unique cell structure results which is comprised of struts (dimensions the order of tenths microns) and window films (microns range). Figure 11.16b shows a microtomography of the typical structure

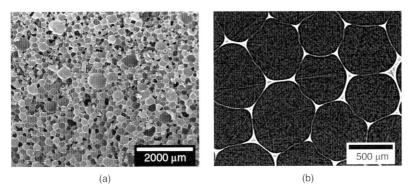

(a)   (b)

Figure 11.16 (a) SEM micrograph of rigid PU foam of density 120 kg m^{-3}. (b) Tomographic slice of a PU foam of density 52 kg m^{-3}.

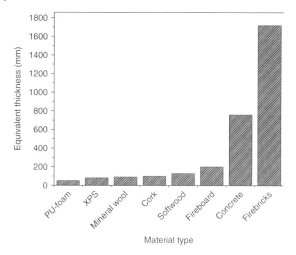

Material type

Figure 11.17 Comparative for different alternative thermal insulating materials to rigid PU foams of 50 mm thickness.

of a low-density closed cell PU (of density 52 kg m^{-3}). This cell structure is different to that of low-density thermoplastic foam whose cell membranes and struts are of similar and uniform thickness.

Low-density PU foams are the best thermal insulators currently available in the market. To illustrate this, Figure 11.17 compares the thermal insulation performance of rigid PU foam with that of other materials. It shows that other materials need much higher thickness in order to reach the insulation performance of 50 mm rigid PU foam. This performance is primarily due to a combination of blowing agent properties, cell size and cell morphology [51].

Despite its major importance, excellent thermal insulation is not the only property that makes rigid PU foams the right selection in several industrial applications. There are some additional advantages which are often critical for the selection of PU foams [52]. The most important advantages are their easy processing, their high mechanical strength and the possibility of forming very strong adhesive joints with many materials. The combination of all these attributes is particularly apparent in the thermal insulator for buildings, sandwich panels and filler of plastic or metals structures where rigid PU foam is virtually unrivaled [53, 54].

11.6
Thermosets Foams: New Trends, Materials and Technologies

11.6.1
Epoxy Foams

Epoxy foams have been in use for decades, although they still remain a specialty product within the thermoset foam family with a much smaller market than

PU foams. These materials are generally rigid and are used when greater heat resistance, solvent resistance, adhesion or higher mechanical properties are needed. Epoxy foams also reduce any potential health issue associated with the isocyanate sensitization in the processing of urethanes. Inert or reactive blowing agents are usually used to expand the epoxy resin [55].

A variety of epoxy resins and curing agents are available to the epoxy formulator and permit a range of heat resistance, toughness and other properties in the cured material. The chemistry of this extremely versatile class of materials can be found in previous literature [55].

Uncured epoxy formulations typically include one or more epoxy resins, one or more curing agents, and modifiers such as fillers and cure accelerators. In the case of epoxy foams, the formulations would also include a blowing agent and surfactants to control the final foam structure. The curing process in epoxy foams requires a much longer duration than PU and is generally assisted by elevated temperatures.

Foamable formulations are quite limited and the selection of proper combinations of epoxy resin and curing agent is critical. For example, a resin with inadequate viscosity will not retain the blowing agent and will not effectively expand. The stability and uniformity of cellular structures are also strongly depending on the surface tension and viscosity, mostly depending on the curing agent. For this reason, surfactants are commonly added to epoxy foam formulations, similar to PU foams, aiming to homogenize the cell size in the final product. The most commonly used surfactants are silicone resins and copolymers of silicone polymers and polyalkylene oxides.

PBA or CBAs can be used with epoxy formulations depending on the processing requirements. Inert gases such as N_2, CO_2 or air are mixed with uncured epoxy formulations and are especially suitable as physical foaming agents in automated processes. In addition, inert liquids with low boiling temperature such as low molecular weight hydrocarbons, ethers, ketones or halogenated hydrocarbons can be used as PBAs. In these cases, the amount of blowing agent induces changes in density and physical properties. CBAs generating gases such as N_2, CO_2 or H_2 are also used. Especially common are a variety of azodicarbonamides, aromatic sulfonyl hydrazides and carbonate salts. The temperature sensitivity/reactivity of these agents can be adjusted with accelerators similar to other resins. It is important to properly match the curing and foaming kinetics to prevent foam collapse, poor cell morphology or a simple loss of the gaseous phase. A unique blowing chemistry used primarily in epoxies is the reaction of liquid polyalkylhydrosiloxanes, particularly methylhydroxiloxanes, with amine curing agents to generate H_2 [55]. The foaming action is usually very gentle and especially suitable for the encapsulation of electronic assemblies. Figure 11.18 shows the cellular structure of a foam produced using this technique. This particular foaming agent has also been used in the automotive industry and other applications.

Most epoxy foam formulations are specialty products designed to fit performance and processing requirements of individual applications. Thus, they are often formulated by the end user instead of the resin producer and also by small

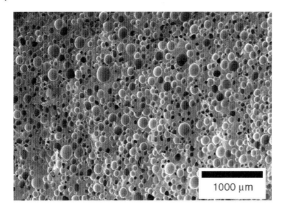

1000 μm

Figure 11.18 Epoxy-based foam with relative density 0.7.

companies that have the flexibility to provide customized solutions. For that reason, only a limited variety of expandable formulated products are available commercially, although companies such as Henkel and Hunstman offer expandable epoxy products. Many of these products are targeted at larger automotive applications.

Commercial applications of epoxy foams are in areas of reinforcement of automotive panels, electronics encapsulation, reparation of damaged buildings and sandwich structures [56]. Epoxy foams have found widespread use in automobiles to add rigidity and vibration damping in flat and hollow panels, to form fiber-reinforced structural panels, and in other areas where their presence provides added strength and performance with little increase in weight [57, 58].

11.6.2
Environmentally Friendly Blowing Agents for PU

High molecular weight gases with low thermal conductivities have been extensively used with PU foams (also XPS foams) offering unique thermal insulation characteristics as they stay within the cells and present extremely low thermal conductivity, in comparison with nitrogen or air, thus reducing the overall conduction through the gas phase. Nevertheless they are volatile chemicals that are prone to diffuse over extremely large time periods (up to 25 years) reaching the atmosphere. During the 1960s, 1970s and 1980s gases used were trichlorofluorocarbones (CFC) and in the later 1990s hydrochlorofluorocarbon (HCFC), having been discovered that they had strong impact on ozone depletion.

The Montreal Protocol led to the reduced use of many chlorine-containing blowing agents, such as CFC-11. Other haloalkanes, such as the 1,1-dichloro-1-fluoroethane (HCFC-141b), were used until their phase out under the IPPC directive on greenhouse gases in 1994 and by the Volatile Organic Compounds (VOCs) directive of the EU in 1997.

By the late 1990s, the use of blowing agents such as pentane, 1,1,1,2-tetrafluoroethane (HFC-134a) and 1,1,1,3,3-pentafluoropropane (HFC-245fa)

Table 11.4 Main characteristics of some of the currently-used blowing agents GWP (Global Warming Potential).

Gas	Mw (g-mol)	Thermal conductivity (mW m^{-1}·K^{-1})	GWP	Flammability limit	Effective diffusivity ($\times 10^{-10}$ cm^2 s^{-1})	Solubility STP (phr atm^{-1})
CFC-12	120.9	9.4	1	NF	5.10	1.5
HCFC-142b	100.5	11.7	0.065	8–16	5.95	6.3
HFC-134a	102.0	13.5	0	NF	8.50	1.0
HFC-152a	66.1	12.6	0	5–30	255	1.8
HFC-32	52.0	14.3	0	13–30	10^5	0.4
CO$_2$	44.0	15.2	0	NF	10^4	0.9

became more widespread, although chlorinated blowing agents remained in use in many developing countries.

Some of the new blowing agents used nowadays are summarized in Table 11.4 [59]. From an environmental point of view and taking into account safety issues, CO$_2$ is the best choice and, in fact, processing technologies in the XPS market tend to use this gas as the main blowing agent. The high diffusivity of this gas has obligated to develop new strategies to reduce the thermal conductivity of plastic foams, such as reducing the heat flow by radiation by reducing the cell size and/or by incorporating infrared blockers (carbon black, graphite, aluminum, titanium oxide, etc.) in the formulation [60, 61]. In the case of PU, the trend is moving towards the use different types of hydrocarbons, such as pentane, which allows retaining good properties with a good environmental behavior; the weak point of this selection yields on the high flammability of this material that complicates the industrial facilities and deteriorates the flame retardancy of the final products.

11.7
Nanocomposite Foams

Solid polymer nanocomposites are those materials with superior properties when compared to their "microcomposite" counterparts. Furthermore, a smaller fraction of nanoparticles can significantly improves a variety of properties. This is the reason why solid polymer nanocomposites are becoming more and more important in recent years covering a vast combination of polymer matrices and nanoparticles [62–65]. This concept has also been translated into foamed-state with the objective of producing nanocomposite foams.

Nanocomposites include at least one phase (filler) that possesses ultrafine dimensions (on the order of a few nanometers). According to their shape, three different nanoparticle types exist, as shown in Figure 11.19. Clay, graphenes,

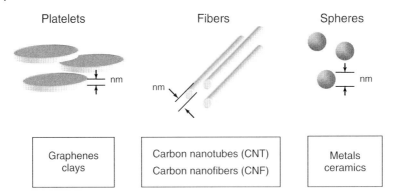

Figure 11.19 Classification of nanofillers according to their shape.

nanotubes and nanofibers, fumed spherical nanosilica, nanocrystals, gold and other metal nanoparticles are examples of these nanofillers.

Achievement of a controlled dispersion of these particles is challenging as they present extremely high surface–volume ratio and tend to agglomerate. Other particles such as clays are layered as stacks and need to be exfoliated, which means in practice a significant amount of technical difficulties. Typical dispersion methods comprise high shear mixing (double screw extrusion, agitation with impellers, etc.) and ultrasonication [66]. In many cases dispersion/exfoliation needs to be assisted via special surface chemistry that promotes "adhesion" between the polymer matrix and the particles. These surface treatments in the particles will also have an important role in the final properties of the nanocomposites foam. The use of surfactants in inorganic nanoparticles and functionalization chemicals in the carbon surface [67] are the most conventional ones. Surfactants in the clay surface increase the hydrophobicity and compatibility through ion exchange reactions with the polymer matrices [68]. Studies on polymer-CNTs and polymer-CNFs composites demonstrate the importance of this functionalization [69, 70].

If particles are well dispersed and there is a good filler–matrix interaction, then properties are expected to be optimum. Thus, the strength of the interaction plays an important role in the final properties, especially in the case of mechanical properties. Elastic modulus, yield stress, ultimate stress and strain-to-failure will mostly depend on particle–matrix interface. Nevertheless, in other transport properties, such as thermal or electrical conductivity, the particle dispersion is the critical variable, apart from the necessity of high nanofiller contents.

On the other hand, foam production in the presence of nanoparticles may imply synergistic effects as particles also have an important role during the foaming step. A small amount of well-dispersed nanoparticles in the polymer may serve as nucleation sites to facilitate bubble nucleation [71–73]. Plate-like nanoparticles can also reduce gas diffusivity in the polymer matrix reducing coarsening effect [74]. Finally, presence of particles and nanoparticles is also associated with an enhanced stability partially inhibiting the cell wall rupture mechanism in the liquid state [72, 73].

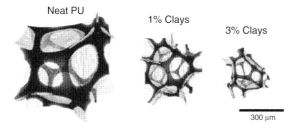

Figure 11.20 Volume rendering of representative unitary cells for different nanoclay-PU foams of density 52 kg m^{-3}.

One existing clear example on the influence of nanoparticles in the final cellular structure of nanocomposites (even when the fractions added are really small) is PU/clay nanocomposite foams [75, 76] in which clay nanoparticles are premixed in the polyol component and later reacted with the isocyanate. Figure 11.20 shows the microtomographic reconstruction of representative unit cells of rigid closed cell PU incorporating different percentages of nanoclays. Final density of the produced foam was not altered by nanoparticle presence although cell size experienced a dramatic reduction that was attributed to an enhanced nucleation effect of these particles in PU systems [77–79].

Under this premise, it could be also possible to obtain microcellular nanocomposites that would combine the advantages of microcellular polymers empowered by the presence of nanoparticles [80–82] An exemplary case is in Figure 11.21 where the compared cellular structures of a microcellular LDPE foam produced by gas dissolution and its equivalent microcellular nanocomposite (LDPE/silica) are shown [74]. Similarly to the previous example, significant cell size reduction is obtained and, in addition, mechanical properties are synergistically enhanced in this case. Both materials had the same relative density: 0.62.

Nanocomposite foams have a significant potential in several industries considering that the addition of controlled and well-dispersed small amounts of nanoparticles promote a significant improvement of the cellular structure and

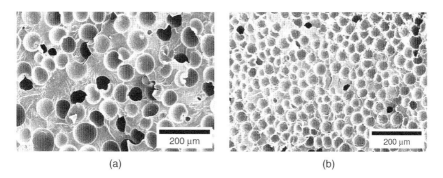

(a)   (b)

Figure 11.21 (a) Microcellular LDPE foam and (b) microcellular LDPE/nanosilica composite foam.

physical properties without modifications on the foaming technologies and at a very low cost. This can be applied, for instance, in the production of better thermal insulators asuming a significant reduction in cell size (PU and PS), in the production of stiffer rigid foams (at constant density) for the core of sandwich panels, in the development of foamed materials with better flame resistance free of halogen compounds and in the production of foamed materials with multifunctional characteristics combining, for instance, low density, good mechanical properties, thermal insulation and electrical conductivity.

11.8
Case Studies

Most of the current processing technologies for polymeric foamed materials and the latest developments and trends are already introduced in the different transportation sectors (automotive, heavy-duty road transportation, aircraft, railway and ships) and, only in the case of the very new developments they are still under consideration for future applications.

These materials are used both in as-produced conditions and either after secondary operations such as cutting and joining or in thermoforming (Figure 11.22) although these particular issues will not be addressed in this chapter.

In this section we will overview all the diverse applications in which foams are present. These applications will be considered in three main groups: structural applications, comfort and security and environmental issues. Polymeric foams are widely present in these applications, although most of them are not structural ones. Figure 11.23 indicates different parts based in polymeric foam present in conventional cars, exemplariness of their abundant presence in transportation.

One of the main characteristics of foams is their multifunctionality that implies these materials usually meet different simultaneous functions (e.g., structural and thermal insulation). In any case, the next section will address each application through the main performing function.

11.8.1
Structural Applications: Sandwich Cores and Structural Foams

The use of polymeric foams in structural applications is intimately associated with their incorporation as core parts in sandwiches or integral structures and,

Figure 11.22 Example of 3D thermoformed cross-linked PE foam.

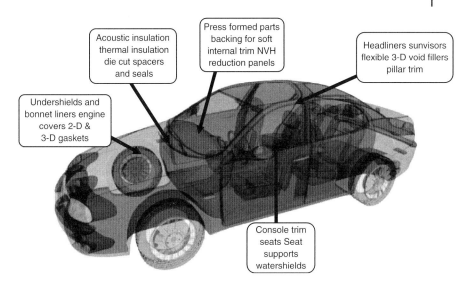

Figure 11.23 Sketch of polymeric foams applications in automotive sector.

in principle, low-density foams are rarely used for structural purposes themselves without any solid outward structure. Thus, hybrid sandwich structures present a unique combination of stiffness and lightness, which puts these materials in direct competition with shape-engineered solid structures (waffled panels, T-shaped crossbeams, etc.).

These structures can be both 2D (panels) and 3D-shaped. One possible technology for 2D structures is the direct coextrusion of a foamed core *in situ* colaminated with a metal or a solid polymer. Alternatively, foam-based sandwich panels can be obtained from already produced foams after cutting and joining operations or after application of a fiber-glass and polyester coating. 3D structural parts can be obtained from shaped panels, chopped foam parts embedded in a fiber glass and polyester or epoxy coating or, more frequently, by any of the injection technologies already described in Section 11.3.1. Size limitations of injection technologies constrain the production of large 3D-shaped parts to only the two first previously mentioned 3D methods.

Finally, foam-based 2D or 3D parts are generally used in those applications under particular lightness demands. Among other applications, we can cite sport ships, light trains and aircraft parts.

The most-used core polymer in the case of multipurpose sandwich panels is PVC that accomplishes a nice stiffness-to-weight ratio, especially considering the rigidity of the base polymers. Other typical PVC-foam parts are extrusion and coextrusion profiles which can be shaped to rather complex sections. Expansion factors considered in the majority of these applications are rather high and density is rarely below 500 kg m^{-3}. Some special cross-linked PVC foam processes reduce the density below 100 kg m^{-3}, these foams are used as core of sandwich panels

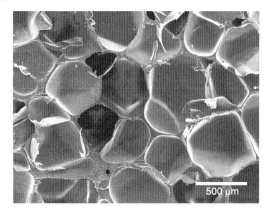

Figure 11.24 Cellular structure of a cross-linked PVC foam with a density of 100 kg m^{-3} currently used as core in sandwich panels for structural applications.

in which the sheets are fiber-glass polyester composites, although these foams are expensive and are only used for very special applications. Figure 11.24 shows the typical cellular structure of these materials.

The second most-used polymer in sandwiches cores after PVC is rigid closed-cell PU. The flexibility in formulation of PU, the wide range of densities, the processing easiness and the low costs associated with it make this material a frequently-used one. Typical core densities are in the range 50–200 kg m^{-3}. If the final foam part has only a structural purpose, rigid PU foams are blown using CO_2 as the blowing agent. The material can be foamed *in situ*, filling cavities in both plastic and metal structures supplying additional stiffness and strength with a low weight increase.

Alternative core cellular plastics such as polyimide, melamine and polysulfone can also be used in very particular sandwich applications, although the reasons for choosing such plastics obey other superimposed functionalities, in many cases related to a need of superior flame resistance and thermal stability (that will be treated in next section).

Known structural applications of polymeric foams in the automotive sector are based on epoxy foams. They were incorporated in hollow sections by setting a solid foam precursor. The precursor was later expanded and cured by the heating conditions experienced in the paint cycle. Application of this technology in Mitsubishi Diamante was an example of the use of such preforms in which the door pillars were filled with a foam insert that expanded and bonded during the paint cycle carried out at elevated temperatures [57, 58].

Fiber-reinforced automotive panels can be also produced by compression molding of foamable epoxies in cavities containing one or more piles of fabric or some type of fiber preform. Such a foaming process provides more complete wetting of the fiber reinforcement and complete filling of the cavity [55].

Finally, nowadays epoxy foams are also meeting important "semistructural" applications in the aeronautics sector by substituting diverse parts that were typically produced by high-density thermoplastic foam injection.

11.8.2
Comfort and Security

The concept of comfort might be directly associated to flexible open cell PU foams. Nevertheless, we will not treat this massive application assuming that it is not novel/interesting by itself although it involves some other important issues that will be addressed in this section.

11.8.2.1 Combustion and Flammability

Development of flame-retardant foams and/or foams with improved response under eventual combustion is becoming more and more important under the requirements of new highly restrictive standards in safety. The use of halogen-free flame-retardant systems (based on aluminum hydroxide or magnesium hydroxide) is not adequate for low-density foams because the high amount of filler needed in these systems [83, 84] hinders foamability. Therefore, halogen compounds are still used to improve the fire behavior of PS-, PU- and polyolefin-based foams, although these compounds release toxic fumes when thermally degrade.

Current standards are particularly oriented to prevent eventual disasters in public transportation systems (road transportation, but especially railway and aircraft) and impose restrictive conditions to any plastic component being part of these vehicles. Therefore, these materials must pass severe combustion and flammability tests. The normative is generally oriented in three aspects: one is associated with the energy that the material releases during its combustion. In this sense, foamed materials present advantages in comparison to their solids assuming that their characteristic fire load is a fraction of that of the solid, proportional to their relative density. Complementarily, resistance to ignition is another important factor, typically measured through the low oxygen index (LOI). Finally, the normative is also focused on minimizing the amount of smoke produced, its opacity and specially to reduce the toxicity of the fumes produced during foam combustion.

This last law requirement will impede the use of the conventional PVC, PU foam boards and sandwiches in these public transport vehicles since the toxicity of their fumes is extremely high. Alternative foamed materials have been successfully developed, such as epoxy foams, melamine foams and polyimide foams. These materials can be frequently found in light trains, airplanes, etc. Nevertheless, the passive protection they offer is subordinated to other major functions such as thermal insulation, noise reduction, vibration damping and/or structural reinforcements. Foam boards present in between the external airplane fuselage and the interior are examples of both functional and passive use of these rather innovative materials.

Law regulations on this issue are continuously being updated and in Europe we are walking into the unique standard direction. Some examples are the CEN/TS 45545–1:2009 and CEN/TS 45545–2:2009 standards which regulate the materials allowed to be introduced in trains and evaluated by different ISO and EN standards: ISO 5660 (Cone Calorimeter), ISO 5658–2 (Spread of flame test), EN 9239 (Spread of flame test – floor coverings) and ISO 5659–2 (Smoke Density Test). Aircraft sector follows the regulations of the US Federal Aviation Administration (FAA) and the European Aviation Safety Agency (EASA) which does not allow any material that keeps flaring a fire if it exceeds 15 s after burning time.

Flexible PU foams present in seats are not exempt of these normatives although there is practically no other substitute for these materials. Therefore, the strategies adopted have been different. Modified PU formulations and the use of flame-retardant additives are continuously being developed. Furthermore, resistance to ignition, low burning rates and reduced smoke emissions can be achieved by a combination of appropriate materials, particularly protective physical barriers over the PU foam. A fire-blocking layer in high performance fabrics (p-aramid with natural or synthetic fibers) is integrated into the design of the seat in order to prevent the smoldering combustion and spread of flame in case of fire.

Finally, combustion and flammability standards for vehicles of individual use (cars, trucks, etc.) are more flexible. The flammability standard adopted by the majority of vehicle manufacturers is the FMVSS 302.

The development of nanocomposite foams opens new possibilities in this aspect. It has been proven that the addition of a small amount of nanoparticles can dramatically improve the flame retardancy of polymeric systems [85, 86]. The next step would be to substitute the halogen compounds used nowadays in many applications with halogen-free flame retardant systems in which the nanoparticles could play a key role.

11.8.2.2 Fogging and Volatiles

The reduction of the amount of VOC emissions from car interior components is one important requirement of the automotive industry. Solid plastics and foams present a natural tendency to release a small amount of organic volatiles as a consequence of its nature and the processing they have suffered. Therefore, producers need to adapt their products in order to fulfill these requirements.

Odor and fogging are also consequences partially dependent on the amount of VOC released [87]. Odor regulation is oriented to limit the negative effects of "new car smell" caused by new materials in cars. On the other hand, additives or constituents in the polymeric materials are the most common source of "fog" in the glasses of new cars which is associated with chemical contamination.

Car manufacturers have created diverse test methods aimed at quantifying these three main subjects. Odor is essentially a subjective judgment but has been regulated under VDA 270. On the other hand, VOC standards are included in VDA 277 and 278 regulations whereas fogging characteristics are quantified under DIN 75201 Methods A and B.

Reduction of VOC emissions in foams is a more complex task in comparison to solids as they present a higher surface-to-volume ratio and the presence of blowing agent residues contributes to VOC emissions. One of the simplest solutions is the use of a PBA such as nitrogen or CO_2 and eliminating the presence of CBAs that are known to be one important source of volatiles. Nevertheless, producers have adopted many other different solutions such as heat treatments, forced ventilation and as a last option, modifications in the formulations.

11.8.2.3 Thermal Insulation

Thermal insulation is necessary in any means of transport with independency of the season or planet location. Thus, considering that polymeric cellular materials are the best heat insulators, it implies that they are deeply used in cars, trains, planes, etc. Cross-linked polyolefin foams (because of the possibility of thermoforming) are currently used to improve thermal insulation in cars. Polyolefin foams are, for instance, used in door panels whereas rigid PU is used in roofs, floors and trays in cars. Insulation is especially important in planes were external temperatures are frequently below −50 °C. In this particular case, thermal insulation needs to be combined with flame retardancy properties.

Thermal insulation is also important for the transportation of refrigerated and frozen food (trucks, vans, etc.). Heat insulation in this application is obtained by using closed cell PU foam the cells of which usually contain a high molecular weight gas of low thermal conductivity (Section 11.5). The foam can be either injected in a walled cavity or, alternatively, sandwich panels can be used to create the insulated room. Figure 11.25 shows an example of a panel for that application.

11.8.2.4 Sound Proofing and Vibration Damping

Open cell foams present excellent sound absorption properties and are used in numerous applications in different means of transport. Car, trains and airplanes intensively use foam based-solutions to accomplish this function, although generally these noise absorption and transmission attenuation involucrate multiwall parts. Foams are generally settled in the first or second layer between the noise emitting part and the traveler/driver. On the other hand, noise reduction is frequently combined with other applications (structural, heat insulation, etc.) that strongly modify the optimum design. Thus, closed cell foams are also present in these designs and even, sometimes, open cell foams are not present as transmission losses are sufficiently high and there is no need for sound absorption.

Figure 11.25 PU foam sandwich used for thermal insulation and structural applications in refrigerated trucks.

Vibration damping is another feature of polymeric cellular materials which is generally associated with the previous one (acoustic absorption). Interior padding applications usually combine damping, noise attenuation and structural functions.

11.8.3
Environmental Issues

Worldwide legislations on fuel economy, tailpipe emission standards, materials recyclability, pedestrian safety and noise reduction are directly or indirectly associated with the materials that are used in the different vehicles. This normative has become particularly restrictive in recent years for personal-use vehicles but it is also currently affecting other means of transport such us planes and heavy road vehicles.

11.8.3.1 Energy Saving and Raw Materials Reduction

Reduction of energy consumption in any means of transport can only be achieved by two additive (nonexclusive) options. The first one considers the increase of efficiency for the system that generates the movement (motor), the reduction of losses in the kinetic energy (friction) or efficient recovery of casual energy losses (braking). Alternatively, the second option passes through diminution in the overall weight of the means of transport which is directly proportional to the energy needed to move the vehicle.

The incorporation of foamed cellular polymers with reduced density would enable this possibility. One of the most interesting alternatives consists in substituting the actual solid materials by those partially foamed (10–20% porosity) without any reduction in the properties of the substituted part. This possibility can be realized by using microcellular nanocomposites. The use of a reinforced polymer in the matrix increases the properties of the equivalent solid, in such a way that the final foamed part presents the same properties as the substituted part.

This concept has not still been translated into a manufactured vehicle foam-part because of higher costs and processing issues but will probably be accomplished in the following years similar to the progressive substitution of solid composites by solid nanocomposites that have recently started in diverse applications [88, 89].

On the other hand, if we consider that foams require less amount of materials to produce a given part we can assume that, somehow, these materials contribute to find more sustainable solutions in a world with increasing demand of feedstocks. The reduced use of polymer quantities during production has the collateral advantage of cost reduction. Owing to this, there is a clear trend of substituting solid parts by foamed parts in the automotive sector, with highly competitive prices, and also in any other of the transportation sectors. For instance, the Mucell Technology (Section 11.4.1) is being used to produce plastic parts with weight reduction up to 20% and costs reduction of 10–15%.

11.8.3.2 Recyclability and Biodegradability

Different types of recyclable and nonrecyclable polymeric foams are present in cars. In principle, cross-linked and thermosets foams are nonrecyclable materials they cannot be remelted but can at least recovered and re-evaluated in different ways.

Non cross-linked polyolefin foams can be readily recycled by granulation and re-extrusion. There are many patents describing different ways for grinding foams and later pelletizing the material [90–92]. Such pellets can be used to produce foams for different types of products. For instance, the recycled material can be incorporated with virgin polymer at levels of 10–20% to produce a new kind of foams with a small effect on its properties.

On the contrary, cross-linked and thermosets foams cannot be easily recovered and used to manufacture new products, as cross-linking interferes with melt flow. There are four main solutions that are used in this case: The first one is to obtain energy by incineration of the polymer; the second one tries to depolymerize the material by chemical means; another interesting solution is to powder the foam and use the powder as an additive for the production of new materials. This solution has been used, for instance, for flexible PU foams – the foam is cryo-powdered and the resultant powder is used as a filler and in small proportions to produce new PU foams [48, 93]. Finally for cross-linked foams, it is possible to produce new products by using a thermoforming process of the residues of the initial materials [94–96].

On the other hand, the incorporation of biodegradable polymers and polymer foam in transport vehicles is still inexistent but in recent decade, the tendency points to the possibility of a partially compostable car in the coming years. Actually, this concept has been recently released and car manufacturers have started their R&D on this possibility. In this sense, polylactic acid (PLA) and polyhydroxybutyrate (PHB) are promising materials for the production of foamed parts for the transport industry and are being largely studied. They have properties comparable to those of petroleum-derived resins and they have enough life time (degradation under normal conditions is almost nonexistent) to be used in this application.

11.9
Summary and Outlook

Polymer foams nowadays have many different applications in the transport industry. The cellular plastic, depending on the part produced, can have many different functions: structural, cushioning, thermal insulation, acoustic absorption and weight reduction, among others. In most of the applications the foam accomplishes several purposes simultaneously.

Polymer foams are used in the core of structural low-weight sandwich panels in railway, naval, aeronautic and automotive industry. Cross-linked PVC and rigid PU are the materials currently used for this type of structures. The current trend towards further weight reduction implies that the use of this type of sandwich construction will become more important in current and future designs. The development of alternative foams for the core of these panels is underway and new foamed cores based on PET, PP and epoxy are ready to be commercialized in the near future. Additionally, these cores could be produced using nanocomposites – novel materials that have improved cellular structures and mechanical properties.

An additional advantage of sandwich constructions is the thermal insulation offered, which cannot be obtained using metal parts. On the other hand, the main drawbacks are the complicated processing (especially when glass fibers composites are used as sheets) and the poor flame retardancy of these materials. Intense research work is being done to improve the fire behavior of these materials by replacing the current halogen compounds that produce toxic fumes with halogen-free solutions. Once again the use of nanoparticles is one interesting possibility to overcome this issue.

Another interesting trend is the substitution of thermoplastic solid parts that need to have certain mechanical properties with cellular plastics or microcellular plastics with low porosity. Different variants of the injection molding technology (low pressure, high pressure, gas-counter pressure, coinjection or microcellular) are being used. The key objective here is to reduce weight and cost and at the same time maintain good performance. This performance can be accomplished because all the previous technologies allow the production of material with a skin-core morphology that optimizes the relative stiffness and strength at low weights. Significant efforts have been done and the trend will continue in the next few years trying to improve the previous technologies or develop new ones with the idea of overcoming some of the main disadvantages of current production methods: poor surface quality, low weight reduction (usually not higher than 25%), parts with a limited size and significant investments for some of the previous technologies (high pressure, gas-counter pressure, coinjection or microcellular).

Two new trends that could significantly modify the type of materials used nowadays are the development of bioplastic foams and the possibility of producing submicrocellular foams.

Foams based on natural and biodegradable materials are already in use in other sectors such as in packaging. The development of semidurable parts is underway and it is almost a certainty that they could be used in the transport industry in about 10–15 years. The key advantage of using these types of materials is the excellent life cycle they have, a property that is now demanded in some transport industries, such as automotive.

On the other hand, the development of cellular materials with cell sizes well below 1 µm would introduce in the market cellular plastics with properties not accomplished by any other material. These properties are very high specific mechanical properties, low thermal conductivity and even transparency (if the polymer matrix is amorphous) would make these materials ideal for structural applications in which thermal insulation and/or transparency are needed.

11.10
Further Reading

The amount of scientific and technical literature on the field of polymeric foams is vast. Probably, the text book that has a more comprehensive treatment of the structure–property relationships for cellular materials is the one from Gibson and Ashby [97]. Although the book is focused on all types of cellular material, most of the scientific knowledge explained can be directly applied to polymer foams. In a similar way, the book by Cunningham and Hilyard [98] is very interesting for those readers who need to go into the details of the structure–property relationships for polymer-based foams.

On the other hand, if the interest is focused on the production technologies of different types of polymer foams, the most comprehensive literature is included in several handbooks [1–3, 99]. These books have chapters devoted to rigid and flexible PU foams, polyolefin-based foams, polystyrene, PVC, epoxy foams, etc., in which the main technologies to produce the materials and the typical structure and properties obtained are summarized. These handbooks also have chapters on blowing agents for foaming and on the fundamentals of the foaming process, that is, the chemical and physical mechanisms involved during the processing of the materials.

The previous literature mainly covers ''classical foams''; these are products developed several years ago and that are produced on an industrial scale. However, new topics such as microcellular foams, nanocellular foams, nanocomposites foams or bioplastic foams that have developed recently can only be found in more specific literature, that is, books on the specific topic of papers published in recent years.

In the case of microcellular foams a few books have been published recently [29, 30]. These books explain the technologies in detail to produce these materials and summarize the hundreds of papers published on the structure and property of foams having cells in the microcellular range. The review published by Kumar [25] is also very informative .

Nanocomposite, nanocellular and bioplastic foams are right now on the frontier of science and many papers on these areas appear each year. The review by L.J. Lee *et al.* [80] is a good starting point to understand the topic of nanocomposites foams. On the other hand, Refs. [37–39] explain the current state of the art in the topic of nanocellular foams produced by gas dissolution. A good summary of the field of bioplastic foams is given in Ref. [40] of the chapter. All these topics evolve very quickly and it is necessary to read the latest papers published to really understand the current situation of each area.

In this sense, there are several conferences in which the latest advancements in these topics are presented every year. The most important conferences are the International Conference on Foams and Foams Technology (FOAMS conference) organized by the Society of Plastic and Engineers (SPE), the Blowing Agents and Foaming Processes conference organized by Smithers Rapra and the Annual Technical Conference (ANTEC) organized by SPE.

Acknowledgments

Financial support from the Spanish Ministry of Science and Innovation and FEDER (MAT2009-14001-C02-01 and MAT2012-34901), the European Spatial Agency (Project MAP AO-99-075), and Juan de la Cierva contract of E. Solórzano by the Ministry of Economy and Competitiveness (JCI-2011-09775) are gratefully acknowledged.

References

1. Business Communications Company Inc. (2004) RP-120X Polymeric Foams – Updated Edition, USA.
2. Eaves, D. (2004) *Handbook of Polymeric Foams*, Rapra Technology, Shrewsbury.
3. Klempner, D. and Sendijarevic, V. (2004) *Handbook of Polymeric Foams and Foam Technology*, 2nd edn, Hanser Publishers, Munich.
4. Khemani, K.C. (1997) Polymeric foams: an overview, in *Polymeric Foams: Science and Technology*, ACS Symposium Series (ed K.C. Khemani) Chapter 1, American Chemical Society, Washington, DC.
5. Glicksman, L.R. (1994) in *Low Density Cellular Plastics: Physical Basis of Behaviour* (eds N.C. Hilyard and A. Cunningham), Chapman & Hall, London.
6. Hurnik, H. (1998) Proceedings of Addcon World'98, Additives of the new millenium, London, UK, 1998, Paper No.16.
7. Quinn, S. (2001) *Plast. Addit. Compd.*, **3**(5), 16.
8. Xu, X. and Park, C.B. (2009) Injection foam molding, in *Injection Moulding Technology and Fundamentals* (eds M.R. Kamal, I. Avraam and L. Shih-Jung), Carl Hanser Verlag, Munich.
9. DaJliels, V.A. (1999) US Patent 6,000,925.
10. L. J. Lee, in *Fundamentals of Reaction Injection Moulding*, Chapter 4, Comprehensive Polymer Science, Vol. 4 (2) (eds G Allen, J. C Bevingto). Pergamon Press, Oxford, New York (1989).
11. Scmerdiiev, S. (1982) *Introduction to Structural Foam*, Society of Plastics Engineers, p. 33.
12. Carveth, P., Stone, A. and White, L. (1987) US Patent 4,657,152.
13. Eckardt, H. and Davies, S. (1979) *Plast. Rubber Int.*, **4**(2), 72–74.
14. Pierick, D. and Jacobsen, D.K. (2001) *Plast. Eng.*, 46–51.
15. Knight, M. (2000) *Plast. Technol.*, 40–49.
16. Jacobsen, K. and Pierick, D. (2000) SPE ANTEC Technical Papers, pp. 1929–1933.
17. Hengesbach, H.A. and Egli, E. (1979) *Plast. Rubber Process*, **1**(2), 56–60.
18. Lassor, R.D. and Caropreso, M. (1985) *Plast. World*, 51–52.
19. Thomas, N.L. (2004) in *Handbook of Polymeric Foams* (ed D. Eaves) Chapter 6, Rapra Technology, pp. 123–154.
20. Park, C.B. (2000) Continuous production of high density and low density microcellular plastics in extrusion, in *Foam Extrusion* (ed S.T. Lee), Technomic Publishing Co, Lancaster, PA.
21. Eaves, D. (2004) in *Handbook of Polymeric Foams* (ed D. Eaves) Chapter 8, Rapra Technology, pp. 173–206.
22. Martínez-Dez, J.A., Rodríguez-Pérez, M.A., de Saja, J.A., Arcos y Rábago, L.O. and Almanza, O.A. (2001) *J. Cell. Plast.*, **37**, 21–42.
23. Puri, R.R. and Collington, K.T. (1988) *Cell. Polym.*, **7**, 219.
24. Rodriguez-Perez, M.A., Hidalgo, F., Solorzano, E. and de Saja, J.A. (2009) *Polym. Test.*, **28**, 188–195.
25. Kumar, V. (2004) in *Handbook of Polymeric Foams* (ed D. Eaves) Chapter 610, Rapra Technology, pp. 243–268.
26. Almanza, O.A., Rodríguez-Pérez, M.A. and de Saja, J.A. (2001) *Polymer*, **42**, 7117.

27. Martini-Vvedensky, J.E., Suh, N.P., and Waldman, F.A. (1984) Inventors; Massachusetts Institute of Technology, assignee. US Patent 4,473,665.

28. Suh, N.P. (1996) in *Innovation in Polymer Processing* (ed J.F. Stevenson) Chapter 3, Hanser/Gardner Publications, Cincinnati, pp. 93–149.

29. Xu, J. (2010) *Microcellular Injection Moulding*, John Wiley & Sons, Inc., Hoboken, NJ.

30. Okamoto, K.T. (2003) *Microcellular Processing*, Carl Hanser Verlag, Munich.

31. Shimbo, M., Nishida, K., Nishikawa, S., Sueda, T., and Eriguti, M. (1998) in *Porous, Cellular and Microcellular Materials*, MD Series, Vol. **82** (ed V. Kumar), ASME, p. 93.

32. Pfannschmidt, O. and Michaeli, W. (1999) SPE ANTEC, Technical Papers 2100–2103.

33. Witzler, S. (2001) Injection Molding Magazine (Dec.), p. 80.

34. Defosse, M. (2009) Modern Plastics Worldwide (Dec. 14–15).

35. Yokoyama, H., Li, L., Nemoto, T., and Sugiyama, K. (2004) *Adv. Mater.*, **16**, 1542.

36. Hedrick, J.L. *et al.* (1996) *React. Funct. Polym.*, **30**, 43–53.

37. Otsuka, T., Taki, K., and Ohshima, M. (2008) *Macromol. Mater. Eng.*, **293**, 78–82.

38. Reglero-Ruiz, J.A., Dumon, M., Pinto, J., and Rodriguez-Perez, M.A. (2011) *Macromol. Mater. Eng.*, 296–300.

39. Thermoplastic nanocellular foams with low relative density using CO_2 as the blowing agent. (2011) Proceedings of the SPE FOAMS Conference, Iselin, New Jersey, 2011.

40. Tarng-Lee, S., Park, C.B., and Ramesh, N.S. (2007) *Polymeric Foam Series, Science and Technology* Chapter 8, Taylor & Francis, pp. 165–193.

41. Nabar, Y., Raquez, J.M., Dubois, P., and Narayan, R. (2005) *Biomacromolecules*, **6**(2), 807–817.

42. Guan, J. and Hanna, M.A. (2005) *Ind. Eng. Chem. Res.*, **44**(9), 3106–3115.

43. Di, Y.W., S.I., and Di, M.E. (2005) *J. Polym. Sci., Part B: Polym. Phys.*, **43**(6), 689–698.

44. Vercelheze, A.E.S., Fakhouri, F.M., and Dall'Antonia, L.H. (2012) *Carbohydr. Polym.*, **87**, 1302–1310.

45. Salgado, P.R., Schmidt, V.C., and Ortiz, S.E.M. (2008) *J. Food Eng.*, **85**, 435–443.

46. Ghosh, S., Gutierrez, V., Fernández, C., Rodriguez-Perez, M.A., Viana, J.C., Reis, R.L., and Mano, J.F. (2008) *Acta Biomater.*, **4**, 950–959.

47. Saunders, J.H. and Frisch, K.C. (1962) *Polyurethanes: Chemistry and Technology*, Applied Science Publishers, London.

48. Woods, G. (1987) *Flexible Polyurethane Foam: Chemistry and Technology*, Applied Science Publishers, London.

49. Herrington, R.M. and Hock, K. (1991) *Flexible Polyurethane Foams*, The Dow Chemical Company, Midland, MI.

50. Zhang, X.D., Neff, R.A., and Macosko, C.W. (2004) in *Polymeric Foams Mechanisms and Materials* (eds S.T. Lee and N.S. Ramesh) Chapter 5, CRC Press LLC, pp. 148–181.

51. Grünbauer, H.J.M., Bicerano, J. *et al.* (2004) in *Polymeric Foams Mechanisms and Materials* (eds S.T. Lee and N.S. Ramesh) Chapter 7, CRC Press LLC, pp. 262–318.

52. Britton, DJ (1989) *Cell. Polym.*, **8**, 125.

53. Oertel, G. (ed.) (1985) *Polyurethanes Handbook*, 1st edn, Carl Hanser Verlag, Munich.

54. Gum, W.F., Riese, W., and Ulrich, H. (eds) (1992) *Reaction Polymers Chemistry: Technology, Applications, Markets*, 1st edn, Carl Hanser Verlag, Munich.

55. Domeier, L.A. (2004) in *Handbook of Polymeric Foams and Foams Technology* Chapter 10, Carl Hanser Verlag, Munich, pp. 347–366.

56. Bledzki, A.K., Kurek, K., and Gassan, J. (1998) *J. Mater. Sci.*, **33**, 3207.

57. Komai, K., Minoshima, K., Tanaka, K., and Tokura, T. (2002) *Int. J. Fatigue*, **24**, 339.

58. Kawaguchi, T. and Pearson, R.A. (2003) *Polymer*, **44**, 4229.

59. Costeux, S., Vo, C.V. and Lawrence, S.H. (2011) Long term performance of insulation foams. Proceedings of the International Conference on Foams and Foams Technologies, Seattle, September, 2010.

60. Vo, C.V., Bunge, F., Duffy, J., and Hood, L. (2011) *Cell. Polym.*, **30**, 3.

61. Kaemmerlen, A., Vo, C., Asllanaj, F., Jeandel, G., and Baillis, D. (2010) *J. Quant. Spectrosc. Radiat. Heat Transfer*, **111**, 865–877.

62. Giannelis, E.P. (1996) *Adv. Mater.*, **8**(1), 29.

63. Alexandre, M. and Dubois, P. (2000) *Mater. Sci. Eng., R.*, **R28**(1–2), 1.

64. Vaia, R.A. and Giannelis, E.P. (2001) Polymer nanocomposites: status and opportunities. *MRS Bull.*, **26**(5), 394.

65. Hammel, E., Tang, X., Trampert, M., Schmitt, T., Mauthner, K., Eder, A. et al. (2004) *Carbon*, **42**(5–6), 1153.

66. Sandler, J., Shaffer, M.S.P., Prasse, T., Bauhofer, W., Schulte, K., and Windle, A.H. (1999) *Polymer*, **40**, 5967.

67. Gong, X., Liu, J., Baskaran, S., Voise, R.D., and Young, J.S. (2000) *Chem. Mater.*, **12**, 1049.

68. Theng, B.K.G. (1974) *The Chemistry of Clay-Organic Reactions*, John Wiley & Sons, Inc., New York.

69. Zhu, J., Kim, J., Peng, H., Margrave, J.L., Khabashesku, V.N., and Barrera, E.V. (2003) *Nano Lett.*, **3**, 1107.

70. Liao, K. and Li, S. (2001) *Appl. Phys. Lett.*, **79**, 4225.

71. Zhai, W., Yu, J., Wu, L., Ma, W., and He, J. (2006) *Polymer*, **47**, 7580–7589 R. Sandler *et al.*, op.cit.

72. Zhai, W., Park, C.B., and Kontopoulou, M. (2011) *Ind. Eng. Chem. Res.*, **50**, 7282–7289.

73. Zhai, W., Kuboki, T., Wang, L., Park, C.B., Lee, E.K., and Naguib, H.E. (2010) *Ind. Eng. Chem. Res.*, **49**, 9834–9845.

74. Saiz-Arroyo, C., Escudero, J., Rodríguez-Pérez, M.A., and de Saja, J.A. (2011) *Cell. Polym.*, **30**, 63–78.

75. Cao, X., Lee, L.J., Widya, T., and Macosko, C. (2004) *Annual Technical Conference, Society of Plastics Engineers*, 62nd edn, Vol. **2**, p. 1896.

76. Cao, X., Lee, L.J., Widya, T., and Macosko, C. (2005) *Polymer*, **46**(3), 775.

77. Pardo-Alonso, S., Solórzano, E., Estravís, S., Rodriguez-Perez, M.A., and de Saja, J.A. (2012) In situ evidence of the nanoparticle nucleating

effect in polyurethane–nanoclay foamed systems. *Soft Matter*, **8**, 11262.

78. Pardo-Alonso, S., Solórzano, E., and Rodriguez-Perez, M.A. (2013) Time-resolved x-ray imaging of nanofiller-polyurethane reactive foam systems. *Colloids and Surfaces A: Physicochem. Eng. Aspects* doi. 10.1016/j.colsurfa.2013.01.045.

79. Pardo-Alonso, S., Solórzano, E., Brabant, L., Vanderniepen, P., Dierick, M., Van Hoorebeke, L., and Rodríguez-Pérez, M.A. (2013) 3D Analysis of the progressive modification of the cellular architecture in polyurethane nanocomposite foams via X-ray microtomography. *Eur. Polym. J.* doi 10.1016/j.eurpolymj.2013.01.005.

80. Lee, L.J. *et al.* (2005) *Compos. Sci. Technol.*, **65**, 2344–2363.

81. Tomasko, D.L. (2003) *Curr. Opin. Solid State Mater. Sci.*, **7**, 407–412.

82. Cao, X., Lee, L., Widya, T., and Macosko, C.W. (2005) *Polymer*, **46**, 775–783.

83. Hull, T.R. and Price, D. (2003) *Polym. Degrad. Stab.*, **82**, 365.

84. Hull, T.R. and Woolley, W.D. (2000) *Polym. Int.*, **49**, 1193.

85. Verdejo, R., Barroso-Bujans, F., Rodriguez-Perez, M.A., de Saja, J.A., Arroyo, M., and Lopez-Manchado, M.A. (2008) *J. Mater. Chem.*, **18**, 3933–3939.

86. Verdejo, R., Saiz-Arroyo, C., Carretero-Gonzalez, J., Barroso-Bujans, F., Rodriguez-Perez, M.A., and Lopez-Manchado, M.A. (2008) *Eur. Polym. J.*, **44**, 2790–2797.

87. Guadarrama, A., Rodriguez-Mendez, M.L., and De Saja, J.A. (2002) *Anal. Chim. Acta*, **455**, 41–47.

88. Krüger, P. (2007) Nanocompostes for Automotive Applications, *http://www.speautomotive.com/ SPEA_CD/SPEA2007/pdf/e/nano composites_part1_paper1_kruger_bms.pdf* (accessed 21 December 2012).

89. Patricia Tibbenham, Ford Motor Company Developing Polymer Nanocomposites for Automotive Applications, *http://www.speautomotive.com/ SPEA_CD/SPEA2006/PDF/h/h1.pdf* (accessed 21 December 2012).

90. Kühnel, W. (1981) US Patent 4,246,211.

91. Tatsuda, N., Fukumori, K., Sato, N., Sahara, S., and Ono, H. (2000) US Patent 6,090,862.

92. Bourland, L., Freundlich, R.A., Nwana, R.G., and Herrant, V.W. (2002) US Patent 6,384,093 B1.

93. Estravís-Sastre, S. and Rodríguez-Pérez, M.A. (2011) Polyurethane Foam Composites: Relationship between Mechanical Properties and Filler-Matrix Chemical Interaction SPE Eurotec® Conference, Barcelona, Spain, November 2011.

94. Svirklys, F. and Shanklin, D. (2005) US Patent 6,932,540 B2.

95. Svirklys, F., McGregor, W., Marsh, R., and Shanklin, D. (2003) US Patent 6,558,548 B2.

96. Saiz-Arroyo, C., Rodríguez-Pérez, M.A., and de Saja, J.A. (2012) Production and characterization of crosslinked low density polyethylene foams using as raw materials waste of foams with the same composition. *J. Appl. Polym. Sci.*, **52**, 751–759.

97. Gibson, L.J. and Ashby, M.F. (1997) *Cellular Solids – Structure and Properties*, 2nd edn, Chapter 1, Cambridge University Press, Cambridge.

98. Cunningham, A. and Hilyard, N.C. (1994) NC physical behavior of foams – an overview, in *Low Density Cellular Plastics: Physical Basis of Behaviour* (eds N.C. Hilyard and A. Cunningham) Chapter 1, Chapman & Hall, London.

99. Rodríguez-Pérez, M.A. (2005) *Adv. Polym. Sci.*, **184**, 87–126.

12
Metal Foams
Joachim Baumeister and Jörg Weise

12.1
Introduction

In comparison to foams based on polymers, glasses, or other matrices, metal foams are a relatively young material. Although some older metallurgical processes could result also in porous metal structures, first specifically intended trials for the production of porous metals were carried out in the 1920s [1]. As for foams, in general, there exist two border cases of metal foam structures [2]: foams in which the pores are all connected to each other and with the environment (open porous foams) and foams in which every single pore is completely enclosed by the (metal) matrix (closed porous foams). Most technical metal foams can approximately be assigned to one of those groups, although there are also foam types such as advanced pore morphology (APM) or sintered hollow sphere structures, which combine considerable fractions of open and closed porosities (Figure 12.1).

For functional applications, open porous metal foams are dominant. Several products can be mentioned such as oil–mist-separator structures in civil airplanes [3] or porous surface coverings of base frames of prototype pantographs of Japanese high-speed trains for noise reduction [4]. However, the most prominent example is the application of INCO nickel foams in pasted-electrode batteries with a production volume of about 3 000 000 m^2 foam per year [5] (Figure 12.2).

In the field of structural applications, closed-pore metal foams are the more important group. This is related to the fact that most closed-pore structures show, in principle, higher strengths in comparison to open porous structures with the same matrix and density as was shown, for example, by theoretical calculations of Gibson and Ashby [2]. In the following, this chapter, therefore, concentrates mainly on this group of metal foams.

First activities for the production of closed-cell metal foams already started in the 1920s and were continued in the late 1940s. However, the activities were really pushed in the 1990s. In this time, various national and international public and industrial projects were funded for the development and optimization of production processes for metal foams and the evaluation of their property spectrum and potential applications. In the course of those projects, a diversity of processes and

Structural Materials and Processes in Transportation, First Edition.
Edited by Dirk Lehmhus, Matthias Busse, Axel S. Herrmann, and Kambiz Kayvantash.
© 2013 Wiley-VCH Verlag GmbH & Co. KGaA. Published 2013 by Wiley-VCH Verlag GmbH & Co. KGaA.

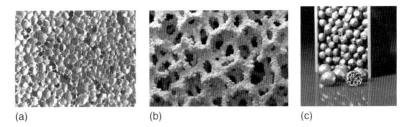

(a)   (b)   (c)

Figure 12.1 (a) Closed porous Foaminal foam and (b) open porous Alantum foam. (c) APM foam with combined open and closed porosities.(Courtesy of Alantum.)

Figure 12.2 INCOFOAM Ni foam coil. (Courtesy of V. Paserin, INCO.)

foam types were developed and brought up to industrial production. Examples of industrially produced (mainly closed-cell) metal foams are Foaminal and Hollomet hollow metal foam structures in Europe, Alporas aluminum foams in Japan, and stabilized aluminum foams in North America.

The interest in closed-cell metal foams is based on the fact that this material group offers properties such as [6]

- low density,
- high specific stiffness,
- high thermal stability,
- very good ability to absorb energy during deformation,
- good acoustic and vibration damping.

In several of these properties, they resemble the group of polymer foams – a nonsurprising fact because many of the specific foam properties are fundamentally based on the specific pore structure and not on the matrix properties (subchapter 7.1). In comparison to polymer foams, however, metal foams stand out by their high thermal stability, higher thermal and electrical conductivities, and (mostly) higher mechanical strength.

For the understanding of the application potential of metal foams, their strengths, and their weaknesses, it is important to highlight some fundamental relationships of the mechanical behavior of foams. In Figure 12.3, typical stress–strain curves

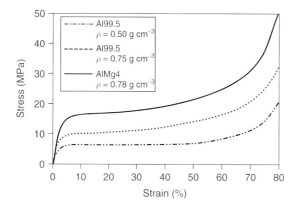

Figure 12.3 Compression stress–strain curves for closed-cell Foaminal aluminum foams.

for several aluminum foams are shown. In all curves, a typical stress plateau can be observed. This plateau is the basis for the very good ability of such (not only metal) foam structures to absorb deformation energy. The more pronounced this plateau is the better is the energy absorption efficiency of the material. As can be seen, the stress level in the plateau region depends on the alloy type, the density, and the respective porosity of the foam. The latter parameter exhibits a dominating influence on most properties of foams. Some properties such as the compression strength, the mean plateau stress level, and the Young's modulus depend approximately quadratically on the relative density of the foam (see Eqs. (12.1) and (12.2) and also introductory subchapter 7.1).

However, when looking at the specific stiffness of constructions, the low elastic modulus of foams is compensated by their low density, as with this, section area and thus moments of inertia can be increased leading to the high specific stiffness of metal foam structures. The increase of stiffness by replacing solid material by porous material depends on the basic component shape and is especially significant for platelike components [2, 6].

$$\sigma_{foam} \approx \sigma_{matrix} \left(\frac{\rho_{foam}}{\rho_{matrix}} \right)^n \quad 1.5 \leq n \leq 2.5 \tag{12.1}$$

$$\frac{E_{foam}}{E_{matrix}} = \left(\varphi \frac{\rho_{foam}}{\rho_{matrix}} \right)^2 + (1 - \varphi) \left(\frac{\rho_{foam}}{\rho_{matrix}} \right),$$
$$\varphi = \text{parts of the overall mass, which is in the struts} \tag{12.2}$$

Owing to the above-mentioned general correlations between the mechanical properties and the relative density, some advantages are offered by foams, which are based on matrix alloys with high specific Young's moduli and specific strengths similar to, for example, aluminum. This is one of the reasons why the majority of scientific and industrial structural foam activities were in aluminum foams. Other foams made, for example, from steel alloys offer, however, specific advantages when high operation temperatures or good ductility are needed.

Most types of metal foams exhibit a rather low ductility so that in the case of tensile or bending loads, foams should be used in combination with compact materials, for example, in the shape of sandwiches, filled tubes, or other filled hollow structures (Figure 12.4). The foam core acts mainly as a spacer between the compact face sheets, which bear the in-plane tensile or compression stresses. The foam is subjected mainly to shear stresses; therefore, the shear modulus is the dominating foam property for this type of application. In comparison to other sandwich materials, metal foam sandwiches offer several advantages such as a more isotropic core structure, higher possible application temperatures, and an improved shaping ability.

Metal foams have high-vibration damping abilities (see, e.g., [6]) and resemble in this way other heterogeneous materials such as gray cast iron, which is heavily used in mechanical engineering. In cast iron, the damping is caused by interactions of the metal matrix and the graphite and easy dislocation movement in graphite [7], whereas in metal foams such as Foaminal, the damping mechanism is dominated by friction in small cracks and dislocation movement due to stress concentration in certain areas of the foam structure [8].

It is important to stress that intensive investigations of metal foams showed that most of the mechanical properties are dominated by the properties of the matrix alloy and the relative foam density and not by the pore structure. Although there is some small influence of the foam structure, this is a second-order effect and can be neglected in most cases as long as no so-called monster-pores occur, which have to be regarded as material faults.

In summary, the potential of closed-cell metal foams lies in a good combination of high specific stiffness and good deformation and vibration-energy-absorbing capabilities, whereas the main weaknesses of foams are the quadratic correlation of most mechanical properties to the relative density and the reduced ductility. Therefore, main research and industrial activities have been related to applications in which the *combined properties of foams* can offer advantages in comparison to alternative material solutions (a survey of research projects is, e.g., given in [9]). For some applications, the rather isotropic structures and properties of foams can be important, for example, in comparison to honeycomb structures.

Not only the properties of the final product but also all aspects related to the production itself such as process control, costs, and quality management

Figure 12.4 Examples for combinations of aluminum foam with compact structures.

are essential. Those different come to the fore for the different foam types. One main concern should be mentioned, however, at this place – the mostly stochastic character of foams and their rather complex structure. Both aspects delimitate the applicability of various standard inspection methods such as, for example, radioscopy and are important drawbacks for applications in safety-sensitive domains such as aircraft production.

Production processes for metal foams for structural applications are dominated by two groups: melt and powder metallurgical approaches. In the following, the main technical developments of the last few years will be illustrated and evaluated for both groups. This description will be complemented with another – rather new – development, which is currently the focus of various research groups, porous structures based on wires and other half-finished parts, such as, for example, Kagome structures.

12.2
Foams Produced by Means of Melt Technologies

There exist several technologies to produce metal foams – mostly aluminum foams – by feeding gas into the molten metal. The different approaches developed differ mainly in the sources of the gas and the foam stabilization. For example, in the ALCAN process, gas is blown directly into the melt by means of rotating impellers and the foam is continuously skimmed from the melt surface to a conveyor belt, solidified, and cut later into simple shapes as blocks and plates. In comparison, Alporas foams are produced by introducing gas-releasing solid agents such as TiH_2 into a melt, which has been thickened before by means of the addition of calcium and subsequent stirring [6]. This approach seems to allow a more homogeneous dispersion of gas throughout the melt.

Similar to aqueous foams, also foams produced from metal melts need foam stabilization. In comparison to aqueous systems, only a limited number of stabilizing effects can be used in metal melts; therefore, the main approaches are to introduce or generate a multitude of small solid particles such as oxides, carbides, ceramics, or intermetallic phases in the melt. Fraction of particles in the resulting foam structure can be up to 20 vol% (Figure 12.5), which can cause increased brittleness of foams and also increased tool wear during the subsequent machining of the foams.

Advantages of melt processes for the production of metal foams are the comparatively low costs and large possible foam dimensions; drawbacks are the limitation to simple product shapes and a very rough surface appearance due to pores cut open at the surfaces. On the other hand, the foams produced by melt processes are often used in architecture and design [11, 12] exactly because of this unique surface appearance.

Melt-metallurgical processes for foam production are well established. So, development in the last few years has concentrated on process variants to facilitate the production of more complex shapes and on subsequent processes such as bonding.

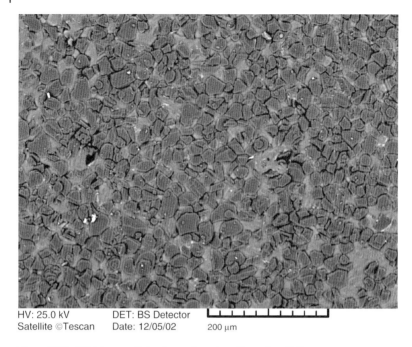

HV: 25.0 kV   DET: BS Detector
Satellite ©Tescan   Date: 12/05/02   200 μm

Figure 12.5 SEM image of aluminum foam lamella surface [10].

However, one relatively new method, developed by the company Alcoa, USA [13, 14], should be mentioned: in this process, calcium carbonate is added to a melt of an Al–Si–Mg alloy as well for the generation of stabilizing particles as for the generation of blowing gases, although for the latter also other chemical agents such as hydrides can be employed.

Several methods have been developed in order to facilitate the production of more complex foam components than blocks and plates. All approaches have in common that the shaping is done when the foam is in the liquid or semiliquid state by casting or injecting the foam into molds or by other shaping measures such as rolling [15, 16].

Development of subsequent processes has mainly been aimed at the combination of foam semifinished products with other materials such as face sheets or pipes to constitute the final product. One example is shown in Figure 12.6, in which aluminum foam plates were combined with meandering cooling pipes for improved and more homogeneous heating in comparison to conventional dry-wall solutions. Recently, sandwich structures were developed by GLEICH Aluminiumwerk GmbH & Co. KG, Kaltenkirchen, which consist of an Alporas foam core and adhesive-bonded aluminum EN AW 5754 (AlMg$_3$) surface sheets of the thickness of 1–3 mm. They have dimensions of up to 1500 mm × 3000 mm, are regarded as noncombustible (following German standard DIN 5510), and can be stud- and spot-welded. Applications are mostly for fixture constructions but first automotive application studies were already done (subchapter 5).

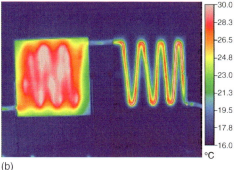

(a)   (b)

Figure 12.6 (a) Metal foam plates with integrated pipes and (b) comparison of surface temperature distribution of aluminum foam panel and dry wall after heating for 15 min with 40 °C water. (Courtesy of Institute of Materials and Machine Mechanics SAS, Bratislava, and IWE GmbH & Co. KG, Greifswald.)

A quite different type of foams produced by means of melt-metallurgical methods is the group of the so-called syntactic metal foams. Syntactic foams are composite materials produced by filling a metal, polymer, or ceramic matrix with hollow particles, for example, such as hollow glass or metal spheres, polymer or carbon microballoons, and foam granules (see, e.g., Figure 12.7) or natural materials such as cenospheres. The filler material is retained in the finished foam structure. Polymer-based syntactic foams were developed in the early 1960s as buoyancy-aid materials for marine applications and a large variety of applications have been realized since then. The development of metal syntactic foams followed quite a bit later and concentrated on alloys with comparably low melting temperatures such as aluminum, magnesium, and titanium because metal melts commonly do not wet the hollow particles and pressure-assisted infiltration techniques such as gas-pressure infiltration or squeeze-casting had to be used (see, e.g., [17–19]). Most investigations have addressed aluminum syntactic foams mainly because of the easy processing of aluminum alloys.

Besides the very homogeneous and assured closed porosity (which is especially important for submarine applications), syntactic foams offer a unique spectrum of properties, determined and adjustable by the combination of the properties of the hollow elements and the matrix. In comparison to other foams such as Foaminal, the compressive and tensile properties of syntactic metal foams cannot simply be related to the relative density. Compressive properties of syntactic foams depend more on the properties of hollow elements; however, tensile properties depend more on the matrix material, although the complex nature of mechanical interactions of the constituents of syntactic foams makes the prediction of the composite behavior very difficult.

Because of the strength contribution of the integrated hollow elements and their relative density, which is higher in comparison to most other metal foams, syntactic foams show superior compression strengths of up to 200 MPa. Therefore, potential

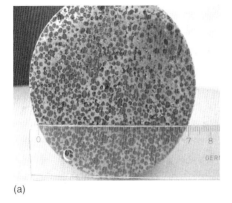

(a)

(b)

Figure 12.7 Syntactic aluminum foams: (a) foam produced by means of gas-pressure infiltration of structures made from glass foam granules. (Courtesy of TU Bergakademie Freiberg.) (b) Metallographic section of foam made from infiltration of preforms of micro-hollow glass spheres.

applications will be found in components in which very high local deformation energy densities have to be absorbed, for example, in connection members of components in cars, railway cars, or trucks.

Owing to their closed porosity, syntactic foams have good corrosion properties and can be subjected to various surface treatments such as anodizing or copper and chromium plating (Figure 12.8).

A very recent development of syntactic foams was made by Fraunhofer IFAM, Germany: the elements integrated into the aluminum matrix were not simple iron metal hollow spheres, but spheres that had been partly filled with loose ceramic powder (Figure 12.9). This concept leads to enormously increased vibration dampening of the syntactic foam (Figure 12.10).

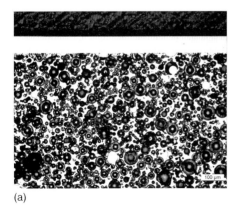

(a)

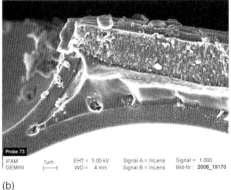

(b)

Figure 12.8 (a) Metallographic section of copper- and chromium-plated zinc–glass bubble composite and (b) SEM image of anodized AlSi9Cu3–S60 glass bubble composite.

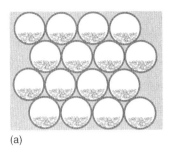 

(a)   (b)

Figure 12.9 (a) Sketch of syntactic foam with partly filled hollow spheres and (b) laboratory example (aluminum matrix) [20].

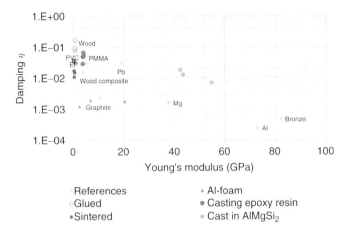

References   ▲ Al-foam
Glued   ● Casting epoxy resin
●Sintered   ● Cast in AlMgSi$_2$

Figure 12.10 Damping of particle-filled steel hollow spheres compared with reference materials for lightweight construction.

12.3
Foams Produced by Means of Powder Metallurgy (P/M)

There exist a variety of technologies for the production of closed-cell metal foams, which start from metal powder: production of compacted foamable precursors that are foamed subsequently in the molten state (e.g., Foaminal), metal powder slurry foaming, syntactic foams made by Powder Metallurgy (P/M) techniques, production of hollow metal sphere assemblies, and other techniques such as self-propagating high-temperature synthesis (SHS) [6]. Although a lot of those technologies are still in the laboratory stage of development, some have been commercialized, for example, Foaminal aluminum foams and hollow metal sphere foams.

The Foaminal technology was developed at Fraunhofer IFAM, Germany, in 1990 and the following years [21, 22] and consists of mixing commercially available metal powders with small quantities of a foaming agent (especially metal hydrides), compaction of the mixture by, for example, hot extrusion, co-extrusion or hot pressing, further shaping steps (sheets, profiles, etc.), and finally the foaming at temperatures around the melting point of the corresponding alloy. In comparison to melt technologies, no stabilizing additives are needed as the foaming melt is stabilized by a fine distribution of oxide fragment stemming from the powder particle surfaces. The Foaminal process that has been commercialized by the company Alulight, Austria, allows the near-net-shape production of foam parts with complex three-dimensional geometries and a closed surface skin. Furthermore, this process can also be used for the production of foam-reinforced hollow structures and sandwiches. The main disadvantage of Foaminal as one of the powder-metallurgical processes for foam production is the high cost of the metal powder. Therefore, the development efforts of the last few years have been aimed to reduce these material costs, for example, by improving process output or using cheaper raw material. Furthermore, alternative foaming agents have been tested intensively and the stabilization of the foams was investigated and optimized in the scope of several projects.

Primarily, the application of the Conform technology for the precursor compaction by Alulight led to significant improvements [23]. In this continuous process, the powder mixture is fed into the profiled groove of a wheel, the groove being closed by a close-fitting shoe. The material is prevented from continuing its passage around the wheel by means of an abutment. High temperatures and pressures arise in the material, which becomes plastic and emerges from the machine through an extrusion die. The precursor products can take shapes such as simple rods, tubes, and complex profiles.

Because of the high cost of aluminum powder, the substitution of the powder by, for example, recycling cutting chips was an issue of several investigations. One approach developed at IFAM Bremen is based on briquetting of chips mixed with a foaming agent and stabilization powders and thixocasting of the briquets to the final foaming precursor. This technique was tested successfully at semi-industrial level [24]. Challenges of this approach were ensuring a homogeneous mixture of chips and powder additives and optimizing the casting process and the foam stabilization (as oxides from the surfaces of the chips have less stabilization effectivity in comparison to oxides from the surfaces of the powder particles). Several 100 kg of precursor material could be produced in very complex shapes (Figure 12.11) using untreated industrial AlMg4.5Mn sawing chips. Only limited amounts of stabilizing additives (e.g., 5 wt% of 3 μm Al_2O_3 or 3 wt% CaO) were needed to reach a good foaming behavior.

Foam stabilization is a critical point for process and quality control. Several research projects were aimed at a better understanding of the mechanisms and the optimization of the stabilizing additives (e.g., stabilization of AlSi11 foam by surface allocation of oxide particles; Figure 12.12). Considerable progress could be observed in the recent years, leading, indeed – in combination with other measures – to an improved quality and property reproducibility of aluminum foam products.

Objective of the PolyFoam concept, developed by Fraunhofer IWU, Germany, is the reduction of handling efforts and expenditures using standardized small

(a) (b)

Figure 12.11 Thixocasting of recycling chips for the production of foamable aluminum precursor. (a) Precursor component after casting and (b) foam component.

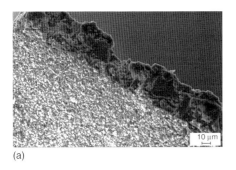

(a) (b)

Figure 12.12 Pore surfaces of AlSi11 foams (a) with 1 wt% TiH_2 and 5 wt% Al_2O_3 (3 µm) and (b) with 1 wt% TiH_2 and 5 wt% Al_2O_3 (12 µm) [25].

foamable precursor geometries to facilitate and automate mold filling before the actual foam process [9] (Figure 12.13).

Other important cost factors of Foaminal aluminum foams are the molds necessary for the near-net-shape production and a quite complicated process control for complex-shaped foam components. Therefore, a new concept (see, e.g., [26]) for cost reduction is based on the separation of the main steps of the foaming process:

1) foam expansion
2) foam part shaping.

In the first step, standardized foam granules are produced in a mass production process feeding continuously granulated foamable precursor material through a belt furnace to effect the foaming. Owing to their simple geometry, the foaming process can easily be controlled and automated. The granules are then coated with thermally activated adhesives and subsequently thermally bonded to larger structures (Figures 12.14 and 12.15). This technology offers significant cost-reduction potential in the case that hollow structures have to be filled as here

Figure 12.13 Mold filling with different precursor material shapes and resulting foam components. (Courtesy of Fraunhofer IWU.)

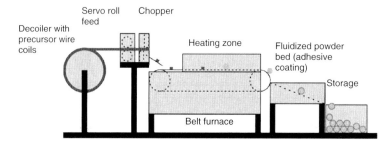

Figure 12.14 Process route for the production of adhesive-coated "APM" aluminum foam granules.

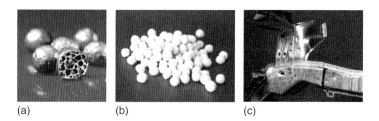

(a)   (b)   (c)

Figure 12.15 (a) APM foam granules, (b) coated foam granules, and (c) component with integrated "APM" foam.

no extra molds are necessary. Furthermore, the producer of the component has not to occupy with the foaming technology, as the aluminum foam granules are prefabricated with reproducible and narrowly defined parameters.

Such "APM" aluminum foam granules can also be used for the production of hybrid aluminum–polymer foams (Figure 12.16). The aluminum foam granules are coated with thermally activated adhesives, which contain chemical foaming agents. The coated granules are subsequently poured into the hollow structures, which have to be (locally) filled, and during a subsequent heat treatment at moderate

Figure 12.16 Polymer–aluminum hybrid foam sample.

temperatures ($<180\,^\circ$C), the adhesive melts, foams, and cures to a polymer foam with integrated aluminum foam granules.

Considerable progress was achieved in the state of the art and the commercialization of aluminum foam sandwiches (AFSs). Main objective was to optimize material and process parameters in order to improve the quality and reproducibility of the foam core and the overall sandwich structure. Parameters considered were, for example, the thermal pretreatment of the foaming agent TiH_2, the composition of the foam alloy, and the use of fast infrared heating concepts in contrast to the earlier predominant air-circulation furnaces. Further challenges were to ensure more strict surface qualities and dimensional accuracies as before [27].

The focus of the research activities of the last few years have also been other foaming techniques such as, for example, slurry foaming and foam production by means of SHS [28]. This method comprises the steps to prepare a slurry from metal powders and a suitable carrier fluid (typically a solvent or a polymeric binder), foam the carrier fluid by mechanical agitation, by foaming agents or by chemical reactions, remove the carrier fluid again, and sinter the remaining fragile metallic foam structure under inert or reducing atmospheres. Slurry foaming has advantages when materials with high melting points such as iron, steels, and also nickel-based alloys have to be processed, but the technique has also been applied to other alloys such as aluminum. Several detailed techniques were developed, for example, slip reaction foaming by IEHK, Aachen University of Technology, Germany [28]. The slip reaction foaming process (Figure 12.17) leads to a foam structure that is characterized by two kinds of porosity: primary pores with the diameter of 0.1–3.5 mm and secondary pores, which can be considered as cell wall porosity (Figure 12.18). Such foams are – strictly speaking – open porous foams, although the primary pore structure resembles more the closed-pore type.

Structures consisting of P/M-produced metal hollow spheres also exhibit different kinds of porosity: internal voids of the hollow spheres, sintering porosity of the sphere walls, and open porosity in the interspace between the spheres.

Raw materials

Room-temperature process

High-temperature process

Products

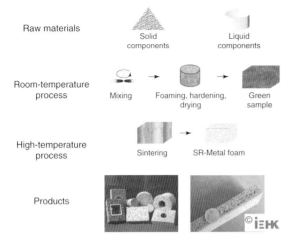

Solid components

Liquid components

Mixing

Foaming, hardening, drying

Green sample

Sintering

SR-Metal foam

Figure 12.17 Process scheme of the slip reaction foaming technology. (Courtesy of IEHK.)

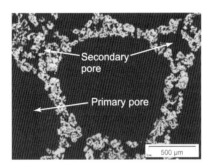

Secondary pore

Primary pore

500 µm

Figure 12.18 Pore structure of foam produced by means of the slip reaction technology. (Courtesy of IEHK.)

Because of the fact that both structural (e.g., crash energy absorber elements and sandwich structures; Figure 12.19) and functional applications (exhaust structures and carriers for catalysts) were considered for this type of foam.

Recent developments were the production of powder-filled hollow spheres (subchapter 2) and the use of hollow metal spheres in hybrid ceramic–metal foam structures. Hollow sphere structures have recently been commercialized by the company Hollomet GmbH, Dresden.

P/M techniques have also been developed for the production of syntactic foams (already introduced in subchapter 2). P/M techniques become interesting for the materials with high melting points – such as iron or steel alloys, for which melt infiltration methods are difficult. The approach is quite similar to the techniques that are used for polymer syntactic foams: hollow elements such as microhollow glass spheres or cenospheres are used as simple additives in conventional P/M production processes such as, for example, metal powder injection molding

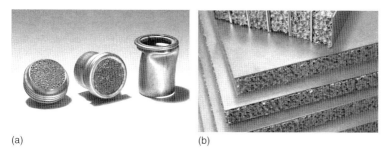

(a)   (b)

Figure 12.19   (a) Crash energy absorber elements and (b) sandwich structures made from hollow metal spheres.

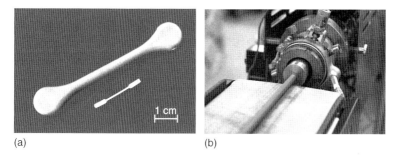

(a)   (b)

Figure 12.20   (a) MIM-produced iron foam components and (b) extrusion of invar foam feedstock.

(MIM), uniaxial pressing, feedstock extrusion, or slurry casting [29] (Figure 12.20). Conventional machinery can be used, although process parameters have to be adapted. In comparison to polymer syntactic foams, a sintering step at high temperatures ($>800\,^{\circ}$C) is needed, which constitutes a critical stage of the overall production process, especially, if hollow spheres with a limited thermal stability similar to microglass bubbles (e.g., 3MS60HS spheres $T_{\mathrm{softening}} \approx 600\,^{\circ}$C) are used. Nevertheless, the feasibility of production could be shown, as demonstrated in Figure 12.21.

12.4
Porous Structures for Structural Applications Produced from Wires and Other Half-Finished Parts

A very different approach for the production of porous structures for structural applications is the shaping and bonding of semifinished parts such as wires or sheets to three-dimensional configurations. Advantages of this approach, which recently found intensive research interest, are

- that relatively cheap and standardized raw material is available,
- the structures are characterized by high reproducibility,
- high-strength alloys and widespread commercial shaping techniques can be used.

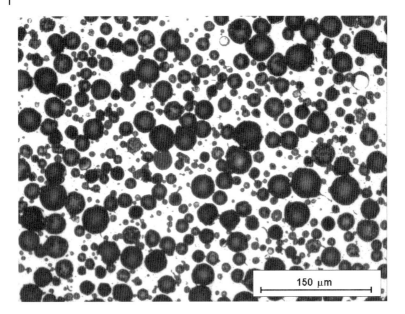

Figure 12.21 Fe foam with 10 wt% glass bubbles, MIM-produced.

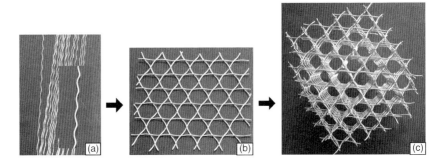

Figure 12.22 WBK structures. (Courtesy of Chonnam National University, Korea.)

For bonding often soldering is used. Various structures were developed, a special form are the so-called Kagome structures. These are structures made from wires, which intersect at an angle of 60° or 120°, with each intersection consisting of three wires. Wire-woven bulk Kagome (WBK; Figure 12.22) shows a very high specific strength (coming very close to the ideal Kagome structures and their mechanical properties) and shows more isotropic behavior in comparison to other structures such as honeycombs. Besides wires, also thin sheets were used as semifinished material for the production of nearly closed hollow spheres and structures made from those [30].

12.5
Case Studies

One of the first serial products has been developed by Alulight together with Alcoa for the high-premium cars, Ferrari 360 and 430 Spider (Figure 12.23). The addition of aluminum foam into the door sill increased the stiffness as well as the behavior in case of a side crash. The production of the components lasted from 1999 to 2009 with approximately 5000 pieces per year.

The applicability and the efficiency of such (local) metal foam reinforcements of hollow structures of cars had also been shown with outstanding results during a European project in which the A-beam of a Ford passenger car was used as demonstrator component. The position of the reinforcement foam is shown in Figure 12.24a and the improvement of crash behavior (40% higher energy absorption at only 3% weight increase) in Figure 12.24b.

The first high-volume production was established with a small crash-energy-absorbing element for the AUDI Q7 sports utility vehicle (SUV) passenger car (Figure 12.25). An important safety requisite is the separation of the passenger

Figure 12.23 Stiffener and crash absorber in the Ferrari 360 and 430 Spider. (Courtesy of Alulight and Alcoa.)

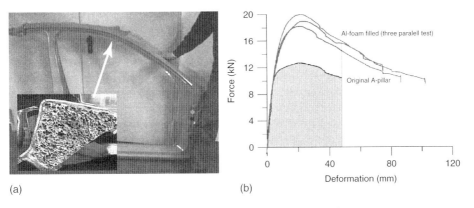

(a)   (b)

Figure 12.24 (a) Local foam reinforcement of A-pillar, (b) comparison of force–deformation curves for nonreinforced with three reinforced A-pillar test structures.

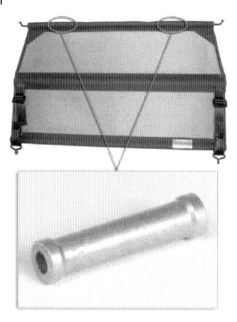

Figure 12.25   Safety net with integrated Foaminal foam energy absorber for AUDI Q7 SUV.

area from the baggage compartment. Otherwise, in case of a crash, unsecured luggage items could be a deadly danger. Therefore, AUDI together with REUM, a subsupplier of automotive interior components, developed a safety net for the protection of the passengers and the load. The specific advantage of the construction with integrated aluminum foam over other solutions is the pronounced and constant stress plateau without a stress peak in the initial stage of deformation.

The serial production of this component started in 2006 (with meanwhile more than 250 000 parts produced) using an automated production line with a capacity of over 100 000 parts per year. Foaming is done by means of an inductive heating system. Every component is weighted automatically for density control and checked visually.

An interesting automotive serial application of AFS has been realized by the company Teupen Maschinenbau GmbH, Germany, in a welded support structure of a mobile telescope working platform (Figure 12.26) in cooperation with Pohl metalfoam, Germany (earlier: Alulight Sandwich). The AFS panels have a very high bending stiffness-to-weight ratio, which allows the increase of the maximum lifting height of the telescope arm to up to 25 m while still complying with the mass restrictions for the Euro-B standard driving license (3500 kg).

AFSs provided by GLEICH Aluminiumwerk GmbH & Co. KG are currently used in the floor structure of the EWE E3 electric vehicle (Figure 12.27). This structure has to bear the weight load of the battery (approximately 300 kg), increase the car body stiffness, and provide some protection of the battery against mechanical

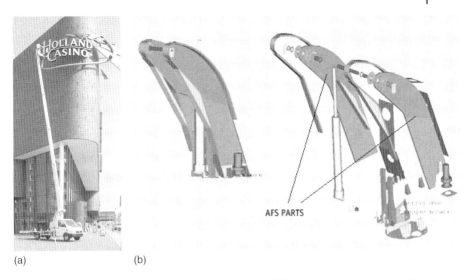

(a)   (b)

Figure 12.26 Mobile telescope working platform (a) with AFS-welded support structure (b). (Courtesy of Pohl metalfoam.)

(a)   (b)

Figure 12.27 (a) EWE E3 electric vehicle and (b) aluminum foam sandwich for the attachment of the battery. (Courtesy of EWE AG and Volkswagen Osnabrück GmbH.)

damage from below. In order to ensure a good chassis clearance, the possible height of the sandwich structure is limited. Because of the limited number of the cars produced, material costs were not the most critical aspect.

The use of AFS panels has also been considered in research projects for the production of railway car constructions (e.g., front module in Figure 12.28). Objectives for the application of AFS were improving crash behavior and easy recycling, as well as reducing weight, fuel/energy consumption, vibration, noise dampening, and flammability.

In addition to AFSs, other aluminum (Foaminal) foam components have been commercialized in crash-energy-absorbing structures of railway cars, for example, in the Siemens Combino tram (Figure 12.29). A soft crash energy absorption is important for the safety of passengers in trains and cable cars, especially for low velocities (at maximum 10 km h^{-1}).

Figure 12.28 Welded AFS prototype front structure for railway vehicles. (Courtesy of Wilhelm Schmidt GmbH, Groß Kienitz.)

Figure 12.29 Aluminum foam part for a front crash element of Combino city trams. (Courtesy of Alulight International.)

Another example of a commercial metal foam product for the increase of the passive safety of sprinter light train (SLT) railway carriages in the Netherlands is shown in Figure 12.30 (development and production by Institute of Materials and Machine Mechanics SAS, Bratislava, and Gleich GmbH, Kaltenkirchen; production started in 2008, approximately 500 parts per annum). The aluminum foam box is used in order to avoid high crash forces in the case that the absorbing capability of existing (conventional) damping elements is exhausted. The foam part is designed to absorb an amount of energy which equals the kinetic energy of a 22 ton carriage moving at 8 km h^{-1} within only 30 mm deformation length. Heat-treated aluminum foam with high compression strength was used with a three-dimensional shaping of the part focused on right deformation progress to reach this target.

Furthermore, several small foam components for bicycles, golf caddies, or motorbikes had been developed and partly commercialized, for example, bicycle seat posts and downfall support pads for motorbikes made from fiber-reinforced polymers with integrated aluminum foam cores [31]. In these examples, the polymer structure is the load-bearing component of the construction, whereas the foam core serves mainly for misuse/impact protection. Another example is – though not yet commercialized – aluminum handle bars of bicycles, where a partial foam reinforcement is applied in order to increase the specific bending strength. The

(a)   (b)

Figure 12.30 (a) Aluminum foam part (volume: 1926 cm³, 1250 g, compression strength: ˜17 MPa) and (b) assembled aluminum crash box for SLT railway vehicle. (Courtesy of Institute of Materials and Machine Mechanics SAS, Bratislava.)

efficiency of this approach was demonstrated for an Al 6060(T6) tube with local foam reinforcement; bending strength increased by 60%, whereas the weight was only increased by 25%.

Cost issues are still a main hindrance for the large-scale implementation of closed-pore metal foams. The importance of costs is instructively demonstrated by two examples.

Objective of an earlier study was weight reduction for brake pistons (Figure 12.31a), as such pistons form part of the unsprung mass and any reduction in weight achieved here is improving driving comfort. In production, there were, among others, at that time, three types of pistons:

1) pistons with the standard design based on forged ferritic steel chromium-plated with a wall thickness of 5 mm (360 g),
2) aluminum pistons with a wall thickness of 8–10 mm, anodized (160 g),
3) polymer pistons promising a weight reduction of nearly 50% but with restrictions from general use because of the properties at high temperatures.

The new design concept (Figure 12.31b) was based on thin sheets of high-strength austenitic stainless steel supported by an integrated aluminum foam core. The reduction in thickness of the steel shell made possible by the foam core served to facilitate and cheapen the shaping process (deep drawing and flow forming). Further cost reduction was achieved by the possible abandonment of surface corrosion protection treatments. In the scope of the study, all relevant issues of the process and product optimization could be settled and all benchmarks be reached. However, because of a marginal increase of costs, the component never reached serial production.

The second example is the Artega GT sports car in which sandwich panels are used in the floor section (Figure 12.32). The sandwiches are bonded with structural adhesives into the body structure and locally reinforced with additional profiles.

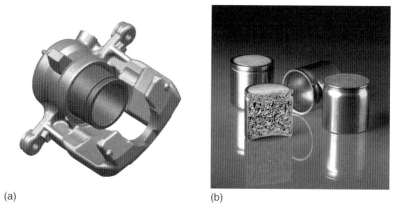

(a)   (b)

Figure 12.31   (a) Design of brake structure and (b) foam-filled brake pistons.

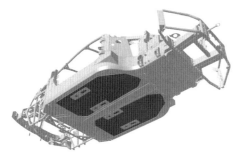

Figure 12.32   Position of sandwich components at the Artega GT car body. (Courtesy of Artega.)

This design concept ensures a good aerodynamic under-floor effect and offers cost reductions, as no special tools and molds are needed. At first, AFSs (ALPORAS 8 mm, with 1 mm EN AW 6082 T651 face sheets) had been used. However, in later constructions, the metal foam sandwich was replaced by Plascore honeycomb sandwich panels because of their lower weight and costs.[1] In this example, the advantages of AFSs (such as, e.g., higher isotropy, see subchapter 1) were not needed for the application, leading to the replacement by more cost-efficient solutions.

As one can see in the above-mentioned case studies, most developed applications can be found in the field of ground transport, that is, automotive and railway systems. Only a few technological studies regarding the application of foam in naval architectures have been done so far. Materials considered were, for example, foam sandwiches for lightweight constructions and syntactic metal foams and sandwiches made from those in the construction of submarine vessels. The main problem encountered is the fact that the sizes of components and raw materials

1) Information of the Artega Automobil GmbH & Co. KG.

needed for shipbuilding are considerably larger than the typical sizes of foams produced so far. Furthermore, shipbuilding is still steel-dominated, but steel foams are more difficult to produce. Last but not least, the welding practice for, for example, foam sandwiches is more complicated and differs from the predominant welding techniques used in the naval industry.

The same limited application of structural foam can be seen in the aerospace industry (although examples of applications of open porous functional foams can be found) [3]. A few research projects were aimed at the production and test of components such as radome shields [32] or maintenance steps. In the case of aerospace components, cost issues are not so dominating but safety considerations are. Foams are *per se* very complicated spatial patterns and most foams are stochastically originated structures (by pore nucleation and growth), which both make quality control more complicated and present an important disadvantage of foam-based materials in comparison to other construction materials such as honeycomb sandwiches.

12.6
Summary and Outlook

Following the intensive research and development activities in the last 20 years, several metal foam components – based on open as well as closed porous structures – have found entrance into industrial production. Fields of application can be found not only in transport and subsidiary industry but also in engine building, fixture construction, design and architecture, sport goods, and implant medicine. Most prominent examples of porous metal applications are, surely, the open porous nickel foams used in batteries (3 million square meters per year), the crash absorber in the AUDI Q7 car (about 250 000 parts), and titanium alloy hip prosthesis structures (about 10 000 implanted), although the use of metal foams improves the performance of several other small-scale and even niche products as well. Nevertheless, it has to be stated that the number of products and the scale of production lag behind the high expectations, which initially led to the intensive public and industrial research efforts.

At the beginning when metal foams just emerged in the literature and in small laboratory and pilot productions, the most important hindrances for the market development were

• missing or incomplete material property databases,
• missing design guidelines,
• lacks in the understanding of production process mechanisms and the influence of process parameters leading to not fully optimized production processes,
• processes that were still not cost-optimized,
• missing experience in quality control and quality management,
• industrial base that was too lean – the fact that production was done only by research institutions or only too few industrial suppliers were available in order

to be accepted, for example, by automotive original equipment manufacturer (OEM),
- metal foam products that still had to show practicability in the everyday use of real components in order to be accepted by following customers.

Most of the above-mentioned aspects are typical for all new-developed materials and lead to a generally slow market introduction of new material concepts and products.

Owing to the intensive research and development activities especially in the time after 1990, a significant progress could be achieved: foam quality and reproducibility could be improved, design and process guidelines and testing standards were developed, and several (though small) industrial suppliers are active in the market backed by an established international research infrastructure. Running metal foam products demonstrate the capability of the material and help to develop a deeper understanding of the material in everyday use. With enduring production and still-increasing number of products (not only in transport industry), further technological progress was achieved and can furthermore be expected.

However, the number of structural metal foam applications is still limited. Two main reasons can be held responsible for this:

1) Metal foams in structural applications are facing a strong competition with well-developed and -established conventional materials in markets such as automotive, which are characterized by intensive cost pressure. In comparison, the cost issue is not that critical for functional materials (although not negligible).
2) The strong correlation and decrease of mechanical properties such as Young's modulus and strength to the relative density and the reduced ductility of most metal foams constitute disadvantages of the foams even if other properties such as specific stiffness, vibration dampening, and crash energy absorption capacity are improved.

Not surprisingly and as demonstrated in above-mentioned case studies, the first applications of foams were realized in components for which the advantages of the structural metal foams overcompensated their drawbacks. Nevertheless, in order to push the market situation for structural metal foams especially, the cost issue had and still has to be considered. As shown in subchapters 2–4, some progress has already been achieved, for example, new concepts were developed to reduce raw material costs: foams based on aluminum foam granules (APM) and foamable precursor produced by means of technologies with higher output (Conform) or produced from recycling material.

In order to evaluate the possible future development of structural metal foams in the transport market, the expected trends of both the technology and the market have to be taken into account. In the following, the situation will be considered for different modes of transport.

12.6.1
Automotive Passenger Cars

Important market trends are as follows:

1) Sales will be growing dominantly in markets such as China and India and also in other emerging markets in Asia and Africa. Those markets are characterized by a very high cost pressure for the final product. Here a trend for low cost and easy engineering can be observed for a number of technical products (e.g., in medical technology) and these trends will flash over to the automotive industry.
2) Electromobility will have a growing share in the overall production with new technological challenges to be considered: battery technology, battery safety, electromagnetic compatibility (EMC), and car climatization.
3) Tightened specifications for car occupant and pedestrian safety.
4) Increased use of polymers and especially fiber-reinforced polymers in structural components of the cars and compatibility to crash behavior specifications.

The first trend will make the market situation for innovative nonconventional materials such as metal foams with a higher cost level clearly more difficult. Other trends such as e-mobility might even establish increased market potential for metal foams and are subject of currently running projects. Also, the future specification of more complex crash scenarios can offer new application chances for quasi-isotropic materials such as metal foams.

12.6.2
Commercial Vehicles

For commercial vehicles, mass reduction due to lightweight construction concepts offers higher financial benefits in comparison to passenger cars, as decrease of car weight increases freight load capacity. Therefore, somewhat higher costs for mass reduction are accepted. As for passenger cars, further tightening of crash behavior specifications for commercial vehicles can be expected and is already under way. For example, intensive activities are currently conducted for the improvement of under-ride barriers. This offers some potential for metal foams. However, several alternative and quite effective solutions are in competition.

Lightweight design of superstructural parts of commercial vehicles is also gaining more importance. Metal foam solutions, for example, based on sandwich structures can possibly benefit from the experiences gained because of their increasing use in other application fields such as fixture constructions. However, in the last few years, a variety of other innovative sandwich structures have been also developed and commercialized; therefore, a tough material competition can be expected.

Regarding military vehicles: with currently increased asymmetric war scenarios, more attention is given to the safety of trucks and other cars against blast impacts. If a military vehicle is subjected to explosions of landmines or improvised explosive devices (IEDs), modern outer metal structures often protect the occupants from injuries caused by shrapnel penetration. However, serious injuries can result from the shock waves of the explosion. In this respect, metal foams can offer interesting features to improve the safety of passengers by reducing the load and can complement the existing ballistic protection.

12.6.3
Railway Transportation

Railway transport industry is rather conservative in the application of new materials as railway cars have to endure long service times and specification procedures are extensive. Nevertheless, as for commercial road-based vehicles, mass reduction of the cars directly results in increased load capacity. This applies especially for public metropolitan and suburban commuter railway systems as here start and stop cycles occur more often and are, therefore, more energy consuming.

Furthermore, means of transportation such as trams are more prone to be involved into crash scenarios in comparison to long-distance trains. This and the specific velocity (and energy) situation of crash scenarios such as tram versus passenger car led already to the application of metal foams in crash-energy-absorbing elements of such rail vehicles. Because of the first successful foam applications in future, an increasing acceptance of metal foam-based constructions in railway transportation can be expected.

12.6.4
Marine Transport

No actual trend is evident, which will cause a large increase of the application potential for metal foams in structural components of ships, ferries, and so on. Applications for superstructures of ships with improved fire protection have been discussed. However, the large sizes of components and semifinished materials common in shipbuilding are one main hindrance against their introduction into this market.

12.6.5
Aeronautical Transport

One major trend of aircraft industry is the increase of the use of fiber-reinforced polymers on the expense of the application of aluminum components. This trend will also negatively affect the market potential of the dominating closed-pore foam type – aluminum foams. In combination with the other above-mentioned disadvantages of foams with stochastic pore structures, no increase of the market potential for metal foams can be expected in the years to come. Newly developed porous structures such as Kagome foams might change this situation. However, this material is still quite in its infancy; therefore, application will not be seen presently.

In conclusion, it can be stated that

- Even though nowadays implementation of metal foams into structural applications lags behind the high expectations of 20 years ago, an already significant number of metal foam components have been brought into production.
- Important drawbacks of the technology (missing material data, standards, etc.) that characterized the situation before 10–20 years have been eliminated.
- Some drawbacks still exist and have to be addressed – especially the material costs.
- Other drawbacks of foams such as the reduction of strength with the decreasing relative density have to be considered to be fundamental and cannot be eliminated, limiting the potential applications to cases in which the advantages of foams prevail.
- Capable international industrial and research metal foam infrastructures have formed.
- The successful market introduction of metal foams in the last 10 years will help to improve further market acceptance.
- Interesting new processes and materials have been developed widening the property and potential application spectrum of metal foams.

The largest variety of current foam applications can be found in commercial automotive and railway vehicles. It is expected that also in future the highest potential for metal foam applications will be found there.

12.7
Further Reading

Books

- Gibson, L.J. and Ashby, M.F. (1997) *Cellular Solids: Structure and Properties*, 2nd edn, Cambridge Solid State Science Series, Cambridge University Press.
  The work by Gibson and Ashby is well recognized as the fundamental study on all kinds of cellular solids. It introduced and classifies the different subgroups of cellular materials and provides comprehensive background information on structure-property relations. For a thorough introduction to the field, it remains extremely useful.
- Ashby, M.F., Evans, A., Fleck, N.A., Gibson, L.J., Hutchinson, J.W., and Wadley, H.N.G. (2000) *Metal Foams – A Design Guide*, Butterworth-Heinemann.
  Ashby *et al.* work provides detailed information on how best to apply metallic foams, with their unique combinations of properties, in engineering design. Though it may not include all the latest information in terms of actual material properties, the general principles it suggest do still remain valid.

Journal Articles

- Banhart, J. (2001) Manufacture, characterisation and application of cellular metals and metal foams. *Prog. Mater. Sci.*, **46**, 559.

This extensive review provides an in-depth overview of the various ways of producing metallic foams. It classifies these approaches, provides information on characterization methods as well as the actual properties and discusses application potentials both as structural and functional materials. Furthermore, with its large number of references, it is an entry point for a deeper familiarization specifically with metal foam research in the 1990s.

- Banhart, J. (2013) Light-metal foams – history of innovation and technological challenges. *Adv. Eng. Mater.*, **15**, 82.

This recent work is focussed specifically on metal foams with light metal matrix and is thus narrower in scope than the previous review. However, written by the same author, it takes up the original thread in this field and adds both its own evaluation of a further decade of research and the related, latest references from this field.

Conferences

- Metfoam Conference Series
  Conference/organiser website: www.metfoam2013.org
  As international conference, the Metfoam Conference takes place since 1999 on a biannual basis in uneven years. Beginning in 2005, a regular change of locations from Asia to America and Europe has been introduced. The 2013 event will be hosted from June 23rd–26th, 2013, by the North Carolina State University at Raleigh, North Carolina, USA. 2015 will see the continuation of the series in Europe. The Metfoam, as the International Conference on Porous Metals and Metallic Foams, covers all aspects of this growing class of materials, from processing to characterization, simulation and application. Proceedings of the conference are regularly published in print, and provide an up-to-date overview of research in the field.
- Cellmat Conference Series
  Conference/organiser website: www.cellmat.de
  The Cellular Materials (Cellmat) Conference series is organized on a biannual basis in even years as a DGM conference and specifically by Fraunhofer IFAM, Dresden, Germany, where it regularly takes place. The scope is generally wider than that of the Metfoam, as not only metallic foams are covered.
- Syntactic and Composite Foams Conference Series
  Conference/organiser website: http://www.engconfintl.org/14af.html (2014 event)
  The event takes place every three to four years, with the next, fourth event in the series to be held at Santa Fe, New Mexico, USA, from November 2nd–7th, 2014. The focus is on syntactic foams both with polymeric and, to an increasing degree, metallic matrix.

Internet Resources

- www.metalfoam.net

The website, which is managed by the Technical University of Berlin's working group on metallic foams under the responsibility of Prof. John Banhart, offers information about recent developments in the field, future and past events, companies working in the field etc.

Acknowledgments

The authors want to acknowledge the amicable help of many colleagues and partners for the collection of the presented data, especially ALANTUM Europe GmbH, Alulight International, Artega Automobil GmbH & Co, Chonnam National University Korea, EWE AG, Fraunhofer Institute IWU, GLEICH Aluminiumwerk GmbH & Co. KG, Helmholtz-Zentrum Berlin, IEHK (RWTH Aachen), INCO, Institute of Materials and Machine Mechanics SAS, IWE GmbH & Co. KG, Pohl metalfoam, TU Bergakademie Freiberg, Volkswagen Osnabrück GmbH, and Wilhelm Schmidt GmbH.

References

1. de Meller, A. (1926) Produit métallique pour l'obtention d'objets laminés, moulés ou autres, et procédés pour sa fabrication. French Patent FR000000615147A, Dec. 30, 1926.

2. Gibson, L.J. and Ashby, M.F. (1997) *Cellular Solids – Structure and Properties*, Cambridge Solid State Science Series, 2nd edn, Cambridge University Press.

3. Cellmet-News 2007-II (2007) Environment well served with RE-CEMAT rotary oil mist separators *http://www.metalfoam.net/CELLMET-NEWS-2007-II.pdf*, p. 3 (accessed January 2013).

4. Ikeda, M., Sueki, T., and Takaishi, T. (2010) *Q. Rep. RTRI*, **51**(4), 220–226.

5. Open Cell Nickel Foam *http://www.novametcorp.com/pdf/Datasheets/Novamet%20Open%20Cell%20Nickel%20Foam%202005-12.pdf.*(accessed January 2013)

6. Ashby, M.F., Evans, A., Fleck, N.A., Gibson, L.J., Hutchinson, J.W., and Wadley, H.N.G. (2000) *Metal Foams – A Design Guide*, Butterworth-Heinemann.

7. Riehemann, W. (1994) Metallische Werkstoffe mit Extremer Innerer Reibung und Deren Messung, Habilitationsschrift TU Clausthal.

8. Banhart, J., Baumeister, J., and Weber, M. (1996) *Mater. Sci. Eng., A*, **205**(1–2), 221–228.

9. Hipke, T., Lange, G., and Poss, R. (2007) *Taschenbuch für Aluminiumschäume*, Aluminium-Verlag Düsseldorf.

10. Babcsan, N., Leitlmeier, D., Degischer, H.-P., and Banhart, J. (2004) *Adv. Eng. Mater.*, **6**(6), 421–428.

11. Aussenfassade Wohnu (2003) Gewerbehaus Kochanowicz, Bochum, Design: Architekturbüro Kochanowicz, Bochum.

12. Furnstahl & Simon Architects Memorial of Service Employees International Union, Dedicated to its Members Lost in the World Trade Center Attack of September 11, 2002.

13. Bryant, J.D. *et al.* (2006) US Patent Application 20060243094.

14. Bryant, J.D. *et al.* (2008) in *Proceedings of International Conference on Porous Metals and Metallic Foams* (eds L. Lefebvre, J. Banhart, and D. Dunand), DEStech Publications, Inc. p. 19.

15. Kenny, L.D. and Thomas, M. (1994) PCT Patent WO 94/09931.

16. Sang, H. *et al.* (1992) PCT Patent WO 92/21457.

17. Balch, D.K., O'Dwyer, J.G., Davis, G.R., Cady, C.M., Gray, G.T., and Dunand, D.C. (2005) *Mater. Sci. Eng., A*, **A391**, 408.

18. Rawal, S.P. and Lanning, B.R. (1994) Composite Materials for Advanced Submarine Technology, National Technical Information Service (NTIS) Report DARPA Research Project MDA-972-89-C-0044.

19. Weise, J., Zanetti-Bueckmann, V., Yezerska, O., Schneider, M., and Haesche, M. (2007) *Adv. Eng. Mater.*, **9**(1–2), 52–56.

20. Jehring, U., Weise, J., Wöstmann, F.-J., and Stephani, G. (2009) *Lightweight DES.*, **6**, 24–27.

21. Baumeister, J. (1990) German Patent 40 18 360.

22. Baumeister, J. and Schrader, H. (1991) German Patent DE 41 01 630.

23. Schäffler, P. and Rajner, W. (2003) in *Proceedings of International Conference on Cellular Metals and Metal Foaming Technology MetFoam* (eds J. Banhart, N.A. Fleck, and A. Mortensen), MIT-Verlag, pp. 39–46.

24. Weise, J., Wichmann, M., Lehmhus, D., Haesche, M., and Cristina Magnabosco, I. (2010) *J. Mater. Process. Technol.*, **26**(9), 845–850.

25. Haesche, M., Weise, J., Baumeister, J., Garcia-Moreno, F., and Banhart, J., Proceedings of International Conference on Porous Metals and Metallic Foams MetFoam 2010, to be published in 2013.

26. Stöbener, K., Baumeister, J., Rausch, G., and Busse, M. (2008) *Adv. Eng. Mater.*, **10**(9), 853.

27. Seeliger, H.-W. Proceedings of International Conference on Porous Metals and Metallic Foams MetFoam 2010, to be published in 2013.

28. Angel, S., Bleck, W., Scholz, P.-F., and Fend, T. (2004) *Steel Res. Int.*, **75**(7), 479–484.

29. Weise, J., Baumeister, J., Yezerska, O., Beltrame, G., Silva, D., and Salk, N. (2010) *Adv. Eng. Mater.*, **12**(7), 604.

30. Uchida, N., Mihara, Y., Shinagawa, K., and Yoshimura, H. (2010)Proceedings of International Conference on Cellular Materials CellMat2010, pp. 327–333.

31. Baumeister, J. and Lehmhus, D. (2003) in *Proceedings of International Conference on Cellular Metals and Metal Foaming Technology MetFoam* (eds J. Banhart, N.A. Fleck, and A. Mortensen), MIT-Verlag, pp. 13–18.

32. Hanssen, A.G., Girard, Y., Olovsson, L., Berstad, T., and Langseth, M. (2006) *Int. J. Impact Eng.*, **32**(7), 1127–1144.

Part V
Modeling and Simulation

Kambiz Kayvantash

Today's engineering practice cannot be dissociated from simulation technology. Virtual prototyping and virtual testing (also referred to as V&V–standing for validation and verification–or analytical substantiation) is at the heart of reliable and cost efficient engineering design. From a mechanical point of view (which is the standpoint of this book), this comes down to relating loading and displacements or strains and strains in a structure or an infinitesimal element of it. Any structure under design needs to be tested under various loading conditions, and at the early stages of the design, it has to be modeled. In essence, a model is a function that needs to take into account the variables (loads) and the parameters (geometry and material properties). Since the loads are assumed and the geometry is proposed in a given design, the key missing information is often the material characteristics.

There are many ways for identifying the material characteristics. One can of course perform laboratory tests and obtain the so-called stress–strain curves, and define a relation between them, at the tested sample level. The key issue here is that the sample itself is a structure, has a given scale, and may therefore influence the outcomes, hence the very nature of the material "law." At which scale should we define this law? Additionally, the laboratory conditions are supposed to be deterministic and known, which is not always the case. There is a need to investigate other methods apart from traditional curve-fitting approaches in order to obtain characteristic laws at various scales ranging from microscopic (molecular bonds) to macroscopic (sample size and beyond).

This part will review three ways (out of a few more) to obtain the material characterization for a given design. The assumption here is that material models are sought for integration into finite element codes, which have become the dominant design too for engineers and analysts.

The first section (Hohe) explores a midway approach, basically concerned with homogenization issues: how do we transfer information at the microscopic level to macroscopic theory of continuum mechanics? As we said earlier, any information (or observation) includes the uncertainties associated with it at the observation scale. This explains the term "homogenization" since at the final instance what the characterization is all about is obtaining a homogeneous (all over) material

Structural Materials and Processes in Transportation, First Edition.
Edited by Dirk Lehmhus, Matthias Busse, Axel S. Herrmann, and Kambiz Kayvantash.
© 2013 Wiley-VCH Verlag GmbH & Co. KGaA. Published 2013 by Wiley-VCH Verlag GmbH & Co. KGaA.

property set for a give material for a given component. This approach is of extreme importance for nonhomogeneous materials such as composites, knowing that a finite element discretization of a given structure may not be representing a full geometrical description of the real layout of the material involved. This is, in particular, the case for nanocomposites or even layered composites that need to be modeled via one single material set of data.

The second section (Kayvantash) assumes a "structural" level identification and tends to ignore the scale issue. It combines all "information" available concerning the material used in a design and along with the finite element formulation of the strains and stresses, which should be selected with care to reflect the complexity of the loading. It essentially uses data mining (not to be confused with fitting) technology to obtain the required material data to be input to a model and the corresponding solver. In this approach, the material model is a holistic set of information, which, combined with the geometry and discretization of the structure, should reproduce all known observations as closely as possible. It is essentially a machine learning approach applied to material characterization. This approach is pragmatic and serves essentially engineering modeling activities but lacks deep insights into what the real material properties are at an infinitesimal scale. It is of course possible to include information at all levels in one sweep, but the outcome is an averaging of the behaviors at various scales providing the information set.

The third section (Friak et al.) is an "ab initio" approach, essentially going back to basics at a nano- or microscopic level (first principles of continuum mechanics at a molecular level). This approach can be considered via a bottom–up or inversely top–down strategy exploring the nature of the material bonds and nano (phase, thermodynamics) level, extending it to micro- (homogenization and finally to macroscopic level (material design). The key investigation here is to substantiate relations at each scale and relate to each other, allowing for obtaining relations for design models (material laws for FE – finite element codes). This justifies the title "multiscale" approach and allows for obtaining the right material data or the right material for a given design starting from first principles. This approach may be considered as optimal if information are sought at an "element" level but relaxes the structural or geometric considerations.

The reader is invited to consider the three approaches as problem-oriented solutions. There is probably no unique tool (or no need for) for material characterization, which explains also why this particular issue remains at the heart of a good modeling practice resulting in predictive outcome.

13
Advanced Simulation and Optimization Techniques for Composites

Jörg Hohe

13.1
Introduction

Composite materials and other materials with distinct microstructure are becoming increasingly popular in modern transport technologies. On the basis of the microstructural geometry, the composite materials can be classified into particle-reinforced materials, short-fiber-reinforced materials with either aligned or randomly orientated fibers, long-fiber-reinforced composites, and (quasi-)infinite-fiber-reinforced composites. An additional class is formed by composites with interpenetrating microstructures, where no differentiation between matrix and inclusions can be made. A special case of composite materials are cellular solids, where inclusions with zero stiffness (the void volume) are embedded into a matrix with finite stiffness formed by the cell walls. To some extent, even classical metallic materials with a polycrystalline microstructure can be considered as composites consisting of an agglomerate of single crystals.

The main advantage of composite materials is their enhanced capability for a design of materials to fit with any kind of prescribed requirements in structural application. Typical requirements in the transport sector are the combination of low weight with reasonable stiffness and strength. Further requirements may include inherent good thermal and acoustic insulation properties or superior vibration damping capabilities. The classical field for the application of composite materials is the aerospace sector. Nevertheless, because of the decreasing expenses in the manufacture and design of composite materials, composite and cellular materials today can be found in a wide range of applications in road, rail, and maritime transport as well.

For reasons of numerical efficiency, the numerical analysis of components consisting of composite material is preferably performed in terms of macroscopic "effective" properties rather than by a direct model of the actual microstructure. In general, the corresponding properties are determined experimentally. Nevertheless, because of the large number of microstructural design options and the large number of material properties, the design and optimization of composite materials

Structural Materials and Processes in Transportation, First Edition.
Edited by Dirk Lehmhus, Matthias Busse, Axel S. Herrmann, and Kambiz Kayvantash.

require a considerable experimental effort. Complementing the experimental characterization by numerical analyses might save significant amounts of experimental expenses. Furthermore, the numerical analysis of the material behavior on the microstructural level may provide a deeper insight into the microstructural mechanisms and material properties, which cannot be measured directly. Finally, the numerical simulation for the prediction of the effective material response provides an efficient tool for parametric studies regarding the effect of the microstructural design on the macroscopic material behavior. It forms the base for any optimization of the material using rigorous mathematical procedures.

Different schemes for the numerical prediction of the effective material behavior of microheterogeneous materials have been provided. The early studies by Voigt [1] and Reuss [2] were concerned with rough upper and lower bounds. Later, Hashin and Shtrikman [3] as well as Walpole [4] or Talbot and Willis [5] established improved bounds. As an alternative to these approaches providing rigorous bounds on the effective properties, methods based on the stress and deformation analysis of a representative volume element for the microstructure are frequently employed in the mechanics of materials [6]. Overviews on these methods and other homogenization techniques are given, for example, by Walpole [7] or Nemat-Nasser and Hori [8].

Actual trends in the numerical analysis in the mechanics of materials are the development of multiphysics homogenization schemes considering mechanical and nonmechanical objectives together. For the assessment of the scatter in the effective properties of microheterogeneous materials with randomly disordered microstructures, probabilistic homogenization techniques are developed. The optimization of materials with respect to different mechanical and nonmechanical functions requires the development of multiobjective optimization procedures in order to comply with in many cases contradictory requirements.

13.2
Multiphysics Homogenization Analysis

13.2.1
Concept of the Representative Volume Element

In the mechanics of materials, the numerical prediction of the macroscopic response of microheterogeneous materials is performed by a "homogenization" analysis. Mathematically, the homogenization problem can be defined as follows: consider a body Ω (structure or structural component) according to Figure 13.1. The body is bounded by a boundary $\partial\Omega = \partial\Omega^u \cup \partial\Omega^t$, where the displacements u_i are prescribed on $\partial\Omega^u$, whereas the surface traction vector $t_i = \sigma_{ij} n_j$ with the Cauchy stress components σ_{ij} and the outward normal unit vector n_j are prescribed on $\partial\Omega^t$. In addition, Ω might be loaded by distributed body forces f_i. Implicitly, it is understood that $i = 1-3$ and Einstein's convention of summation over repeated indices is employed. For an efficient numerical analysis in the industrial design process, the body Ω is to be substituted by a similar body Ω^* with the same size and external shape subject to similar external loading conditions. In contrast to the original body Ω, the substitute body Ω^* is assumed to consist of a quasi-homogeneous medium with yet unknown properties.

If the microstructure of Ω does not explicitly depend on the macroscopic spatial position within the body, a representative volume element Ω^{RVE} according to Figure 13.1 can be considered. The characteristic length l of Ω^{RVE} has to be much smaller than the characteristic length L of the entire body Ω. Nevertheless, it is not infinitesimally small:

$$L \gg l \gg dl \tag{13.1}$$

During the homogenization analysis, the representative volume element Ω^{RVE} is substituted by a similar volume element $\Omega^{RVE\,*}$ consisting of the effective medium. The mechanical, thermal, and other physical properties of the effective medium have to be determined such that the material response of both volume elements is equivalent on the mesoscopic level.

13.2.2
Equivalence Conditions

For the definition of the mesoscopic equivalence of Ω^{RVE} and $\Omega^{RVE\,*}$, different types of equivalence conditions have been proposed in the literature. For the definition of the equivalence of the strain or deformation state in mechanical analyses, most authors employ the kinematic equivalence condition

$$\bar{\varepsilon}_{ij} = \frac{1}{V^{RVE}} \int_{\Omega^{RVE}} \varepsilon_{ij}\, dV = \frac{1}{V^{RVE*}} \int_{\Omega^{RVE*}} \varepsilon_{ij}^*\, dV = \bar{\varepsilon}_{ij}^* \tag{13.2}$$

requiring that the volume average of the strain components ε_{ij} in both volume elements is equal [9]. In geometrically nonlinear analysis, a similar equivalence condition may be used, where the infinitesimal strain components $\varepsilon_{ij} = 1/2(u_{i,j} + u_{j,i})$ are replaced by the components $F_{ij} = u_{i,j} + \delta_{ij}$ of the deformation gradient as the

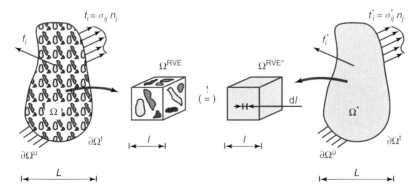

Figure 13.1 Concept of the representative volume element.

basic measure of deformation in continuum mechanics [10]. As usual in index notation, $u_{i,j}$ denotes the partial derivative $\partial u_i/\partial x_j$, whereas δ_{ij} are the components of the unit tensor.

Alternatively, an asymptotic two-scale expansion of the displacement field is frequently used in the mathematically oriented literature [11, 12]. In this approach, the displacement field u_i of the body Ω consisting of the real microstructure is decomposed by

$$u_i = \overline{u}_i^* + \widetilde{u}_i\left(\overline{\varepsilon}_{kl}^*\right) \tag{13.3}$$

into the effective displacement field $\overline{u}_i^*$ and an additional fluctuating part $\widetilde{u}_i\left(\overline{\varepsilon}_{kl}^*\right)$ with a magnitude scaled by the local effective stress level $\overline{\varepsilon}_{kl}^*$. This approach has been extended to large deformation theory by Willis [13] as well as to nonlocal elasticity by Gambin and Kröner [14].

The equivalence of the stress state can be defined in a similar manner as the kinematic equivalence by a volume-averaging approach

$$\overline{\sigma}_{ij} = \frac{1}{V^{\text{RVE}}} \int_{\Omega^{\text{RVE}}} \sigma_{ij}\,\mathrm{d}V = \frac{1}{V^{\text{RVE}*}} \int_{\Omega^{\text{RVE}*}} \sigma_{ij}^*\,\mathrm{d}V = \overline{\sigma}_{ij}^* \tag{13.4}$$

considering the Cauchy stress σ_{ij} or any other stress measure forming the energy conjugate to the employed strain or deformation measure in the kinematic equivalence condition. Alternatively, the surface average

$$t_i = \frac{1}{A^{\text{RVE}}} \int_{\partial\Omega^{\text{RVE}}} t_i\,\mathrm{d}V = \frac{1}{A^{\text{RVE}*}} \int_{\partial\Omega^{\text{RVE}*}} t_i^*\,\mathrm{d}V = t_i^* \tag{13.5}$$

of the traction vector $t_i = \sigma_{ij}n_j$ on opposite surfaces of the representative volume element is often employed for the definition of the equivalence of the load level of the two volume elements Ω^{RVE} and $\Omega^{\text{RVE}*}$. In two-scale expansion approaches for periodic microstructures, a decomposition of the stress field similar to the decomposition of the strain field in Eq. (13.3) is employed.

Irrespective of the choice of the equivalence conditions, it has to be assured that the basic principles of physics, especially the conservation of energy, are satisfied by the homogenization approach as it has been pointed out by Bishop and Hill [15]. In order to satisfy the Hill condition inherently, energy-based equivalence conditions instead of the stress- or traction-based conditions in Eqs. (13.4) and (13.5) are becoming increasingly popular. In this type of approach, the energetic equivalence condition

$$\overline{w} = \frac{1}{V^{\text{RVE}}} \int_{\Omega^{\text{RVE}}} w\,\mathrm{d}V = \frac{1}{V^{\text{RVE}*}} \int_{\Omega^{\text{RVE}*}} w^*\,\mathrm{d}V = \overline{w}^* \tag{13.6}$$

with the strain energy density w substitutes the stress-based conditions [16]. Using this type of approach implies the deformation of a representative volume element Ω^{RVE} for the given microstructure according to a prescribed effective strain state using Eq. (13.2) and computation of the effective strain energy density using Eq. (13.6).

Appropriate boundary conditions for the numerical analysis of the representative volume element under a prescribed effective strain state can be obtained by

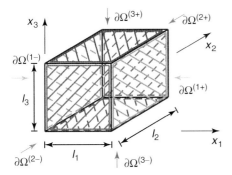

Figure 13.2 Boundary conditions.

transforming the volume integral in Eq. (13.2) into a surface integral using Green's theorem resulting in

$$\int_{\partial\Omega^{RVE}} \frac{1}{2}\left(u_i n_j + u_j n_i\right)\ \mathrm{d}A = \bar{\varepsilon}^*_{ij}$$

(13.7)

where u_i are the displacements on the surfaces of the volume element Ω^{RVE}, whereas n_j are the components of the outward normal unit vector. By Eq. (13.7), a relation between the prescribed effective strain components $\bar{\varepsilon}^*_{ij}$ and the average displacement difference across the surfaces of the representative volume element is established. If constant reference strain states are considered, and if the microstructure is either periodic or idealized by a periodic geometry, periodic boundary conditions can be applied. Periodic boundary conditions require that the deformation of opposite boundaries of the representative volume element must deform to a similar shape in order to ensure that neighboring volume elements fit even in the deformed configuration. In this case, the displacement difference between two corresponding spatial points on opposite surfaces of the volume element has to be constant over the respective pair of external surfaces, resulting in

$$\frac{1}{2}\left(\frac{u_i^{(j+)} - u_i^{(j-)}}{l^{(j)}} + \frac{u_j^{(i+)} - u_j^{(i-)}}{l^{(i)}}\right) = \bar{\varepsilon}^*_{ij}$$

(13.8)

where $u_i^{(j+)}$ and $u_i^{(j-)}$ are the displacements of two corresponding points on the external surfaces $\partial\Omega^{RVE(j+)}$ and $\partial\Omega^{RVE(j-)}$ according to Figure 13.2. An alternative interpretation of these boundary conditions is that the partial displacement derivatives $u_{i,j}$ in the definition of the strain components are substituted by the respective numerical derivatives $\Delta u_i / \Delta x_j$ with respect to the dimensions of the representative volume element Ω^{RVE}.

Once the effective strain energy density $\bar{w}^*$ for a prescribed effective strain state $\bar{\varepsilon}^*_{ij}$ is determined, the effective stresses can be computed by using their definition in terms of strain and strain energy density

$$\sigma_{ij}^* = \left.\frac{\partial \overline{w}^*}{\partial \overline{\varepsilon}_{ij}^*}\right|_{d\overline{\varepsilon}_{kl}^{\mathrm{pl}*}=0} \approx \left.\frac{\Delta \overline{w}^*}{\Delta \overline{\varepsilon}_{ij}^*}\right|_{\Delta \overline{\varepsilon}_{kl}^{\mathrm{pl}*}=0} \tag{13.9}$$

on the macroscopic level.

The advantage of the energetic homogenization approaches is their general formulation based on one of the basic conservation laws in physics. They are applicable under small and finite deformations, irrespectively of the material behavior. Furthermore, these concepts can be extended to a variety of mechanical and nonmechanical problems for multiphysics approaches in a straightforward manner by substituting the strain energy in Eq. (13.6) with the respective energy type and the displacement gradient in Eq. (13.2) with the respective transport equation in the physical problem considered.

For thermal analyses, the effective displacement gradient and the strain energy are substituted with the effective temperature gradient $\overline{T}_i$ and the effective heat flux $\overline{\dot{q}}_i$, respectively. Hereby, Eqs. (13.2) and (13.6) are substituted with

$$\overline{T}_i = \frac{1}{V^{\mathrm{RVE}}} \int_{\Omega^{\mathrm{RVE}}} T_i \mathrm{d}V = \frac{1}{V^{\mathrm{RVE}*}} \int_{\Omega^{\mathrm{RVE}*}} T_i^* \mathrm{d}V = \overline{T}_i^* \tag{13.10}$$

and

$$\overline{\dot{q}}_i = \frac{1}{V^{\mathrm{RVE}}} \int_{\Omega^{\mathrm{RVE}}} \dot{q}_i \mathrm{d}V = \frac{1}{V^{\mathrm{RVE}*}} \int_{\Omega^{\mathrm{RVE}*}} \dot{q}_i^* \mathrm{d}V = \overline{\dot{q}}_i^* \tag{13.11}$$

[17]. Other physical effects such as the effective acoustic or electrical conductivity can be treated in a similar manner. Owing to the comprehensive formulation, energy-based homogenization schemes are easily used for an analysis of inter-actions between the different material properties such as the thermomechanical coupling effects through thermal expansion and thermal dissipation of plastic work or the electromechanical coupling through the Piezo effect. In the combined approaches, the different physical fields and energy components have to be treated separately. The same type of energetic separation may be employed in purely me-chanical analyses involving kinetic energy as well as energy dissipation by plastic deformation or internal damping.

The energetic homogenization procedure provides a general tool for the pre-diction of all types of mechanical and nonmechanical material properties in multiphysics analyses, which can efficiently be employed in the design and mathematical optimization of composite or other microheterogeneous materials. Especially in the design of multifunctional materials, these numerical techniques can be helpful. In addition to the prediction of the macroscopic properties, the nu-merical analysis of microheterogeneous materials enables a deeper insight into the underlying microstructural mechanisms and processes responsible for the effects observed on the macroscopic level. In a reverse approach, they can be employed for an analysis of the microscopic conditions at "hot spots" in macroscopic structures detected during the macroscopic analysis in the design process using the effective homogenized properties.

13.3
Probabilistic Homogenization Approaches

13.3.1
Assessment of Material Uncertainties

The homogenization techniques described in Section 13.2 are rigorously deterministic approaches based on the analysis of a representative volume element with well-defined microstructural geometry. On the other hand, many microheterogeneous materials in structural application possess irregular random microstructures with uncertain local geometries. Examples are short-fiber-reinforced composites where the local orientation of the individual fibers is unpredictable or solid foams where the local cell size and cell geometry of individual cells cannot be predicted. The uncertain microstructural parameters such as the fiber orientation or the cell size can either be completely random (i.e., all fiber orientations may occur with equal probability) or the values of the uncertain parameters may be described by a probability distribution such as the cell size distribution of structural foams.

For such types of material, the numerical prediction of the effective material properties using volume-element-based homogenization techniques requires the analysis of a large-scale representative volume element in order to include a sufficient number of constituents in order to be statistically representative. Depending on the volume fraction of the second phase, the representative volume element does need to include not only the constituents with all possible values of the random variables (e.g., fiber orientations) but also a sufficient number of inclusions to account for all possible types of interaction between neighboring inclusions. Analyses of the effective elastic moduli of two-dimensional model foams by Silva *et al.* [18] show that even for two-dimensional volume elements with more than 500 cells in the plane, the elastic moduli for the two spatial directions are not equal. The missing isotropy of the cellular microstructure indicates that a single volume element of this size may still not be large enough in order to be statistically representative.

For microheterogeneous materials with large characteristic length l of the microstructure, the required size of the representative volume element may be in the same order of magnitude or even beyond the characteristic length L of the entire structure (Figure 13.1). Hence, the requirement of Eq. (13.1) is not satisfied and no well-defined representative volume element exists. Thus, a probabilistic analysis is required. Examples are thin-walled structures made from short-fiber-reinforced composites or sandwich cores made from large cell metallic foams. In both the cases, only a small number of inclusions or pores will be found through the thickness of the structure. Probabilistic homogenization procedures can also be useful in the analysis of the effective properties for randomly microheterogeneous materials, where statistically representative volume elements can be identified because probabilistic procedures enable the analysis of the scatter in the effective material response in addition to the mean effective properties.

For the probabilistic homogenization of the effective material response of microheterogeneous materials based on the numerical analysis of volume elements of the microstructure, two different approaches have been established. In the first type of approach, numerical experiments on a large number of small-to-medium-scale volume elements with different microstructure – often termed *testing volume elements* – are performed. Although the individual testing volume elements are not statistically representative, the entire set of testing volume elements has to be representative for the given microstructure. For each of the testing volume elements, an individual homogenization analysis is performed. Subsequently, the entire set of homogenization results is assessed by stochastic methods in the same manner as in the stochastic assessment of experimental data sets.

The microstructure of the testing volume elements is generated computationally by a random positioning and reorientation of the second-phase inclusions such as short fibers or pores. For foam microstructures, procedures based on the Voronoï [19] process and its variants are frequently used. In its basic form, the process consists in a random definition of n nuclei p_i within the considered volume. Subsequently, cells belonging to the individual nuclei are defined as the sets of spatial points p with an Euclidean distance $r(p, p_i)$ to the respective nucleus, which is less than the distance to all other nuclei:

$$\Omega^{\text{cell}}\left(p_i\right) = \left\{p \middle| p \in R^3, r\left(p, p_i\right) < r\left(p, p_j\right), j \neq i\right\}, \quad i, j = 1, \ldots, n \quad (13.12)$$

Enhancements of the standard or Γ-Voronoï process defined by Eq. (13.12) include the δ-Voronoï process, where a minimum distance between neighboring nuclei is introduced or the Voronoï process in Laguerre geometry, where the nuclei are supplied with surrounding, nonoverlapping spheres and the Euclidean distance is substituted with the Laguerre distance measured to the closest points on the sphere surfaces [20]. The Voronoï process in Laguerre geometry enables a more direct control of the cell size distribution via a prescribed size distribution of the spheres surrounding the individual nuclei. A competitive assessment may be found, for example, in a recent contribution by the present author [21]. As an alternative to Voronoï procedures, perturbation methods can be used, where an originally regular microstructure is perturbed by a random repositioning of the cell wall intersections within prescribed windows [22].

Computational models for polycrystalline microstructures can be generated by Voronoï-type approaches in the same manner as cellular microstructures. Models for particle- and short-fiber-reinforced composites can be generated by a similar strategy as the Voronoï process in Laguerre geometry, where the nonoverlapping spheres are substituted with the inclusions. In all cases, periodic idealizations of the respective microstructure may be advantageous in order to enable the use of periodic boundary conditions in the homogenization analysis and thus to avoid pathological perturbations of the local fields near the testing volume element boundaries.

Examples for testing volume elements with periodically idealized microstructure generated by Voronoï processes are presented in Figure 13.3. In this context, the model for the short-fiber-reinforced composite includes three complete ellipsoidal

Random short fiber reinforced microstructure

Matrix   Inclusions

Random cellular microstructure

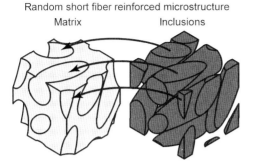
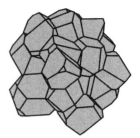

Figure 13.3 Testing volume elements (finite element meshes removed).

inclusions, which are intersected by the plane testing volume element boundaries. The testing volume element with the cellular microstructure has nonplane surfaces in order to avoid the intersection of individual cells with the testing volume element boundaries. Owing to the periodicity of microstructure and boundary conditions, the testing volume element definition with plane or nonplane (but parallel) external boundaries yields equal homogenization results.

Alternatively to the multiple analyses of small- or medium-scale testing volume elements, the stochastic information can be generated from a single analysis of a large-scale, statistically representative volume element, where the homogenization analysis is performed on subelements as testing volume elements. In this case, both the effective local strain state and the effective local stress state have to be determined using an appropriate pair of equivalence conditions, as described in Section 13.2.2. In most cases, "moving window"-type approaches are employed, where the testing volume element is a spatial element with prescribed size and shape, which is moved through the representative volume element [23]. Alternatively, microstructure-related spatial subsets such as the individual cells of a foam microstructure may be employed as testing volume elements.

13.3.2
Statistical Assessment of the Homogenization Results

The results of the testing volume element analyses are evaluated by stochastic methods. The simplest type of stochastic assessment is the determination of the statistical mean or expectation value

$$E\left(\bar{z}^*\right) = \sum_{i=1}^{n} \bar{z}^*\left(y_i\right) p\left(y_i\right) \tag{13.13}$$

of the effective property $\bar{z}^*$. In Eq. (13.13), y_i is a random variable such as the local inclusion volume fraction varied during the analysis of the testing volume elements, whereas $p(y_i)$ is the individual probability of occurrence for the underlying microstructure with $\sum_{i=1}^{n} p(\gamma_i) = 1$. In the simplest case, where a number of n numerical experiments on nominally equal testing volume elements are performed, the random variable γ_i is just the number i of the actual numerical experiment and $p(\gamma_i) = 1/n$. The effective property $\bar{z}^*$ can be any effective material property determined in the homogenization analysis such as the effective Young's modulus, the effective yield stress, and the effective thermal conductivity. The scatter in the effective property $\bar{z}^*$ can be assessed in terms of the variance

$$V\left(\bar{z}^*\right) = \left(\bar{z}^*\left(\gamma_i\right) - E\left(\bar{z}^*\right)\right)^2 p\left(\gamma_i\right) \tag{13.14}$$

or the standard deviation $\sigma\left(\bar{z}^*\right) = \sqrt{V\left(\bar{z}^*\right)}$. In addition to the expectation value $E\left(\bar{z}^*\right)$ and the variance $V\left(\bar{z}^*\right)$ as the first stochastic moment and the second central stochastic moment, respectively, higher order stochastic moments may be considered for a more sophisticated characterization of the material uncertainty.

Alternatively, a characterization of the material uncertainty directly in terms of the (cumulative) probability distribution $F(\bar{z}^*)$ or the probability density distribution $f(\bar{z}^*) = \partial F(\bar{z}^*) / \partial \bar{z}^*$ is in many cases more significant. For this purpose, the discrete homogenization results $\bar{z}^*(\gamma_i)$ are rearranged in ascending order. Subsequently, each result is supplied with the cumulative probability

$$F\left(\bar{z}^*\left(\gamma_j\right)\right) = \sum_{k=1}^{j-1} p\left(\gamma_k\right) + \frac{1}{2} p\left(\gamma_j\right) \tag{13.15}$$

where j is the number of the respective numerical experiment after renumbering in ascending order with respect to the respective homogenization result. The distribution $F(\bar{z}^*)$ describes the probability that the random effective property $\bar{z}^*$ has the value $\bar{z}^*\left(\gamma_j\right)$ or less. The probability density distribution is obtained as the derivative $f(\bar{z}^*) = \partial F(\bar{z}^*) / \partial \bar{z}^*$ of the probability distribution.

As an illustrative example for the application of probabilistic homogenization analyses, the effective normal and shear stiffness components $\bar{C}_N^*$ and $\bar{C}_S^*$ in the reduced elasticity law

$$\begin{pmatrix} \bar{\sigma}_{11}^* \\ \bar{\sigma}_{22}^* \\ \bar{\sigma}_{12}^* \end{pmatrix} = \begin{pmatrix} \bar{C}_N^* & \bar{C}_C^* & 0 \\ \bar{C}_C^* & \bar{C}_N^* & 0 \\ 0 & 0 & \bar{C}_S^* \end{pmatrix} \begin{pmatrix} \bar{\varepsilon}_{11}^* \\ \bar{\varepsilon}_{22}^* \\ 2\bar{\varepsilon}_{12}^* \end{pmatrix} \tag{13.16}$$

are determined for a two-dimensional Al model foam with a relative density of $\bar{\rho} = 5\%$ [24]. The results for the probability distributions at three different levels of microstructural disorder characterized by the variance V_{cellvol} of the cell size are presented in Figure 13.4. It can be observed that increasing geometric scatter on the microscopic level causes increasing uncertainties in the effective properties. In this context, the microstructural geometric uncertainty does not only affect the scatter in the effective properties but also affect the median value at $F\left(\bar{C}_S^*\right) = 0.5$ of the effective shear stiffness $\bar{C}_S^*$. Especially in this case, distinctively asymmetric shapes of the probability distribution $F\left(\bar{C}_S^*\right)$ develop, indicating that

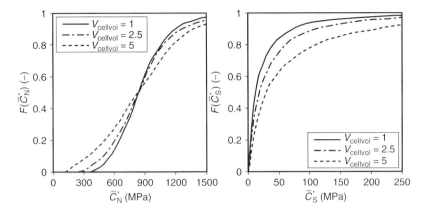

Figure 13.4 Probabilistic analysis of the effective stiffness of two-dimensional model foams.

a probabilistic assessment in terms of the basic stochastic parameters $E\left(\overline{C}_S^*\right)$ and $V\left(\overline{C}_S^*\right)$ according to Eqs. (13.13) and (13.14) alone might be insufficient.

13.3.3
Probabilistic Material Models

With the methods presented in Section 13.3.2, the scatter in the macroscopic material properties to be expected at a specific material point can be predicted. Care has to be taken in all cases, where more than one variable is required to characterize the material because the different stochastic constitutive parameters will be cross-correlated rather than being independent from each other. Furthermore, the values of the stochastic material parameters at neighboring spatial positions, in general, will be correlated. Hence, the effect of stochastic material properties in the macroscopic probabilistic analysis of a body Ω according to Figure 13.1 has to be described in terms of random fields [25]. Within this approach, it is assumed that the value of an effective material property $\overline{z}^*$ at a spatial position x_i can be described by

$$\overline{z}^* \left(x_i\right) = E\left(\overline{z}^*\right)\left(1 + f^{\text{rnd}}\left(x_i\right)\right) \tag{13.17}$$

where $f^{\text{rnd}}(x_i)$ is a random field with zero mean. In order to account for interrelations between different effective material properties at the individual spatial positions, tensor-valued random fields may be employed instead of the scalar-valued random field in Eq. (13.17) [26].

The characteristics of the random field are defined by the autocorrelation function $\Psi_{\overline{z}^*\overline{z}^*}$ with the coefficient of variation $\sigma_{\overline{z}^*\overline{z}^*}$ and the correlation length $l_{\overline{z}^*}$. The autocorrelation function describes the correlation of the material properties of neighboring spatial points, whereas the correlation length describes the distance over which the correlation between the stochastic variables at two spatial points decays to a vanishing value. In many cases, approximate formulations are employed for the autocorrelation function because of the lack of experimental data on the material uncertainty. Supplementing the experimental characterization by stochastic microstructural simulations may be useful in order to gain the required correlation data.

13.4
Optimization

The numerical models and simulation strategies presented in the previous sections enable the optimization of a material by means of rigorous mathematical optimization strategies. The aim of an optimization process is the design of structures or structural components with the best possible properties for the respective application. In this context, the performance depends on both the effective properties of the material used and the global properties such as the size, shape, and loading conditions of the structural component. The global geometric and loading parameters govern the local loading conditions; however, the local stress–strain response affects the deformation of the components, resulting in distinct interactions of the material level and the structural level. Hence, the structural mechanics problem and the material design in most cases cannot be separated, even in cases where either the structure geometry or the material to be used is fixed.

Mathematically, the optimization problem is defined by an objective function, design variables, constraint conditions, and the optimization algorithm. The objective function defines the dependence of the objective in the optimization problem as a function of design variables. Objective functions frequently used are the overall weight of a structure, its stiffness, its performance for crash energy absorption, its acoustic or thermal insulation properties, and so on. The objective function has to be either minimized or maximized by the appropriate choice of the design variables. The design variables are the variables that may be changed in order to improve the performance as the variables defining the overall shape of the component, its loading conditions, or parameters defining the microstructure of the material. In most cases, the objective function cannot be defined in closed form but has to be evaluated numerically, for example, using output data from a finite element analysis of a given design of the component under consideration.

In order to ensure manufacturability of the material and the component, the space of the design variables is usually bounded, that is, the design variables are not permitted outside prescribed ranges. These ranges are defined by constraints, usually in terms of inequalities for the respective design variable. For the optimization process itself, a variety of numerical algorithms have been developed [27]. In general, these algorithms enable the determination of an improved design from the actual design by some kind of descent technique. Hence, the improved designs are determined in a sequential process starting from an initial or reference design of the component under consideration.

Present research fields include the development of enhanced optimization algorithms in order to improve the computational performance or to reduce the risk for the algorithm to get trapped in local suboptima and thus to ensure a reliable identification of the global optimum design. A research field important especially in the design of materials is the multicriteria optimization, where the design has to be optimized with respect to different objectives, which might even be contradictory (e.g., structures with minimum weight and maximum stiffness). Approaches in this field range from a simple recombination of the different objective functions into a single one to more sophisticated Pareto-optimization procedures. Multiobjective optimization procedures are especially important in conjunction with multiphysics homogenization analyses for the design of multifunctional materials. An exhaustive discussion on multiobjective optimization procedures may be found, for example, in the book by Eschenauer and Koski [28].

13.5
Summary and Conclusions

The numerical simulation of the effective material response of microheterogeneous materials in homogenization analyses forms an important tool for complementing the experimental characterization of new materials. In addition to the possible reduction of experimental expenses, the numerical simulation of microheterogeneous materials in many cases enables a deeper insight into the micromechanical processes, causing the material response observed on the macroscopic level. In the recent time, the increasing demand for an integrated design of structures causes an increasing demand for multifunctional materials requiring multiphysics homogenization approaches.

The increasing use of probabilistic methods such as the stochastic finite element method in structural analysis enables an analysis of structural components accounting for all kinds of material uncertainties. Driven by these developments, probabilistic methods are increasingly used in the mechanics of materials. These methods allow a prediction of not only the mean effective properties of a material but also an assessment of the scatter to be expected. In this context, large efforts are required in order to properly specify all cross-correlations between different stochastic material parameters as well as the autocorrelation of the material properties at neighboring spatial positions. Owing to the general lack of experimental data on the scatter of the material data on length scales below the respective specimen level and the large experimental effort necessary to obtain reliable data in this sense, only approximate stochastic material laws are currently used in stochastic structural mechanics. Here, the probabilistic homogenization analysis provides a tool to fill the gap between the experimental data available at reasonable efforts and the requirements for the structural analysis on the macroscopic level.

Finally, the numerical analysis of the effective material behavior enables the use of rigorous mathematical optimization procedures for the design of optimum materials fitting with all kinds of prescribed requirements in structural application. By this means, the exploitation of the potential of materials can be increased and over-conservatisms can be reduced. The possible reduction of weight in transport might help to save the limited natural resources. Finally, the rigorous mathematical optimization of structural materials provides a powerful tool in order to deal with contradictory requirements in the design of multifunctional structures.

References

1. Voigt, W. (1889) Ueber die Beziehung zwischen den beiden Elasticitätskonstanten isotroper Körper. *Ann. Phys. Chem.*, **274**, 573–587.
2. Reuss, A. (1929) Berechnung der fließgrenze von mischkristallen auf grund der Plastizitätsgrenze von einkristallen. *Z. Angew. Math. Mech.*, **9**, 49–58.
3. Hashin, Z. and Shtrikman, S. (1962) On some variational principles in anisotropic and nonhomogeneous elasticity. *J. Mech. Phys. Solids*, **10**, 335–342.
4. Walpole, L.J. (1966) On bounds for the overall elastic moduli of inhomogeneous systems I. *J. Mech. Phys. Solids*, **14**, 151–162.
5. Talbot, D.R.S. and Willis, J.R. (1985) Variational principles for inhomogeneous non-linear media. *J. Appl. Math.*, **35**, 39–54.
6. Hohe, J. and Becker, W. (2002) Effective stress–strain relations for two-dimensional cellular sandwich cores: Homogenization, material models and properties. *Adv. Appl. Mech.*, **55**, 61–87.
7. Walpole, L.J. (1981) Elastic behaviour of composite materials: theoretical foundations. *Adv. Appl. Mech.*, **21**, 169–242.
8. Nemat-Nasser, S. and Hori, M. (1993) *Micromechanics: Overall Properties of Heterogeneous Materials*, North Holland Publishing, Amsterdam.
9. Hill, R. (1963) Elastic properties of reinforced solids: Some theoretical principles. *J. Mech. Phys. Solids*, **11**, 357–372.
10. Hohe, J. and Becker, W. (2003) Effective mechanical behavior of hyperelastic honeycombs and two-dimensional model foams at finite strain. *Int. J. Mech. Sci.*, **45**, 891–913.
11. Bakhvalov, N. and Panasenko, G. (1989) *Homogenisation: Averaging Process in Periodic Media*, Kluwer Academic Publishers, Dordrecht.
12. Bensoussan, A., Lions, J.L., and Papanicolaou, G. (1978) *Asymptotic Analysis for Periodic Structures*, North Holland Publishing, Amsterdam.
13. Willis, J.R. (1987) in *Homogenization Techniques for Composite Media* (eds E. Sanchez-Palencia and A. Zaoui), Springer-Verlag, Berlin, pp. 297–336.
14. Gambin, B. and Kröner, E. (1989) Higher-order terms in the homogenized stress–strain relation of periodic elastic media. *Phys. Status Solidi.*, **B151**, 513–519.
15. Bishop, J.F.W. and Hill, J.R. (1951) A theory of the plastic distortion of a polycrystalline aggregate under combined stress. *Philos. Mag.*, **42**, 414–427.
16. Hohe, J. and Becker, W. (2001) An energetic homogenization procedure for the elastic properties of general cellular sandwich cores. *Composites B*, **32**, 185–197.
17. Hohe, J. and Gumbsch, P. (2010) On the potential of tungsten–vanadium composites for high temperature application with wide-range thermal operation window. *J. Nucl. Mater*, **400**, 218–231.
18. Silva, M.J., Hayes, W.C., and Gibson, L.J. (1995) The effects of non-periodic microstructure on the elastic properties of two-dimensional cellular solids. *Int. J. Mech. Sci.*, **37**, 1161–1177.
19. Voronoï, G. (1908) Nouvelles applications des paramètres continus à la théorie des formes quadratiques. *J. Reine Angew. Math.*, **134**, 198–312.
20. Fan, Z., Wu, Y., Zhao, X., and Lu, Y. (2004) Simulation of polycrystalline structure with Voronoi diagram in Laguerre geometry based on random closed packing of spheres. *Comput. Mater. Sci.*, **29**, 301–308.

21. Hardenacke, V. and Hohe, J. (2010) Assessment of space division strategies for generation of adequate computational models for solid foams. *Int. J. Mech. Sci.*, **52**, 1772–1782.

22. Li, K., Gao, X.L., and Subhash, G. (2006) Effects of cell shape and strut cross-sectional area variations on the elastic properties of three-dimensional open-cell foams. *J. Mech. Phys. Solids*, **54**, 783–806.

23. Baxter, S.C., Hossain, M.I., and Graham, L.L. (2001) Micromechanics based random material property fields for particulate reinforced composites. *Int. J. Solids Struct.*, **38**, 9209–9220.

24. Hardenacke, V. and Hohe, J. (2009) Local probabilistic homogenization of two-dimensional model foams accounting for micro structural disorder. *Int. J. Solids Struct.*, **46**, 989–1006.

25. Charmpis, D.C., Schuëller, G.I., and Pellissetti, M.F. (2007) The need for linking micromechanics of materials with stochastic finite elements: a challenge for materials science. *Comput. Mater. Sci.*, **41**, 27–37.

26. Soize, C. (2008) Tensor-valued random fields for meso-scale stochastic model of anisotropic elastic microstructure and probabilistic analysis of representative volume element size. *Probab. Eng. Mech.*, **23**, 307–323.

27. Spillers, W.R. and McBain, K.M. (2009) *Structural Optimization*, Springer-Verlag, Berlin.

28. Eschenauer, H. and Koski, J. (1990) *Multicriteria Design Optimization: Procedures and Applications*, Springer-Verlag, Berlin.

14
An Artificial-Intelligence-Based Approach for Generalized Material Modeling

Kambiz Kayvantash

14.1
Introduction

Automotive materials have become so diverse that we see a new material available for improving the product nearly every day. In general, these materials could be considered as either load bearing, for which we need some characterization of their behavior, or styling or isolating for which some specific characteristics may be needed, depending on the application. The discussions in this chapter refer to the first category but could easily be extended to the second as well.

In recent years, the process of identification of load-bearing characteristics has become a rather complex issue because of the impact of the manufacturing and geometry of the material on its mechanical behavior. This is especially the case for nonmetallic materials, which are used with great success for, among other applications, vehicle weight reduction purposes. In order to define a mechanical material model for a load-bearing material, we essentially need to know two things. First, we need to know what we measure as the "quantity" representing the response of the material (ordinate or function). This corresponds to what we actually observe as a physical behavior. Second, we need to know along which abscise (variable) we observe the change in behavior. This corresponds to what we willingly impose as state change on the structure, which brings us to the fundamental question of what is the best choice for the measured quantity (stress) and the variable (strain). There is a third difficulty, often not mentioned or underestimated in many studies proposing material laws – this is the transition of the observed axes (strain and stress or variable and function) from single-dimensional characteristic to three-dimensional (3D) space, which usually is not obvious. It is often (wrongly) assumed that a curve representing a given behavior in one dimension can be easily extended to 3D, based on some, often unjustified, assumption on *Poisson's effect*. Indeed, material identification extends even beyond the three geometrical axes, although these are the only ones a material model in a software (finite element (FE) code) has access to. All other information needs to be combined and expressed in terms of the geometrical measures of change (strains or rates).

Structural Materials and Processes in Transportation, First Edition.
Edited by Dirk Lehmhus, Matthias Busse, Axel S. Herrmann, and Kambiz Kayvantash.

The implementation of a 1D material model expressed in terms of a 1D formula or experimental 1D data immediately poses the problem of summing the effects of a general loading in terms of a combination of equivalent pure deformations. These may include uniaxial, equi-biaxial, planar (pure shear), and volumetric modes. Frequently, no shear tests are available (at least for high strain rates), especially for large deformations (such as foams or some low-density composites where damage in material or structural bonding precedes large elongations). In these cases, the natural choice of expressing the material law in the principal axes of the deformation is attractive and has frequently been adopted, the reason being that only 1D information is collected in order to establish a 3D behavior. Another important aspect is the correct choice of strains and stresses that correspond to the abscissa and the ordinate of the experimental curve, namely the engineering (or nominal) stress and strain. In addition, the extension of 1D–3D formulation requires an efficient treatment of the coupling effects that may be introduced via the introduction of a variable Poisson's ratio. Finally unloading, creep, and hysteretic behaviors must also be treated in a numerically stable manner.

The essential choice when considering large nonlinear elastic deformations as in the case of foams and rubbers is the strain measure (this also governs the choice of an objective stress measure). This choice should first be "measurable" within an experimental setup and, second, should give enough information on the nature of the nonlinearities. Finally, a corresponding objective definition of the stresses must be available in such a way that a unique monotonous function of the strain could be defined for the stress computation. In the next paragraphs, we shall present a detailed discussion of the above-mentioned problems and the solutions adopted. We shall give particular attention to the definition of strains as it underlies the whole methodology of this work. Note that many standard textbooks have already treated this important subject in nonlinear elasticity and we shall therefore only summarize the definitions concerned. It may seem redundant to an experienced reader although essential to our work.

In the following section, we shall present a summary of few available (and relevant) strain and stress measures applicable to large inelastic deformations. The interested reader may find many of the details not presented here in Ogden's classic reference book on the theory of large elastic deformations [1].

14.2
Strain Measures

Physically speaking, the strain of a material point is a measure of the stretch of that point when moving in space, excluding rotations and rigid body movements (which involve uniform movement of all material particles and not a group of them only). Note that the strain needs to be pure, that is, independent of and excluding any other change of state. Turning around the point where you stand or moving (rigidly) to another point does not stretch your body. It simply changes your viewpoint which is why we have to remove it, via appropriate filtering or transformation, from our observation.

From a mathematical point of view, the deformation of a continuum at a given point is defined by the second-order deformation gradient tensor $\mathbf{A}$ where

$$\mathbf{A}_{ij} = \frac{\partial x_j}{\partial X_j}$$

where $X_j(X_j, t)$ denotes the coordinates of a point with initial coordinate $x_j(t = 0)$ at time t. These deformations include both the stretch (elongation) and the rotation of an arbitrary line element. The quantity referred to as the *deformation gradient A* is a non-singular tensor and J (=det $\mathbf{A}$), which represents the volume change in the medium in consideration, is nonzero. $\mathbf{A}^T\mathbf{A}$ is symmetric and positive definite and $\mathbf{A}^T\mathbf{A}$ and $\mathbf{A}\mathbf{A}^T$ are called the *right Cauchy–Green deformation tensors* and *left Cauchy–Green deformation tensors*.

The material is said to be unstrained at X if the length of an arbitrary line segment dX is unchanged after deformation. This implies that for material to remain unstrained at X, $\mathbf{A}^T\mathbf{A} = \mathbf{I}$, which includes the case of a rigid body rotation accompanied by a rigid translation for the most general case. If the material does not satisfy the above condition, then it is said to be strained and the tensor $\mathbf{A}^T\mathbf{A} - \mathbf{I}$ can be regarded as a strain tensor as it provides a measure of the change in length of an arbitrary line segment of material. However, various strain definitions may be defined depending on the standpoint of the observer.

The Green (Lagrange) strain tensor $\mathbf{E}$ is defined by

$$\mathbf{E} = \frac{1}{2}(\mathbf{A}^T\mathbf{A} - \mathbf{I})$$

The displacement of a particle X from reference configuration to the current one is defined by the point difference $u(X) = x - X$ and the displacement gradient $\mathbf{D}$ tensor is given by

$$\mathbf{D} = \text{grad } u(X)$$

It follows that

$$\mathbf{E} = \frac{1}{2}\left(\mathbf{D} + \mathbf{D}^T + \mathbf{D}^T\mathbf{D}\right)$$

Recall that, in general, $\mathbf{A}$ includes both elongations and rotations, whereas only elongations are relevant for a material behavior observation. We, therefore, need to identify and remove the rotations from $\mathbf{A}$. The *polar decomposition theorem* states that for any non-singular second-order tensor $\mathbf{A}$, there exists unique positive definite symmetric second-order tensors $\mathbf{U}$ and $\mathbf{V}$ and an orthogonal second-order tensor $\mathbf{R}$ such that

$$\mathbf{A} = \mathbf{R}\mathbf{U} = \mathbf{V}\mathbf{R}$$

and that

$$\det A = \det \mathbf{U} = \det \mathbf{V}$$

(recall that det **A** represents the change in volume of the original element)

$$\mathbf{R}^T\mathbf{R} = \mathbf{I}\mathbf{U} = \mathbf{R}^T\mathbf{V}\mathbf{R}$$

The deformation gradient **A** represents a rigid rotation if and only if **U** = **V** = **I**.
If **R** = **I**, then **A** = **U** = **V** and the deformation is referred to as *pure strain*.

The *Lagrangian* and *Eulerian strain* tensors can be written as $\frac{1}{2}(\mathbf{U}^2 - \mathbf{I})$ and $\frac{1}{2}(\mathbf{I} - \mathbf{V}^2)$, respectively. The tensors **U** and **V** are called, respectively, the *right stretch tensor* and *left stretch tensor* (they correspond to the *right* and *left Cauchy strain tensors*).

Let λ_i be the principal values of **U** corresponding to the principal directions $u^{(i)}$ such that

$$\mathbf{U}u^{(i)} = \lambda_i u^{(i)}$$

We have also

$$\mathbf{V}\mathbf{R}u^{(i)} = \mathbf{R}\mathbf{U}u^{(i)} = \lambda_i \mathbf{R}u^{(i)}$$

This shows that λ_i are also the principal values of **V** corresponding to the principal directions $\mathbf{R}u^{(i)}$. Thus the deformation rotates the principal directions of **U** into those of **V**. If rotations are zero, **U** = **V** is valid. Note that $J = \rho_0/\rho = \det \mathbf{A} = \lambda_1\lambda_2\lambda_3$ (J = relative volume, ρ = density, ρ_0 = initial density). Finally, the *Lagrangian* and *Eulerian* strain tensors may generally be defined as

$$\frac{1}{m}\left(\mathbf{U}^m - \mathbf{I}\right), \quad \frac{1}{m}\left(\mathbf{V}^m - \mathbf{I}\right), \quad m \neq 0$$

$$\ln \mathbf{U}, \quad \ln \mathbf{V}, \quad m = 0$$

where m is usually an integer (this constraint may even be released).

The above strain measures are coaxial with **U** and **V** and have the principal values

$$\frac{1}{m}\left(\lambda_i^m - \mathbf{I}\right), \quad m \neq 0$$

$$\ln \lambda_i, \quad m = 0$$

Some special cases are of particular interest in elasticity, namely

$m = 0$	Hencky (logarithmic – this is often referred to as *true* strain as it may be shown to correspond to a measure of the change in elongation with respect to the actual configuration/length of material segment)
$m = 1$	Biot (nominal – this is sometimes referred to as *engineering* strain as it refers to the initial configuration/length of material segment)
$m = 2$	Green
$m = -2$	Almansi

For further reading, we refer to [1] where most of the above-mentioned formulations are developed. For our interests, it is this last concluding paragraph which we shall retain for the rest of this chapter as it is of utmost importance to our approach. Identifying a material "law" is equivalent to establishing a relation (linear or else) between a function of the stretch λ_i and a suitable stress (force divided by an appropriate representative surface normal to the stretch measured. In simple terms, the strain may be defined via different expressions (functions) of the stretch λ_i ranging from linear to higher order definitions. There is no reason to prefer one strain measure to another except for practical material model coding purposes relating to the function (stress) that we measure, which should correspond to the directly related measure of stretch (this is referred to as the objectivity of the strain–stress relationship). Indeed, for small strains most definitions are equivalent. We conclude that any large deformations of foams, composites, plastics, or rubbers may be formulated in terms of any appropriate above-mentioned measures of the deformations. The choice of the strain is, however, governed by the definition of the stress–strain relationship and the corresponding experimental measurements available.

The fundamental issue to be considered here is that the definition of strain is only a convention. It simply defines the axis along which the observed phenomena (usually stress) should be presented and considered for defining a material behavior.

14.3
Stress Measures

In large deformations, the choice of the stress undergoes the same limitations as the deformations. Once the abscissa of the characterization is defined, we need to define the ordinate. However, the choice is not arbitrary and needs to correspond to the abscissa we are referring to (in continuum mechanics, this requirement is formulated by stating that the strain and stress measures need to be conjugate). In the differential formulation, the two measures must be objective (one should cause and explain the other), and their definitions must be coherent with respect to the principle of virtual work. Frequently, the equilibrium equations of a solid are expressed in terms of the *Cauchy* stress T. The components of the *Cauchy* stress tensor may be obtained from direct experimental measures and have a clear interpretation. However, the use of this stress measure is complicated by the fact that the rate of *Cauchy* stress is not objective. In short, the Cauchy stress refers to the force per actual unit surface, whereas its change refers to two different stages, and therefore two different unit surfaces.

One solution to this problem consists of writing the material law in terms of other objective derivatives of the Cauchy tensor (e.g., Jaumann, Truesdell, and Naghdi–Green) as rates refer to small time increments during which adequate assumptions on the reference unit surface may be made. However, these measures only give good estimates of the stress rate if the pure deformations between two consecutive configurations can be considered as small. This condition is usually satisfied sufficiently if the time steps are small and if the constitutive material law is presented in a differential form. This is fortunately the case in explicit time integration. Recall that some FE codes use Jaumann stress rates and the true (logarithmic) strain measure, which is related to the measure of "velocity strain" which has limitations for large deformations (elastomers and foams), although more efficient for metal elastoplasticity and quite fast to compute (as there is no need for polar decomposition of the deformation gradient).

Another solution that seems to be better suited for some material characterizations (such as foams) is to use another objective stress tensor. For example, the *second Piola–Kirchhoff stress* $T^{(2)}$ is such a tensor defined with respect to the initial configuration. The rate of this stress tensor in a differential formulation may be related to the *Green strain tensor*. After the computation of the total stress components (based on the 1D formulae), the *Cauchy stress tensor* in the new configuration may be obtained by the transformation:

$$T = \frac{1}{J}\mathbf{A}T^{(2)}\mathbf{A}T$$

In the formulation of a material law (i.e., a 1D formula often using nominal stress measures), we actually define an objective stress tensor $\mathbf{S}$, whose inverse is often referred to as the *first Piola–Kirchhoff tensor*. However, as we have already chosen the measure of strain in the deformed configuration, the tensor $\mathbf{S}$ should also be defined in the same configuration but as in the *second Piola–Kirchhoff tensor*, per unit of the reference surface. The principal directions of the strains coincide with those of the tensor $\mathbf{S}$ (also referred to as the nominal stress tensor) as well as the *Cauchy tensor* $\mathbf{T}$. As the *Cauchy tensor* is a measure of the stress per unit of the deformed surface, the relation between the two stress tensors in the principal directions may be expressed as

$$\mathbf{S}_i = \mathbf{T}_i \lambda_j \lambda_k$$

It should be noted that the tensor $\mathbf{S}$ and the *second Piola–Kirchhoff* tensor have different eigenvalues. The following relationship exists which allows for transformations from one definition to another depending on the material model to be established:

$$\mathbf{T}_i^{(2)} = \frac{\mathbf{S}_i}{\lambda_i}$$

We can define a conjugate set of stress and strain measures as those which are equivalent to the above equation. We state that the stress power density is objective and consequently any set of conjugate stress–strain measures. More generally, we can associate with $\mathbf{E}^{(m)}$ a symmetric stress tensor $\mathbf{T}^{(m)}$ such that $\mathrm{tr}(\mathbf{T}^{(m)}\mathbf{E}^{(m)})$ gives the power density independently of m where each of $\mathbf{T}^{(m)}$ and $\mathbf{E}^{(m)}$ is a Lagrangian tensor.

In general, consider the general constitutive relation of an elastic material, which can be written in terms of $\mathbf{T}^{\hat{}} = F(\mathbf{U})$, where $\mathbf{T}^{\hat{}} = \mathbf{R}^\mathsf{T}\mathbf{T}\mathbf{R}$. The tensor $\mathbf{T}^{\hat{}}$ is called the *rotated-stress tensor* (*Kirchhoff tensor*). It shows that the strain determines the stress tensor, through the constitutive equation, only to within the rotation of the material particle. In other words, a constituent relation may only relate stresses to a strain tensor, which does not contain any rotational components. What we should retain from the above discussion is that as for strain the definition of stress is not unique and should be adapted (conjugate) to the strain observed [1]. Additionally, it is not sufficient to establish a relation between the imposed load or displacement and observed reaction or deformation to conclude on a material law relating directly the observed quantities. As a final conclusion, we state that a material law may be expressed as a function of stress (or its rate) of the deformation (or its rate) as long as the function respects uniqueness and objectivity. From a mathematical point of view (related to our approach developed in the following sections) as long as $\mathbf{U}$ and $\mathbf{T}'$ are known, selecting a material law may be considered as the identification of the function F. The relation is only valid if and only if the relevant variables of the functions are identified in advance. This statement is central to our approach presented hereafter and may be extended to a nonlinear case.

14.3.1
One-Dimensional Material Models

Once a set of objective strain–stress tensors are chosen (this may depend on many factors such as the choice of the experiments available or the code into which the material laws are to be introduced), a material law may be devised (often in form of a 1D formula). Generally, three types of material models may be considered and are of major interest to us. Either an "energy functional" of some measure of the strain is available in which case a stress–strain relationship may be obtained by deriving the functional with respect to the strain measures or this is not the case and a phenomenological model based on some kind of separation of the components of the complex behavior must be alternatively considered (Figure 14.1).

In the case of hyperelastic materials such as rubbers or some plastics given an appropriate measure of strain, the energy functional is defined and derived in order to arrive at the stresses (stress – per unit area – is the increment of work needed to generate the increment of stretch):

$$W = f(\lambda_i)$$
$$\sigma = \frac{\partial W}{\partial \lambda}$$

where λ is a principal stretch calculated from the strains as explained previously. Note that the stress needs to be scaled (proper conjugate pairs of the strain–stress measure need to be adopted).

The incremental formulation of a typical material law (e.g., viscoelastic) may be as follows:

$$\sigma = s + p$$
$$\dot{s} = E\dot{e} + \frac{(E + Et)}{\mu}s + \frac{(EEt)}{\mu}e$$
$$p = f(\varepsilon m)$$

It is a 1D phenomenological description of the mechanical behavior. The first approach is well suited and historically tested for rubberlike materials. We have presented these models in terms of the more general Ogden formulation, which also includes the case of the widely used Mooney–Rivlin models. The second one is used for the formulation of the viscoelastoplastic models for foams (and to a lesser extent rubbers). Note that any other 1D formula, which fits the experimental data, may be used instead (Figure 14.2).

A third approach may also be adopted via direct imposition of a stress–strain curve obtained during some simple experimental procedure. This approach assumes that this simple test underlies the global behavior of the material in a general loading state (e.g., RADIOSS law 38). It may, however, be considered as a subset of the second method and is essentially a phenomenological approach in the sense that the phenomena under observation are represented by a digitalized set of values instead of a 1D formula fitted to these data. Both the second and the third approach rely heavily, of course, on the availability and reliability of the data. In all the cases,

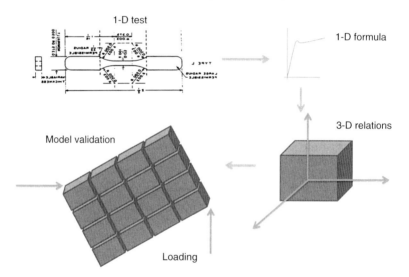

Figure 14.1 Transformation steps needed from 1-D testing to 3-D element formulation.

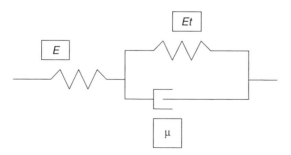

Figure 14.2 A simple maxwell - Kelvin model with three material parameters E, Et, μ.

the choice of the strain and stress measures is predominant for a practical (and objective) determination of the material constants.

14.3.2
General Material Models in Three Dimensions

The majority of material models used in FE codes applied in automotive industry are phenomenological laws. They decompose the macroscopic behavior of the material into separate components and subsequently use a curve fitting procedure in order to derive the material constants. The fit parameters are not unique and must be obtained step-by-step for different strains and strain rates (Figure 14.3).

A typical material law based on a 1D compression observation needs to be extended and represented in a general 3D configuration (Figure 14.4).

The load–deflection curve is then extracted and normalized (with respect to original surface and original length). This suggests that a typical axial cube compression test may be used in order to formulate the essential part of the 3D behavior of the material (because of the fact that most foams exhibit little if any dilatation under pure compression). Recall that foamlike materials are indeed microscopic

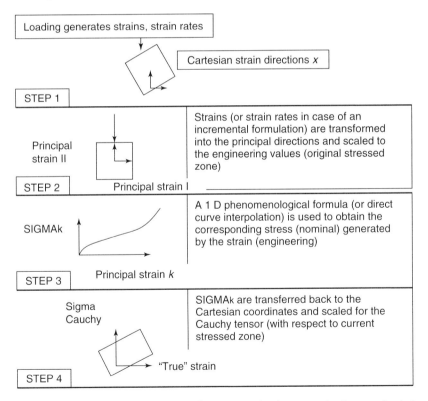

Figure 14.3 Material formulation steps from a generalized cartesian loading to a local element level coordinate system and back.

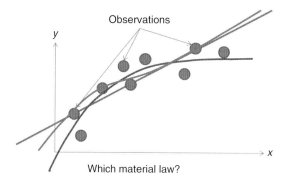

Figure 14.4 A material law is not unique. A physical behavior may be interpreted in many ways, depending on the physical parameters that define the conditions during which the characterization is performed. A judgment is necessary.

structures and we are speaking here of a macroscopic phenomenological behavior law. The 1D stress–strain curve (or the normalized force deflection curve) may thus be used as the basis of the material law, as long as it is defined in a principal direction. Recall that a principal direction corresponds to a direction for which the shear stresses are zero (obtained theoretically via an eigenvalue transformation of the Cartesian strain tensor). A typical procedure would then be the following:

14.3.3
A Generalized Material Model

The concept of a material model is a simplification of number of phenomena, which may lead, on a macroscopic level, to the identification of the stress tensor (**Y**) as a function of the strain tensor, strain rate tensor, or other material macroscopic variables (*X*), or vice versa, the deformations as a result of loadings on the structure. In reality, the abscissa is not only the strain, and the function may involve more than just stress values. Other quantities such as temperature or humidity could be used as measures of loading change and other changes of material state such as color or radiation could also be used as "response" measures. Apart from strict laboratory conditions in which a material is observed or a deterministic computer FE model with predefined material formulation and parameters, real world presents few conditions where loading and response are completely known or even understood. Today's design solutions are permanently in quest of quality and robustness, which requires the engineer to take into account not only the "average" but also the deviations and in some cases even the outliers.

There is no reason why the loading versus material response relationship should always be explainable via a deterministic analytical formula as is often suggested by the very concept of the material law. In this section, we shall explore the options and devise a more generic approach for the identification of the material response (cf. law) subject to an imposed loading or displacement field. In our approach, the cause and effect (material law) is not always explained (formulated) using an algebraic formula but rather via a collection of cause-and-effect data, which are used to explain the behavior.

In the following section, bold letters represent vectors and bold italic letters represent matrices. Let us assume in a most general manner that the material model may be constructed via a relationship $Y = H(A, X)$. We shall call Y the descriptors, X the explanatory variables, and A the parameters of the model, described by H. Note that the material model is not only a matter of X and Y (e.g., strains and stresses) but also of A, which we can interpret as the *parameters* or conditions of the laboratory test in general. This is of extreme importance and is unfortunately often ignored or too simplified.

The main difference between this generic approach and the previous "macroscopic" approach is twofold. First, we do not only relate the Y and the X via a mathematical formula but via an explainable relation or "rule," which can be interpreted as the material relation as observed. Second, we do not necessarily predetermine what is X and what is Y. In this sense, the Y (stresses or loads or anything else) are observed, whereas X (strains, rates, temperatures, etc.) are assumed or imposed. Note that the relation may easily be inverted, which is the case in many material tests as the loading is imposed and displacements (deformations) are observed.

In the next section, we shall consider the material model to be represented by a set of observations (sites or "load cases"), a number of descriptors (e.g., loads), and a set of explanatory variables (e.g., strains). Let us also consider another set of "system," known or assumed parameters, which we shall call regulatory parameters (e.g., experimental setup or error).

Let us assume that we are looking for a relation of the form $Y = H(A, X)$, where all quantities are vectors or matrices. We may represent the above-mentioned formula in a partitioned matrix form, which for simplicity, we shall call a "data set" representing all observation sites. We have

$$\mathbf{F} = (Y|A|X)$$

where $\mathbf{F}$ is the matrix of sites. It is to be noted that in a general sense, neither $\mathbf{F}$ nor A or X need to belong to the set of real numbers as they may equally.

Presented in this manner, the task of establishing a material model becomes one of extracting the "rules" explaining the hidden patterns or relationships in the data set, which also include the strains and stresses. We shall handle this issue via application of some basic machine learning techniques and present in the following. Note that the techniques themselves are not fully covered here but only their application. The interested reader may refer to many interesting textbooks on the topic including [2].

One very interesting outcome of the above-mentioned generalized approach is that more than one definition of strain can be used for identification purposes in the X partition of the data set as long as a relevant stress description (conjugate) is available on the Y partition. Indeed, if a material law is physically and theoretically correct, one would expect the X columns and the corresponding Y columns "fit" with more or less same order of accuracy.

14.3.4
Reliability and Robustness

In the following section, we shall explain why we have introduced the A partition in the data set. In general, in data mining techniques, there is no distinction of A. It is considered as part of either Y (as constants or weights) or X (as variables). A material identification technique may refer to this as secondary (cf. X as primary explanatory variables). In the simplest case, A may contain simply weights expressing confidence we have in the particular line of data in the F matrix. However, in a rule-based characterization, it may play a major role in the way it relates different sets of data to be related to each other.

Let us assume that we have a set of data including tension, shear, and hydrostatic compression tests for an elastic material. Note that from a theoretical point of view, one of the tests is redundant (only two Lamé constants necessary to characterize the material). $F1$, $F2$, and $F3$ are considered as "pure" cases, that is, providing only "diagonal" information on the descriptors. The important issue to consider is that in nearly all "fitting" procedures, the A part is completely forgotten (or simplified by some reference to norms – not always respected). Usually $F1$, $F2$, or $F3$ represent averages of many Fs with certainly different As, which is not correct. The machine learning approach is primarily aiming to remedy this, thus allowing for improved rating (precision), reliability (controlled statistics), and robustness (reduced surprises!).

The idea is simply to include the As in the rule extraction relating Ys and Xs. The outcome, (i.e., the stress–strain relationship) is different if we do or do not include the As. Obviously, inclusion of the As provides valuable information in terms of the exactness of the relationship we can extract from the data set.

It is obvious that the choice of the material law or formula depends not only on the real (physical) relationship between the X and Y but also on the way we include these in the expression $F(X) = Y$. The function F includes (in a combined or compressed manner) all the influence of As. Another more appropriate way to define the law is indeed to include the As arriving at $F(A, X) = Y$. This simply means that the material law is not unique but depends simply on the choice we make in order to represent the influence of As. The choice is indeed a decision and not a rule or a law itself, which is often ignored.

In the ever expanding literature, we are frequently confronted with contributions, where a material law is reduced to a curve fitting practice, stating, for example, that as the relation between X and Y is quadratic, therefore is the material law. This is wrong first since our choice of X itself is not unique (we could have explored the relation between X^2 and Y or $\ln(X)$ and Y, arriving at another formula. Whichever decision we make, the As are of fundamental importance and should not be neglected. This brings us to the concept of material behavior "learning" instead of material law fitting.

14.3.5
Machine Learning Techniques and Rule

In general, one can distinguish between two categories of identification techniques. The first one is a direct method. In this approach, an a priori relationship is assumed (a formula) via observation, and specific tests are devised in order to bring into evidence (or measure) the influence of the model components. In a sense, this is a real-physics-based approach, which justifies its strength. Its major disadvantage lies in the fact that the model needs to be assumed in advance and that it does not fully exploit modern data collection and generation techniques. One could say that it is best reserved for experts!

All the methods introduced hereafter belong to the category of inverse methods. In this approach, data is first collected not always with the objective of explaining the model components. At a later stage, data is exploited (learnt) in order to explain the assumed variables and parameters (which explains its relationship to the direct methods). The idea here is not to be comprehensive but present methods which the author has employed successfully for material characterization. Many other techniques are available (radial basis function (RBF), support vector machine (SVM), Kernel techniques, etc.), but which are either too complex to be described here or are simply too complicated for our applications. Similarly, we do not intend to be anyway comprehensive in presenting machine learning technique; therefore, we shall briefly introduce the following techniques:

1) Regression (least square, minimization, etc.)
   This is an indirect method, although it has a close link to the direct method in the sense that it is the most traditional (classical analysis of variance (*ANOVA*) type) "curve fitting" approach well adapted to situations where relationships between site descriptors Y and explanatory variable X are known to be of a low-order polynomial (of usually first or second order), while system or environmental parameters A are assumed to be constants (to be evaluated by the algorithm). Higher order relations may also be established via the replacement of X by some function of X expressed via A. Note that A is the outcome of the process and not an input, which means that the A partition of the data set is either considered as a part of X or ignored simply via the implicit form of the function to be fitted.
   The idea is to obtain the relationship using a classical least square or any other adapted minimization technique (especially if constraints are to be taken into account) in order to minimize the distance between the observed results and those calculated using the polynomial description. The advantage of this method lies in its simplicity of implementation as well as availability of stand-alone tools. The major disadvantage of the method is that it requires an a priori assumption on the algebraic form of the relationship, which makes it unsuitable for non-numerical observations to be taken into account. Another sometimes negative side effect lies in the fact that it has a strong "filtering" effect on the data not taking into account specific details of the data set.

2) Kriging

The Kriging method has become very popular because of its generality, robustness, and sensitivity to details of the data set. It can be considered as a general-purpose interpolation (extrapolation) technique or simply or more appropriately as a complex surface fitting technique. Its major advantage is that it does not assume any imposed characteristic (shape) on the relationship apart from the fact that it can be explained via the combination of a polynomial (usually of low order) explaining the general features of the data set (basis) with a more localized (point-wise) function explaining each data in the data set. It can easily be applied to any number of descriptors, variables, and parameters. The major disadvantage of this method lies in the fact that it can only handle real numbered data sets and that a matrix inversion is necessary, which makes it sometimes inefficient.

3) Neural networks

This is the simplest classical method with "reasoning" as it allows for some sort of iterative "trial-and-error" logic. Its clear mathematical foundations are yet to be established but one can say that it belongs in a sense to the category "nonlinear principal component analysis (PCA)" methods (see later sections). The principle is simple and is explained in the following (we shall avoid formulas but only consider the logic behind the reasoning):

Assume you have a set of descriptors Y and a set of explanatory variables X. Starting with a set of initial values of X, one can suggest a simple relation between the X and the Y (let us call this functions). If the relation explains Y, the shape function is rated high, else it is penalized. That is all there is to it. This process may be automated in order to obtain the best shape functions, which explain Y given X (describe Y as a function of X).

4) PCA

This is our favorite method because of its simplicity as well as its "visibility." The method is generally used to visualize the data, allowing to extract simple rules expressed in a set (basis) of independent orthogonal vectors $\mathbf{U}$ (eigenvectors of the correlation matrix $\mathbf{R}$ constructed from the standardized components of Y (or X or A), which are linear (in case of linear PCA) combinations of the descriptors Y. Once the principal components $\mathbf{U}$ are obtained, any element of the data set (F, Y, A, X) may be represented on the $\mathbf{U}$ space and relations obtained. From a graphical point of view, the PCA is simply a rotation in the multidimensional space of the original system of axes. The principal coordinates may also be obtained from the dispersion matrices $\mathbf{S}$ (*not standardized, therefore unit dependent*). Note that these are NOT the same as those obtained from the $\mathbf{R}$ matrix and should, in general, not be used for material identification. In this case, the descriptors are standardized. It follows that the distances among objects. The major idea is then to establish relations between the Y, the X, and the A. The only disadvantage of this approach lies in the fact that the relationship expressed in terms of $\mathbf{U}$ may prove to be difficult to interpret physically (as $\mathbf{U}$ is in fact a combination of the descriptors).

14.3.6
How to Extract the Generalized Material Law Extraction

Using one of the techniques mentioned above, we can now proceed to analyze the data set. The trick in analyzing the data in Table 14.1 is in the fusion of the As in an additional vector of **X**. In simple terms, we proceed with a separation of variables, replacing $F(A, X) = Y$ by $G(A)*H(X) = Y$ in which the operation "*" needs to be determined. In the following, we shall use a PCA method for the analysis.

Proceeding with a PCA analysis on the $Y|A$, we obtain a relationship that allows us to describe all As in a new representation orthogonal space (**Ua**), which is in turn considered as a new vector (additional column) contributing to X. We arrive at an augmented matrix (vector **Ua** is added), which allows us to represent the relationship $F(A, X) = Y$ by $F(X_{new}) = Y$. The rest is straightforward using another PCA (or nonlinear PCA) in order to explain the relationship between the X_{new} and Y. One should observe that our intention is not always to express the Y in terms of an approximate polynomial of function of the (A, X) (or X_{new}) but always be able to identify the position of a new couple (a, x) in a set of a data set A, X. Indeed this approach may best be called material classification rather than material identification because we intend to include the probability aspects of the material behavior in our formulation.

The data in Table 14.1 now becomes that in Table 14.2.

The reader may realize that the "pure" tests (i.e., diagonal cases) are no more pure and include a constant term, which needs to be taken into account as a new variable. This may complicate the fitting because it adds to complexity. However, there is a way to remove this difficulty if one uses a constrained optimization procedure rather than a simple least-square-type algorithm as the **X4** columns may simply be considered as an equality constraint (the Gs). One could also consider

Table 14.1 Hypothetical test results (data set).

Sites	Descriptors			Parameters (secondary variables)			Explanatory (primary) variables		
	Y_1	Y_2	Y_3	A_1	A_2	A_3	X_1	X_2	X_3
	Normal stress	Shear stress	Pressure	Confidence	Temperature	Machine precisions	Elongation	Shear	Volumetric strain
F1	YP1	0	0	AW1	AT1	AM1	XV1	—	—
F2	0	YP2	0	AW2	AT2	AM2	—	XV2	—
F3	0	0	YP3	AW3	AT3	AM3	—	—	XV3
F4	YP4	YP5	YP6	AW4	AT4	AM4	XV4	XV5	XV6
F5	YP7	YP8	—	AW5	AT5	AM5	XV7	XV8	XV9
F6	YP9		YP10	AW6	AT6	AM6	XV10	XV11	XV12
....	....	....	....	....	....	....	....	....	....

Table 14.2 Condensed test results (condensed data set).

Sites	Descriptors			Explanatory variables			
	Y_1	Y_2	Y_3	X_1	X_2	X_3	$X_4 = G(A)$
	Normal stress	Shear stress	Pressure	Elongation	Shear strain	Volumetric strain	Parameters
F1	YP1	0	0	XV1	—	—	XA1
F2	0	YP2	0	—	XV2	—	XA2
F3	0	0	YP3	—	—	XV3	XA3
F4	YP4	YP5	YP6	XV4	XV5	XV6	XA4
F5	YP7	YP8	—	XV7	XV8	XV9	XA5
F6	YP9	—	YP10	XV10	XV11	XV12	XA6
....	....	....	....	....	....	....	.....

the *A*s as Lagrange multipliers or as an additional unknown if the least-square solution is the only one at hand.

The introduction of *A*s as constraints has an immediate consequence and advantage because it allows for "stochastic" analysis to be conducted simultaneously during the fitting allowing for "relaxations" of the constraint, generating a cloud of possible solutions instead of a single material law. Finally, remember that *A*s can only be "learnt" and not always computed because of the fact that original data given in Table 14.1 may not be complete (missing data), which justifies a machine learning approach rather than a straightforward regression-type procedure.

Another major advantage of the above approach is to identify all unknowns simultaneously, using all the data sets by exploiting all (or available) strain and stress measures. This has the advantage of adding generality to the material behavior formulation.

14.4
Example

In the following example, we shall apply the method to the material (elastic properties) identification of a PCB (printed circuit board) made from FR4 (fiber-reinforced, FR) material (courtesy of French Fonds Unique Interministériel (FUI) 13 projects O2M, SP8, reported by Melanie Le and Kambiz Kayvantash). A material law is sought in order to relate the elastic (Young's) modulus to the temperature and the orientation of the fibers. The results are presented in the following graphs (all results are normalized). At first sight, it is difficult to establish a relation between *E* and *T* (Figure 14.5). Any curve fitting process would have delivered nonphysical relations because of overfitting caused by attributing effects to only two variables (temperature and fiber orientation).

The learning process would much improve this interpretation via considering not only the obvious but also the less apparent environmental and process parameters such as:

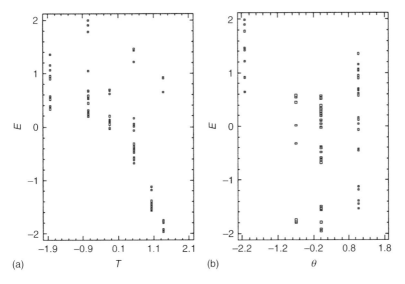

Figure 14.5 (a) Graph of *E* versus *T*. (b) Graph of *E* versus θ.

- material treatment (process)
- temperature
- age
- loading velocity
- operator direction of sample cut.

The results (material laws) in the form of a polynomial material law are given in the following for an AI (artificial intelligence or machine learning) based approach and a classical-regression-based solution. Good correlation is found between the two approaches for both aged and new materials. It is to be noted that the results based on aged (long term) materials are much closer than new materials, where the environmental process has a more immediate (short term) impact on the tested materials. This clearly indicated the superiority of the AI-based approach capable of taking into account not only the observed data but also the environmental and process factors.

References

1. Ogden, R.W. (1984) *Non-Linear Elastic Deformations*, Ellis Horwood Ltd, Chichester.

2. Bishop, C.M. (2006) *Pattern Recognition and Machine Learning*, Springer, New York.

15
Ab Initio Guided Design of Materials

Martin Friák, Dierk Raabe, and Jörg Neugebauer

15.1
Introduction

Quantum-mechanical (so-called *ab initio*) calculations have achieved remarkable accuracy in predicting physical and chemical properties and phenomena in engineering materials. Owing to their universality and reliability, they are becoming increasingly useful when designing new alloys or revealing the origin of phenomena in existing materials. Having succeeded in revealing the atomistic origin of specific material properties, this deeper insight and understanding allow for (i) fine-tuning of material characteristics, (ii) enhanced compositional tailoring of materials that are designed specifically for given industrial applications, or, last but not least, (iii) identifying more accessible or more affordable solutes as alternatives to existing ones. On the basis of the universality and validity of a consistent quantum-mechanical approach, these calculations accurately predict basic material properties without any experimental input. Importantly, the quality of the quantum-mechanical description is not limited to ground-state parameters – material responses to rather extreme external conditions can reliably be determined as well. Owing to the facts that (i) such theory-guided material design can significantly reduce both time and costs when developing new alloys and (ii) increasingly faster massively parallelized computers are becoming extensively available, design strategies involving theoretical modeling tools are expected to be extensively used in upcoming decades. This chapter provides selected examples of quantum-mechanical approaches to the theory-guided material design. Specifically, the development of ultralightweight Mg-based alloys intended for aerospace and automotive applications is discussed.

15.2
Top-Down and Bottom-Up Multiscale Modeling Strategies

Global challenges regarding energy savings and CO_2 reduction coming together with the worldwide trend toward an ever increasing demand for global mobility are

Structural Materials and Processes in Transportation, First Edition.
Edited by Dirk Lehmhus, Matthias Busse, Axel S. Herrmann, and Kambiz Kayvantash.
© 2013 Wiley-VCH Verlag GmbH & Co. KGaA. Published 2013 by Wiley-VCH Verlag GmbH & Co. KGaA.

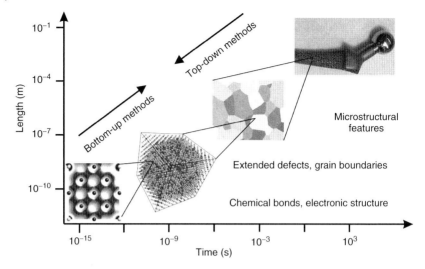

Figure 15.1 Schematic visualization of the hierarchical nature of structural engineering materials and the relevant simulation approaches for each respective level. The example shown here is a Ti-based medical implant for hip replacement, exposed to long-term cyclic loading conditions. Similar levels of hierarchy exist for most industrial products intended for engineering applications. Note the many orders of magnitude that have to be bridged to go from the smallest level, the atomistic scale, to the macroscopic scale both on length and time.

among the most essential tasks in material science and engineering. Any successful attempt to achieve both goals requires the development of new metallic materials with excellent mechanical and functional properties. Computer-based simulation techniques have become increasingly important in developing and designing new materials, reducing development times, identifying potential failure mechanisms, or addressing lifetime issues.

A major challenge in developing and applying techniques to design and optimize structural engineering materials is their inherent complexity and hierarchical nature. This is schematically shown for the example of a medical implant in Figure 15.1. This figure shows a hip replacement exposed to long-term cyclic loading conditions. From an application/engineering point of view, the main interest is on the behavior at the macroscale, that is, which parts of the implant become deformed when loaded, how is the energy absorbed, and so on. However, the mechanisms causing the specific mechanical behavior are at much smaller length scales and are determined by the specific microstructure such as size and orientation of grains, formation of precipitates or multiple phases, dislocations, and their mutual interaction. The structure, the energetics, and the dynamics of all these features is a direct consequence of the chemical interactions, that is, of the formation and breakage of chemical bonds at the atomic scale.

Owing to this complexity, simulation techniques employed for structural materials had been commonly restricted to the macroscale. In these techniques, the underlying material parameters such as elastic and plastic quantities are obtained by careful and extensive experimental measurements and interpolation techniques to obtain desired compositions. This single-scale approach works well when restricting the simulation to a narrow class of materials with similar properties and to materials where extensive experimental data is available. However, when trying to extrapolate to new material compositions or other microstructures, the predictive power of these approaches often drops dramatically or even fails.

In order to improve the predictive power and broaden the applicability range, methods have been developed, which allow to include also physical effects and mechanisms down to a certain length scale. These so-called top-down approaches allow, for example, to improve the description of plastic flow or strain hardening of real materials significantly by including the character of the grains, the relevant glide planes, and dislocation reactions. A prominent example is the texture-based crystal plasticity approach, which significantly improved existing constitutive models in finite element simulations (see, e.g., [1, 2]).

While top-down approaches remarkably improved the realism and the accuracy in simulating engineering materials, a potential shortcoming is that many of the relevant input parameters are on the microscale. These parameters are thus difficult or only indirectly accessible by experiment. Therefore, like for the single-scale methods, the availability of experimental data for the specific material system is crucial and cost expensive. Rather recently, a fully new approach has attracted a lot of attention in metallurgical simulations. The key idea is to turn the simulation process upside-down.

Rather than starting at the macroscopic scale, one starts at the most fundamental level, that is, quantum-mechanical level, which describes the interaction of the fundamental constituents of any material. More specifically, the laws of quantum mechanics as expressed in the Schrödinger equation in combination with electrodynamics describe the interaction of electrons and atomic cores precisely, without needing any material-specific experimental input parameters.

The only parameters entering the quantum-mechanical calculations are fundamental physical constants such as the mass and charge of the electron, Planck's constant, or the nucleus number of the specific chemical element. The important difference of this approach compared to a single-scale or top-down approach is that it is completely free of any material-specific parameters. Thus, simulations solving the quantum-mechanical equations start from identical principles and building blocks as used by nature to construct materials. Therefore, these methods are called *first principles methods* or using the corresponding Latin term *ab initio*. The most important and successful quantum-mechanical approach to address material science issues is the density-functional theory (DFT) [3, 4], which has been invented and pioneered by W. Kohn who had received in 1998 the Nobel Prize for its development.

Ab initio methods opened the way to a completely new generation of simulation techniques and promise a number of key advantages. First, new materials and their properties can be modeled solely on the computer without having to perform expensive and time-consuming experiments. This opened for the first time the possibility to perform combinatorial material design over the whole periodic table on the computer. Second, this approach is free of any adjustable or empirical parameters and thus allows an unbiased and highly accurate prediction of material properties. Third, because it accurately reproduces the interactions and constituents of nature, it provides a hitherto not possible insight into the mechanisms actually controlling the material properties or guiding chemical trends. Finally, it can be used as an ultimate starting point for multiscale techniques, which calculate material parameters at the lower scale and transfer them to the succeeding next scale. These schemes significantly broaden the use of *ab initio* methods well beyond the atomistic level (for a combination with finite element simulations (FEM – finite element method), see, e.g., [5, 6]).

Despite various computational and implementational challenges, *ab initio* based multiscale techniques have been developed and successfully applied for a wide range of materials. Areas in which these methods are often nowadays employed in industrial research and development are semiconductor technology, catalysis, or pharmacy. In such multiscale modeling of materials or scale-hopping approaches, the role of *ab initio* calculations is mostly twofold: (i) to study those systems and phenomena where the electronic effects are crucial and must be treated quantum-mechanically and (ii) to provide data for the generation of interatomic potentials with an extended range of transferability (see, e.g., [7]). Let us note that there exists a vast amount of literature devoted to multiscale modeling of materials (recent reviews may be found, e.g., in [8–11]).

The introduction of *ab initio* based multiscale techniques in metallurgy of metallic alloys has been largely hampered by the complexity of the material system with respect to microstructure and the large number of chemical elements and phases involved. In the following, a brief and, by no means, complete review is given on how the above-mentioned challenges can be addressed and what is presently achievable when using *ab initio* methods. The main focus will be on elastic properties and a few examples of how such methods can be applied to address practical material science problems are given. Specifically, the example of theory-guided development of ultralightweight Mg–Li alloys is discussed (see the following text).

15.3
Ab Initio Based Multiscale Modeling of Materials

In order to carry out an *ab initio* guided material design strategy that is able to predict material properties at the macroscale, a multidisciplinary approach is crucial. Figure 15.2 schematically shows an example of such a material design approach aiming at specific elastic properties of alloys. First, the thermodynamic stability for a variety of phases is determined in order to identify the stable one(s) as well as their volumetric ratio in a multiphase alloy, if necessary. Together with the thermodynamic stability of phases, the mechanical stability is tested by computing the elastic tensor of the single crystal. Second, employing linear-elasticity homogenization techniques that allow scale-bridging between atomistic and macroscopic

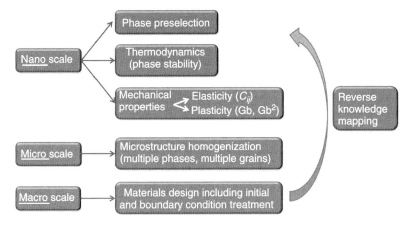

Figure 15.2 Schematic overview of the proposed multiscale and multidisciplinary strategy combining (i) thermodynamic phase stability with (ii) single-crystalline elasticity data. The latter is obtained at atomic level by first-principles calculations and is combined with self-consistent homogenization techniques in order to bridge scale differences.

levels, polycrystalline elastic moduli and other engineering parameters measurable at macroscale are predicted.

Starting from an initial composition and based on the residuum/deviation of the properties on the macroscale, a new atomic composition is suggested and studied. This cycle is repeated until the desired properties are obtained. If this turns out to be impossible because the targeted properties are outside this material class, the targets have to be revised, new phases/compositions have to be considered, or the condition of an exact match is replaced by minimizing the difference between actual and targeted properties. Following this strategy, an optimum alloy composition is obtained.

Such a theory-guided material design has successfully been applied to a wide range of materials. When aiming at materials with specific elastic properties, a few recent cases can be mentioned: (i) the development of new Ti-based biocompatible Ti–Nb and Ti–Mo alloys [12] intended for medical application (implant materials), (ii) ultralightweight Mg–Li alloys (see a detailed discussion below), and (iii) elasticity tuning of the so-called MAX phases (ternary nanolaminated transition metal (TM) carbides or nitrides) [13, 14].

The actual application of *ab initio* techniques to model and predict material properties faces a number of serious challenges. First, while quantum mechanics is principally exact for the energy scale relevant for materials, a direct solution of the underlying quantum-mechanical equations – the Schrödinger equation – would be restricted to a few electrons. Thus, for simulating actual materials, one has to rely on alternative formulations, which are typically restricted to specific cases and/or rely on approximations. While DFT has been proven to be mathematically exact for the electronic ground state, for practical realizations, assumptions regarding the so-called exchange-correlation (xc) functional have to be made. Owing to their construction, these functionals are universal, that is, can be applied to principally any material system. However, no practical approach exists yet to systematically improve their accuracy and it is therefore crucial to carefully check accuracy and predictive power for a given material system and property.

Second, because of the complex physics involved, quantum-mechanical techniques are extremely demanding with respect to computational effort and also with respect to code development and maintenance. A particular challenge is that, because of the quantum-mechanical nature, each electron has to be represented by a three-dimensional object – the wave function – which extends over the entire simulation cell. As each atom has typically a few dozen electrons, this limits the system size for routine calculations to a few hundred atoms even on high-performance computer clusters.

Third, while *ab initio* techniques are principally ideally suited to construct multiscale simulation techniques, a direct connection to the next higher scale turns out to be often impractical because of the huge number of possible configurations at the atomic scale. Therefore, the development of advanced filter techniques is crucial to reduce the data flow from the lower to the higher scale without losing accuracy and predictive power.

In this chapter, we use our recent development of ultralightweight Mg-based alloys as an example of quantum-mechanical calculations applied to reveal the atomistic origin of phenomena in existing materials as well as to design new alloys. In the case of Mg–Li alloys, an optimum alloy was sought for, which represents a compromise between two conflicting criteria: (i) specific Young's modulus as a measure of stiffness and (ii) the bulk-over-shear modulus ratio as an approximate indicator [15] of either brittle or ductile behavior (see, e.g., [16–25]).

15.4
Modeling of Ultralightweight Mg–Li Alloys

Magnesium is the sixth most abundant element occurring in the Earth's mantle and because of its low density, it can potentially be more often used as a structural material in automotive and aerospace applications where weight savings are crucial [26–30]. A wider use of wrought Mg and Mg alloys is limited because of the fact that these materials are difficult to deform at room temperature and when formed they have undesirable mechanical properties.

The very low room-temperature ductility in Mg alloys is primarily caused by the hexagonal closed-packed (hcp) crystal structure of Mg, where the critical resolved shear stress of the two basal slip systems is much lower than that of other slip systems. The two active slip systems are not enough to fulfill the von Mises criterion (requiring at least five independent slip systems) to accommodate an arbitrary deformation state via dislocation glide. The absence of room-temperature ductility and the evolution of strong deformation textures are then inherent limitations of the hcp Mg crystal structure, which have not been completely overcome [31–35].

Changing the hcp crystal structure to a more symmetric cubic one, such as a body-centered cubic (bcc), is one potential option of making Mg more suitable for processing. Cubic Mg is expected to be more workable at room temperature, because of a higher number of available slip systems, and possibly also less prone to form disadvantageous deformation textures. Leaving aside pressure-induced phase transitions (see, e.g., [36]) or epitaxy-induced transformations (e.g., [37]), a convenient metallurgical way of changing the crystal structure into a different one is by alloying with an element that stabilizes the desired phase (see, e.g., [38]). An element stabilizing the bcc phase is a so-called bcc-stabilizer. Specifically, in case of stabilizing the bcc phase of Mg alloys, the lightest known metal, lithium, is a convenient candidate.

Li is experimentally known to stabilize the bcc structure over the hcp one with as little as 30 at% Li [39, 40]. As an additional benefit with respect to automotive applications, Mg–Li alloys can potentially become one of the lightest possible metallic alloy systems because of the fact that the density of Mg–Li alloys is expected to range between that of Mg (1740 kg m^{-3}) and that of Li (580 kg m^{-3}). While there is little doubt that bcc Mg–Li alloys will be lightweight, it is unclear whether these alloys will offer enough advantageous elastic properties, such as the stiffness required by transportation applications. Therefore, a systematic material design search is needed, which accomplishes an optimization with respect to multiple and often contradicting material criteria.

In order to systematically analyze properties of Mg–Li alloys, we have determined both the thermodynamic stability and single-crystal elasticity of Mg–Li alloys using DFL at $T = 0$ K and compared with experimental data available for both room and lower temperatures. On the basis of DFT-determined properties, engineering parameters such as the ratio of bulk modulus over shear modulus (B/G) and the ratio of Young's modulus over mass density (Y/ρ) are calculated. The former may be used as an approximate indicator of either ductile or brittle behavior (with the 1.75 threshold value) and the latter can be used as a measure of strength of the studied materials.

Binary alloys were described by supercells consisting of $2 \times 2 \times 2$ elementary cubic (Figure 15.3) or hexagonal unit cells with a total of 16 atoms (see details in Refs. 17 and [18]). Fifteen alloy compositions were studied by systematically replacing Mg and Li atoms. Specifically in case of bcc Mg–Li alloys, the single-crystalline elastic constants of the calculated compounds were determined from the changes of the total energy as a function of the applied strain (see Ref. 41). As all the computed compounds were of cubic symmetry, three elastic constants were determined from three independent cell shape deformations, specifically (i) volumetric, (ii) tetragonal, and finally (iii) trigonal (see Ref. 41). In order to determine polycrystalline moduli, the self-consistent Hershey's [42] homogenization scheme for texture-free cubic aggregates (see also, e.g., Ref. 25) based on the single-crystalline elastic constants C_{ij} was employed and compared with the values within the Reuss [43] and Voigt [44] schemes constituting the lower and upper bounds, respectively, of the polycrystalline shear modulus. All modulus values are representative of polycrystalline materials with a random texture. After computing compositional trends of polycrystalline shear moduli, the corresponding trends were also derived for Young's moduli (Figure 15.3).

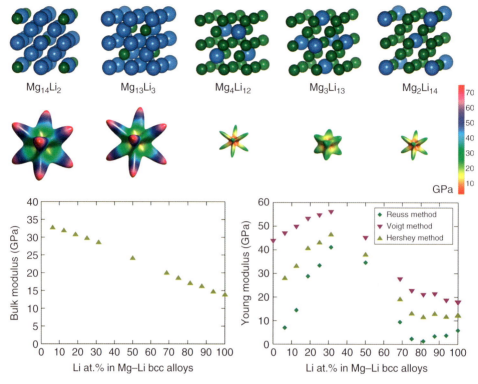

Figure 15.3 Schematic visualization of five bcc-based 16-atomic supercells used in *ab initio* calculations of Mg–Li alloys (upper row) accompanied by directional dependences of their respective single-crystalline Young's moduli (middle row). *Ab initio* calculated single-crystalline elastic constants were further employed to predict compositional trends of both bulk modulus (bottom left) and homogenized polycrystalline Young's modulus (bottom right) using Reuss [43], Voigt [44], and self-consistent Hershey [42] homogenization methods.

The predicted properties of these alloys compare well with available experimental results available in the literature. The bulk modulus predictions were within 20% (or 2–5 GPa) from experimental data [45]. In contrast to the nearly linear compositional trend of the bulk modulus, the other polycrystalline elastic moduli, such as Young's modulus Y shown in Figure 15.3, show strongly nonlinear variations as a function of alloy composition. Alloys with 30 at% Li were found to be the stiffest and alloys with >70 at% Li were found to be the softest. Importantly, without performing a quantum-mechanical study, an experimental determination of such strong nonlinear trends would require expensive and lengthy casting and testing of numerous Mg–Li compositions spanning the whole concentration range.

Using the DFT-calculated data, an Ashby map containing Y/ρ versus B/G was constructed (Figure 15.4). Plotting the Reuss, Voigt, and self-consistent Hershey values of various local arrangements together in this map resulted in a universal master curve. Such a universal curve exists indicates that it is not possible to increase

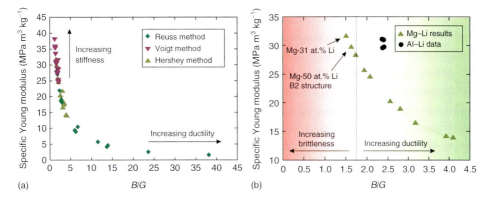

Figure 15.4 Ashby map of specific Young's modulus (Y/ρ) as a function of the ratio of the bulk and polycrystalline shear moduli, B/G, for bcc Mg–Li alloys employing Reuss, Voigt, and self-consistent Hershey homogenization methods to determine polycrystalline moduli (a). The region close to the B/G = 1.75 threshold value is magnified in (b) displaying only the results based on the Hershey's homogenization method and comparing them with data collected in literature for fcc Al–Li alloys [46]. Red and green backgrounds indicate either brittle or ductile alloys, respectively.

both the stiffness (represented here by Y/ρ) and the ductility (approximately quantified by the B/G ratio) by changing only the composition of a binary alloy.

The DFT-based Ashby map allowed revealed that alloys with 30–50 at% Li offer the most potential as lightweight structural material. The B/G ratio for these alloys is close to B/G = 1.75 threshold value, separating expected ductile and brittle material behaviors. The specific modulus (Y/ρ) for these bcc Mg–Li alloys compares favorably to that of fcc Al–Li alloys, indicating that these alloys could offer potential weight savings in the future. Importantly, the predicted Ashby map containing Y/ρ versus B/G shows that it is not possible to increase both Y/ρ and B/G by changing only the composition of a binary alloy.

15.5
Ternary bcc MgLi–X Alloys

In order to overcome the limitations found in binary Mg–Li alloys, which are clearly expressed by the universal master curve, we extended the study to ternary alloys. To do this, the optimum Mg–Li binary candidate with respect to stiffness and ductility (stoichiometric MgLi) was chosen as a starting point to construct ternary alloys. MgLi–X ternaries with solute additions were modeled using 16-atomic supercells (Figure 15.5) formed from $2 \times 2 \times 2$ MgLi (CsCl, B2) cubic unit cells, in which one (either Mg or Li) atom was replaced by one solute atom. The considered solutes include the third-row (Na, Al, Si, P, S, and Cl) and fourth-row TM elements (Sc, Ti, V, Cr, Mn, Fe, Co, Ni, Cu, and Zn), totaling to 16.

Repeating the computational procedure applied to Mg–Li binaries, we determine basic mechanical and physical properties, such as lattice parameters

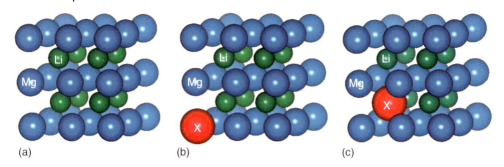

(a)   (b)   (c)

Figure 15.5 Schematic visualization of the 16-atomic 2 × 2 × 2 supercell used in calculations to model stoichiometric MgLi with the B2 (CsCl) structure (a) as well as a substitution of either one Mg atom (b) or one Li atom (c). The considered solutes include the third-row (Na, Al, Si, P, S, and Cl) and fourth-row TM elements (Sc, Ti, V, Cr, Mn, Fe, Co, Ni, Cu, and Zn).

and single-crystal elastic constants at $T = 0$ K using DFT. The single-crystal elastic constants are used to calculate polycrystalline elastic moduli using the self-consistent linear-elasticity Hershey technique [46]. The homogenized polycrystalline elastic constants are then used to calculate the B/G ratio and specific Young's modulus (Y/ρ). Using the DFT-calculated data, an Ashby map containing Y/ρ versus G/B was constructed for both types of substitution in MgLi and compared with the results obtained earlier for the Mg–Li alloys. Using this approach, we systematically search for compositions that increase *both* specific Young's modulus and ductility, for example, we look in the Ashby map for data points that are above the universal master curve of the binary Mg-Li system (Figure 15.4a).

As shown in Figure 15.6, none of the 16 chosen elements has the desired impact. We, therefore, conclude that none of the studied ternary elements is able to overcome the limitation of the binary. This result holds for any type of substitution (Mg or Li atoms). It should be noted that the results in Figure 15.6 provide a combined perspective on the elastic properties of the MgLi–X alloys. In contrast to the studied individual elastic constants where alloying causes both positive and negative effects [19, 20], the Ashby maps reveal that the improved bulk, Young's, and shear moduli values come at the expense of making the MgLi–X alloy more brittle.

The predicted trends can be partly explained by the fact that all the studied ternary elements have higher mass than either Li or Mg. Thus, the selected solutes increase the density of the studied alloys. This increase of the density is unfortunately not fully compensated by the change of the elastic moduli (and/or their ratios). The right choice of the alloying element, specifically with respect to an increasing density and its relation to the solid–solution strengthening, is clearly of paramount importance as elements in the periodic table that would be yet lighter than the constituents of the binary matrix, here Mg and Li, are lacking. As *ab initio*

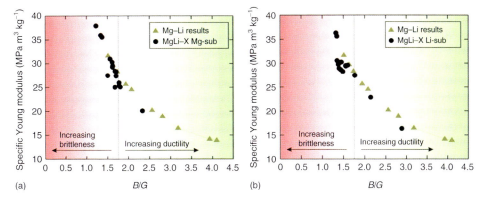

Figure 15.6 Ashby maps of specific Young's modulus Y/ρ versus B/G ratio for MgLi crystals with either Mg (a) or Li (b) atoms substituted by third-row (Na, Al, Si, P, S, and Cl) and fourth-row TM elements (Sc, Ti, V, Cr, Mn, Fe, Co, Ni, Cu, and Zn). The vertical lines indicate the threshold separating the ranges with either brittle or ductile behavior. The values for substituted MgLi–B2 are compared with the values predicted earlier for a set of Mg–Li bcc compounds (these data points are connected by a line to guide the eyes).

calculations can reliably predict changes of both density and elastic parameters, their use may shorten and optimize this search.

When searching for an optimum alloying element, it is also demonstrated [19, 20] that there is no straightforward relation between the properties of the studied solutes in their elemental ground-state phase and properties that they can induce in an alloy when used as alloying elements. The search combining both quantum-mechanical approaches and experimental techniques should therefore be extended to an even broader set of solutes (see, e.g., [19, 20]) and/or multiphase higher-order alloys. Specifically, in case that both options fail in increasing simultaneously ductility and strength, alternative approaches are possible. For example, it is often worthwhile to explore whether other mechanisms improving ductility can be activated, such as, for example, twinning-induced plasticity (TWIP) or transformation-induced plasticity (TRIP) known in steels (see, e.g., [47–51]).

15.6
Summary and Outlook

Ab initio based multiscale techniques have emerged in the past few years as powerful tools in metallurgy and material design. While still being in their infancy, enormous progress in the available *ab initio* computer codes, in computer power, and in coupling *ab initio* techniques to larger length and timescales makes these approaches more and more interesting to address and solve pressing questions related to the design and performance of modem structural materials. A major advantage of *ab initio* based techniques is that they are inherently free of empirical, fitted, or experimental input parameters. Therefore, they enable for the first time an accurate and unbiased screening of principally any material system and its properties solely on the computer, that is, without the need to cast/synthesize it beforehand. The *ab initio* approaches can be thus used within so-called theory-guided material design to preselect promising alloys and composites with properties dictated by specific (industrial) needs.

The *ab initio* modeling tools are not going to replace existing experimental and theoretical simulation techniques but will rather complement and enhance them. Examples are related to generate input for existing top-down or single-scale approaches to address questions that are out of reach for experiment or to identify fundamental mechanisms and chemical trends. We therefore expect that in metallurgy, as in other material science topics such as semiconductor technology, catalysis, or pharmacy, *ab initio* based simulation techniques will be quickly adopted. Challenges specifically related to metallurgy are that (i) the composition of metallic alloys is often less precisely defined when compared to, for example, Si-based semiconductor technologies, (ii) the complexity of the underlying microstructure is substantially higher, and (iii) the complexity of physics at the atomic scale due to the often strong interaction between magnetic, chemical, elastic, and electronic contributions.

The ongoing development in both quantum-mechanical methods and atomistic and/or continuum approaches to predict material properties shows many promising trends toward further, extending the above-mentioned multiscale and multidisciplinary schemes beyond the elasticity of materials at $T = 0$ K. Examples cover a wide range of fields such as pressure-induced behavior (see, e.g., [52, 53]), complex plasticity (see, e.g., [54–56]), and/or finite temperatures via taking into account, for example, vibronic degrees of freedom (see, e.g., [57–59]) or magnetic excitations (see, e.g., [60–72]).

15.7
Further Reading

In order to learn more about DFT, which is the most common *ab initio* approach in material science, see, for example, the Nobel lecture of Professor Kohn [73] or reviews on DFT (e.g., Refs. [74–76]). Comprehensive books focusing on the electronic structure of materials are, for example, those by Harrison [77], Martin [78], or Kohanoff [79], for calculations of thermal properties see, e.g., Refs. [80, 81].

Acknowledgments

We would like to acknowledge financial support of the collaborative research center SFB 761 "Stahl – *ab initio*" of the Deutsche Forschungsgemeinschaft as well as funding by the Interdisciplinary Centre for Materials Simulation (ICAMS), which is supported by ThyssenKrupp AG, Bayer MaterialScience AG, Salzgitter Mannesmann Forschung GmbH, Robert Bosch GmbH, Benteler Stahl/Rohr GmbH,

Bayer Technology Services GmbH, and the state of North-Rhine Westphalia, as well as the European Commission in the framework of the European Regional Development Fund (ERDF). We thank Dr D. Ma and Dr A. Dick (MPIE) for their helping us with figures. We also acknowledge extensive contributions of Dr W.A. Counts (formerly MPIE, currently QuesTek Innovations LLC) to *ab initio* calculations of Mg–Li and MgLi–X alloys. Figures depicting supercell details have been created by employing the VESTA visualization tool [82].

References

1. Zhao, Z., Roters, F., Mao, W., and Raabe, D. (2001) *Adv. Eng. Mater.*, **3**, 984.

2. Raabe, D. and Roters, F. (2004) *Int. J. Plasticity*, **20**, 339.

3. Hohenberg, P. and Kohn, W. (1964) *Phys. Rev.*, **13**, B864.

4. Kohn, W. and Sham, L.J. (1965) *Phys. Rev.*, **140**, A1133.

5. Ma, D., Friák, M., Raabe, D., Neugebauer, J., and Roters, F. (2008) *Phys. Status Solidi B*, **245**, 2642.

6. Roters, F., Eisenlohr, P., Hantcherli, L., Tjahjanto, D.D., Bieler, T.R., and Raabe, D. (2010) *Acta Mater.*, **58**, 1152.

7. Music, D., Basse, F.H.-U., Haßdorf, R., and Schneider, J.M. (2010) *J. Appl. Phys.*, **108**, 013707.

8. Mattsson, A.E., Schultz, P.A., Desjarlais, M.P., Mattsson, T.R., and Leung, K. (2005) *Modell. Simul. Mater. Sci. Eng.*, **13**, R1.

9. Moriarty, J.A., Benedict, L.X., Glosli, J.N., Hood, R.Q., Orlikowski, D.A., Patel, M.V., Söderlind, P., Streitz, F.H., Tang, M., and Yang, L.H. (2006) *J. Mater. Res.*, **21**, 563.

10. Yip, S. (ed) (2005) *Handbook of Materials Modeling*, Springer, Dordrecht, Berlin, Heidelberg, and New York.

11. Gumbsch, P. (ed) (2006) *Proceedings of the 3rd International Conference on Multiscale Materials Modeling*, Fraunhofer IRB, Stuttgart.

12. Raabe, D., Sander, B., Friák, M., Ma, D., and Neugebauer, J. (2007) *Acta Mater.*, **55**, 4475.

13. Sun, Z., Music, D., Ahuja, R., Li, S., and Schneider, J.M. (2004) *Phys. Rev. B*, **70**, 092102.

14. Schneider, J.M., Sigumonrong, D.P., Music, D., Walter, C., Emmerlich, J., Iskandar, R., and Mayer, J. (2007) *Scr. Mater.*, **57**, 1137.

15. Pugh, S.F. (1954) *Philos. Mag.*, **45**, 823.

16. Friák, M., Counts, W.A., Raabe, D., and Neugebauer, J. (2008) *Phys. Status Solidi B*, **245**, 2636.

17. Counts, W.A., Friák, M., Raabe, D., and Neugebauer, J. (2009) *Acta Mater.*, **57**, 69.

18. Counts, W.A., Friák, M., Raabe, D., and Neugebauer, J. (2009) in *Magnesium, 8th International Conference on Magnesium Alloys and their Applications* (ed K.U. Kainer), Wiley-VCH Verlag GmbH, Weinheim, p. 133.

19. Counts, W.A., Friák, M., Raabe, D., and Neugebauer, J. (2010) *Adv. Eng. Mater.*, **12**, 572.

20. Counts, W.A., Friák, M., Raabe, D., and Neugebauer, J. (2010) *Adv. Eng. Mater.*, **12**, 1198.

21. Friák, M., Deges, J., Krein, R., Frommeyer, G., and Neugebauer, J. (2010) *Intermetallics*, **18**, 1310.

22. Friák, M., Deges, J., Stein, F., Palm, M., Frommeyer, G., and Neugebauer, J. (2009) in *Advanced Intermetallic-Based Alloys for Extreme Environment and Energy Applications*, Materials Research Society Symposium Proceedings, Vol. 1128 (eds M. Palm, B.P. Bewlay, M. Takeyama, J.M.K. Wiezorek, and Y.-H. He), Materials Research Society, Warrendale, PA, p. 59.

23. Counts, W.A., Friák, M., Battaile, C.C., Raabe, D., and Neugebauer, J. (2008) *Phys. Status Solidi B*, **245**, 2630.

24. Liu, J.Z., van de Walle, A., Ghosh, G., and Asta, M. (2005) *Phys. Rev. B,* **72,** 144109.

25. Zhu, L.-F., Friák, M., Dick, A., Grabowski, B., Hickel, T., Liot, F., Holec, D., Schlieter, A., Kühn, U., Eckert, J., Ebrahimi, Z., Emmerich, H., and Neugebauer, J. (2012) *Acta Mater.,* **60,** 1594.

26. Mordike, B.L. and Ebert, T. (2001) *Mater. Sci. Eng., A,* **302,** 37.

27. Mordike, B.L. (2002) *Mater. Sci. Eng., A,* **324,** 103.

28. Hort, N., Huang, Y.D., and Kainer, K.U. (2006) *Adv. Eng. Mater.,* **8,** 235.

29. Potzie, C. and Kainer, K.U. (2004) *Adv. Eng. Mater.,* **6,** 281.

30. Gehrmann, R., Frommert, M.M., and Gottstein, G. (2005) *Mater. Sci. Eng., A,* **395,** 338.

31. Agnew, S.R., Yoo, M.H., and Tome, C.N. (2001) *Acta Mater.,* **49,** 4277.

32. Agnew, S.R., Tome, C.N., Brown, D.W., Holden, T.M., and Vogel, S.C. (2003) *Scr. Mater.,* **48,** 1003.

33. Agnew, S.R. and Duygulu, O. (2005) *Int. J. Plast.,* **21,** 1161.

34. Yoo, M.H., Agnew, S.R., Morris, J.R., and Ho, K.M. (2001) *Mater. Sci. Eng., A,* **319,** 87.

35. Galiyev, A., Kaibyshev, R., and Gottstein, G. (2001) *Acta Mater.,* **49,** 1199.

36. Friák, M. and Šob, M. (2008) *Phys. Rev. B,* **77,** 174117.

37. Friák, M., Schindlmayr, A., and Scheffler, M. (2007) *New J. Phys.,* **9,** 5.

38. Zhu, L.-F., Friák, M., Dick, A., Grabowski, B., Hickel, T., Liot, F., Holec, D., Schlieter, A., Kühn, U., Eckert, J., Ebrahimi, Z., Emmerich, H., and Neugebauer, J. (2012) *Acta Mater.,* **60,** 1594.

39. Jackson, R.J. and Frost, P.D. (1967) *Properties and Current Applications of Magnesium-Lithium Alloys,* NASA SP-5068, NASA, Washington, DC.

40. Shen, G.J. and Duggan, B.J. (2007) *Metall. Trans. A,* **38,** 2593.

41. Chen, K., Zhao, L.R., and Tse, J.S. (2003) *J. Appl. Phys.,* **93,** 2414.

42. Hershey, A.V. (1954) *J. Appl. Mech.,* **21,** 236.

43. Reuss, A.Z. (1929) *Z. Angew. Math. Mech.,* **9,** 49.

44. Voigt, W. (1928) *Lehrbuch der Kristallphysik, Leipzig Germany,* B.G. Teubner, Stuttgart.

45. Khein, A., Singh, D.J., and Umrigar, C.J. (1995) *Phys. Rev. B,* **51,** 4105.

46. Taga, A., Vitos, L., Johansson, B., and Grimvall, G. (2005) *Phys. Rev. B,* **71,** 14201.

47. Frommeyer, G., Brux, U., and Neumann, P. (2003) *ISIJ Int.,* **43,** 438.

48. Fischer, F.D., Reisner, G., Werner, E., Tanaka, K., Cailletaud, G., and Antretter, T. (2000) *Int. J. Plast.,* **16,** 723.

49. Grassel, O., Kruger, L., Frommeyer, G., and Meyer, L.W. (2000) *Int. J. Plast.,* **16,** 1391.

50. Sugimoto, K., Usui, N., Kobayashi, M., and Hashimoto, S. (1992) *ISIJ Int.,* **32,** 1311.

51. Dick, A., Hickel, T., and Neugebauer, J. (2009) *Steel Res. Int.,* **80,** 603.

52. Holec, D., Rovere, F., Mayrhofer, P.H., and Barna, P.B. (2010) *Scr. Mater.,* **62,** 349.

53. Holec, D., Franz, R., Mayrhofer, P.H., and Mitterer, C. (2010) *J. Phys. D: Appl. Phys.,* **43,** 145403.

54. Steinbach, I. (2009) *Modell. Simul. Mater. Sci. Eng.,* **17,** 073001.

55. Miller, R.E. and Tadmor, E.B. (2009) *Modell. Simul. Mater. Sci. Eng.,* **17,** 053001.

56. Belytschko, T., Gracie, R., and Ventura, G. (2009) *Modell. Simul. Mater. Sci. Eng.,* **17,** 043001.

57. Debernardi, A., Alouani, M., and Dreyss, H. (2001) *Phys. Rev. B,* **63,** 064305.

58. Moruzzi, V.L., Janak, J.F., and Schwarz, K. (1988) *Phys. Rev. B,* **37,** 790.

59. Grabowski, B., Hickel, T., and Neugebauer, J. (2007) *Phys. Rev. B,* **76,** 024309.

60. Nolting, W., Vega, A., and Fauster, T. (1995) *Z. Phys. B: Condens. Matter,* **96,** 357.

61. Kübler, J. (2006) *J. Phys. Condens. Matter,* **18,** 9795.

62. Kübler, J., Fecher, G.H., and Felser, C. (2007) *Phys. Rev. B,* **76,** 024414.

63. Halilov, S.V., Perlov, A.Y., Oppeneer, P.M., and Eschrig, H. (1997) *Europhys. Lett.,* **39,** 91.

64. Rosengaard, N.M. and Johansson, B. (1997) *Phys. Rev. B*, **55**, 14975.

65. Rusz, J., Bergqvist, L., Kudrnovský, J., and Turek, I. (2006) *Phys. Rev. B*, **73**, 214412.

66. Ležaić, M., Mavropoulos, P., and Blügel, S. (2007) *Appl. Phys. Lett.*, **90**, 082504.

67. Ruban, A.V., Khmelevskyi, S., Mohn, P., and Johansson, B. (2007) *Phys. Rev. B*, **75**, 054402.

68. Gao, G.Y., Yao, K.L., Şaşioğlu, E., Sandratskii, L.M., Liu, Z.L., and Jiang, J.L. (2007) *Phys. Rev. B*, **75**, 174442.

69. Pajda, M., Kudrnovsky, J., Turek, I., Drchal, V., and Bruno, P. (2001) *Phys. Rev. B*, **64**, 174402.

70. Körmann, F., Dick, A., Grabowski, B., Hallstedt, B., Hickel, T., and Neugebauer, J. (2008) *Phys. Rev. B*, **78**, 033102.

71. Körmann, F., Dick, A., Hickel, T., and Neugebauer, J. (2010) *Phys. Rev. B*, **81**, 134425.

72. Hallstedt, B., Djurovic, D., von Appen, J., Dronskowski, R., Dick, A., Körmann, F., Hickel, T., and Neugebauer, J. (2010) *Calphad*, **34**, 129.

73. Kohn, W. (1999) *Rev. Mod. Phys.*, **71**, 1253.

74. Burke K. *et al.* The ABC of DFT, *http://www.chem.uci.edu/~kieron/dftold2/*.

75. Perdew, J.P. and Kurth, S. (2003) in *Density Functionals for Non-relativistic Coulomb Systems in the New Century*, Lecture Notes in Physics, Vol. 620 (eds C. Fiolhais, F. Nogueira, and M. Marques), Springer-Verlag, Berlin, Heidelberg, pp. 1–55.

76. Klaus, C. (2006) *Braz. J. Phys.*, **26**, 1318.

77. Harrison, W.A. (1989) *Electronic Structure and the Properties of Solids*, Dover Publications, New York.

78. Martin, R.M. (2004) *Electronic Structure: Basic Theory and Practical Methods*, Cambridge University Press, Cambridge.

79. Kohanoff, J. (2006) *Electronic Structure Calculations for Solids and Molecules: Theory and Computational Methods*, Cambridge University Press, Cambridge.

80. Narasimhan, S. and de Gironcoli, S. (2002) *Phys. Rev. B*, **65**, 064302.

81. Xie, J., de Gironcoli, S., Baroni, S., and Scheffler, M. (1999) *Phys. Rev. B*, **59**, 965.

82. Momma, K. and Izumi, F. (2011) VESTA 3 for three-dimensional visualization of crystal, volumetric and morphology data. *J. Appl. Crystallogr.*, **44**, 1272–1276.

Part VI
Higher Level Trends

Dirk Lehmhus

The present book has been structured according to the main material classes. This concept leaves little room for topics that either explicitly bridge these subdivisions or go beyond them. For capturing such higher level trends, part VI has been introduced, featuring contributions on hybrid design, sensorial materials, and additive manufacturing techniques.

There is a statement by Professor Werner Hufenbach of the Technical University of Dresden's Institute of Lightweight Structures Engineering and Polymer Technology, saying that lightweight design should aim at using "the right material at the right place at the right cost and with the right ecological impact" ("Der richtige Werkstoff an der richtigen Stelle zum richtigen Preis mit der richtigen Ökologie"). In shorter terms, what is implied is multi-material or hybrid design of engineering structures. Needless to say, the potentials of this concept have been realized in transportation too. Chapter 16 contrasts current single-material dominated approaches such as the various ultra light steel solutions with such a multi-material design as investigated, for example, in the recent SuperLight-Car project funded by the European Union. While this project is taken as a ready example to explain the general potential, the present situation in the automotive sector is outlined and expected future trends, including the possible transfer to other transport industry sectors, are discussed.

A very special case of hybrid designs are the so-called sensorial materials; in this case, hybrid does refer not only to the use of different structural materials in different places, but to the integration of sensing, data evaluation, decision-making, and communication capabilities within them. The vision behind these materials is related to concepts such as ubiquitous computing and ambient intelligence, as materials and components thus equipped would represent a considerable amount of distributed computing power. On the level of the individual part, their use would be equivalent to providing engineering parts with a technical nervous system: They would gain the ability to feel, and they could use it to monitor their life and their own state from birth (i.e. from production) to end of life. In this, they would represent a further development of sensor-equipped materials and structures, as discussed in the context of structural health monitoring in Chapter 8. In the foreseeable future, realization of such materials must be based on a top-down methodology, that

Structural Materials and Processes in Transportation, First Edition.
Edited by Dirk Lehmhus, Matthias Busse, Axel S. Herrmann, and Kambiz Kayvantash.

is, by integrating the various functionalities of a network of autonomous sensor nodes in a material via embedding miniaturized microsystems, microelectronics, and peripheral components, including communication paths into the material. Chapter 17 follows this line of thought by introducing first the general vision; then describing related developments in various contributing areas of research such as sensors, sensor integration, or data processing; and closing with an application scenario that was, in this case, borrowed from the robotics rather than from the transportation field.

Although extremely versatile now, many production processes still face limitations on the basis of their generic principles. An example is the problem of realizing undercuts or internal cavities in permanent mould casting or forming processes. At the same time, advanced optimization techniques may advocate such geometrical features. Additive manufacturing processes face much less restrictions in realizing them, as their approach is to build up the part in a layer-by-layer manner, relying on 3D numerical models in doing so ("art-to-part"). Process variants that adhere to this general description have been developed for polymeric, ceramic, and metallic materials. Examples range from stereolithography to 3D printing and laser sintering techniques. Optimization of these processes has progressed in parallel. As a consequence, additive manufacturing of final components has become a reality in first niche areas, and is now being considered for an increasing number of areas, including the aerospace sector, whereas before, the available methods were typically limited to deliver prototypes or tools for prototype series. Chapter 18 sketches the past as well as expected future developments, describing capabilities as well as the remaining challenges.

16
Hybrid Design Approaches

Daniele Bassan

16.1
Introduction

Over recent years, total vehicle weight has risen significantly. Owing to its direct influence on the power demand of vehicle, weight reduction is one of the most important concerns among other measures in order to decrease the fuel consumption and CO_2 emission. When aiming at a weight reduction for the body-in-white (BiW) of a compact passenger car, one promising approach is the multi-material design. By this, the most suitable material is chosen for every component of the body structure based on the criteria such as energy absorption, structural integrity, and stiffness Figure 16.1.

Demands on modern vehicle grow from generation to generation. This mainly concerns the features such as safety, comfort, driving performance, roominess, variability, and quality. In general, these requirements are fulfilled by the use of heavy engine, the improvement of the chassis, a higher stiffness in BiW, and more package parts. These implementations influence and enforce each other, so that the vehicle weight increased in the past years.

16.2
Motivation

Lightweight design is a good technique to reduce the fuel consumption and the CO_2 emission. A contribution to achieve the CO_2 targets, which were decided by the European Union (EU) in December 2008, should be possible. Figure 16.2 illustrates the current state of the regulation for new vehicles regarding the CO_2 emission, starting with the European Automobile Manufacturers' Association (ACEA) negotiated agreement since 1995. According to this, the emission of all new vehicles inside the EU shall be reduced to 130 g CO_2 km^{-1} until 2012. For vehicles starting in 2012 there are penalties between €5 and €95 per gram CO_2 if the values are not kept.

Structural Materials and Processes in Transportation, First Edition.
Edited by Dirk Lehmhus, Matthias Busse, Axel S. Herrmann, and Kambiz Kayvantash.
© 2013 Wiley-VCH Verlag GmbH & Co. KGaA. Published 2013 by Wiley-VCH Verlag GmbH & Co. KGaA.

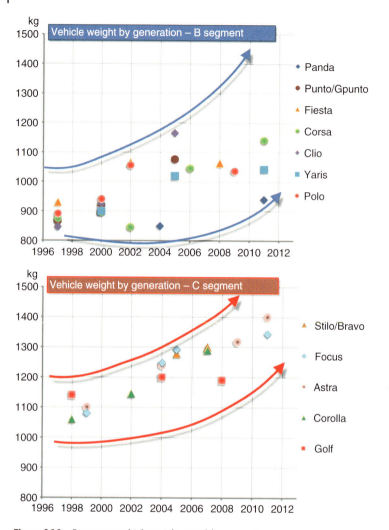

Figure 16.1 Passenger vehicle weight trend by generation.

For the development of a C-class compact car, a weight scenario can be defined, which is based on the ACEA-negotiated agreement. Hereby, the CO_2 emission should be reduced by 20 g CO_2 km^{-1}. Half of this reduction should be achieved by weight reduction. Assuming that 0.3 l 100 km^{-1} fuel is saved by saving 100 kg, and a reduction of fuel consumption of 0.1 l 100 km^{-1} equals to a CO_2 reduction of 2.5 g km^{-1}, a weight reduction of 130 kg is needed. Although the BiW has a high contribution to the gross weight, it is not the only part because there are some nonstructural-relevant parts such as doors and closures. These parts allow a reduction of approximately 45 kg using plastics right away. Hence for the BiW, a reduction of about 85 kg is aspired. *This represents 30% of the weight of a current midsize car.*

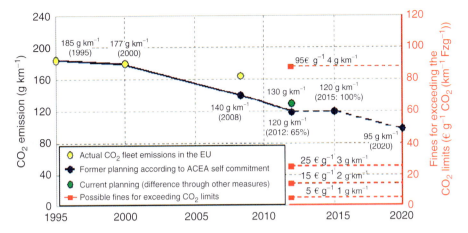

Figure 16.2 CO$_2$ emission – evolution in the EU [1].

Besides a life cycle assessment (LCA) on a per-vehicle basis, the potential contribution of lightweighting to a reduction of global transport energy consumption and greenhouse gas (GHG) emissions has also been estimated in the study of Helms and Lambrecht [2]. In 2000, all modes of transport were responsible for GHG emissions (referring to CO$_2$eq) of about 7600 Mt. As shown in Figure 16.3, by far, the highest saving potential lies in the weight savings of passenger vehicles, while global GHG emissions of vehicles will decrease as weight reduction takes place in the developed nations.

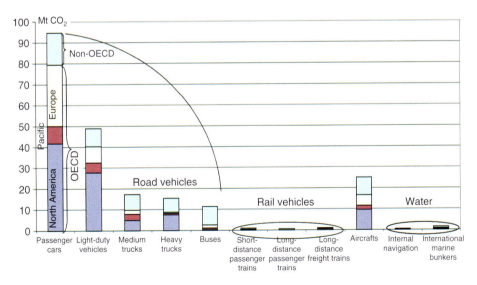

Figure 16.3 Influence of the different types of vehicles on the global annual potential of greenhouse gas savings [3].

16.3
From Monomaterial to Hybrid Multi-Material Design Approach in the Automotive Sector

In recent years, a considerable progress in automotive lightweight design was made by the implementation of new materials such as high-strength steels and a partial use of aluminum as well as plastic. However, in recent century, the steel-intensive usage in car body manufacturing was a fact (Figure 16.4).

Over the years, carmakers have exploited steel application on BiW at its maximum potential. In recent years, because of also the emerging role of aluminum, the steel producers have focused their development efforts to new high-strength steel grades and have highlightened the current steel design potential in different concepts.

16.4
ULSAB AVC Project/FSV Future Steel Vehicle Project

Primary to *ULSAB-AVC*'s design development was the use of the steel materials and manufacturing process that reflect state-of-the-art or future trends. The need to reduce the added mass required to satisfy future safety mandates presented the opportunity to also consider the application of newer types of high-strength steels. Advanced high-strength steels (AHSS) were considered to assist in achieving the overall aims of the program. As this is a concept program focusing on manufacturability in the year 2004 and beyond, it provides an opportunity to expand the list of candidate steels to those steels that are currently available and those under development that would become available by 2004 (ULSAB-AVC project ended in 2002).

The body structure analysis was run using dynamic high-strain rate properties provided by the ULSAB-AVC member companies. The closure designs suggested in ULSAB-AVC utilize the designs and findings from the UltraLight Steel Auto Closures (ULSAC) program concept and validation phases completed in May 2001.

The use of advanced technologies, such as tailored blanks, hydroformed tubes, tailored tubes, and laser assembly welding, were also basic considerations toward

Figure 16.4 First steel monocoque [2].

Figure 16.5 ULSAB-AVC C-class family sketches [4].

achieving a safe, lightweight vehicle. Proven technologies were maximized to benefit the overall vehicle goals. For example, when laser welding was needed for some application, the use of the required laser weld station was then maximized and applied to eliminate other spot welds (Figure 16.5).

Within the project, the USLAB-AVC BiW was just one example of how advanced high-strength steel and advanced automobile design could be combined in a cutting-edge structure to achieve the delicate balance of safety and mass efficiency, yet remaining an affordable solution to mass reduction challenges.

The overall weight achieved was, for a C-class BiW, 210 kg, according to the crash safety requirements for 2004.

The Future Steel Vehicle (FSV) program has been completed recently and the main target was to demonstrate the exploitation of new-steel-based BiW performance improvement and weigh reduction. The overall target was to achieve a mass reduction of 35% compared to a benchmark vehicle. In the project were considered all steel alloys available and also in development, ready by 2020. The final result was a concept body weighing 188 kg, almost all steels adopted belong to the high-strength grades.

16.5
S-in Motion – Steel BiW Project

ArcelorMittal's S-in motion project was first rolled out to customers in November 2010. It represents another intensive effort to meet the current and future requirements surrounding the production of electric vehicle while developing innovative, cost-effective coatings for a new generation of hot stamping materials.

The lightest S-in motion solution reduces BiW and closures weight by 57 kg, while the lightest chassis offering nets 16 kg in weight savings compared to a baseline C-segment vehicle.

Nevertheless, for an advanced weight reduction with the aim to reverse the trend of rising vehicle curb weight and to achieve a more efficient contribution to fuel consumption reduction, a *transition to new body structure design is needed*. One key to super lightweight design could be the multi-material mix. In this process, the material for a special body part is used where it is needed in order to fulfill the requirement and to minimize the weight in parallel.

16.6
Multi-Material Hybrid Design Approach

Hybrid technology opens up great innovation potential for the automotive industry, particularly for producing assembly support parts for extensive system modules that feature a high level of integration. Principal characteristics are excellent structural properties and good overloading behavior at low weight, high integration possibilities for functional elements, as well as high precision in production and in practical use. Hybrid parts can be produced in high volumes by the most economical methods.

There are structural components that can be implemented in hybrid technology. These include, among others, the crossbeam beneath the dashboard, structural components in elaborately furnished vehicle seats, and structures for the central console and the rear hat rack in the vehicle.

Material lightweight design uses material such as light metals of plastics. A combination of these methods is possible. The idea behind multi-material design is to choose the best material for each part of the BiW, which fulfills the given requirements by minimal weight. Some R&D-funded projects use this approach.

16.7
Optimum Multi-Material Solutions: The Reason for Hybrid Design Approach

New vehicle applications show how both aluminum and steel can contribute viable material solutions. Modular front-end crash management systems frequently use aluminum extrusion for bumper support sections and longitudinal arms combined with high-strength steel chassis legs. These are frequently formed by the use of tailored blanks using different steel compositions to allow controlled energy absorption by progressive crumpling. In other applications, aluminum honeycomb or convoluted tubular structures are used to dissipate energy. The use of aluminum is increasing in bonnet area and front end, where the combination of weight saving and its ability to progressively deform under pedestrian impact makes it a good choice. As alternative, more and more nonmetal exterior panels are going to be introduced in hang-on parts.

Steel can also be used in combination with aluminum roof panels by the adoption of proper joining technologies.

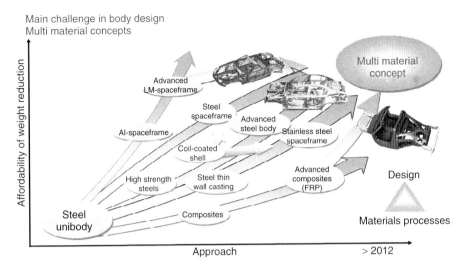

Figure 16.6 Multi-material concepts [5].

The right approach is to adopt the most relevant material where needed, considering the peculiar characteristic of each automotive material (e.g., mechanical performances, lightweight, manufacturing cost, and complex shape manufacturability).

Complex aluminum casting and extrusion are increasingly being considered for the use in hidden structures such as cockpit carriers, or instrument panel beams, while floor pan and sill areas remain largely the preserve of steel.

Even steel itself is progressing by improving its properties by the development of new steel grade or new formable steel by the major steel producers.

On the other hand, the continuing development of sophisticated joining technologies is crucial for the development of *multi-material body shell* (Figure 16.6). The mass-market motor industry demands solutions that are compatible with high-volume production, assembly, and paint application processes.

16.8
SuperLIGHT-Car Project

The idea behind multi-material design is to choose the best material for each part of the BiW, which fulfills the given requirements by minimal weight. The SuperLIGHT-Car project was sticked to this approach in the development process. Especially in the field of multi-material design, there are additional requirements regarding joining, costs, and producibility. Material distribution in the final concept is shown in Figure 16.7. In the total weight of a SuperLIGHT-Car, aluminum makes up to 53%, the most used material. Besides aluminum sheet and cast, also extrusion profiles are used. Steel is used up to 36%, including hot formed steel. With 7 wt%, magnesium has a high share. A special part is the magnesium die-cast strut tower. The last 4% of structural minor relevant parts are composed of plastics.

High-strength steel
Hot-formed steel
Aluminum sheet
Aluminum cast
Aluminum extrusion
Mg-sheet
Mg-diecast
Fiber-reinforced plastic

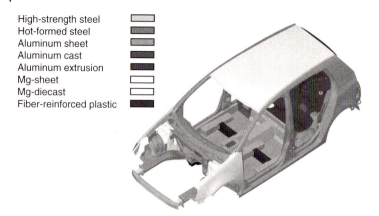

Figure 16.7 SuperLIGHT-Car body concept [5].

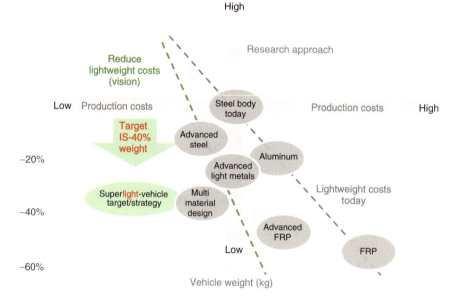

Figure 16.8 Research approach [5].

The overall mass reduction by the proper adoption of a multi-material solution was 30% of the BiW excluding closures. It has been evaluated that the achievable overall mass reduction including closures would be 40% (Figure 16.8).

16.9
Hybrid Solutions: Overview of Current Automotive Production

Despite the multi-material approach seems to be the most profitable in terms of weight reduction, the management of different material mix in a BiW, together with the available joining techniques that restrict the adoption of welding technologies (standard resistant spot welding (RSW)), moving toward the adoption of adhesives or mechanical joints, affects the introduction of huge hybrid design in BiW for high-series production volume cars.

Even today, the materials used in the construction of vehicle bodies are mainly various grades of steel. Mass production vehicle aluminum structural parts are included mainly as bumper, bolted to the front-end rails in steel. Although aluminum-intensive body concepts were restricted to the executive class cars to start with, they were later applied to other classes as well, including the 3 l car concept. Plastics mainly dominate the vehicle interior, their external application being chiefly limited to nonload-bearing components. The utilization of fiber-reinforced composite materials for supporting body parts has been limited to special series, as well as premium and racing models.

Tier 1 material suppliers and Tier 2 product and service companies are developing more flexible materials to make more unique vehicle structures. As a result, new lightweight materials such as aluminum, magnesium, plastics, and carbon fiber have been introduced to help make better lightweight BiW designs that are comparable to the strength of a typical steel structure. Therefore, new structural designs such as unibody design, space frame design, and multi-material design have been proposed in such a way that less content of the material needs to be used providing savings in weight. For example, in 2006, the BiW structure consisted of 85% steel and up to 10% aluminum, whereas in 2015, depending on the general development of these new light materials, the BiW structure is likely to consist of 60% steel and 40% demanding alternative lightweight material.

An increasing proportion of the material used in the chassis frame is made by light metals, particularly aluminum alloys such as cast or wrought alloy. The proportion of steel and iron materials in the chassis frame is decreasing correspondingly.

The great number of different materials used in the power train is appropriately described by the term *hybrid multi-material design*. Crank case may be made of high silicon content aluminum or cast iron, intake pipes may be made of glass-fiber-reinforced plastic (GFRP), and crank shafts may be made of welded steel, for instance. Ceramic materials in catalytic converters are state-of-the-art, as well as ceramic preforms for low-wear cylinder surfaces as insert in aluminum crank cases. Magnesium is the chief competitor of GFRPs and aluminum is expected to become relevant as casing material in the future.

Another example of *hybrid* approach can be found in the *structural reinforcement materials* (Citroen C5, but also other plastic/steel-hybrid-reinforced structures). The adoption of plastic reinforcement into hollow structural parts contributes to the attainment of higher energy absorption level during structure collapse, giving higher mechanical stability to the structure. The reinforcements are typically adopted in rails, but also in nodes, where there is conjunction of different structures (Figure 16.9).

Structural foams are typically polyurethane or epoxy resin based, and can have different range of densities, giving different energy absorptions and stiffening effects. These foams can replace conventional steel reinforcement giving advantages

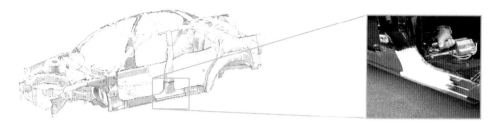

Figure 16.9 Structural foam application areas [6].

Terocore® Hybrids

Metal / plastic hybrids

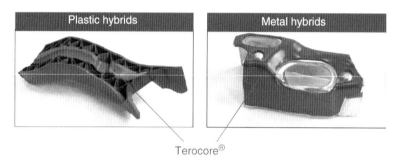

Figure 16.10 Structural foam inserts [7].

in terms of performance and weight. Usually are placed in Body Shop and then expand during electro coating curing process. The structural foam has, usually, a plastic "carrier" that facilitates its positioning into BiW hollow structure. The energy absorption and stiffening effect with the structural foams can increase up to three or four times with respect to the hollow base structure, even if with weight increment (Figure 16.10).

Another application of structural foams is for the Noise, vibration, and harshness (NVH) performance: localized reinforcement can contribute to improve the overall NVH behavior, changing the resonance frequency.

An evolution of simple structural hybrid reinforcement insert is the hybrid metal–plastic or metal–metal structures. The synergic action of dissimilar material reinforcement gives higher overall structure performance (restriction of local buckling).

Bimetallic structures have been developed by some Tier 1 companies. The principle is to adopt medium pressure die cast of semisolid aluminum casting and "bimetallic" structural automotive components comprising of steel tubes and cast aluminum nodes (Figure 16.11).

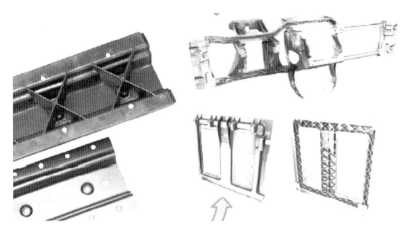

Figure 16.11 Joining without joining process – Vario Struct [8].

The main advantages are reduced assembly and machining, comparable performance, and weight reduction (20–30%) with respect to the original steel solution. Applications are engine cradles, suspension links, and knuckles.

The multi-material hybrid approach is adopted currently for the premium segment, where we consider sport cars. The idea is to adopt functionally integrated hybrid lightweight structures. This solution can be used to manufacture higher mechanical performance structures such as cockpit, seat frame, and structural nodes (Figure 16.12).

Another solution of hybrid application in BiW is on Jaguar Land Rover (JLR) B Post composite, which consists of steel, plastic-reinforced structures, and expanding foam and gives an overall mass reduction of 20% with respect to the original monosteel solution.

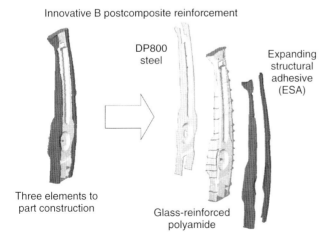

Figure 16.12 Use of lightweight materials in JLR body structures and closures [9].

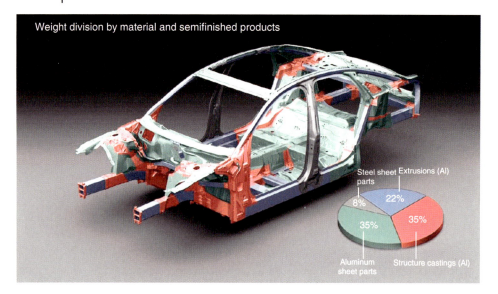

Figure 16.13   Use of metal hybrid approach [10].

The premium segment is the most suitable for a large implementation of hybrid design solution (Figure 16.13).

Different scenarios can be identified when we consider other parts such as *front end*. As this subassembly is mounted after BiW assembly, it is possible to avoid most of the critical topics that encounter the BiW during manufacturing, primarily, the e-coat oven (which generates different material dilatation hard to be managed in a mixed multi-material BiW) and the identification of proper joining technologies (Figure 16.14).

The Ford Focus (C170) was one of the first mass-produced vehicles to have a front end (grill-opening reinforcement, GOR) as a structural component that was developed consistently in hybrid technology.

The application of hybrid technology can reduce component weight, increase quality, and lower production costs. Dependable computational methods employing the aid of finite elements support the design of these complex components during the design stage. The pathbreaking hybrid technology combines the advantages possessed by two completely different materials and the corresponding production methods and opens up totally new possibilities in structural development.

Hybrid designs allow the production of complex, ready-to-assemble components in a few working steps and thus combine the economic process of sheet metal

Figure 16.14   Hybrid front end [11].

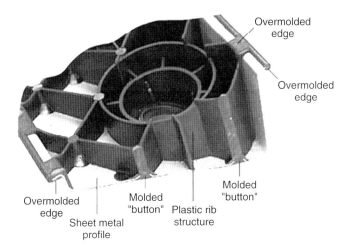

Figure 16.15 Profile cross section of a hybrid structure with special structural parts labeled [11].

stamping with injection molding. In such case, a plastic structure is injection molded to a stamped, perforated sheet metal profile. The plastic melt passes through the punched openings of the profile and around its peripheral edges. It forms molded buttons and overmolded edges (Figure 16.3) between the mold cavity and the inserted metal part (Figure 16.15).

Additional subassembled profiles are no longer necessary. This fact decisively improves the cost-efficient production of profiled structures using hybrid. Even if, in some loud cases, the plastic ribs cannot fully cope with the rigidity of a closed profile; a smart hybrid design always raises the overall performance with lower weight and cost.

Another example of hybrid structure is represented by BMW 5-series, model year 2003 (Figure 16.16).

Several applications of hybrid design approach can be found at vehicle component level.

In the hybrid brake pedal, the metal insert is overmolded at the plastic injection stage via a process using water injection technology to fuse GFRP with metal to create a tubular body. Every vehicle component is subjected, as never before, to the analysis of potential mass, with weight-saving solutions applied whenever feasible and cost-effective, in areas that even a decade ago would not have been considered.

The pedal blends GFRP with metal. The result is a weight savings over a conventional metal pedal of at least 30% and up to 50%. The process delivers the complete and final assembly in one step.

Another example of hybrid subassembly is represented by a bimetallic engine cradle, which has steel crossmembers with aluminum A356 cast over the tube ends.

The bimetallic engine cradle has been design-validated for durability and corrosion at the joint level. The bimetallic cradle can almost reach the low mass

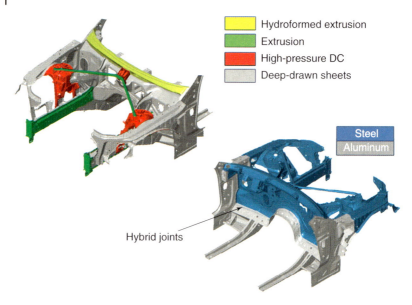

Hydroformed extrusion
Extrusion
High-pressure DC
Deep-drawn sheets

Steel
Aluminum

Hybrid joints

Figure 16.16 Front structure hybrid concept [12].

of an all-aluminum cradle but at a lower cost. During the casting process for the bimetallic cradle, aluminum is injected into the die containing steel crossmembers. The process produces a mold over the tube ends. The differential in coefficient of thermal expansion between steel and aluminum creates a shrink fit at the joint.

This solution represents a multimetal concept, which will make available a new generation of cradle.

16.10
Trends in Automotive Materials and Structural Design

Mild steel is due to see a major decline from 40% in 2006 down to 15% in 2015 as it will be taken over by higher strength steels [13]. As aluminum is predicted to be the main competitor of steel, steel is developing to meet certain demands of the automotive market and, therefore, is offering different steel grades from 300 MPa up to more than 1000 MPa in the form of advanced high-strength steel, expected to account for 10% of the vehicle structure by 2015.

The demand for lightweight materials is expected to increase with the materials such as aluminum that is already found in vehicles such as the Audi R8 and the Jaguar XJ. Aluminum is expected to grow from 10% in 2006 to 30% in 2015. An earlier example of the use of carbon fiber in the BiW structure has been witnessed in the Mercedes-Benz SLR. The use of carbon fiber is expected to be increasingly used in luxury-sport vehicles by 2015, increasing from 2% in 2006 to 5%.

Owing to the development of new technologies, there is more potential for alternative structural designs. Steel unibody has currently the highest number of production as steel is the most popular BiW material with a usage of 99% of a typical basic-segment vehicle structural mass. Through the demand of lightweight materials, Monocoque design has almost doubled in production from approximately 600 000 in 2006 to approximately a million vehicles in 2015. Multi-material design is not as popular in 2006, as it only accounts for 18% of vehicles produced. This is because of extra cost and related technology product and services that companies have to provide to accommodate to build this structural design. A further step in manufacturing and joining technologies is the obliged way to enable the application of these materials into a large BiW volume production.

Aluminum space frame design is going to be the most used structural design with 34% of vehicle production, as this design is mostly favored for lightweight materials.

16.11
Hybrid Solutions in Aircraft, Rail, and Ship Market

The *aircraft market* is the most sophisticated with new alloys being developed, some containing exotic metals such as scandium (which costs more than gold). For instance, Al–Mg–Sc and revised Al–Cu–Li alloys are currently under development.

These require ultrapure primary aluminum to produce them and there are, therefore, purity issues regarding the recycling of machined scrap and end-of-life scrap. A key factor when using these more exotic and costly aluminum alloys was to achieve a high "fly-to-buy" ratio. At present, typically 95% of the material supplied as plate or rolled section is machined away to reduce weight, leaving stiffening ribs. To make a typical spar for a midsized business jet, a 1000-kg rolled section is machined resulting in a final component weighing just 47 kg. New joining technologies will enable these ribs to be welded to thinner plate to achieve a similar result but with a much higher "fly-to-buy" ratio.

There is an increasing challenge from composite materials in the aircraft market. For example, the Airbus A350 presently at the design stage is to contain 37% by weight of composites including the floor and frames of the fuselage and wing (with metal ribs), while the Boeing 787 *Dream-liner* now under construction was announced at its launch to consist of 50% composite materials including the fuselage and only 20% of aluminum alloys. New aircraft in the design stage include the Airbus A30X short-range plane, which will use a composite wing with aluminum ribs (Figure 16.17).

A typical example of hybrid structure is represented by the fiber metal laminate (FML) [14]. It is a class of metallic materials consisting of a laminate of several thin metal layers bonded with layers of composite material. This allows the material to behave much as a simple metal structure, but with considerable specific advantages regarding the properties such as metal fatigue, impact, corrosion resistance, fire resistance, weight savings, and specialized strength properties. This material is widely used also for structural parts such as fuselage panels and leading-edge horizontal/vertical tailplane.

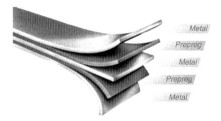

Figure 16.17 Hybrid aeronautical fiber metal laminate structure [14].

In the *railway* and *shipbuilding* industries, increasingly wide hollow extruded sections are being used. Taking into account the reduced operating costs, the overall costs of building a train from aluminum rather than steel is some 30% less.

Very large hollow-section extrusions are being produced for the rail industry allowing a single section to span the whole length of a carriage. Adjacent extrusions are welded together to attain the width required, for example, to fabricate a complete carriage floor.

In shipbuilding, high-speed ferries offer an obvious choice for aluminum where, for example, bulb flats can be integrated with plate by extruding them, thereby saving weight and fabrication costs.

The use of aluminum in a high-speed ferry was first introduced in 1975 in the Incat Tasmania. Likewise, some aluminum superstructures on ships have been replaced by high-strength steel following adverse experiences brought about by fire. Over the time, composites have been widely introduced in the design because of the advantages of shape structure complexity, which can be achieved with those manufacturing technologies. Hybrid design is widely adopted in shipbuilding industries also mainly in yacht design, where cost is not so a stringent issue.

16.12
General Aspects on Joining Technologies for Multi-Material Mix

Despite the fact that a hybrid approach on material utilization is the best way to fully exploit structure properties, designers have to consider the question of how to join altogether these different materials.

In particular, there are some aspects that should be considered also in the design phase because of the different material responses to the environment: corrosion phenomena and thermal expansion.

Considering the first aspect, corrosion phenomena on steel unibody are well known and properly managed; thanks to the standard protective coating applied on BiW, which starts with zinc coating on steel coil up to electrophoresis deposition on BiW in the paint shop.

Nevertheless, the adoption of different materials could enhance the corrosion phenomena, in particular, when we mix together steel and aluminum with their different electrochemical potentials. The better solution is to apply an interface between them in order to insulate someway them from the electrical point of view, and to avoid welding solution preferring mechanical joints or adhesives.

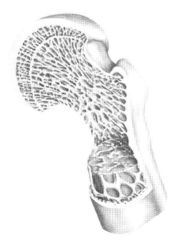

Figure 16.18 Bone structure.

Same situation has to be faced with the adoption of Carbon Fiber Reinforced Plastic (CFRP)–titanium aerospace structures. Actually this subject is matter of research projects aimed at the development of innovative coating process and materials, which fulfill, also, the new, more stringent regulation on toxic substances (Figure 16.18).

The other big aspect that has to be considered is the different thermal expansion coefficients. In particular, car body of different material mixes is subjected in the same manufacturing process steps to heat sources, which generate different elongations on the structures (e.g., after electrophoresis deposition, the BiW passes through the oven, which is used to "cure" the coating, and usually the process takes 15–20 min at a temperature in the range 150–180 °C). This process step is also a limitation to the introduction of plastic parts because of their specific glass temperature. It is possible, in effect, to adopt in this stage a thermoset plastic material with a higher glass transition temperature. In case of thermoplastic ones (where adoption is increasing because of the end of life and recycling requirement), it is possible to adopt them after the curing process, even if it represents a complication for BiW parts but is usual for closures.

16.13
Conclusion

Considering recent developments in material science and engineering, we are assisting to an era where the different material solutions are going to be fully exploited to their limit. Further, thanks to the high level of design optimization reached in the industries, we can assist the evaluation of best material usage in the structures.

Besides, as has been outlined in the previous examples, the extreme monomaterial exploitation cannot reach the ultimate result that a good combination of different materials, in synergic action, can enable.

The availability of different material properties can provide designers the capability to adopt the proper material where it is really needed and confer a tailored and functionally graded structure that cannot be reached by the extreme usage of a single material.

In fact, the nature itself has reached the same conclusion after 1000 years of evolutive optimization. Bone structure is a typical example of composite structure in nature where materials with different properties are located where they are really needed.

References

1. Sahr, C. (2009) Concept tools and simulation for lightweight body design. RWTH Aachen, International Conference: Innovative Development of Lightweight Vehicle Structure, Wolfsburg, May 26–27, 2009.

2. Helms, H. and Lambrecht, U. (2004) First Steel Monocoque, *http://lowreysautorestoration.com/products/ brook/coupebody.html* (accessed 21 December 2012).

3. Helms, H. and Lambrecht, U. (2004) Energy Savings by Light-Weighting. Part II. IFEU Institute for Energy and Environmental Research.

4. Ulsab-AVC (2002) Ulsab-AVC Advanced Vehicle Concept. Overview Report, January 2002. http://www.autosteel.org/∼/media/Files/ Autosteel/Programs/ULSAB-AVC/avc_overview_rpt_complete.pdf

5. SLC Conference TRA (– SLC_presentation_TRA08_VW_March08.ppt), *http://www.superlightcar.com/public/docs/ SLC_presentation_TRA08_VW_March08. ppt*

6. Kriescher, M. (DLR) (2010) Metal-hybrid structure for an improved crash behaviour of car body structures. WAMM Conference.

7. Engels, T. (2010) New application for structural organic materials. Joining Carbody Conference.

8. Roth, T. (2010) Joining without joining process. Joining in Car Body Engineering Conference.

9. Black, S. and Lidgard, B. (2010) Use of lightweight materials in JLR body structures and closures. Jaguar & Land Rover Cars, Materials in Car Body Engineering Conference.

10. Fidorra, A. and Baur, J. (Audi AG) (2010) The new Audi A8. Eurocarbody Conference.

11. LANXESS Deutschland GmbH (2008) Hybrid-Front End Ford Focus.

12. Hanle, U. and Lief, K. (BMW AG) (2003) BMW hybrid concept. EuroCar-Body Conference.

13. Frost & Sullivan (2008) The Next Generation: Lightweight Materials. Poonam Tamana, Frost & Sullivan Report, *http://www.frost.com/prod/servlet/market-insight-top.pag?docid=128138913* (accessed 21 December 2012).

14. Botelho, E.C., Almeida, R.S., Pardini, L.C., and Rezende, M.C. (2006) A review on the development and properties of continuous fiber/epoxy/aluminum hybrid composites for aircraft structures. *Mater. Res.*, **9**, 247–249. http://www.utwente.nl/ctw/pt/education/ Master%20Assignment%20and%20 Projects/Previous%20Master%20Projects/ bestanden/Wiering%20Rolf%20-%20 Damage%20development%20 of%20Glare.doc/

17
Sensorial Materials

Dirk Lehmhus, Stefan Bosse, and Matthias Busse

17.1
Introduction

By our definition, sensorial materials are materials that are able to feel, that is, materials that can gather and evaluate sensorial information. The definition is wide enough to accommodate different kinds of physical or chemical signals, but it does call for integration of the associated sensor nodes or networks in the material, and for additional data-processing capabilities. Naturally, aspects such as provision for (internal and external) communication as well as a reliable energy supply need to be added [1, 2].

The natural equivalent of such a technical system is the nervous system of the various animal genera. For them, it is the basis of one of the defining characteristics of life itself, irritability or the response to stimuli. The technical motivation to reproduce nature's invention is manifold, as are the envisaged applications. Within the context of this book, we will confine ourselves to advantages foreseen in the field of load-bearing structures, and thus naturally to usage in structural health monitoring (SHM). SHM is well established in civil engineering [3] and currently being discussed for aerospace applications. In the latter field, the transition from metal to composite aircraft structures reflected in several chapters of this work has raised concerns with respect to these materials' typical response to impact loading: Optically non-detectable failure may occur, calling for a constant monitoring of the state of the material. The transition from the metal-dominated "silver eagle" to the composite-based "blackbird" is therefore expected to provide a major boost for SHM not only in terms of market penetration, but also technologically, as it is likely to fuel developments toward sensor-integrated and sensorial materials [4, 5]. Aerospace structures call for component sizes orders of magnitude smaller than what is acceptable for bridges, the present mainstay of SHM applications.

Safety of potentially impact-loaded composite structures is one motivation for implementing SHM or, to begin with, load-monitoring systems. There are others, too. Knowing the exact state of a structure at any moment in time allows timing of maintenance work according to actual needs. Need-based, as opposed to regular,

Structural Materials and Processes in Transportation, First Edition.
Edited by Dirk Lehmhus, Matthias Busse, Axel S. Herrmann, and Kambiz Kayvantash.
© 2013 Wiley-VCH Verlag GmbH & Co. KGaA. Published 2013 by Wiley-VCH Verlag GmbH & Co. KGaA.

maintenance can reduce operating costs specifically in industries where high costs are incurred either by the maintenance work itself or by taking the object of maintenance out of service. Offshore wind energy plants and commercial aircraft are examples in this respect. Further advantages may be realized wherever damage-tolerant or fail-safe as opposed to safe-life dimensioning is applied as design philosophy. Here, constant monitoring in conjunction with an understanding of the development of damage over time will enable predictive maintenance strategies as well as weight savings based on an adaptation of safety factors. The latter would be justified by the greater proximity to damage development achieved by means of an integrated system. Current developments in the field of load and SHM systems for fiber-reinforced composites are discussed in Chapter 8 on polymer matrix composites.

Having thus defined sensorial materials, a distinction has to be made with respect to certain types of stimuli-responsive materials sometimes referred to as *self-X materials*. The best known among these and probably the example developed furthest are self-healing materials [6]. In these materials, aspects such as structure, constitution, and composition are predefined in a way that one specific stimulus will result in one specific response of the material. Thus the link between stimulus and response is hard corded/hard wired. In this way, a crack in a typical type of self-healing polymer will, when reaching an embedded microsphere, induce it to crack and release its liquid content, which fills the crack tip region and hardens under the supporting influence of a catalyst embedded in the polymer matrix surrounding the microsphere. Thus the main difference between sensorial and self-X materials lies in the fact that the former are flexible in their reaction to a stimulus because they have the potential to actively evaluate it, they can adapt their knowledge about themselves and in consequence their response based on sensory information, and they can communicate the associated knowledge to other entities. In a sense, they have a potential for spatial and temporal self-awareness, and in contrast to self-X materials, they respond consciously, not spontaneously.

Basically, two ways may be envisaged to realize sensorial materials: one is top-down and the other is bottom-up [7]. The top-down approach develops sensor-equipped structures along the lines of miniaturization, compliant solutions for sensors, embedding techniques, and so on toward the point where they may be considered sensorial materials. The threshold implied between structure and material is hard to pinpoint: one indirect, practical approach at a definition is by stating that in a true sensorial material, the dimensioning to the primary (in our case mechanical) role should not be affected by any provision for the sensorial capabilities of the material. This is contradictory because it may be read to say there is no benefit to structural design brought about by sensorial materials. What is meant, however, is that the influence of embedded microsystems on mechanical performance should not have to be explicitly modeled in the sense of geometrically representing heterogeneity. Instead, sensorial materials should allow being treated as homogeneous materials rather than as hybrid structures. Homogeneous in this sense may include providing for local, stochastic property variations introduced or altered by embedded systems in a way similar to provisions of this kind nowadays made for composite materials.

It is obvious from this description that there is a close link between the "smart dust" concept and sensorial materials [8]. The envisaged smart dust particles, when embedded in, for example, a structural material, may be identified with the individual sensor nodes present in a sensorial material, while the envisaged size of a dust particle would allow fulfilling the homogeneity principle. Challenges, too, are very similar: in both cases, further miniaturization is an issue, and sensor nodes, such as smart dust particles, need to get by on as little energy as possible without losing their ability to communicate and process data [9]. Some aspects, although may favor the development of sensorial materials, for example, depending on host material and application, size reduction in one dimension may suffice in their case. Furthermore, the spatial relation between sensor nodes will be fixed in most sensorial materials, whereas it is dynamic in smart dust. Besides, the distinct position of components in a product made of a sensorial material will facilitate placement of energy-harvesting elements and render their potential yield more predictable in relation to known service cycles.

Summing up, sensorial materials thus turn out to be highly integrated, self-sufficient variants of smart or rather intelligent structures. Not surprisingly, this implies that development of true sensorial materials is highly interdisciplinary and involves several subordinate aspects. As a consequence, there is a multitude of research organizations that address one or more of the central aspects, but do not cover the entire field.

17.2
Components

Considering sensorial materials from the top-down perspective, development becomes a jigsaw puzzle in which different components have to be made to fit together on a new level of miniaturization and compliance with each other as well as the surrounding material or structure. Which components these are is exemplarily depicted in Figure 17.1.

Currently, sensorial materials are a vision with top-down approaches toward their development most likely to succeed within the next 5 or 10 years. Development of these materials will thus greatly depend on development of the individual components.

This is the justification of the chosen chapter substructure, which addresses the sectors sensors, component integration, data processing, and energy supply, including management and storage separately. Owing to the width of the field and the many disciplines involved, the level of detail will necessarily be limited and some highlights only brought to the full attention of the reader. These will be selected based on the specific relevance for the encompassing field of interest.

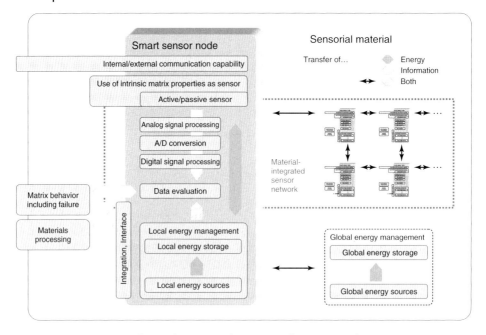

Figure 17.1 Fundamental structure of a sensorial material, perceived as combination of a host material and an integrated network of smart sensor nodes. Added global energy supply, storage, and management are optional, as they constitute a deviation from the concept of homogeneity.

17.2.1
Sensors

For the purposes of structural monitoring, sensors that detect damage or mechanical loads-the latter either directly or by their effects-, are of primary interest. Furthermore, for a true sensorial material, these sensors must provide a response that is accessible to further processing. Thus material concepts that rely on transferring the information of damage via "bruises" do not fall in this category, as they leave the interpretation of the – in this case, optical – signal to an external observer, be it a human being or an external technical system. In general, those types of damage sensors reporting the event in a binary manner via being destroyed themselves are of less interest to sensorial material technology. Here, data acquisition becomes a one-time event and quantitative interpretation at least difficult as the sensor response can only be interpreted in terms of a threshold value having been passed, but not in terms of at which height. Thus, indirect inference of damage, for example, based on quantitative strain sensing, and an internal description of the structure that relates sensor signal patterns to structural state are preferable.

Detection of mechanical strain fulfilling these requirements is commonly achieved by means of piezoelectric, piezoresistive, or optical sensors. Table 17.1 gives an exemplary overview of relevant optical sensor principles and relates some of their major characteristics based on a collection by Lopez-Higuera *et al.* [10].

Table 17.1 Overview of commercially available optical displacement, strain, and pressure-sensing systems suitable for structural monitoring tasks with some characteristics, primarily based on Lopez-Higuera *et al.* [10].

Basic principle	Sensor type	Resolution	Frequency/speed
Fiber Bragg grating (FBG)	Strain	$0.2\,\mu\varepsilon$	Maximum 10 kHz
SOFO V	Displacement	$2\,\mu m$	Hz
SOFO dynamic	Displacement	$0.01\,\mu m$	0–10 kHz
Brillouin scattering	Strain	1 m	—
Fabry–Perot interferometry	Strain	$0.1\,\mu\varepsilon$	Maximum 500 Hz

Naturally, the concrete choice of sensors depends very much on host material and application requirements, such as required accuracy, sampling frequency, and long-term stability under service conditions. The standard choice for strain sensing today are either conventional strain gauges or fiber optic sensors working according to the principle of fiber-Bragg gratings (FBGs). Both are passive sensors that require a source of energy, either electrical or as light passing through the fiber. In fiber-reinforced polymer matrix composites (FRP), measurement of lamb wave propagation through the material with excitation based on piezoelectric transducers and signal detection realized according to the same principle is another method applicable for damage detection purposes (Chapter 8) [11].

Strain gauges detect strain via the piezoresistive effect, that is, the change in resistivity of a material subjected to mechanical strain. Their sensitivity depends on the material selected. It is lower in metallic conductors and significantly increased in semiconductor materials, the drawback being that in the latter case, temperature sensitivity is also increased, meaning that compensation of its effects must be considered. For this reason, alloys such as Cu55Ni44Mn1 are favored as they are temperature insensitive over a considerable temperature range around the ambient. They usually come on foil to be adhesively bonded to the structure to monitor. An alternative, more flexible approach is maskless printing [12], which can also be based on CuNi alloy inks [13, 14].

FBG-type sensors are easily integrated in carbon and specifically long glass fiber-reinforced composites. They can provide distributed, localized information about strain based on the possibility to integrate a multitude of sensors in a single fiber and still read them out individually. However, conventional solutions are limited to detecting only longitudinal strain, and besides need compensation for temperature-induced strain. The latter problem can be solved either by integrating additional, mechanically decoupled sensors of identical type or by introducing a secondary sensor network to capture the temperature field and use this information in data evaluation. The former aspect is addressed in present research efforts, which combine special, polarization-maintaining fiber buildups with a cross-sectional shape in which the inner core and thus the actual waveguide move out of the neutral axis in bending. This way, considering intensity and wavelength shift, both bending radius and orientation can be fixed. As a side aspect, the actual number of sensors needed for a given task that involves measurement of complex strain fields can be reduced, and spatial separation of the exact point of measurement of individual strain components avoided [15].

17.2.2
Component Integration

Sensorial materials require sensors and peripheral devices, including energy supply and communication lines to be integrated into the material. This raises several technological issues, which can be roughly associated to the ability of devices to withstand the thermal and mechanical loads during production of the host material and its processing into a component, compliance with host material properties and, last but not least, the cost of integration versus its benefit for the product.

Integration is greatly facilitated if a sensorial material can be assembled from components for which sensor integration solutions exist. Examples are technical textiles, in which fiber-type sensors can be integrated either in the making of the textile, for example, during weaving [16], or by following this step, for example, by stitching [17]. The final material can then be realized as fiber-reinforced composite, for example, by resin infusion in processes such as resin transfer molding. Solutions of this kind have been discussed in the context of SHM in Chapter 6. As the references show, these techniques can provide for the sensor, but usually not for the peripheral elements.

The latter can be achieved if layered structures are integrated in a material that has, in itself, a layered buildup. Again this is the case with several types of fiber-reinforced composites. Examples exist for both dry fiber- and prepreg-based processes. The sensor elements themselves are commercially available, for example, by companies such as Accellent (SMART layer™), which rely on piezoelectric devices produced via printed circuit techniques for this purpose [18, 19]. Similar "smart patch" solutions exist for optical sensors, too [20].

An approach that goes beyond the SMART layer™ technique has recently been suggested by Hufenbach *et al.* for fiber-reinforced composites based on thermo-plastic matrices. Piezoelectric sensors and actuators are provided on thermoplastic foils that match the composition of the matrix of the composite. In the composite buildup, these elements are placed between fiber layers (e.g., prepregs). During hot pressing, the supporting foil partially melts and guarantees a material integration, which is shown to outclass adhesive bonding in terms of transmission of strain [21].

For connecting such elements to form a sensor network, additional efforts are needed, irrespective of the communication being envisaged as wireless or based on physical connections. Printing techniques may provide one common solution for this problem, provided that a continuous layer exists within the material, which is accessible to such functionalization. In sheet-type materials, this is usually the case with the outer surface, leading to sensor application rather than to integration. In layered materials, however, true sensor integration may be achieved if the multilayer built-up is used to apply the sensor network to one of the internal layers.

Basic printing processes for sensor integration include maskless (aerosol jet printing and inkjet printing) as well as mask-based processes (screen printing and offset printing). Printing of various kinds of sensors, including strain gauges for detection of mechanical strain, has been demonstrated for both inkjet and aerosol jet processes as well as for screen printing. Differences lie in the achievable resolution, which goes down to approximately $10\,\mu m$ and slightly below using aerosol jet printing, $20-50\,\mu m$ for inkjet printing, and millimeter-sized structures for screen printing of metallic pastes – for other types of functional materials, screen printing too can reach higher resolutions. Later development in inkjet printing technology promises even higher resolutions of few micrometers and is reported to enable true three-dimensional (3D) structures, too [22, 23]. Substrates can be metals (if conductive structures are printed, this requires application of an insulating layer first, which is also possible), ceramic, and even textiles in the case of aerosol jet printing [12]. Roll-to-roll processes are of special interest when it comes to realizing large-area structures with high degree of detail, but nevertheless at low cost. Recently, offset printing has been applied to produce large-area electronic structures, including sensors, conductive paths, and data evaluation electronics. The advantageous flexibility of printing processes is not limited to freedom in geometry implied by their classification as direct-write processes ("art-to-part," i.e., direct realization of computer-generated models), but extends to the wide range of materials that can be processed, and based on it, a similar scope in terms of components. For example, energy storage devices such as batteries and capacitors as well as RFID antennae have been realized in this way [23, 24].

Generally, compliance of sensors with production and use is an important aspect. In terms of production processes, sensors need to be able to withstand the considerable thermal and mechanical loads associated with the making of a structural material. This explains why composites are presently leading the field. Here, both the thermal and the mechanical load are limited, and in many processes, the initial shape is also the final, as no subsequent forming is foreseen. However, first experiments with metal matrices have also been made. This includes rapid manufacturing approaches, as presented in Chapter 18, which have the general potential to facilitate component integration based on the layerwise buildup of structures typical for these techniques.

Besides, fiber metal laminates, which consist of alternating layers of sheet metal and fiber-reinforced composites, bear great promise for extending sensor integration to metallic materials, as they combine certain metallic performance characteristics with a consolidation process that essentially relies on polymer processing techniques: The thermal and mechanical loads which the integrated components have to endure and are thus greatly reduced.

How severe these conditions can be, and, on the other hand, the fact that they are nevertheless manageable, is exemplified by first approaches to integrate sensors, actuators, and energy-harvesting devices in light metal castings [25]. Metal integration of sensors generally tends to be the greater challenge as thermal and/or mechanical loads exerted on components to be integrated are usually higher. If casting is the prime example in this respect for the thermal case, so is metal

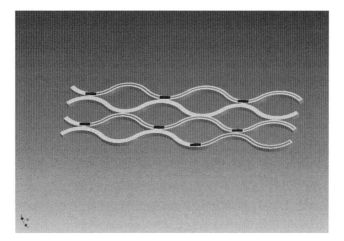

Figure 17.2 A thermogenerator realized as functional net, shown here to illustrate this principle (image provided by Institute of Microsensors, Actuators, and Systems, University of Bremen). The wavy shape of remaining substrate and active components is meant to provide stretchability. Source: Image courtesy of Institute of Microsensors, Actuators, and Systems, University of Bremen.

forming with respect to mechanical stress. If the latter is dominant, flexibility and stretchability of all embedded elements is needed – or, as this may not be achievable in most cases, at least of their interconnects. Positioning the components in the neutral plane can be an alternative, but depends on processing and service loads allowing such measures. Besides, it may not be advisable for the sensor itself, which is expected to see some signal after all. Progress in this field will provide further support for sensorial material development. Several approaches to realize compliance have been discussed by Lang *et al.* [7]. Besides materials characteristics, in many cases, geometries are adapted. Figure 17.2 depicts a thermogenerator for integration with castings, which represents this approach. Here, the aim is to accommodate thermal expansion or rather contraction during cooling of the cast part. Solutions for sheet metal integration of piezoelectric sensors and actuators based, for example, on microforming of cavities in sheet metals, meant to take up the active elements, have been suggested by Neugebauer *et al.* [22], and their capability to serve as a semifinished product that can sustain limited degrees of forming to generate a product shape demonstrated.

Had all integration-related issues been solved, material-embedded sensor networks would still remain a wound within the host material. For this reason, reduction in size is required. Lang *et al.* [7] have coined the term *function scale integration* for this approach. The size of sensors and peripheral elements and thus the footprint they leave in the material are to be reduced to the absolute minimum needed to guarantee the respective element functionality. Progress in this area is very much influenced by packaging technology, and the development of RFID techniques and smart card solutions have brought about several breakthroughs in this respect. Table 17.2 provides an overview of sensor "footprints,"

Table 17.2 Footprint associated with pressure sensor integration as a function of progress in packaging technology [7].

Packaging technology/housing type	Volume (mm³)	Footprint (μm)	Approximate year of introduction
Silicon chip in TO8 housing	300	10 000	1990
Dual-in-line (DIL) package	100	500	↓
Chip size package	20	3000	↓
Capacitive pressure sensor chip, surface micromachining	0.2	500	↓
			↓
RFID chip, thin chip technology	0.025–0.2	400–2000	↓
Perforated chip functional net	0.05–0.4	40	2010

comparing similar functionality – in this case pressure sensing – and the influence of packaging technologies that reflect the development progress in recent decades and years. Note that *footprint* in this case is defined as the greatest continuous extension of the component in a critical spatial dimension.

Looking at the sensor and classical microsystem technology approaches, the next step will be to further reduce substrate volume to create functional nets, as depicted in Figure 17.2 [7].

17.2.3
Data Processing: Algorithms and Hardware Architectures

The major motivation behind the defining requirement of internal data evaluations toward sensorial materials is the understanding that only concepts based on localized data processing will be able to provide real-time evaluation of a structure's state. Besides, distributing data processing may reduce vulnerability to damage and increase robustness and fault tolerance of a sensorial material.

Recently, emerging trends in engineering and microsystem applications such as the development of sensorial materials pose a growing demand for active networks of miniaturized smart active sensors embedded in technical structures [23, 24]. Each sensor node consists of some kind of physical sensor, electronics, data processing, and communication, providing a certain level of autonomy, dynamic behavior, and robustness [29].

With increasing miniaturization and sensor density, decentralized network and data-processing architectures are preferred or required. Spatial data fusion can be used to extract desired structural or temporal information, rather than to collect a large set of data, finally processed by a central data-processing unit.

Initially, independent data-processing nodes of a sensor network can be coupled using message-based communication enabling distributed computing capabilities. In the context of sensorial materials, 2D grid networks are suitable, shown in Figure 17.3. This network topology connects each node with up to four neighbor

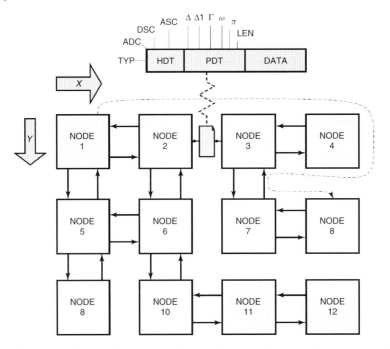

Figure 17.3 An example of an irregular two-dimensional network using point-to-point link connections for message transfer. Each node is a service endpoint and router, too [30]. A message consists of a header descriptor HDT, a packet descriptor PDT, and the data part.

nodes. Interconnection dependencies between nodes are limited to a small local area, a prerequisite for high-density sensor networks embedded in materials. The network topology can be irregular, for example, depending on design or because of temporary failures or permanent physical defects of the existing communication links. Smart routing can be used to deliver messages on different alternative routes around partially connected or defective areas. An example of a fault-tolerant message-based communication system is the scalable local intranet protocol (SLIP) [30], which is discussed later. Nodes are able to send messages to a destination node using relative delta-addressing, avoiding unique node address identifier assignments, not applicable to nodes in high-density sensor networks.

Traditionally, embedded systems are composed of generic processor systems and peripheral components. Usually, generic processor systems have limited concurrency and power management capabilities. Application-specific single-chip digital logic circuits can overcome those limitations. The design and modeling of digital logic on hardware level is limited to less complex systems. Complex algorithms can only be mapped to digital logic systems using high-level synthesis on behavioral programming level.

For example, the ConPro development framework [31] enables high-level synthesis of parallel and distributed embedded systems on behavioral level. An imperative programming model is provided, based on concurrently communicating sequential processes (CSP) with an extensive set of inter-process communication primitives and guarded atomic actions preventing and resolving resource-sharing conflicts. The programming language and the compiler-based synthesis process enable the design of constrained power- and resource-aware embedded systems on register-transfer level (RTL) efficiently mapped to FPGA and ASIC technologies. Concurrency is modeled explicitly on control- and data path level.

There are local and global resources (storage, IPC), accessed by one process and several processes, respectively. Concurrent access of global resources is automatically guarded by a mutex scheduler, serializing access, and providing atomic access without conflicts. Figure 17.4 gives an overview of building blocks and the design flow using high-level synthesis. The multiprocess model provides concurrency on control path level using concurrently executing processes, and on data path level using bounded program blocks. Parallelism available on process level requires inter-process communication and parallelism available on data path level can be applied to arbitrary sequences of data path instructions bounded in basic blocks within a process. In addition, the high-level programming level can be used to synthesize software designs using the same compiler providing the same functional behavior as the hardware design.

The main synthesis flow transforms and maps process instructions to states of clock-synchronous finite-state machines (FSMs) controlling the process RTL data path temporally and spatially, shown in Figure 17.4c.

Abstraction of hardware components, implementing, for example, device drivers, is provided by the external module interface (EMI), closing the gap between the software and hardware levels. The EMI encapsulates hardware components with abstract objects, visible on programming level. Objects can only be used and modified using method operations applied to these objects.

Many systems and algorithms used in sensor networks can be explicitly partitioned into communicating concurrently executing processes using the previously described programming model and design methodology.

Figure 17.5 shows an example, which is the multiprocess architecture of the SLIP protocol stack implementing the router of a sensor node. A sensor node is able to communicate with four neighbor nodes, arranged in the network topology shown in Figure 17.3. Both incomplete (missing links) and irregular networks (with missing nodes and links) are supported for each dimension class using a set of smart routing rules.

The path from a source to a destination node is specified by a delta-distance vector. A delta-distance vector Δ specifies the way from the source to a destination node counting the number of node hops for each dimension (known as *XY routing*).

The SLIP protocol and message format are scalable with respect to network size (extension in each dimension), maximal data payload length, and the network topology dimension size. A message packet contains a header descriptor specifying the type of the packet and the scalable parameters, followed by a packet descriptor containing the actual delta-vector Δ, the original delta-vector Δ^0, a preferred routing direction ω, an application layer port π, a backward-propagation vector Γ, and the length of the following data part.

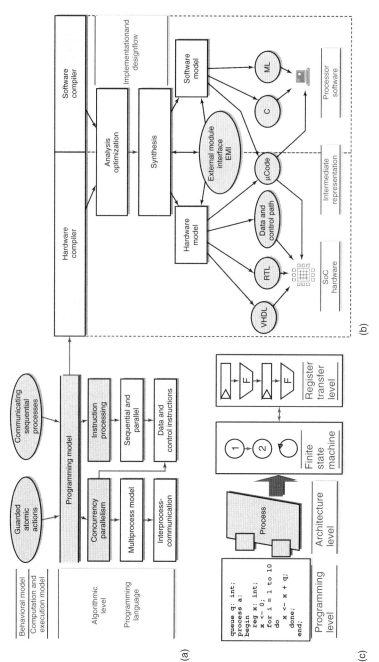

Figure 17.4 Building blocks of the programming model (a), the high-level synthesis (b), mapping the programming level to digital logic using RTL architecture (c).

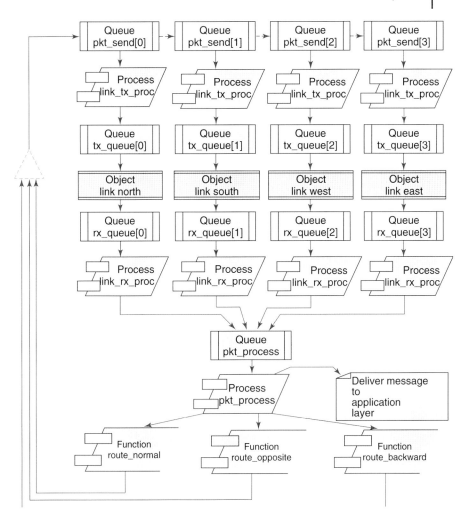

Figure 17.5 Process and inter-process communication architecture of the SLIP protocol stack.

Although the protocol stack is implemented entirely in hardware, it is modeled on high algorithmic level, shown in simplified form in Algorithm 17.1. First, the normal XY routing is tried, where the packet is routed in each direction one after another with the goal to minimize the delta count of each particular direction. If this is impossible (due to missing connectivity), the packet is tried to send to the opposite direction, marked in the message packet descriptor. Opposite routing is used to escape small-area traps; backward routing is used to escape large-area traps or send the packet back to the source node (packet not deliverable). Link and message-processing paths are implemented concurrently using sequential executing processes interchanging data using queues for inter-process communication.

Algorithm 17.1 Smart routing protocol SLIP (simplified)

```
M:   Message(Δ,Δ⁰,Γ,ω,Π,Len,Data)
PRO smart_route(M):
     IF  Δ = 0   THEN   DELIVER(M,π) ELSE
     TRY  route_normal(M)   ELSE
     TRY  route_opposite(M)   ELSE
     TRY  route_backward(M)   ELSE   DISCARD(M);
PRO route_normal(M):
     FORONE δᵢ ∈ Δ TRY minimize δᵢ :
       route(Δ,M)  WITH δᵢ := (δᵢ + 1)|δᵢ<0 ∨ (δᵢ−1)|δᵢ>0;
PRO route_opposite(M):
     FOREONE δᵢ ∈ Δ TRY minimize δᵢ :
        route(Δ,M)   WITH δᵢ := (δᵢ−1)|δᵢ<0 ∨ (δᵢ + 1)|δᵢ>0;
PRO route_backward(M):
     SEND  M (received  from  direction δᵢ)
       back  to  direction −δᵢ WITH Γᵢ = −δᵢ/|δᵢ|;
```

A comparison of traditional XY and smart routing using the routing rules is shown in Figure 17.6. The diagram shows the analysis results of operational paths depending on the number of link failures. A path is operational (reachable) if and only if a node (device under test), for example, node at position (2,2), can deliver a request message to a destination node at position (x,y) with $x \neq 2 \wedge y \neq 2$, and a reply can be delivered back to the requesting node. A failure of a specific link and node results in a broken connection between two nodes. Figure 17.6b shows an incomplete network with 100 broken links.

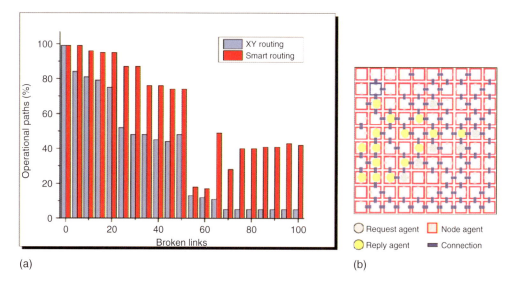

(a)   (b)

Figure 17.6 Robustness analysis with results obtained from simulation (a) and snapshot of sensor network (b).

With traditional XY routing, there is a strong decrease in operational paths, from a specific node (DUT) to any other node, if the number of broken links increases. Using smart routing increases the number of operational paths significantly, especially for considerable damaged networks, up to 50% compared with XY routing providing only 5% reachable paths any more.

The hardware implementation (using ConPro and standard cell ASIC synthesis) requires about 244k gates, 15k FF 2.5 mm^2 assuming ASIC standard cell technology of 0.18 μm. The design is partitioned on programming level in 34 processes, communicating using 16 queues.

17.2.4
Energy Supply and Management

Offering sufficient amounts of energy to the right sensor network node at the right time is a central issue for sensorial materials. Its main aspects are provision, storage, distribution, and management of energy. The supply itself can be based on external sources, energy storage devices, energy harvesting, or a combination of these three. From entirely local to centralized generation and storage, everything is possible, as are the many conceivable intermediary solutions. In practice, combinations of the latter two dominate, which usually implies a need for intelligent management of resources both in time and in space.

Energy storage on the microscale relies mostly on thin film batteries and capacitors. While the former generally provide higher energy densities, the latter excel in power density, but suffer from leakage currents that prevent efficient long-term power storage. Other concepts include microscale fuel cells or even miniaturized internal combustion engines, both of which need a reservoir for the energy carrier and are thus considered more exotic solutions – the more so because in their case the combination with energy harvesting, which typically provides electrical energy (see below), is difficult to realize. An overview covering the full variety of smaller scale energy system concepts has been provided by Krewer [32]. Among the microfuel cells specifically, proton exchange membrane fuel cells (PEMFCs, PEMμFC) and direct methanol fuel cells (DMFCs, DMμFC) prevail. An overview of both is given by Nguyen and Chan [33], while DMFC developments are highlighted by Kamarudin *et al.* [34].

Thus for sensorial materials, the focus is mostly on electrical and electrochemical energy storage systems and thus microbatteries, capacitors, and nowadays certain intermediary or hybrid systems, all of which are usually realized as thin film devices. With typical thicknesses in an order of magnitude below 100 μm, their achievable storage capacity and power output are often expressed per unit area. Table 17.3 provides a comparison of typical systems, adding selected data on fuel cells as further example. The dominant technical solution in terms of batteries are Li-ion-based concepts, which currently provide the best combination of many practical aspects, including besides energy density high numbers of charging/discharging cycles achievable. According to Patil *et al.*, the energy content for Li ion systems will reach saturation levels at approximately 220 Wh kg^{-1} and 530 Wh l^{-1} [35]. Thin

Table 17.3 Battery systems as energy storage solutions for microscale power supply compared. Note that units of power density vary throughout the table, with the heading giving the prevalent one for each case.

Type of storage device	Voltage	Energy density		Power density	References
	(V)	(Wh kg^{-1})	(Wh l^{-1})	(kW kg^{-1})	
Batteries					
Ni–Cd	1.2	40	100	—	[35]
Ni metal hydride (MH)	1.2	90	245	—	[35]
Ag–Zn	1.5	110	220	—	[35]
Li ion	3.6	155	400	—	[35]
Li- polymer	3.6	180	380	—	[35]
Li thin film	3.6	250	1000	—	[35]
Capacitors					
Activated carbon/PbO$_2$/ sulfuric acid	2.25–1.0	15.7	39.2	8.9[a]	[36]
Activated carbon/NiOOH/ KOH	1.6–0.6	13.9	31.5	4.0[a]	[36]
Li$_4$Ti$_5$O$_{12}$/activated carbon/acetonitrile	2.8–1.6	13.8	24	3.8[a]	[36]
Advanced carbon/advanced carbon/acetonitrile	3.3–2.0	13.7	19	1.9[a]	[36]
Activated carbon/activated carbon/sulfuric acid	1.0–0.5	1.72	2.2	1.2[a]	[36]
Activated carbon/activated carbon/acetonitrile	2.7–1.35	5.7	7.6	6.4[a]	[36]
Microfuel cells[b]				mW cm^{-2}	
Proton exchange membrane fuel cell (PEMFC)	1.23	—	—	Up to 315 mW cm^{-2}	[33]
Direct methanol fuel cell (DMFC)	1.21	—	—	Up to 100 mW cm^{-2}	[33]
DMFC	1.21	—	—	43 mW cm^{-2}	[34]

[a]95% efficiency according to [36].
[b]Note that it is difficult to specify energy density for fuel cells, as energy transduction and storage of the energy carrier are separated and thus a larger fuel capacity will push the figure to higher values, with the energy content of the fuel usable based on the efficiency of the process as upper limit.

film systems may exceed these values, as shown in Table 17.3. Note however that for Li ion batteries, improvement of energy and power density usually is conflicting aims.

Recently, 3D structuring of batteries has received renewed interest. This approach should not be misunderstood as a mere stacking of planar thin film batteries in the third dimension. In contrast, structuring takes place within the cell at the nano- and microscopic levels and allows improvement in performance-controlling characteristics such as the electrode–electrolyte interface area or the effective thickness of electrode and electrolyte layers, thus, for example, reducing otherwise power-limiting diffusion path lengths.

The resulting components may remain as thin film, but their energy and power density can be stepped up significantly at the same real footprint. The potential of the method has been repeatedly demonstrated, for example, by Ho *et al.* [23] based on inkjet-printed Ag–Zn microbatteries, in which a comparatively wide spacing of printed silver pillars providing electrode structuring led to an increase in the areal energy density to 3.95 from 2.33 Wh cm^{-2}, observed for a planar-silver-electrode-based buildup produced by means of the same super inkjet printing process. Similarly, increases by an approximate factor of 5 depending on aspect ratio of electrode-structuring trenches and pores have recently been published by Baggetto *et al.* [37] for poly-Si-based negative electrodes to be used in Li ion batteries.

Detailed reviews on recent work and potentials of 3D structuring techniques applied to microbatteries have been published by Long *et al.* [38] and more recently by Roberts *et al.* [39] and Oudenhoven *et al.* [40]. The investigated systems are usually Li ion variants, with both solid and liquid electrolyte concepts addressed.

In terms of systems under study, another special focus is on Li-based solutions using solid-state electrolytes. Challenges in this field have been summarized by Patil *et al.* [35] and include the identification of candidate electrolytes providing sufficient Li ion conductivity. The latter is of specific significance as otherwise the internal resistance of the cell will produce an unwanted limitation to power output. High-conductivity electrolyte materials would thus complement structuring approaches aimed at maintaining the low conductive path length (equivalent to electrode layer thickness) typically realized in planar thin film systems.

Besides structural optimization and search for improvement in active materials, additional efforts are directed at alternative material systems, that is, at alternatives in terms of the fundamental reaction enabling energy storage and retrieval. Because of their promise for high energy density, which has its theoretical limit about an order of magnitude above conventional lithium ion systems, lithium–oxygen batteries receive special attention in this context. However, despite their attractive performance, their applicability to sensorial materials may turn out to be limited as their fundamental principle of operation implies access to air or at least a suitable source of oxygen, which may contradict the need for embedding of systems in a host material to ensure durability. Research on the system as such is driven by the need for breakthrough energy storage solutions for future electric vehicles. Currently, many technological challenges remain, and so far no functional full cell has been realized. The state of research is outlined by Cheng and Chen [41], Christensen *et al.* [42], and Bruce *et al.* [43].

Development of capacitors mirrors exertions in the field of batteries in as far as the alleviation of typical weaknesses is addressed. For capacitors, this is energy rather than power density. The focus is on electrochemical capacitors, which replace the dielectric separating the electrodes in conventional capacitors with an electrolyte. Two fundamental storage mechanisms can be distinguished. In electrochemical double-layer capacitors (EDLCs), charge storage is electrostatic based on reversible adsorption of electrolyte ions on the oppositely charged electrode's active material surface. High levels of capacity are basically enabled by large active material surfaces (e.g., porous or nanostructured carbon- and materials) and by the small distance between opposite charges in the double layers at each electrode. Charging and discharging can be extremely fast, as no chemical reaction kinetics have a part in energy storage. Recent developments to further improve energy density include nanoporous active materials with pore sizes tailored to match sizes of charge-carrying ions as some studies suggest that such morphologies can partially remove solvation shells surrounding these ions in the electrolyte, resulting in smaller effective size and thus a higher density of charger carriers on the double layer's electrolyte side [44].

Redox-based electrochemical capacitors showing pseudocapacitive charge storage form another class of electrochemical capacitors. In their case, fast and reversible redox reactions occur at the active material surface, which is very often a metal oxide such as RuO_2, Fe_3O_4, or MnO_2. Specific capacitance observed exceeds that of EDLCs, but as chemical reactions are involved, reaction kinetics can limit power output, and the advantageous characteristics of electrostatic processes in terms of cycling stability can suffer [44].

As finding the right balance between energy and power density is central for optimized energy storage, attempts at combining the two fundamentally different concepts of batteries and capacitors deserve attention. The resulting hybrid capacitors usually combine a capacitor electrode with a battery one, at the cost of a certain amount of power density, but naturally with benefits for energy storage capacity. The concept can facilitate increased cell voltage, too, which affects both energy and power density favorably. A drawback is the fact that cyclability, one of the mainstays of capacitors besides their superior power density, can suffer in hybrid designs [44].

As future sensorial materials will have to stay functional over a considerable service life without much possibilities for maintenance, energy supply is unlikely to be feasible based on storage and thus primary batteries alone. Therefore, secondary or rechargeable systems have been the focus of the preceding discussion. The answer to the question where the energy for recharge should originate from may be twofold. Need-based recharging from external sources following extended service cycles is an option – however, as outmost autonomy of sensor networks in sensorial materials is a major aim, internal energy generation and thus energy scavenging or harvesting is a central issue in their development. Generally, energy harvesting means tapping energy sources from the environment to match a system's requirements. Major principles that can be used are harvesting of

- thermal energy (e.g., thermoelectrics and Seebeck effect);
- vibration energy (e.g., piezoelectrics);
- light (e.g., photovoltaics);
- radio frequency electromagnetic radiation;
- fluid flow.

Table 17.4 Overview of ambient energy sources here for the tapping by means of energy-harvesting devices, approximate data collected by and according to Yildiz [45] and Valenzuela [46].

Ambient energy source		Technical principle[a]	Approximate power density		Typical efficiency[b]
Class	Subclass		(μW cm^{-2})	(μW cm^{-3})	(%)
Light	Outdoor	Photovoltaic	10^5	n.a.	10–24
	Indoor	Photovoltaic	100	n.a.	10–24
Thermal	Human	Thermoelectric	60	n.a.	0.1
	Industrial	Thermoelectric	10^3–10^4	n.a.	3
	Temperature variation	—	n.a.	10^1	—
RF	Ambient RF	—	1^a	n.a.	—
	GSM 900 MHz	—	0.1^b	n.a.	50
	WiFi	—	0.001^b	n.a.	50
Vibration	Human (Hz)	Microgenerator	n.a.	4	25–50
	Industrial (kHz)	Microgenerator	n.a.	800	25–50
	—	Piezoelectric	n.a.	200^a	25–50
Acoustic	Acoustic noise, 75 dB	—	n.a.	0.003^a	—
	Acoustic noise, 100 dB	—	n.a.	0.96^a	—
Airflow	—	—	1^a	n.a.	—

[a] Exclusive source: [45].
[b] Exclusive source: [46], all others [45] and [46].

Table 17.4 provides a rough estimate of the available energy per unit volume or area for several such solutions.

Energy harvesting solutions often rely on smart materials, such as thermo- or piezoelectric materials, with energy-transduction capability. In many cases, when microdevices are sought for, these are integrated in microsystems using MEMS and related technologies. The active materials themselves profit from new developments in nanostructures and low-dimensional architectures. Thermoelectrics are an example in this respect. In many relevant materials, for example, nanograined structures have been demonstrated to perform favorably in comparison to conventional materials as phonon scattering at grain boundaries allows influencing thermal and electrical conductivities independently. This opens up possibilities to lower the former while retaining the latter, a path that leads to increased efficiency in thermoelectrics.

Furthermore, theoretical work as well as early experimental validations for 2D quantum wells in PbTe-type materials by Dresselhaus *et al.* has long since provided evidence that low-dimensional structures can lead to performance increase [47]. Among many others, Davila *et al.* [48] have demonstrated this fact for Si nanowires in contrast to bulk Si, at the same time realizing major steps from fundamental material characterization toward working microscale thermogenerators using MEMS techniques. Developments in nanocomposite thermoelectric materials are partly aimed at utilizing related effects [49]. In terms of material classes in general, even though highest figures of merit (thermoelectric efficiency) are still documented for classical systems such as PbTe or BiTe, there is a tendency toward identifying alternatives that are less critical in terms of stability at elevated temperatures, environmental concerns, cost and so on, like many oxide materials [50].

Vibration energy harvesting is mostly based on piezoelectric materials and MEMS techniques for production of generators. Highest efficiencies are generally achieved if vibrating systems, for example, comprising a beam with attached mass can be tuned to the excitation frequency to work in resonance mode. However, this implies peak performance at a fixed frequency, which may not be ideal where vibration is randomly distributed over a larger frequency range. For this reason, wide bandwidth or even frequency-adaptable generators are being investigated [51]. Further approaches combine the sensor function with energy supply, that is, by deducing sensory information from the harvested energy flow. Systems of this kind relying on special zigzag metal counter-electrode designs to gather charges from flexed ZnO nanowires have been suggested by Wang *et al.* [52, 53], with improvements expected from transferring the basic concept to nanowires showing higher piezoelectric efficiency and thus a higher current yield than reported by Wang *et al.*, such as $BaTiO_3$ [54].

In terms of materials, the brittle behavior of many piezoelectric ceramics such as lead zirconate titanate (PZT) has fueled interest in polymers with generally comparable properties, and in piezoelectric composites. In the latter, active ceramic particles are embedded in a polymer matrix that addresses the concern in terms of mechanics [55], while at the same time facilitating processing of the materials, including techniques such as mask-based [56], but also maskless printing.

General overviews covering different types of energy-harvesting systems in conjunction with sensor and sensor network energy supply have recently been put together by Vullers *et al.* [57], who add information about power management circuitry, while Bogue [54, 58, 59] includes storage techniques in his reviews, too. Table 17.5 provides a glimpse at energy issues of systems that have recently been proposed in the application context of structural monitoring. A more detailed overview of energy-harvesting systems was provided by Hudak and Amatucci [60] in 2008.

Having thus discussed ambient and local energy availability and associated transduction, the issue of adequately using it in distributed sensor networks, and thus the consumer perspective, remains.

With increasing miniaturization and sensor density, decentralized energy supply using self-powered architectures is preferred. Energy harvesting, for example, using thermoelectrical sources, actually delivers only low electrical power, requiring (i) smart energy management on the consumer side controlling the energy consumption and (ii) low-power capabilities and design of sensor nodes.

Table 17.5 Characteristics of selected sensor nodes suggested for structural monitoring applications, focusing on aerospace, with and without energy harvesting solutions implemented.

Description				Power			References
Application	Harvester	Storage	Sensor principle	System requirement (μW)	Provision[a] (μW)	System efficiency (%)	
Aircraft wireless sensor node for structural load monitoring	TG, combined with PCM	Cap., 1.8–2 F	Crack wire	189[b]	6000	Approximately 0.05[c]	[61]
Credit-card-sized smart tag sensor node for aerospace applications	PEG, thick film printed	Cap., 0.55 F, 4.5 V	Temperature and pressure sensor, accelerometer	88 900[d]	240	—	[62]
Autonomous sensor node in distributed wireless sensor network	None	Primary battery	Impedance sensor	150 (sleep) 18 000 (active)	n.a.	—	[63]
RF-powered wireless sensor node	RFG	Cap., 2.2 mF	Strain gauge	6000	~3801	52[e]	[64]
Wireless sensor node for guided wave-based SHM	None	LiPo primary battery	Guided wave propagation, piezoelectric excitation	1399 mW (active), 50 mW (inactive)	n.a.	—	[65]

[a] Power as supplied by harvesting unit under typical or design conditions of usage, prior to power conditioning an similar measures such as DC–DC conversion.
[b] Excluding communication requirements.
[c] Direct system efficiency, storage efficiency not included.
[d] Including communication and sensor power requirements.
[e] RF-to-DC conversion efficiency only, no full system efficiency.
Harvesting: PEG = piezoelectric Generator, PCM = phase change material for latent heat storage, RFG = radio frequency generator. Electrical energy storage: Cap. = capacitor.

Advanced design methodologies for embedded systems can aid to satisfy low-power requirements targeting self-powered sensor nodes. Contributions to energy management are:

1. *Local smart energy management* performed at runtime using advanced computer science algorithms (artificial intelligence) providing optimization of power consumption.

   In contrast to various other approaches targeting algorithms and architectures with high computational effort, for example [66], *smart energy management* can be performed spatially at runtime by a selection from a set of different (implemented) algorithms classified by their demand of computation power, and temporally by varying data-processing rates. It can be shown that power/energy consumption of an application-specific system-on-chip (SoC) design strongly depends on computation complexity [67].

2. Optimized application-specific SoC design on RTL using high-level synthesis. Low-power systems can be designed on algorithmic rather than on technological level. Averaged SoC cell activity correlates strongly with computation and signal/data flow [67].

3. *Smart energy distribution* management in decentralized self-powered sensor networks using distributed artificial intelligence concepts and algorithms, such as multi-agent systems [68].

Sharing of one interconnect medium for both data communication and energy transfer significantly reduces node and network resources and complexity, a prerequisite for a high degree of miniaturization required in high-density sensor networks embedded in sensorial materials. Point-to-point connections and mesh-network topologies are preferred in high-density networks because they allow good scalability (and maximal path length) in the order of $O(\log N)$, with N as the number of nodes.

Figure 17.7 shows the main building blocks of a self-powered sensor node, the proposed technical implementation of the optical serial interconnect modules, and the local energy management module collecting energy from a local source, for example, a thermoelectric generator, and energy retrieved from the optical communication receiver modules.

The data-processing system can use the communication unit to transfer data (D) and superposed energy (E) pulses using a light-emitting or laser diode. The diode current, driven by a differential-output sum amplifier, and the pulse duration time determine the amount of energy to be transferred. The data pulses have a fixed intensity several orders lower than those of the adjustable energy pulses. On the receiver side, the incoming light is converted into an electrical current using a photodiode. The data part is separated by a high-pass filter, and the electrical energy is stored by the harvester module.

Information and energy is encapsulated in messages routed in the network from a source to a destination node, using a simple delta-routing protocol. An alternative, advanced smart routing protocol, which allows incomplete mesh-networks and compensates link failures using different routing rules, is described in [30].

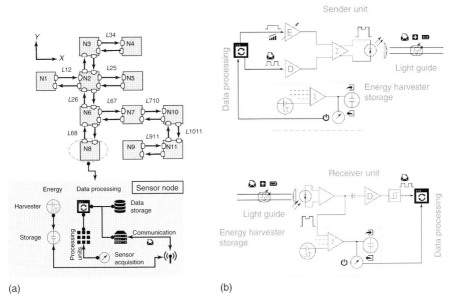

Figure 17.7 Network topology (a) and sender and receiver blocks (b) used for data and energy transmission between neighbor nodes. Each node connects up to four neighbors and uses optical links to transfer data messages and energy.

Using those technical abilities, it is possible to use active messaging to transfer energy from good nodes having enough energy toward bad nodes, requiring energy. An agent can be sent by a bad node to explore and exploit the near neighborhood. The agent examines sensor nodes during path travel or passing a region of interest (perception) and decides to send agents holding additional energy back to the original requesting node (action). Additionally, a sensor node is represented by a node agent, too. The node and the energy management agents must negotiate the energy request.

Some simulation results are shown in Figure 17.8, showing the benefits of energy management using multi-agent systems to negotiate energy demand and distribution using communication links for transmission of both messages and energy [69]. Each node is initially supplied by energy locally harvested and stored in an energy deposit. Without energy management, there are nodes with insufficient energy. Using smart energy management, all nodes can be supplied with enough energy to fulfill data and message processing.

17.3
Case Study

In this section, a case study of the modular robot arm manipulator project Mod-uACT is presented, showing the integration of sensorial materials in robotics [70].

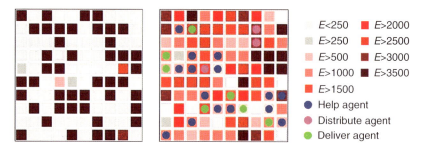

$E<250$		■	$E>2000$
	$E>250$	■	$E>2500$
	$E>500$	■	$E>3000$
■	$E>1000$	■	$E>3500$
■	$E>1500$		

● Help agent
● Distribute agent
● Deliver agent

Figure 17.8 Results from multi-agent simulation without and with energy management. Help, deliver, and distribute agents are used to compensate low-energy nodes (blue color). Sensor nodes can exchange messages (agents) and energy using communication links.

An active sensor network embedded in the connection structure of a robot arm manipulator is used to provide load and touch perception of the environment, shown in Figure 17.9. It features locally tight coupling and integration of sensors, actors, and data processing and communication, and globally decentralized data processing with a network of distributed sensor nodes embedded in the robot arm structure.

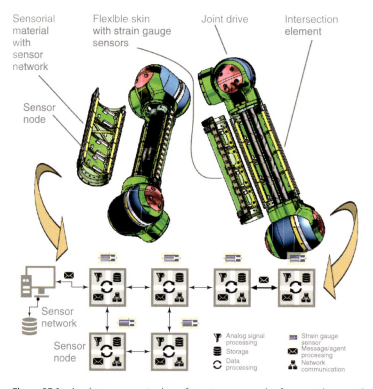

Figure 17.9 A robot arm manipulator featuring a network of sensorial materials and actuator joints. The intersection structures integrate a sensorial material for environmental perception.

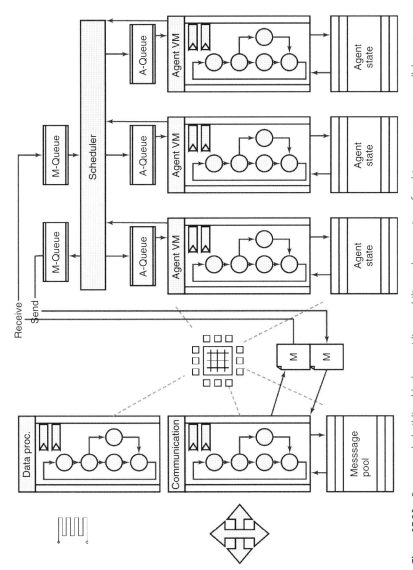

Figure 17.10 Sensor node building blocks providing mobility and processing of multi-agent systems: parallel agent virtual machines, agent-processing scheduler, communication, and data processing [69].

Multi-agent systems are used to implement

- distributed data processing and information exchange [69];
- data fusion and propagation of information;
- finally lower levels of machine learning methods for fast load inference and classification in a discrete position-force space [29].

An agent approach provides stronger autonomy than a traditional object or remote-procedure-call-based approach. Agents can decide for themselves which actions are performed, and they are capable of reacting on the environment and other agents with flexible behavior.

Traditionally, mobile agents are executed on generic computer architectures [71, 72], which usually cannot easily be reduced to single-chip systems as they are required, for example, in sensorial materials with high sensor node densities.

SoC design embedding low- and high-level data processing in a single microchip ensures a high degree of miniaturization and low-power data-processing systems suitable for local energy-harvesting methods. The runtime environment is modeled on behavioral level using the high-level multi-process programming language ConPro with atomic-guarded actions and inter-process communication (communicating sequential processes) [31], introduced in the previous section.

Internode communication is performed with messages encapsulating agents. Smart message routing provides robustness [30]. The hardware architecture is optimized for mesh networks of data processing nodes, which can be organized, for example, in a 2D grid topology with each node having connections to its up to four direct neighbors.

The functional behavior of agents is implemented with finite-state machines on RTL embedded in the sensor node, shown in Figure 17.10. Only the state of the agent (consisting of control and data state variables) is propagated using messages. For example, the agent state machine implementing the behavior of the smart energy management agent consists of only nine control states and nine data state variables.

17.4
Further Reading

As yet no all-encompassing reference work on sensorial materials exists. However, several aspects of them are covered in the literature on smart structures. Besides, scientific journals have recently accepted and published contributions that delineate the concept. Some of these, such as the work by Lang *et al.* [7], have already been cited above.

Furthermore, the general issue of material-integrated intelligence and sensing as expressed in the term *sensorial materials* is covered as a major aspect in several conference series, of which some examples are named below. As some of the conferences are dynamic in terms of their topics, that is, the list of symposia is put together anew for each time, selected events of the past have been named below.

Naturally, the numerous individual aspects that contribute to realizing sensorial materials, such as NEMS/MEMS technology, energy harvesting and storage, and network data evaluation, are covered by specialized events. These have been omitted here to maintain clarity.

Books:

- Wadhawan, W.K. (2007) *Smart Structures-Blurring the Distinction between the Living and the Nonliving*, Oxford University Press, Oxford.
- Gaudenzi, P. (2009) *Smart Structures-Physical Behaviour, Mathematical Modelling and Applications*, John Wiley & Sons, Hoboken.

Conferences:

- SysInt Conference Series, International Conference on System Integrated intelligence
  Conference/organizer web site: *www.sysint-conference.org*
  A biannual international event dealing with all aspects of system-integrated intelligence from product development and engineering design to life cycle aspects and covering the physical realization of intelligent structures and materials, as well as the technologies needed for this purpose. Held for the first time in Hanover, Germany, in June 2012 including a special session on sensorial materials, planned for July $2^{nd}-4^{th}$, 2014 in Bremen, Germany and June/July 2016 in Paderborn, Germany.
- Smart Systems Integration (SSI)
  Conference/organizer web site: *www.smart-systems-integration.com*
  An annual European conference with accompanying exhibition usually held in Spring at changing locations. The wider topic of smart systems integration regularly includes most aspects of sensorial materials. Organization is supported by the European Technology Platform on Smart Systems Integration (EPoSS).
- E-MRS Spring Meeting
  Conference/organizer web site: *www.emrs-strasbourg.com*
  One of the largest annual European conferences on materials science on a fundamental and applied level, organized by the European Materials Research Society, and with very few exceptions regularly held in Strasbourg in May or June.
  Past years have seen several symposia linked to the topic of sensorial materials, that is:
  – E-MRS Spring Meeting 2010,
    Symposium A: From Embedded Sensors to Sensorial Materials, organized by Professor Walter Lang, Institute for Microsensors, Actuators and Systems, University of Bremen, Germany

- E-MRS Spring Meeting 2012,
    Symposium Q, organized by Danick Briand, École Polytechnique Fédérale de Lausanne (EPFL), Switzerland.
- Euromat
Conference/organizer web site: ..., *www.euromat2009.fems.eu*
www.euromat2011.fems.eu, www.euromat2013.fems.eu,...
www.fems.eu
A conference similar in size and diversity to the E-MRS Spring Meeting. Euromat is organized by the Federation of European Materials Societies and held biannually in uneven years in changing countries, with the local materials society hosting the event. As for the E-MRS Spring Meeting, the selection of symposia varies from event to event, with topics linked to sensorial materials regularly covered.
    - Euromat 2011,
        Symposium A53: MEMS/NEMS for Sensorial and Actorial Materials, organized by Dirk Lehmhus, ISIS Sensorial Materials Scientific Centre, University of Bremen, Germany and Professor Jürgen Brugger, École Polytechnique Féderale de Lausanne (EPFL), Lausanne, Switzerland.
- CIMTEC Conference Series
Conference/organizer web site: *http://www.cimtec-congress.org/*
A biannual conference usually held in Italy and for the fourth time in 2012. CIMTEC conferences have two recurring topics. One of these is "Smart Materials, Structures and Systems", the focus in 2012 and thus again in 2016. In 2012, the entire conference held in Montecatini Therme, Italy, from 10 to 14 June, was of relevance for sensorial materials research with a special focus set by symposium G, "Emboding Intelligence in Structures and Integrated Systems."
- The International Workshop on Structural Health Monitoring
Conference/organizer web site: *http://structure.stanford.edu/workshop/*
Rooted in SHM as one of the major fields of application foreseen for sensorial materials, this workshop, which is organized by and held biannually in uneven years at Stanford University, USA, is among the leading international events in the field. It is complemented, in even years, by the "European Workshop on Structural Health Monitoring" (EWSHM) which addresses identical topics.

Acknowledgment

The authors would like to acknowledge support to this work granted by the University of Bremen and the Federal State of Bremen within the framework of the ISIS Sensorial Materials Scientific Centre.

References

1. Lang, W., Lehmhus, D., v. d. Zwaag, S., Dorey, R. (2011) Sensorial Materials – a vision about where progress in sensor integration may lead to. *Sens. Actuators A*, **171**, 1–2.

2. Lehmhus, D., Busse, M. (2012) An introduction to Sensorial Materials. *Proceedings of the 1st Joint International Symposium on System-Integrated Intelligence (SysInt 2012)*, PZH Produktionstechnisches Zentrum GmbH, Hanover, Germany, ISBN: 978-3-943104-59-2, 176–178.

3. Li, H.-N., Li, D.-S., Song, G.-B. (2004) Recent applications of fibre-optic sensors to health monitoring in civil engineering. *Eng. Struct.*, **26**, 1647–1657.

4. Renton, W.J. (2001) (Boeing): Aerospace and structures: where are we headed? *Int. J. Solids Struct.*, **38**, 3309–3319.

5. Peters, C., Zahlen, P., Bockenheimer, C., and Hermann, A.S. (2011) Structural health monitoring needs S3 (sensor-structure-system) logic for efficient product development. *Proc. SPIE*, **7981**, 79815N–79815N-10.

6. Wu, D.Y., Meure, S., Solomon, D. (2008) Self-healing polymeric materials: A review of recent developments. Progress in Polymer Science, **33**, 479–522.

7. Lang, W., Jakobs, F., Tolstosheeva, E., Sturm, H., Ibragimov, A., Kesel, A., Lehmhus, D., and Dicke, U. (2011) From embedded sensors to sensorial materials – the road to function scale integration. *Sens. Actuators, A*, **171**, 3–11.

8. Warneke, M., Last, M., Liebowitz, B., and Pister, K.S.J. (2001) Smart Dust: communicating with a cubic-millimeter computer. *Computer*, **34**, 44–51.

9. Cook, B.W., Lanzisera, S., and Pister, K.S.J. (2006) SoC issues RF smart dust. *Proc. IEEE*, **94**, 1177–1196.

10. Lopez-Higuera, J.M. *et al.*

11. Schubert, K. and Herrmann, A.S. (2011) On attenuation and measurement of Lamb waves in viscoelastic composites. *Compos. Struct.*, **94**, 177–185.

12. Maiwald, M., Werner, C., Zöllmer, V., and Busse, M. (2010) INKtelligent printed strain gauges. *Sens. Actuators, A*, **162**, 198–201.

13. Pál, E., Zöllmer, V., Lehmhus, D., and Busse, M. (2011) Synthesis of $Cu_{0.55}Ni_{0.44}Mn_{0.01}$ alloy nanoparticles by solution combustion method and their application in aerosol printing. *Colloids Surf., A*, **384**, 661–667.

14. Pál, E., Kun, R., Schulze, C., Zöllmer, V., Lehmhus, D., Bäumer, M., and Busse, M. (2012) Composition dependent sintering behaviour of chemically synthesised CuNi nanoparticles and their application in aerosol printing for preparation of conductive microstructures. *Colloid Polym. Sci.* doi: 10.1007/s00396-012-2612-3

15. Kibben, S., Kropp, M., Durnstorff, G., Seefeld, T., Lang, W., and Vollertsen, F. (2011) Fiber optic bend sensor with precise angle resolution and compact evaluation unit. Euromat 2011 Conference, Montpellier, France, September 12–15, 2011.

16. Naumann, S., Lapeyronnie, P., Cristian, I., Boussu, F., and Koncar, V. (2011) Online measurement of structural deformations in composites. *IEEE Sens. J.*, **11**, 1329–1336.

17. Walther, M., Kroll, L., Stockmann, M., Elsner, H., Heinrich, M., and Wagner, S. (2011) Investigation in development of embroidered strain measurement sensors, in *10th Youth Symposium on Experimental Solid Mechanics, Chemnitz, 2011* (eds M. Stockmann and J. Kretzschmar), Universitätsverlag, Chemnitz University of Technology, Chemnitz, Germany, pp. S.117–118, ISBN: 978-3-941003-34-7.

18. Lin, M., Kumar, A., Beard, S.J., and Xinlin, Q. (2001) Built-in structural diagnostic with the SMART Layer™ and SMART Suitcase™. *Smart Mater. Bull.*, 7–11.

19. Qing, X.P., Beard, S.J., Kumar, A., Chan, H.-L., and Ikegami, R. (2006) Advances in the development of built-in diagnostic system for filament wound composite structures. *Compos. Sci. Technol.*, **66**, 1694–1702.

20. Doyle, C., Quinn, S., and Dulieu-Barton, J.M. (2005) Evaluation of rugged 'smart patch' fibre-optic strain sensors. *Appl. Mech. Mater.*, **34–35**, 343–348.

21. Hufenbach, W., Gude, M., Heber, T. (2011) Embedding versus adhesive bonding of adapted piezoceramic modules for

function-intgrative thermoplastic composite structures. *Compos. Sci. Technol.*, **71**, 1132–1137.

22. Murata, K., Matsumoto, J., Tezuka, A., Matsuba, Y., Yokoyama, H. (2005) Super-fine ink-jet printing: toward the minimal manufacturing system. *Microsyst. Technol.*, **12**, 2–7.

23. Ho, C.C., Murata, K., Steingart, D.A., Evans, J.W., and Wright, P.K. (2009) A super ink jet printed zinc–silver 3D microbattery. *J. Micromech. Microeng.*, **19**, 094013 (5 pp.).

24. Pech, D., Brunet, M., Taberna, P.-L., Simon, P., Fabre, N., Mesnilgrente, F., Conédéra, V., and Durou, H. (2010) Elaboration of a microstructured inkjet-printed carbon electrochemical capacitor. *J. Power Sources*, **195**, 1266–1269.

25. Ibragimov, A., Pleteit, H., Pille, C., and Lang, W. (2012) A thermoelectric energy harvester directly embedded into casted aluminum. *IEEE Electron Device Lett.*, **33**, 233–235.

26. Neugebauer, R., Ihlemann, J., Lachmann, L., Drossel, W.-G., Hensel, S., Nestler, M., Landgraf, R., and Rudolph, M. (2011) Piezo-metal-composites in structural parts: technological design, process simulation and material modelling. Proceedings of the CRC/TR39 3rd Scientific Symposium "Integration of Active Functions into Structural Elements", Chemnitz, Germany, October 12–13, 2011.

27. Makarenko, A. and Durrant-Whyte, H. (2004) Decentralized data fusion and control in active sensor networks. Proceedings of the Seventh International Conference on Information Fusion 2004.

28. Levis, P., Gay, D., and Culler, D. (2005) Active sensor networks, in *Proceeding NSDI'05 Proceedings of the 2nd Conference on Symposium on Networked Systems Design and Implementation*, **2**, USENIX Association, Berkeley, CA.

29. Pantke, F., Bosse, S., Lehmhus, D., and Lawo, M. (2011) An artificial intelligence approach towards sensorial materials. Future Computing Conference, 2011.

30. Bosse, S. and Lehmhus, D. (2010) Smart communication in a wired sensor- and actuator-network of a modular robot actuator system using a hop-protocol with delta-routing. Proceedings of Smart Systems Integration Conference, Como, Italy, March 23–24, 2010.

31. Bosse, S. (2011) Hardware-softwareco-design of parallel and distributed systems using a unique behavioural programming and multi-process model with high-level synthesis. Proceedings of the SPIE Microtechnologies 2011 Conference, Session EMT 102 VLSI Circuits and Systems, Prague, Czech Republic, April 18–20, 2011.

32. Krewer U. (2011) Portable Energiesysteme: Von elektrochemischer Wandlung bis Energy Harvesting. *Chem. Ing. Tech.*, **83**, 1974–1983.

33. Nguyen, N.-T. and Chan, S.H. (2006) Micromachined polymer electrolyte membrane and direct methanol fuel cells – a review. *J. Micromech. Microeng.*, **16**, R1–R12.

34. Kamarudin, S.K., Daud, W.R.W., Ho, S.L., and Hasran, U.A. (2007) Overview on the challenges and developments of micro-direct methanol fuel cells (DMFC). *J. Power Sources*, **163**, 743–754.

35. Patil, A., Patil, V., Shin, D.W., Choi, J.-W., Paik, D.-S., and Yoon, S.-J. (2008) Issues and challenges facing rechargeable thin film lithium batteries. *Mater. Res. Bull.*, **43**, 1913–1942.

36. Burke, A (2007) R&D considerations for the performance and application of electrochemical capacitors. *Electrochimica Acta*, **53**, 1083–1091.

37. Baggetto, L., Knoops, H.C.M., Niessen, R.A.H., Kessels, W.M.M., and Notten, P.H.L. (2010) 3D negative electrode stacks for integrated all-solid-state lithium-ion microbatteries. *J. Mater. Chem.*, **20**, 3703–3708.

38. Long, J.W., Dunn, B., Rolison, D.R., and White, H.S. (2004) Three-dimensional battery architectures. *Chem. Rev.*, **104**, 4463–4492.

39. Roberts, M., Johns, P., Owen, J., Brandell, D., Edstrom, K., El Enany, G., Guery, C., Golodnitsky, D., Lacey, M., Lecoeur, C., Mazor, H., Peled, E., Perre, E., Shaijumon, M.M., Simon, P., and Taberna, P.-L. (2011) 3D lithium ion batteries – from fundamentals to fabrication. *J. Mater. Chem.*, **21**, 9876–9890.

40. Oudenhoven, J.F.M., Baggetto, L., and Notten, P.H.L. (2011) All-solid-state lithium-ion microbatteries: a review of various three-dimensional concepts. *Adv. Energy Mater.*, **1**, 10–33.

41. Cheng, F. and Chen, J. (2012) Metal–air batteries: from oxygen reduction electrochemistry to cathode catalysts. *Chem. Soc. Rev.*, **41**, 2172–2192.

42. Christensen, J., Albertus, P., Sanchez-Carrera, R.S., Lohmann, T., Kozinsky, B., Liedtke, R., Ahmed, J., and Kojic, A. (2012) A critical review of Li/air batteries. *J. Electrochem. Soc.*, **159**(2), R1–R30.

43. Bruce, P.G., Freunberger, S.A., Hardwick, L.J., and Tarascon, J.-M. (2012) Li–O2 and Li–S batteries with high energy storage. *Nat. Mater.*, **11**, 19–29.

44. Simon, P. and Gogotsi, Y. (2008) Materials for electrochemical capacitors. *Nat. Mater.*, **7**, 845–854.

45. Yildiz, F. (2009) Potential ambient energy-harvesting sources and techniques. *J. Technol. Stud.*, **35**, 40–48.

46. Valenzuela, A. (2008) Energy Harvesting for No-Power Embedded Systems, *http://focus.ti.com/graphics/mcu/ulp/ energy_harvesting_embedded_systems_using_msp430.pdf* (accessed 15 March 2012).

47. Dresselhaus, M.S., Koga, T., Sun, X., Cronin, S.B., Wang, K.L., and Chen, G. (1997) Low dimensional thermoelectrics. Proceedings ICT'97-XVI International Conference on Thermoelectrics, Dresden, Germany, August 6–29, 1997.

48. Davila, D., Tarancon, A., Kendig, D., Fernandez-Regulez, M., Sabate, N., Salleras, M., Calaza, C., Cane, C., Gracia, I., Figueras, E., Santander, J., San Paulo, A., Shakouri, A., and Fonseca, L. (2011) Planar thermoelectric microgenerators based on silicon nanowires. *J. Electron. Mater.*, **40**, 851–855.

49. Dresselhaus, M.S., Chen, G., Tang, M.Y., Yang, R., Lee, H., Wang, D., Ren, Z., Fleurial, J.-P., and Gogna, P. (2007) New directions for low-dimensional thermoelectric materials. *Adv. Mater.*, **19**, 1043–1053.

50. He, J., Liu, Y.F., and Funahashi, R. (2011) Oxide thermoelectrics: the challenges, progress, and outlook. *J. Mater. Res.*, **26**, 1762–1772.

51. Tang, L., Yang, Y., and Soh, C.K. (2010) Toward broadband vibration-based energy harvesting. *J. Intell. Mater. Syst. Struct.*, **21**, 1867–1897.

52. Wang, Z.L. and Song, J.H. (2006) Piezoelectric nanogenerators based on zinc oxide nanowire arrays. *Science*, **312**, 242–246.

53. Wang, X.D., Song, J.H., Liu, J., and Wang, Z.L. (2007) Direct-current nanogenerator driven by ultrasonic waves. *Science*, **316**, 102–105.

54. Bogue, P. (2009) Energy harvesting and wireless sensors: a review of recent developments. *Sensor Rev.*, **29**(3), 194–199.

55. van den Ende, D.A., Bory, B.F., Groen, W.A., and van der Zwaag, S. (2010) Improving the d33 and g33 properties of 0–3 piezoelectric composites by dielectrophoresis. *J. Appl. Phys.*, **107**, 024107.

56. Dietze, M. and Es-Souni, M. (2008) Structural and functional properties of screen-printed PZT–PVDF-TrFE composites. *Sens. Actuators, A*, **143**, 329–334.

57. Vullers, R.J.M., van Schaijk, R., Doms, I., van Hoof, I., and Mertens, R. (2009) Micropower energy harvesting. *Solid-State Electron.*, **53**, 684–693.

58. Bogue, P. (2010) Powering tomorrow's sensor: a review of technologies – Part 1. *Sensor Rev.*, **30**(3), 182–186.

59. Bogue, P. (2010) Powering tomorrow's sensor: a review of technologies – Part 1. *Sensor Rev.*, **30**(4), 271–275.

60. Hudak, N.S. and Amatucci, G.G. (2008) Small-scale energy harvesting through thermoelectric, vibration, and radiofrequency power conversion. *J. Appl. Phys.*, **103**, 101301.

61. Samson, D., Kluge, M., Becker, T., and Schmid, U. (2011) Wireless sensor node powered by aircraft specific thermoelectric energy harvesting. *Sens. Actuators, A*, **172**, 240–244.

62. Zhu, D., Beeby, S.P., Tudor, M.J., and Harris, N.R. (2011) A credit card sized

self powered smart sensor node. *Sens. Actuators, A*, **169**, 317–325.

63. Zhou, D.O., Ha, D.S., and Inman, D.J. (2010) Ultra low-power active wireless sensor for structural health monitoring. *Smart Struct. Syst.*, **6**, 675–687.

64. Hew, Y., Yu, A., Huang, H. (2011) RF-powered wireless strain sensor, in (ed. Chang, F. K.) *Structural Health Monitoring 2011: Condition-Based Maintenance and Intelligent Structures, Proceedings of the 8th International Workshop on Structural Health Monitoring, September 13–15, 2011*, Vol. **1** and **2**. Stanford University, Stanford, CA.

65. Dürager, C., Heinzelmann, A., Riederer, D. (2011) Wireless sensor network for guided wave propagation with piezo-electric transducers, in (ed. Chang, F.K.) *Structural Health Monitoring 2011: Condition-Based Maintenance and Intelligent Structures, Proceedings of the 8th International Workshop on Structural Health Monitoring, September 13–15, 2011*, Vol. **1** and **2**, Stanford University, Stanford, CA.

66. Nagesh, D.Y.R., Krishna, J.V.V., and Tulasiram, S.S. (2010) A real-time architecture for smart energy management. Conference on IEEE Innovative Smart Grid Technologies (ISGT), 2010.

67. Bosse, S. and Behrmann, T. (2011) Smart energy management and low-power design of sensor and actuator nodes on algorithmic level for selfpowered sensorial materials and robotics. Proceedings of the SPIE Microtechnologies 2011 Conference, Prague, Czech Republic, April, 18–20, 2011.

68. Lagorse, J., Paire, D., and Miraoui, A. (2010) A multi-agent system for energy management of distributed power sources. *J. Renewable Energy*, **35**, 174–182.

69. Bosse, S. and Kirchner, F. (2012) Smart energy management and energy distribution in decentralized self-powered sensor networks using artificial intelligence concepts. Proceedings of the Smart Systems Integration Conference 2012, Session 4, Zurich, Switzerland, March 22–23, 2012.

70. ISIS Sensorial Materials Scientific Centre, University of Bremen, Modu-ACT Case Study, *http://www.isis.unibremen.de/forschung/projekte-casestudies/moduact.html* (accessed 21 December 2012).

71. Peine, H. and Stolpmann, T. (1997) The architecture of the Ara platform for mobile agents, MA '97, in *Proceedings of the First International Workshop on Mobile Agents*, Springer-Verlag, Springer.

72. Wang, A.L., Sørensen, C.F., and Indal, E. (2003) A Mobile Agent Architecture for Heterogeneous Devices, Wireless and Optical Communications.

18
Additive Manufacturing Approaches

Juan F. Isaza P. and Claus Aumund-Kopp

18.1
Introduction

The term additive manufacturing (AM) has to be explained first as there are a lot of definitions and expressions around dealing with this technology. Here we talk of AM in terms of "making parts based on 3D model data by classical layer-by-layer manufacturing techniques," which is most accepted in science and is defined by American Society for Testing and Materials (ASTM) International Committee F42. Other terms often used are

- additive fabrication (VDI 3404);
- freeform fabrication;
- fabbing (mostly used by independent online communities and AM open-source users);
- layer manufacturing;
- 3D printing (mostly used by press, designers, etc.; more specifically, a nozzle-based printing technique);
- rapid manufacturing (outdated meanwhile);
- rapid prototyping (outdated meanwhile).

To give a more detailed impression, the two most common expressions are explained according to ASTM International Committee F42:

Additive manufacturing:
"Process of joining materials to make objects from 3D model data, usually layer upon layer, as opposed to subtractive manufacturing methodologies."
3D printing:
"Fabrication of objects through the deposition of a material using a print head, nozzle, or another printing technology."

According to [1], AM is used "to build physical models, prototypes, patterns, tooling components, and production parts in plastic, metal ceramic, and composite materials." Talking about real manufacturing in terms of large-scale production of large numbers of products, only a few niche products have reached this state.

Structural Materials and Processes in Transportation, First Edition.
Edited by Dirk Lehmhus, Matthias Busse, Axel S. Herrmann, and Kambiz Kayvantash.
© 2013 Wiley-VCH Verlag GmbH & Co. KGaA. Published 2013 by Wiley-VCH Verlag GmbH & Co. KGaA.

First AM approaches started in the mid-1980s. In 1987, the initial commercialization of the plastic processing technique stereolithography (SLA) [1] started. Metal-based processes started in the late 1990s and became mature during 2000–2003 when several companies launched systems for laser melting approaches, which were able to produce dense metal parts directly from the machines. Since then, these systems have become more reliable, more efficient, and the palette of available materials has grown significantly.

According to the definition of rapid manufacturing given in the VDI guideline for additive fabrication (VDI Norm), it is important to understand that the emphasis of this chapter lies on the additive fabrication of end products with all the characteristics of the final good. Some of the processes described in the

Table 18.1 Two-dimensional classification of AM processes.

Method of construction Feedstock state	1D channel	Array of 1D channels	2D channel
Liquid	Polymer[a]	Polymer[b]	Polymer[c]
Discrete particles	Polymer[d] Metal[g] Ceramic[g]	Polymer[e] Metal[e]	Polymer[f]
Molten material	Polymer[h]		
Solid sheets	Paper[i] Polymer[i] Metal[i] Ceramic[i]		

With this classification, it is easy to understand the importance of polymers and metals in the implementation of Rapid Manufacturing techniques. The use of these materials in the production of direct parts is significant because of the diversity of materials on the market, making possible a better achievement of the characteristics needed. Examples of processes listed in the table are:
[a] Stereolithography (SL).
[b] Polyjet modeling (PJM).
[c] Digital light processing (DLP).
[d] Laser sintering (LS).
[e] 3D printing (3DP).
[f] Mask sintering (MS).
[g] Selective laser melting (SLM).
[h] Fused deposition modeling (FDM).
[i] Layer-laminated manufacturing (LLM).

following are also proper for the manufacturing of concept models, geometric- and functional prototypes, products that are used during the first steps of product development.

To get a better idea of the materials used in additive fabrication, it is possible to organize the materials using a two-dimensional (2D) classification based on the method described by Pham and Gault [2] (Table 18.1).

AM, at present, definitely is no means for classical mass production for millions of identical parts. Its advantages are derived from its extremely high flexibility because of the production directly from the computer-assisted design (CAD) model without any necessary tooling.

This also facilitates AM to produce almost any geometry that can be designed. Until now, the outer geometry of a part and its function/strength were of main interest for the user, but AM allows the integration of additional functions and new fields of application of technical parts.

The best known example for the ability of AM systems to integrate functionality is the production of parts with conformal cooling or vacuum channels running directly below the surface of the contour of the die geometry of injection molding or extrusion tools, for example (Figure 18.1). Those channels could not be produced with any other technique.

But even more integration of functionality is possible by producing with AM systems, for example, repeating internal patterns that enlarge inner surfaces for a better energy exchange or mass exchange in devices for heat recovery or filtration. As an example, Figure 18.2 shows these kind of honeycomblike structures, which could also be used for flow guidance.

Internal structures can also have irregular but also detailed forms such as topology-optimized structures (Figure 18.3).

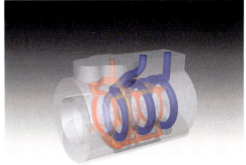

(a)   (b)

Figure 18.1 (a) Additively manufactured calibration tool and (b) scheme of internal vacuum (red) and cooling channels (blue) (designed and built on an EOS M270 system at Fraunhofer-IFAM).

Figure 18.2 Additively manufactured repetitive structures (designed and built on an EOS M270 system at Fraunhofer-IFAM).

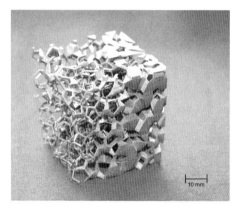

Figure 18.3 Cantilever – topology-optimized structure (calculated and built on an EOS M270 system at Fraunhofer-IFAM).

To use the full potential of AM techniques, complex internal structures can be combined with other geometric features as shells. Separating parts into shell and core volumes enables new solutions not only for lightweight parts, which need a "dense shell" and a "porous core" (Figure 18.4), but also for parts with internal functionality. Both volumes are produced in one manufacturing process, which results in considerable savings in powder material and part weight as well as energy consumption during processing (Figures 18.5 and 18.6).

Applications directly related to transportation purposes can be mentioned as follows:

- production of models and prototypes during a product development phase;
- production of parts for pilot series in automotive and aerospace industry;
- production of parts for very small series products where tooling costs for casting or injection molding would be too high;

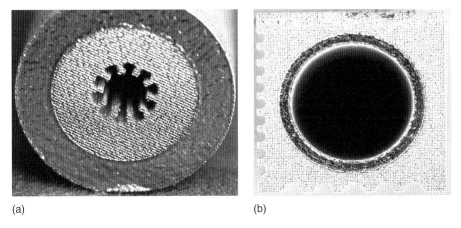

(a)   (b)

Figure 18.4 Additively manufactured study for material combinations of porous and dense areas of a part (a) dense shell – porous core and (b) porous shell – dense core.

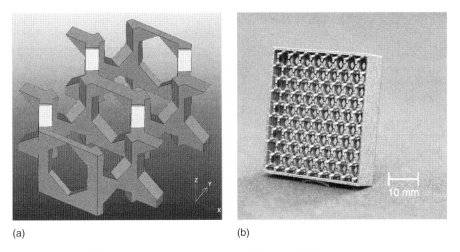

(a)   (b)

Figure 18.5 Double honeycomb structure – (a) base element and (b) test part.

- production of parts of highest geometrical complexity, which could not be realized by means of conventional manufacturing (grinding, milling, and even casting).

Owing to limited building envelopes of AM systems, the production of large structural elements by AM is restricted. Table 18.2 shows some examples of the largest building envelopes for different AM systems:

Sections 18.2 and 18.3 give an overview of different processing variants and available materials of metal- and nonmetal-based AM processes. Section 18.4 shows some specifics about postprocessing of AM parts. The three case studies presented in Section 18.5 give an impression of what AM at present can look like and what results can be achieved.

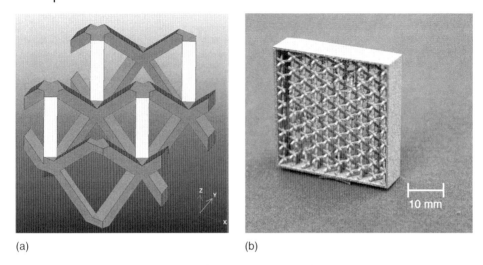

(a)   (b)

Figure 18.6   Reduced honeycomb structure – (a) base element and (b) test part.

Table 18.2   Building envelopes for different AM systems.

Envelope size (L × W × H in mm)	Process	System	Material	Remark
1016 × 508 × 254 (40 × 20 × 10)	3D printing of metal powder	ProMetal R10	Stainless steel + bronze	Parts have to be infiltrated with bronze and tempered after printing; great risk of deformation
280 × 280 × 350	Selective laser melting	SLM 280 HL	Metals	—
2100 × 700 × 800	Stereolithography	MATERIALIZE Mammoth	ABS-like photopolymers	Exclusively operating at MATERIALIZE, Belgium

18.2
Metal Materials

For a better understanding of the use of metals in AM, it is sensible to make a short description of the processes, which use metal as feedstock. Most of these processes work using a point-wise or 1D method (Table 18.1) and nearly all of them are based on metal powder.

The main exception to this approach are the sheet lamination systems, among others the method developed by Neue Materialen Bayreuth GmbH, that uses commercially available steel sheets. The sheets are first coated, and then after cutting out the object outline for the corresponding layer, the sheets are joined by diffusion soldering. This joining process is carried out by a direct contact of the sheets with massive, inductively heated plates. The combination of high build rates and an inexpensive feedstock forms an excellent basis for an economic approach [3].

Concerning the powder-based systems, almost every machine uses a powder deposition method using a coating mechanism to generate a layer onto a substrate plate. Usually the layers have a thickness of 20–100 μm. Once the powder layer is distributed, a 2D slice is selective melted using an energy beam applied on the powder surface. This energy is normally a high-power laser, except in the case of the electron beam melting (EBM) process by the Swedish company Arcam [4, p. 32].

The process is repeated layer by layer until the last layer is melted and the part is complete. These systems are grouped as laser melting processes and depending on the manufacturer, they have specific names, such as selective laser melting (SLM), laser cusing, and direct metal laser sintering (DMLS).

Another approach is the laser-engineered net shaping (LENS) powder delivery system used by Optomec. This method allows the process to be used to add material to an existing part, which means that it can be used for repair of expensive metal components that may have been damaged, such as chipped turbine blades and injection molding tool inserts [4, p. 32].

Common materials used in these processes are stainless steels and titanium alloys. Some machine manufacturers offer their own materials, guaranteeing quality standards for the produced parts. The variety of materials is huge and grows continuously. Taking into account the most important ones for transportation, they can be grouped as shown in Table 18.3 (for detailed specifications of the materials, please see references). The name of the material varies depending on the manufacturer; therefore, the name used here corresponds to the specifications on the material datasheets and, in some cases, the European nomination is given. Nevertheless, there are some efforts in standardization of the processes and the materials, an important step for a better comparison between products.

This wide variety of materials offers the user a good possibility of choosing the right material to achieve the specifications of the product.

Material properties, for example, tensile strength, hardness, and elongation, are important and often used reference points for the decision of the right material. Figure 18.7 illustrates some different alloys in an extensive range of tensile strengths. This diagram helps the user to pick the proper material depending on tensile strength and hardness.

As AM techniques have been already tested in some branches of the manufacturing industry, here are some examples of the properties and the uses given to the materials. System manufacturers are often offering their products (systems and materials) based on successfully produced and tested parts; some of these examples can be found in the recommended further reading.

Table 18.3 List of common metal materials.

	Material	DIN	Manufacturer				
			Arcam	Concept laser	EOS	SLM realizer	SLM solutions
Aluminum alloys	AlSi10Mg	3.2381	—	×	×	—	—
	AlSi12	3.3581	—	×	—	×	×
Cobalt-based alloys	ASTM F75	2.4723	×	—	×	×	×
	EOS MP1	2.4723	—	—	×	—	—
Maraging steel	AISI 420	1.2083	—	×	—	—	—
	Marage 300	1.2709	—	×	×	×	×
Nickel-based alloys	Inconel 718	2.4668	—	×	—	—	—
	Inconel 625	2.4856	—	—	×	—	—
	Hastelloy X	—	—	—	—	—	×
Stainless steel	316 L	1.4404	—	×	—	×	×
	EOS PH1	1.4540	—	—	×	—	—
	17-4PH	1.4542	—	—	×	×	×
Titanium alloys	Titanium Grade 2	3.7035	—	—	—	×	×
	Ti6Al4V	3.7165	×	×	×	×	×
	Ti6Al4V ELI	3.7165 ELI	×	—	×	—	—
	TiAl6Nb7	—	—	—	—	×	×
Tool steel (with carbon)	H13	1.2344	—	—	—	×	×

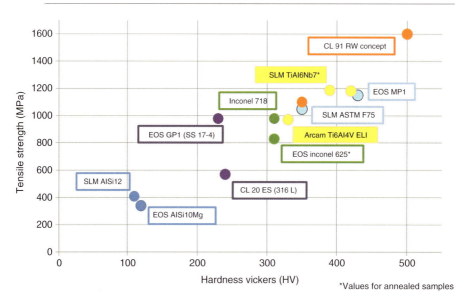

Figure 18.7 Mechanical properties of metal materials.

18.2.1
Titanium Alloys

Owing to the excellent mechanical properties, even in extreme temperatures, high strength-to-weight ratio, and extraordinary corrosion resistance, the use of this material is relevant in aerospace, motor racing, and maritime applications. "In the majority of these and other engineering applications, titanium has replaced heavier, less serviceable, or less cost-effective materials. Designing with titanium, taking all factors into account, has resulted in reliable, economic, and more durable systems and components, which, in many situations, have substantially exceeded performance and service life expectations" [5].

18.2.2
Stainless Steel

This material has a high resistance to oxidation and corrosion in a wide variety of environmental conditions. This characteristic, together with the good mechanical properties, allows the use of stainless steel in direct manufacturing processes for a varied kind of applications in transportation.

18.2.3
Aluminum

The good combination of thermal properties with low density represents aluminum parts as a good option for direct manufacture of functional prototypes, especially in aerospace applications and motor racing. One-off parts and short-run batches can be finished at a fraction of time compared with die-casting techniques. The mechanical properties of the parts are at least equivalent.

18.2.4
Cobalt-Based Alloys

This super alloy offers outstanding properties for high-temperature engineering applications, for example, turbines. Also, filigree parts with small wall thickness or internal structures with thin struts can be manufactured with cobalt-based alloys because of its high strength and stiffness.

18.2.5
Nickel-Based Alloys

Spare parts exposed to high mechanical loads and elevated temperatures in corrosive environments represent a good example for parts that can be manufactured with nickel-based alloys. Some examples are gas turbines, oil, petroleum, and natural gas system parts. This material also shows excellent cryogenic properties.

Parts resulting from AM systems can hardly be introduced in a process chain without being postprocessed. Section 18.4 gives a deeper description of the process used to improve the quality of AM parts. Usual roughness values for metal parts vary between 12 and 17 μm (Rz in X/Y direction).

An interesting approach to improve the surface quality of SLM parts is the LUMEX Avance-25 from Matsuura Machinery Corporation. This "Metal Laser Sintering Hybrid Milling Machine" allows the user to benefit from freeform fabrication by SLM and good surface quality of milled parts, all in one machine. The company presented this interesting hybrid system during the EuroMold 2011, expanding the possibilities for mold fabrication. Complex molds can be formed in a monobloc structure, eliminating errors due to buildup of component parts. Features such as porous molding (for degassing) and conformal cooling are easy to integrate, shortening the resin casting time and increasing the cooling efficiency of the molds. The materials that can be processed with the machine are usual maraging steel alloys, through steel-based powders and titanium alloys, such as Ti–6Al–7Nb. These can be used for the manufacturing of filters, radiators, and frames among others [6].

Over the last decades, scandium and zirconium as alloying elements have become a major interest for researchers all over the world. The improvements of properties that can be achieved by adding these elements to an Al alloy are remarkable. An investigation at EADS Innovation Works in cooperation with the Hamburg University of Technology shows the tremendous results of scandium regarding the combination of high-strength properties with a reduction of density. The rapid solidification offered by the SLM process allows for cooling rates that are sufficient to keep all alloyed scandium in a hypereutectic Al–scandium composition. This produces high strength properties, around 500 MPa, with nearly isotropic results. Furthermore, the ductility of the additively manufactured material, with an elongation of 14% by an area reduction of 20%, is remarkable [7].

Amorphous metallic glasses resulting from the ultrarapid cooling of molten metal alloys have been known for almost 60 years. Metallic glasses have the reputation to be castable in forms such as plastics, harder than steel and titanium, elastic, wear-resistant, and stainless, with the only condition to be cooled fast enough. AM techniques, such as SLM based on the principle of high-energy inputs for short period times, allow for ultrarapid cooling of the melt pool. This offers new interesting possibilities in the manufacturing of bulk metallic glasses.

It is expected that new alloys are found, for example, much harder alloys or alloys with completely new properties. With mixtures of materials (composites) of amorphous and crystalline metals, it could be possible to design materials with desired qualities, developments that could turn automotive and aircraft industry to become AMs as major customers.

18.3
Nonmetal Materials

The history of AM started with the use of polymers, waxes, and paper sheets. Many of those materials were already available on the market before AM processes

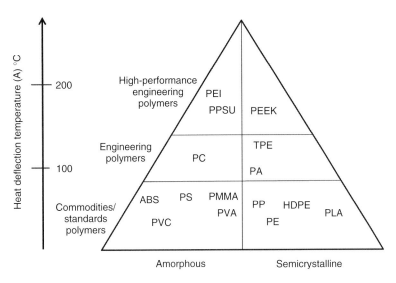

Figure 18.8 Polymers used for additive manufacturing.

were well-known suiting exigencies of other manufacturing processes. After the introduction of SLA in 1986, the necessity of special materials to fit the requirements of prototypes was identified. When it was possible to better understand the market needs, materials were developed to be used to full capacity.

Nowadays, the list of materials positioned on the market fulfilling the demands of the industry is huge. Figure 18.8 illustrates the wide range of polymers, starting with standard polymers until high-performance materials. The performance of both amorphous and semicrystalline polymers varies in relationship to their cost, and not every system can produce high-quality components.

Unlike metal materials, polymer systems are not offered only by high-end technology manufacturers. The sector of the so-called office solutions is remarkable in the polymer part production. Additionally, the high number of open-source systems allows various materials to be in a constant development and is the ambition of many independent researcher groups and end users.

Parts resulting from office and open-source systems do not represent the group of high-performance or engineering materials, but offer a good solution during the first steps of product development. They also open a new sector in the formation of new markets in the transportation sector, comparable maybe with the "massification" of the computer in telecommunication. European R&D projects such as "Directspare" present a new business model for manufacturers to rapidly produce spare parts locally, that is, close to the equipment to be repaired.

Selective laser sintering and extrusion (FDM, fused deposition modeling) based systems cover almost 80% of the materials being commercially offered. This is the result of the variety of materials possible to be manufactured with these systems and also of the final quality of the produced parts. Polymers such as ABS (acrylnitril-butadien-styrol) used to mold dashboards and door panels, PP (polypropylene)

Table 18.4 List of polymer materials grouped by manufacturing process.

Laser sintering	PA	PC	PEEK	PP	PS	TPE	TPU	
Extrusion	ABS	HDPE	PE	PLA	PSU	PVA	PPSU	PEI
3D printing	PMMA	—	—	—	—	—	—	—
Sheet lamination	PVC	—	—	—	—	—	—	—
Stereolithography	Photopolymer		—	—	—	—	—	—

highly resistant to scratches used for rear shelves or trunk linings, or PMMA (polymethyl methacrylate) used in airplane windows are just some examples of the capabilities offered by AM (Table 18.4).

Many polymer-based systems use two materials, one for the final part and the second to support the part overhangs and critical build angles. The support material is removed after production. The part properties correspond to those of the built material.

Multi-material 3D printing systems refer to the application of at least two materials for part production. A combination of materials offers new possibilities, for example, graded designs, expanding the mechanical and physical properties of the manufactured part in a single buildup.

Fraunhofer UMSICHT has developed a new material based on thermoplastic polyurethane (TPU), which extends the range of thermoplastic laser sintering powder materials. The combination of high strength, flexibility, and abrasion resistance of the material opens new possibilities for AM processes. In this way, TPU-sintered powder results in flexible and highly reliable components within a few hours. Application areas such as the manufacturing of flexible tubes or pneumatic structures in the automotive sector will foster the use of this material in transportation [8].

The process of highly filled materials also represents an important sector of the ongoing research efforts. One approach of this interesting area is a process developed by TNO (The Netherlands) called *high filled μSLA*. On the basis of the SLA principle, resins are filled with ceramic materials such as silicon carbide or aluminum oxide, enhancing the properties of the produced parts [9]. Better mechanical properties such as hardness and strength, important for high-performance parts, can be achieved with these kinds of composites.

18.4
Secondary Processes

In order to achieve or improve selected properties such as surface quality, geometrical accuracy, and mechanical properties, it is often necessary to postprocess and finish metal and polymer products resulting from AM techniques. Mostly, physical characteristics can be enhanced by adding well-established fabrication processes at the end of the AM process chain.

Owing to the high quality of metal products from AM processes, it is possible to use every kind of metal-cutting finishes to meet the requirements of surface quality and geometry. Starting with sawing of support structures to separate the parts from the build platform (SLM), the products can be milled, drilled, polished, and so on. Internal surfaces, as in internal/tempering channels, can be polished using abrasive flow machining to improve the stream of the fluid.

Heat treatment is another technique of metal processing, which can be employed, as well as shot peening, used to improve the mechanical properties of the AM-produced parts.

An interesting postprocess for metallic parts is electropolishing. This chemical treatment significantly improves the surface finishing of AM-manufactured parts. Its first objective is to minimize microroughness, thus reducing the risk of dirt or product residues adhering and improving the cleanability of surfaces. But electropolishing can also be used for deburring, brightening, and passivating, particularly for surfaces exposed to abrasive media. As electropolishing involves no mechanical, thermal or chemical impact, small and mechanically fragile parts can also be treated.

With plastic materials, the patented RP-Tempering™ technology is used. Usually, it is a combination of nanomaterial components with special application techniques to produce a composite structure. It is possible to be applied for most plastic-based AM processes, such as FDM, SLA, laser sintering, and multijet solidification, improving the mechanical, electrical, thermal, and chemical properties of common materials. In this way it is possible to specifically enhance the properties for a better performance of end-use parts [10].

18.5
Case Studies

The following three case studies represent examples of how AM is presently in use. On the basis of the SLM technology using metal powder as raw material, these cases show

- the possible performance increase when using the freedom of design offered by AM;
- the geometric complexity of parts AM is able to produce;
- a process chain realized for a multivariant product.

18.5.1
Hydraulic Crossing

This first AM case study emphasizes the enhancement of performance through AM techniques. Here, SLM was used to let designers profit from the use of rapid manufacturing. Significant increases of performance in hydraulics, for example, are possible by exploiting the possibilities of this layer manufacturing process.

(a)   (b)

Figure 18.9 (a) Additively manufactured hydraulic crossing and (b) scheme of internal channel geometry (designed during EU-funded project CompoLight; built on an EOS M270 system at Fraunhofer-IFAM).

Inside the hydraulic crossing, two fluid streams cross within a limited space without mixing. The conventionally produced part consists of a massive metal block where drilled and locked blind holes meet in two levels. The main bore is substituted by four smaller bores to keep the height and the possible mass flow constant. The weight of the conventionally produced pipe crossing is 20 kg; its dimensions are 230 mm × 230 mm × 50 mm.

Without being hindered by the limits of conventional production means, the designer is free to optimize the functionality of an element. In this case, the fluid flow is improved by adapting the internal channel geometry according to flow simulations results. Geometrical changes of cross-section profiles are possible to be manufactured without great effort. Concerning thermodynamics, fins inside the channels are possible to improve thermal exchange processes and to increase the part stiffness at the same time. The newly designed hydraulic crossing (Figure 18.9) was produced from a stainless steel material. Internal fins supported the part during processing, and the dimensions are 80 mm × 80 mm × 50 mm. The total weight is only 0.7 kg. This corresponds to a mass reduction of about 96%.

In this study, the optimization of the channel geometry has an enormous impact on the crossing performance and thus on this entire hydraulic system. At a mass flow of 100 l min^{-1}, the pressure loss of the new design is only 20% of the conventionally designed and manufactured part (Figure 18.10). Additionally, without any postprocessing, the surface quality of the AM part was sufficient enough to be flanged to the connecting piping. Even at a test pressure of 1400 bar, the part showed neither plastic deformation nor leakage.

18.5.2
Wing Profile

This case study also shows geometrical possibilities of AM, here used for a setup for anti-icing tests on aircraft wings. The part shown in Figure 18.11 has two separated chambers for temperature control. For a better distribution of the tempering fluid,

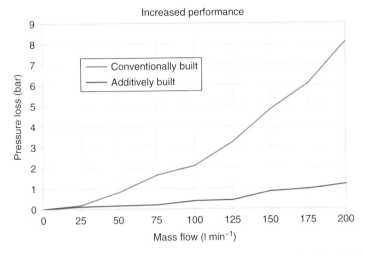

Figure 18.10 Comparison of pressure loss of conventionally and additively built hydraulic crossing.

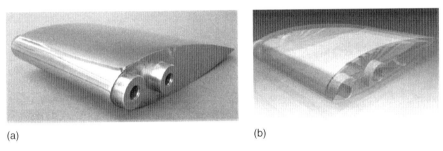

(a)   (b)

Figure 18.11 (a) Additively manufactured wing profile and (b) scheme of internal channel geometry; polished after AM buildup (designed and built on an EOS M270 system at Fraunhofer-IFAM).

the rear chamber is equipped with internal channels spreading from the inlet. The part is hollow; its wall thickness is 3 mm.

18.5.3
Series Production of Individual Parts – Dental Restorations

As there is no present example of an AM process in use to do real series manufacturing for automotive parts, this last case study shows the manufacturing of semifinished parts for the so-called dental restorations done by BEGO Medical, Bremen, Germany. Especially the metal basis of dental crowns and bridges (Figure 18.12) are manufactured by SLM. The product itself is extremely multivariant. It is the production of a high number of one-of-a-kind parts but within a limited product size. A 48 h supply-chain manufacturing by SLM here substitutes a conventional and laborious investment casting process with lost molds.

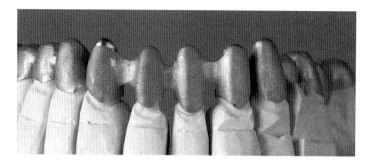

Figure 18.12 Manufactured by SLM – metal "scaffold" of a dental bridge (before ceramic veneering) (BEGO Medical GmbH).

The appropriate process steps are as follows: the dental laboratory produces a so-called positive model of the patient left tooth stumps, which is then 3D-scanned. The resulting data model can be modified as necessary and then be transferred to an SLM manufacturing center. After necessary data verification and preprocessing steps, the generated data then is transferred to the SLM system and afterwards up to 400 individual parts are manufactured simultaneously on one building platform.

After the manufacturing step, all parts are mechanically removed from the building platform and are controlled concerning geometrical accuracy. Subsequently, the parts are sent to the dental technicians for further processing, including ceramic veneering.

The described process chain illustrates that for the realization of an AM approach not only the actual manufacturing step has to be dealt with, but also all pre- and postprocessing steps have to be seen and integrated. The big challenge of a series production of individual parts is to guarantee an identifiability of each part at all times during the process. If one part gets mixed up, the whole building job is lost. Quality assurance of material properties here is realized by building test specimens on each building platform during manufacturing.

18.6
Summary and Outlook

The outstanding feature of all AM techniques is their capability to produce parts of highest geometrical complexity, which cannot be built by any other production technique. This works because of the tool-free layer-by-layer approach of all AM processes. Parts are produced based on 3D CAD model data without any tooling needed.

The number of available materials is still limited compared to other processes such as milling or injection molding, but the number of materials qualified for SLA or SLM is continuously rising. Another limitation still is the part size possible to produce with AM. In SLA, we already talk of axis lengths of 2000 mm. For many other processes, this is one aim that is under way but not yet reached.

Concerning technology readiness, a lot of AM techniques offer part qualities, which are comparable to those of conventional methods. The AM-produced parts can be used and postprocessed (milled, drilled, and coated) as standard parts known to the user. Especially in metal, AM-produced parts often exceed the mechanical property values of those machined from standard bulk material. On the basis of the specific process feature of extremely high cooling rates during solidification, even new materials will become possible. Recently developed special aluminum–scandium alloys (Section 18.2) processed by SLM showed material properties that were significantly better than those of standard alloys.

Another aspect becoming more and more important is connected to the outstanding material efficiency of most AM processes. Almost all raw materials, either powder particles or liquid resin, are used to build the part. Scrap rates are usually below 5%. Compared to scrap rates of more than 90% with many complex milled parts, with declining raw material availabilities or rising costs in the future, this material efficiency will become a major advantage.

The general future of AM will be one of the definitely growing market share in production. It will be reached by faster systems with more powerful lasers and larger building chambers. A significant number of materials will be qualified for AM and – looking at long-term developments – multi-material systems for a lot of processes will become available.

18.7
Further Reading

In this section, some relevant handbooks on AM aiming at readers who need in-depth information are listed.

- *Wohlers Report* (annually published)
- Wohlers, T.(ed.): *Wohlers Report 2011*. Wohlers Associates, Fort Collins, CO, 2011. ISBN 0-9754429-7-X. The annually published *Wohlers Report* is an in-depth global study on the advances in AM technologies and applications with actual examples, *figures on market shares* and new developments of systems and materials from all over the world.
- Gebhardt A (2003) *Rapid Prototyping*, 1st edn, Hanser Gardner Publications, Munich and Cincinnati, 379 pp., ISBN: 10: 156990281X, ISBN: 13: 978–1569902813. This book also describes all major processes and available materials in detail. In addition, it contains chapters dealing with *requirements for series production* as well as aspects of *economic efficiency* and *environmental protection.*
- Gibson I., Rosen D.W., Stucker B., *Additive Manufacturing Technologies: Rapid Prototyping to Direct Digital Manufacturing*, 1st edn, Springer, New York 14 December, 2009 462 pp., ISBN: 10: 1441911197, ISBN: 13: 978-1441911193. This book also describes all major processes and available materials in detail. In addition, it contains chapters dealing with *Design for AM* and offers *Guidelines for Process Selections*. As the book is explicitly meant for students, each chapter ends with some exercises.

- Hopkinson N., Hague R., Dickens P. (eds) *Rapid Manufacturing: An Industrial Revolution for the Digital Age*, John Wiley & Sons, Ltd, Chichester 11 January, 2006, 304 pp., ISBN: 10: 0470016132, ISBN: 13: 978-0470016138. This book also describes all major processes and available materials in detail and is based on various examples. In addition, it contains a chapter dealing with *Automotive Applications.*
- Grimm T., *User's Guide to Rapid Prototyping*, illustrated edition, Society of Manufacturing Engineers, Dearborn, February, 2004, 404 pp., ISBN: 10: 0872636976, ISBN: 13: 978–0872636972. This book, compared to the already mentioned, is a *relatively rough guide to AM*. It contains general information with not many details on systems or materials.

References

1. Wohlers, T. (ed.): *Wohlers Report 2011.* Wohlers Associates, Fort Collins, 2011. ISBN: 0-9754429-7-X.
2. Pham, D.T. and Gault, R.S. (1998) A comparison of rapid prototyping technologies. *Int. J. Mach. Tools Manuf.*, **38**, 1257–1287.
3. Rudnik, Y. (2011) Neues generatives Verfahren zur schnellen Herstellung von Werkzeugen mit komplexer Außen- und Innengeometrie. Internal Presentation Kompetenzzentrum Neue Materialien Nordbayern GmbH, Bayreuth, February 2011.
4. Gibson, I., Rosen, D.W., and Stucker, B. (2010) *Rapid Prototyping to Direct Digital Manufacturing*, Springer, New York, Heidelberg, Dordrecht, London, ISBN: 978-1-4419-1119-3.
5. Arcam Ti6Al4V Titanium Alloy, *http://www.arcam.com/CommonResources/Files/www.arcam.com/Documents/EBM%20Materials/Arcam-Ti6Al4V-Titanium-Alloy.pdf* (accessed 21 December 2012).
6. Matsuura Machinery Corporation (2012) *http://www.matsuura.co.uk/matsuura/lumex-avance-25/*, February 2012 (accessed 21 December 2012).
7. Schmidtke, K., Palm, F., Hawkins, A., and Emmelmann, C. (2011) Process and mechanical properties: applicability of a scandium modified Al-alloy for laser additive manufacturing. Proceedings of the Sixth International WLT Conference on Lasers in Manufacturing.
8. Kupmann, I. (2011) Neuer TPU-Werkstoff für Generative Fertigung, Fraunhofer Institute UMSICHT, Press release, November 2011.
9. Maalderink, H. (2011) Mask Projection Micro Stereolithography, TNO, Precisionfair Veldhoven, The Netherlands, November 30, 2011.
10. RP Tempering™ (2011) Nano-Composite Technology, February 2011, *http://www.rptempering.com* (accessed 21 December 2012).

Index

Structural Materials and Processes in Transportation, First Edition.
Edited by Dirk Lehmhus, Matthias Busse, Axel S. Herrmann, and Kambiz Kayvantash.
© 2013 Wiley-VCH Verlag GmbH & Co. KGaA. Published 2013 by Wiley-VCH Verlag GmbH & Co. KGaA.